Lecture Notes in Computational Science and Engineering

38

Springer
Berlin
Heidelberg
New York
Hong Kong
London
Milan
Paris
Tokyo

Silviu-Iulian Niculescu
Keqin Gu
Editors

Advances in Time-Delay Systems

Springer

Editors

Silviu-Iulian Niculescu

Université de Technologie de Compiègne
UMR CNRS 6022
Centre de Recherche de Royallieu
60205 Compiègne, France
e-mail: silviu@hds.utc.fr

Keqin Gu

Southern Illinois University
Department of Mechanical
and Industrial Engineering
Edwardsville, IL 62025-1805, USA
e-mail: kgu@siue.edu

Library of Congress Control Number: 2004102413

Mathematics Subject Classification (2000): 34K05, 34K20, 93D06, 93D20, 34K35, 35B37, 34K40, 34K50, 34K60, 65P30, 65L99, 65M99

ISSN 1439-7358
ISBN 3-540-20890-9 Springer-Verlag Berlin Heidelberg New York

Springer-Verlag is a part of Springer Science+Business Media
springeronline.com

Printed in Germany

Cover Design: Friedhelm Steinen-Broo, Estudio Calamar, Spain
Cover production: *design & production*
Typeset by the authors using a Springer TEX macro package

Printed on acid-free paper 46/3142/LK - 5 4 3 2 1 0

Preface

In the mathematical description of a physical or biological process, it is a common practice to assume that the future behavior of the process considered depends only on the *present* state, and therefore can be described by a finite set of ordinary differential equations. This is satisfactory for a large class of practical systems. However, the existence of *time-delay elements*, such as material or information transport, often renders such description unsatisfactory in accounting for important behaviors of many practical systems. Indeed, due largely to the current lack of effective methodology for analysis and control design for such systems, the time-delay elements are often either neglected or poorly approximated, which frequently results in analysis and simulation of insufficient accuracy, which in turns leads to poor performance of the systems designed. Indeed, it has been demonstrated in the area of automatic control that a relatively small delay may lead to instability or significantly deteriorated performances for the corresponding closed-loop systems.

In order to reliably analyze and design feedback controls for such systems, it is necessary to consider the fact that the system's future behaviors depend on not only the *current value* of the state variables, but also some *past history* of the state variables. These systems are called *time-delay systems* (also known as *hereditary systems* or *systems with aftereffects*). Examples of system elements which cause time delays include: the measurements of system variables (engineering process), the physical nature of some system's components (engineering systems, such as hydroelectric power systems; biological and ecological systems, such as population dynamics) or signal transmissions (power and communication systems). According to the causes of delays, we may roughly classify them as *physically inherent delays* (physical or biological systems), *technological delays, transmission delays,* and *information delays.*

To effectively deal with time-delay systems, the researchers and practitioners in the systems and control fields are faced with three natural issues: 1) How to mathematically describe such systems in a form which is convenient for analysis and design? 2) How to analyze a given system to extract some fundamental properties such as stability? and 3) How to design a control via feedback principle to achieve stability and satisfy prescribed performance requirements?

Regarding the first issue, there are mainly three ways of representing dynamical systems with time delays based on the differential interpretations of the "states": differential equations on abstract spaces of infinite dimensions, functional differential equations, and differential equations over rings and modules. Each representation has its advantages and disadvantages, and the choice often depends on the specific problem to be solved. The readers will see various examples regarding the second and the third issues in the chapters to follow.

Background, purpose, and intended audience

In recent few years, we have witnessed substantial increase of research activities in the field of time-delay systems, as evident from the large volumes of publications and conference presentations. The purpose of this book is to bring together specialists working on various aspects of time-delay systems with different background such as mathematics, engineering, control systems theory, and present some of the important progress of research on the theory, methodology, new applications as well as overviews of background materials. Most of the chapters are based on the materials presented in the CNRS-NSF Workshop "Advances in Time-Delay Systems" held in Paris in January 22–24, 2003. We believe the interdisciplinary nature of the subject can benefit greatly from contributors of diversified background.

Similar to robust stability and control theory for finite dimensional systems, the analysis and design of time-delay systems have benefited substantially from the availability of amazing computational power due to the breathtaking pace of the development of modern computers as well as modern numerical procedures such as semidefinite programming. On the other hand, important numerical issues continue to emerge to challenge the researchers in this area. For example, an implementation of distributed feedback control may lead to instability even for very small error, as discussed by Michiels, et al (the last chapter of Part IV). It is also desirable to evaluate the design algorithms on the growth of the computational requirements as the size of the system grows. This book is published as a volume in the series of Lecture Notes in Computational Science and Engineering (LNCSE) with a hope that it will raise the awareness in the systems and control community on the computational issues involved in the subject. At the same time, we hope to call to the attention of the experts in the computational area regarding the challenges and opportunities in the area of time-delay systems.

The book should be of interest to both researchers and practitioners in this area, as well as those new to the area. The reporting of the newest progress in various fronts of this rich area is clearly of interest to researchers and practitioners who are concerned with the newest tools available and remaining challenges. For those new to the area, the overviews of background materials will allow them to develop a sense of the issues involved, and familiarize themselves on the topics studied in this area, and find from the chapters the specific topics they are interested in.

Book outline

We have divided the book into seven parts, with topics ranging from stability analysis, numerical analysis, to applications. In soliciting the contributions, we have sought to present a wide range of topics which we feel important and avoid redundancy. We have also provided a table of index to allow the readers to quickly locate the specific topics discussed. It should be pointed out that the subjects discussed in many chapters are relevant to more than one part, and indeed can be naturally put in a different part from where they are now. Therefore, in the following, we will provide a brief description of each chapter to highlight the main topics discussed.

PART I. BASIC THEORY

This part contains one chapter contributed by *Verduyn-Lunel*. It introduces many basic concepts and important results of time-delay systems by considering linear systems with a single delay. Many of these concepts are used repeatedly in the subsequent chapters. An abstract differential equation approach is used. It is a nice illustration that some of the properties may be easily obtained which would be very difficult to obtain using other approaches.

PART II. STABILITY AND ROBUST STABILITY

This part discusses the stability problem of time-delay systems. The stability criteria may be either dependent or independent on delays. It contains four chapters. Since stability is one of the most important properties of systems theory, it is a recurring scene in other parts of the book as well.

The first chapter, contributed by *Kharitonov*, discusses the construction of complete quadratic Lyapunov-Krasovskii functionals for linear systems. The result has important implications in robust stability under uncertain perturbations, as well as effective numerical implementations of Lyapunov approach to the stability and robust stability problem.

The next two chapters use the other main approach frequently used in the stability problem—frequency domain approach. The second chapter of this part, contributed by *Chen and Niculescu*, deals with the robust delay-independent stability of quasipolynomials in the form of frequency-sweeping tests. The third chapter, contributed by *Sipahi and Olgac*, discusses a method of checking the stability of quasipolynomials by considering the crossing of its roots as the delays increase.

The last chapter of this part, contributed by *Bliman*, attempts to build links for both Lyapunov and frequency domain approaches for delay-independent stability. It provides a unified view on various approaches for studying stability of time-delay systems, parameter-dependent systems, and structured singular values.

PART III. CONTROL, IDENTIFICATION, AND OBSERVER DESIGN

This part discusses the design of controllers to render the system stable and satisfy performance specifications, the identifiability and identifier design of system parameters, as well as the observations of system states.

The first three chapters consider the feedback control using inputs which are subject to delays. As is well known, this scenario occurs repeatedly in practical applications. The first chapter in this part, contributed by *Mondié and Loiseau*, discusses finite spectrum assignments. The next chapter, contributed by *Răsvan and Popescu*, uses an elementary approach based on a variant of the Smith predictor, and discusses some implementation schemes. The implementation issue will be revisited in Part IV.

The third chapter in this part, contributed by *Mazenc, Mondié, and Niculescu*, discusses the possibility and specific control design for systems which are subject to arbitrarily large delays, and the open-loop systems are oscillators.

The next chapter, contributed by *Belkoura, et al.*, discusses the identifiability of system parameters using the measured data, and the design of parameter identification schemes.

The last chapter in this part, contributed by *Fattouh and Sename*, discusses the design of robust observers for uncertain linear systems.

PART IV. COMPUTATION, SOFTWARE, AND IMPLEMENTATION

This part tackles the numerical, computations and implementation issues. The first chapter, contributed by *Bellen and Zennaro*, discusses the numerical solution of delay-differential equations using a variable stepsize Runge-Kutta method.

The next chapter, contributed by *Roose, et al.*, describes a Matlab software package "DDE-BIFTOOL," which is capable of carrying out the stability and bifurcation analysis for parameter-dependent systems of delay-differential equations.

The third chapter, contributed by *Datko*, proposes two schemes of checking the stability of systems, based on variants of Lyapunov and analytic function approaches, and take advantage of existing numerical packages such as Matlab. The fourth chapter, contributed by *Louisell*, discusses the numerical calculation of exact upper bound of real parts of system eigenvalues.

The last chapter in this part, contributed by *Michiels, et al.*, discusses the implementations of distributed delay control laws, which is often arrived at by a finite spectrum assignment. It illustrates the mechanism of losing stability in point-wise approximation and proposes safe implementations of distributed delay control laws.

PART V. PARTIAL DIFFERENTIAL EQUATIONS, NONLINEAR AND NEUTRAL SYSTEMS

This part considers the time-delay problem in the partial differential equations, nonlinear systems, as well as systems of neutral type.

The first chapter, contributed by *Hale*, illustrates the possibility of synchronization of systems defined by partial differential equations through the boundary interaction. This is carried out by considering lossless transmission lines which interact through resistive coupling at the end of the lines. An equivalent formulation in terms of a set of partial neutral functional differential equations is used to solve the problem.

The next chapter, contributed by *Fridman*, generalizes the output regulation of nonlinear systems to neutral time-delay systems. The condition is in the form of a set of partial differential and algebraic equations.

The third chapter, contributed by *Bonnet and Partington*, discusses the robust stability and stabilization of time-delay systems, including those of neutral type, using a BIBO stability framework. The fourth chapter, contributed by *Rabah, Sklyar and Rezounenko*, addresses the stability and stabilization problem based on the framework of the strong stability.

The last chapter in this part, contributed by *Rodriguez, Dion, and Dugard* discusses the delay-dependent robust stability of neutral time-delay systems with a norm-bounded nonlinear and time-varying uncertainty using a Lyapunov approach.

PART VI. APPLICATIONS

This part presents some applications of time-delay systems. The first chapter, contributed by *Yuan, Efe, and Özbay*, discusses modeling and control of cavity flow. The second chapter, contributed by *Annaswamy*, deals with the active-adaptive control of acoustic resonance flows related to propulsion and power generation systems.

The third chapter, contributed by *Lozano, et al.*, discusses a robust prediction-based control for time-delay systems with unstable open-loop systems. The method is implemented in the real-time control of the yaw angle displacement of a 4-rotor mini-helicoptor.

The fourth chapter, contributed by *Taoutaou, Niculescu, and Gu*, discusses the stability analysis of a teleoperation system subject to constant or time-varying communication delays.

The fifth chapter of this part, contributed by *Tarbouriech, Abdallah, and Ariola*, discusses bounded control of systems with multiple delays with specific applications to ATM networks.

The last two chapters, contributed by *Birdwell, et al.*, and *Hayat, et al.*, respectively, discuss dynamic time-delay models of load balancing in parallel computing. One of them uses deterministic models, and the other carries out a stochastic analysis of the effect of delay uncertainty.

PART VII. MISCELLANEOUS TOPICS

The last part of this book discusses a number of other topics of interest in time-delay systems. The first chapter, contributed by *Verriest*, is an overview of stochastic time-delay systems. The last chapter, contributed by *Haddad and Chellaboina*, presents an extension to time-delay systems of the stability and Dissipativity theory for nonnegative and compartmental systems.

Acknowledgements

The idea of an edited book was formed through a series of exchanges of e-mails and face-to-face discussions between two of us when Keqin visited HeuDiaSyC in

France as part of implementation of the project "Time-delay systems: analysis, computer aided design, and applications," started in 2002 and jointly funded by CNRS (France) and NSF (USA), and for which we served as the coordinators of the two countries. Indeed, many contributors are among the participants of the project. All the French participants also belong to *CNRS GDR Automatique: Delay systems*, a research network in Automatic Control in existence since 1994. However, as can be seen from the list of contributors, we have also invited researchers from outside of these research teams in order to provide a more balanced representation of the area.

We would first like thank all the contributors of the book. Without their encouragement, enthusiasm and patience, this book cannot be imagined. A list of contributors is at the end of the book. We would also like to thank Professors Mark Spong and Michel Fliess, who made stimulating presentations in the CNRS-NSF Workshop in Paris mentioned earlier. Indeed, we feel very privileged at the opportunity of working with this group of outstanding scholars in this project.

We would also like to thank CNRS and NSF, especially the program managers Claire Giraud, Jean-Luc Clément (CNRS) and Rose Gombay, and Kishan Baheti (NSF) for funding the joint research project which made this book possible. Thanks also go to *GDR Automatique* (France), *MENRT* (France), *AS 01 (CNRS-STIC) Auto-Télécoms*, the laboratory *HeuDiaSyC*, as well as the grants *ACI: Application de l'Automatique en algorithmique des télécommunications* and *ANVAR (Gradient-UTC): Modélisation et contrôle des congestions dans des réseaux*; these organizations and grants also helped fund the Paris workshop and provided various assistance in the editing process.

Part of editing work is completed when Keqin is on sabbatical leave in HeuDiaSyC, France, in Fall 2003. The granting of sabbatical leave by Southern Illinois University at Edwardsville and the financial support of CNRS are gratefully acknowledged.

We would also like to thank Springer for agreeing to publish this book. Especially, we would like to express our gratitude for Editor-in-Chief Dr. Martin Peters and the Series Editors T. Barth, M. Griebel, D. E. Keyes, R. M. Nieminen, D. Roose, and T. Schlick for their careful considerations and helpful suggestions regarding the format and organization of the book. Among them, Professor Dirk Roose provided especially detailed and helpful suggestions. Thanks also go to Ms Thanh-Ha Le Thi, Mathematical Editorial IV, who handled many of our communications with patience and courtesy, and Ms Leonie Kunz in the production department who helped with our LaTeX issues.

Compiegne, France,
October 2003

Silviu-Iulian NICULESCU
Keqin GU

Contents

Part VI Applications

Part VII Miscellaneous Topics

Part I

Basic Theory

Basic Theory for Linear Delay Equations

Sjoerd M. Verduyn Lunel

Mathematisch Instituut, Universiteit Leiden, P.O. Box 9512, 2300 RA Leiden, The Netherlands
verduyn@math.leidenuniv.nl

Summary. For dynamical systems governed by feedback laws, time delays arise naturally in the feedback loop to represent effects due to communication, transmission, transportation or inertia effects. The introduction of time delays in a system of differential equations results in an infinite dimensional state space. The solution operator becomes a nonself-adjoint operator acting on a Banach space of segments of functions. In this chapter we discuss the state space approach, the solution operator and its spectral properties for differential delay equations. As an application we present strong convergence results for series expansions of solutions and construct examples of solutions of delay equations that decay faster than any exponential.

1 Introduction

The aim of this chapter is to present the basic theory for linear autonomous delay equations. The topics include the very definition, the state space approach, the solution operator and an analysis of the spectral properties of autonomous differential delay equations. As motivation for the theory that we develop in this chapter, we briefly discuss some examples.

In the implementation of any feedback control system, e.g., the control of partial differential equations through the application of forces on the boundary, it is very likely that time delays will occur. Therefore, it is of importance to understand the sensitivity of the control system with respect to the introduction of small delays in the feedback loop. For some systems, small delays lead to destabilization while other systems are robust with respect to small time delays. In [11] a first attempt was made for a unifying theory which explains the underlying mechanisms in terms of spectral properties of the solution operator. See also [12,29] for applications towards stabilization of neutral differential delay equations.

The second example concerns the identifiability of unknown parameters that appear in differential delay equations. Parameter identifiability is concerned with the question whether the parameters of a specific model can be identified from knowledge about certain solutions of the model, assuming perfect data. Using the solution operator approach conditions for identifiability of parameters and time delays in linear differential delay equations, assuming knowledge of particular solutions on bounded time intervals, can be given. Completeness of the eigenvectors and generalized eigenvectors of the solution operator plays a crucial role in these results ([26,28]).

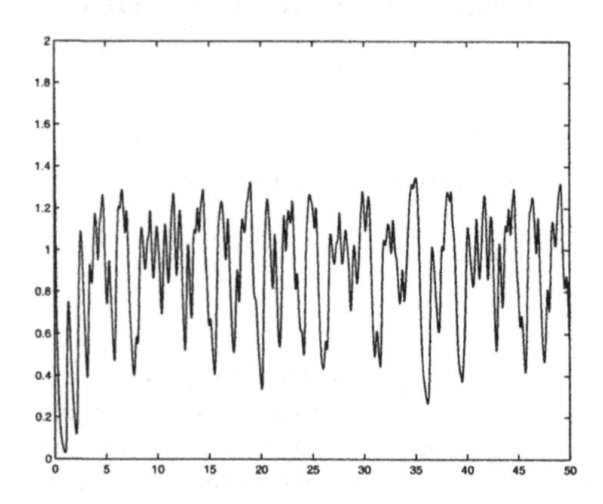

Fig. 1. Chaotic behaviour of the solutions of (1) for the initial data $\varphi(\theta) = 2$, $-1 \leq \theta \leq 0$.

To illustrate how complicated the behaviour of scalar differential delay equations can become, we plot the solution of

$$\dot{x}(t) = -5x(t) + 10\frac{x(t-1)}{1 + x(t-1)^8}, \qquad x_0 = \varphi, \tag{1}$$

with initial data $\varphi(\theta) = 2$ for $-1 \leq t \leq 0$ in Fig. 1. Equation (1) was introduced and studied by Mackey [7] as a model to describe different periodic diseases. Note that the solution intersects the steady state $x(t) = 1$, but this turns out not to be a contradiction to the uniqueness of solutions because the states of the solutions are different (see Section 2).

We conclude this introduction with an outline of the chapter. In Section 2 we define the state of a solution and discuss existence and uniqueness of solutions. In Section 3 we discuss a functional analytic approach toward differential delay equations based on semiflows which is essential in the modern treatment of differential delay equations. In Section 4 we discuss the spectral theory for autonomous delay equations. For autonomous equations the theory is well developed, see [10], but since we restrict our attention to the special case of one point delay – relevant in applications as we have seen – it is possible to present a novel simpler approach based on resolvent estimates for the characteristic matrix and series expansions of solutions. We discuss completeness and noncompleteness results, the existence of solutions that decay faster than any exponential, and the convergence of series of spectral projections. Our results extend earlier results in [1, 2, 18]. Further extensions to linear periodic delay equations and to neutral equations can be found in [27, 29].

2 Basic Theory for Delay Equations

In this section we present a short introduction to the basic theory for differential delay equations. We start to discuss the existence and uniqueness of solutions and then continue with a discussion of linear autonomous delay equations.

2.1 Existence and uniqueness of solutions

Since all our examples are of the form

$$\dot{x}(t) = F(x(t), x(t-\tau)), \qquad t \geq 0, \tag{2}$$

where $F : \mathbb{R}^n \times \mathbb{R}^n \to \mathbb{R}^n$ is a Lipschitz continuous vector field and the time delay $\tau \geq 0$ is a fixed real number, we shall restrict the introduction to the basic theory of equations of this type.

The solution of (2) is defined to be a vector-valued function $x : [-\tau, \infty) \to \mathbb{R}^n$ satisfying (2) for positive time. A moment of reflection indicates that, in order for (2) to have a unique solution, the minimal amount of information of initial data is given by a continuous function on the interval $[-\tau, 0]$.

In fact, if $x(t) = \varphi(t)$ for $-\tau \leq t \leq 0$, where φ is a given continuous function, then the solution $x(t)$ for $0 \leq t \leq \tau$ satisfies the nonautonomous ordinary differential equation

$$\dot{x}(t) = F(x(t), \varphi(t-\tau)) \qquad \text{for } 0 \leq t \leq \tau, \quad x(0) = \varphi(0). \tag{3}$$

This equation has a unique solution and the solution of (3) on $[0, \tau]$ coincides with the solution of (2) on $[0, \tau]$ with initial data φ on $[-\tau, 0]$. Once the solution x is known on $[0, \tau]$, we can repeat the same procedure, starting with the solution on $[0, \tau]$, to find the solution $x(t)$ for $\tau \leq t \leq 2\tau$, etc. This process is called *the method of steps* and yields a unique, globally defined, solution $x(\,\cdot\,; \varphi)$ of (2), given an initial condition φ on $[-\tau, 0]$. Note that the solution becomes smoother in t as t increases. If we start with a continuous function on $[-\tau, 0]$, then the solution $x(\,\cdot\,; \varphi)$ is continuously differentiable on $(0, \tau)$, twice continuously differentiable on $(\tau, 2\tau)$, etc.

From the method of steps, it also follows that, in general, solutions of (2) are only defined on $[-\tau, \infty)$. Backward continuation of solutions of differential delay equations for negative time requires additional smoothness of the initial function φ on $[-\tau, 0]$. Furthermore, there is the additional problem that, if there exists a backward continuation, then this backward continuation is not necessarily unique, see Theorem 5.

Since there is only one time delay in equation (2), we can rescale time and assume that $\tau = 1$. To understand the large time behaviour of solutions of (2), one begins with an analysis of the behaviour of solutions of (2) near steady states. A steady state of (2) is a solution of the form $x(t) = \overline{x}$, where $\overline{x}$ is a solution of the algebraic equation $F(\overline{x}, \overline{x}) = 0$. To understand the behaviour of solutions near a steady state, one has to linearize around the steady state $\overline{x}$. This results in an autonomous linear delay equation

$$\dot{x}(t) = B_0 x(t) + B_1 x(t-1), \qquad t \geq 0 \tag{4}$$

with $B_0 = D_1 F(\overline{x}, \overline{x})$ and $B_1 = D_2 F(\overline{x}, \overline{x})$. Here D_i denotes the partial derivative of the function F with respect to the ith variable.

Next we present the basic theory for linear autonomous delay equations.

2.2 Linear autonomous equations

For autonomous linear delay equations the theory is well-developed, using principles from the theory of Laplace transformation. If we define $\mathcal{L}(x)$ to be the Laplace transform of the function x, i.e.,

$$\mathcal{L}(x)(z) = \int_0^\infty e^{-zt} x(t)\, dt$$

and apply the Laplace transform to the autonomous equation (4) with initial data $x(\theta) = \varphi(\theta)$ for $-1 \leq \theta \leq 0$, we obtain

$$\Delta(z)\mathcal{L}(x)(z) = \varphi(0) + B_1 \int_0^1 e^{-zt} \varphi(t-1)\, dt, \tag{5}$$

where $\Delta(z)$ denotes the characteristic matrix of (4) and is given by

$$\Delta(z) = zI - B_0 - B_1 e^{-z}. \tag{6}$$

The idea is to obtain an explicit representation of x by using the inverse Laplace transform, the Cauchy theorem and a residue calculus. In order to follow this approach, we need good estimates for $\Delta(z)^{-1}$ when z approaches infinity.

Since the zeros of the determinant of $\Delta(z)$, i.e., the roots of the equation

$$\det \Delta(z) = \det\bigl[zI - B_0 - B_1 e^{-z}\bigr] = 0 \tag{7}$$

satisfy a transcendental equation, there are, in general, infinitely many zeros. Characteristic equations of type (7) are well-studied (cf. [2,20]).

The asymptotic behaviour of the solutions of (4) as t tends to infinity is completely controlled by the behaviour of the roots of the characteristic equation (7).

Theorem 1. *Let x be the solution of* (4) *corresponding to an initial function φ. For any $\gamma \in \mathbb{R}$ such that* (7) *has no roots on the line* $\operatorname{Re} z = \gamma$*, we have the asymptotic expansion of the solution*

$$x(t) = \sum_{j=1}^{m} p_j(t) e^{\lambda_j t} + o(e^{\gamma t}) \quad \textit{for}\; t \to \infty, \tag{8}$$

where $\lambda_1, \ldots, \lambda_m$ are the finitely many roots of (7) *with real part exceeding γ and where $p_1(t), \ldots, p_m(t)$ are polynomials in t.*

Corollary 1. *All solutions of* (4) *converge to zero exponentially as* $t \to \infty$ *if and only if* (7) *has no roots in the right half-plane* $\{z \mid \operatorname{Re} z \geq 0\}$.

Therefore questions about exponential stability of steady states of (2) can be reduced to questions about the location of the zeros of the entire function $\det \Delta(z)$ where $\Delta(z)$ is given by (6).

The proof of Theorem 1 is based on the representation of the solution x of (4) using the inverse Laplace transform. From (5) and properties of the characteristic roots, it follows that

$$x(t) = \frac{1}{2\pi i} \int_{\gamma_0 - i\infty}^{\gamma_0 + i\infty} e^{zt} \Delta(z)^{-1} \left[\varphi(0) + B_1 \int_0^1 e^{-zt} \varphi(t-1)\, dt\right] dz. \tag{9}$$

Next the idea is to shift the line of integration to the left, while keeping track of the residues corresponding to the singularities of $\Delta(z)^{-1}$ that we pass (see [10, 10] for details).

What happens when we let $\gamma_0 \to -\infty$ in the series expansion (8)? Do we get a convergent infinite series and if so, is it a faithful representation of $x(t)$ or does $x(t)$ contain a component which goes to zero (as $t \to \infty$) faster than any exponential? A solution which does go to zero faster than any exponential is called a *small solution*. How do we recognize from the characteristic matrix $\Delta(z)$ whether or not small solutions exist? And how do we recognize whether the infinite expansion in polynomial-exponential functions is convergent and represents the solution for arbitrary initial data or, otherwise, how can we characterize those initial data for which it does?

For autonomous functional differential equations the importance of these questions have already been addressed in [1,3,4,19].

In the next section we shall outline this approach which is based on analyzing solution operators acting on function spaces of initial data rather than analyzing $\mathbb{C}^n$-valued solutions. We explain the natural connection between the questions above and very interesting questions about the completeness of systems of eigenvectors and generalized eigenvectors of nonself-adjoint operators.

3 A Functional Analytic Approach

The dynamical system approach to differential delay equations is to associate with (2) a semiflow acting on the space of initial data, defined by the time evolution of segments of solutions (see Fig. 2).

Let C denote the Banach space of continuous functions defined on the interval $[-1, 0]$ with values in $\mathbb{C}^n$, provided with the sup-norm $\|\varphi\| := \sup_{-1 \leq \theta \leq 0} |\varphi(\theta)|$ for $\varphi \in C$. If the solution of (2) with initial data $x(\theta) = \varphi(\theta)$, $-1 \leq \theta \leq 0$, is denoted by $x(\,\cdot\,; \varphi)$ and the state of the solution $x(\,\cdot\,; \varphi)$ is defined by

$$x_t(\theta; \varphi) = x(t + \theta; \varphi), \qquad -1 \leq \theta \leq 0,$$

then the semiflow $\Sigma(t;\,\cdot\,) : C \to C$ is defined by

$$\Sigma(t;\varphi) = x_t(\,\cdot\,;\varphi), \qquad t \geq 0.$$

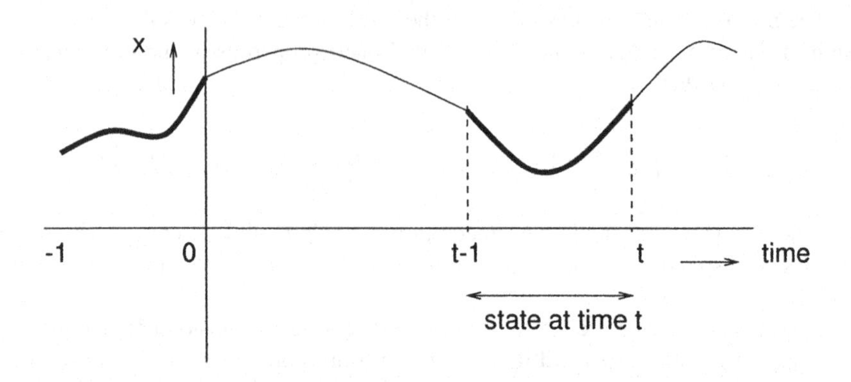

Fig. 2. The state space approach

3.1 Linear autonomous equations

The linearization of the semiflow $\Sigma(t;\,\cdot\,)$ around an equilibrium leads to a strongly continuous semigroup $T(t)$ defined by translation along the solutions of (4), i.e., $T(t)\varphi = x_t$, where x the solution of

$$\dot{x}(t) = B_0 x(t) + B_1 x(t-1), \qquad x_0 = \varphi, \quad \varphi \in C. \tag{10}$$

An important observation is that the evolution of the state x_t is given by an abstract ordinary differential equation in the infinite dimensional state space C. Given equation (10) this abstract ordinary differential equation can be computed explicitly and is given by

$$\frac{du}{dt} = Au, \qquad u(0) = \varphi,\ \varphi \in C, \tag{11}$$

where $A(C \to C)$ is an unbounded operator defined by

$$\begin{cases} A\varphi &= \frac{d\varphi}{d\theta} \\ \mathcal{D}(A) &= \{\varphi \in C \mid A\varphi \in C \text{ and } \frac{d\varphi}{d\theta}(0) = B_0\varphi(0) + B_1\varphi(-1)\} \end{cases} \tag{12}$$

and $u : [0,\infty) \to C$ is given by $u(t) = x_t$ for $t \geq 0$. Thus, the system of differential equations (10) with time delays can be viewed as a transport equation with nonlocal boundary conditions. Furthermore, the solutions of (10) are in one-to-one correspondence with the solutions of the infinite dimensional ordinary differential equation (11) and this correspondence is given by

$$u(t)(0) = x(t).$$

This observation originated with Krasovskii and has been crucial in the development of the qualitative theory of differential delay equations.

The semigroup $T(t)$ satisfies the following basic properties

(i) $T(0) = I$;
(ii) $T(t+s) = T(t)T(s)$ for $t, s \geq 0$;
(iii) for every $\varphi \in C$, the maps $t \to T(t)\varphi$ are continuous.

Furthermore, the operator A defined by (12) is the infinitesimal generator of the semigroup $T(t)$ and is formally defined by

$$A\varphi := \lim_{t \downarrow 0} \frac{1}{t}\left[T(t)\varphi - \varphi\right] \quad \text{for every } \varphi \in \mathcal{D}(A), \tag{13}$$

where $\mathcal{D}(A)$ consists of all $\varphi \in C$ for which the limit in (13) exists.

The growth bound of $T(t)$ is defined by

$$\omega_0 := \lim_{t \to \infty} \frac{1}{t} \log \|T(t)\|.$$

It is known that $\omega_0 < \infty$ and that, for $\operatorname{Re} z > \omega_0$, the resolvent of A equals the Laplace transform of the semigroup $T(t)$

$$(zI - A)^{-1} = \int_0^{\infty} e^{-zt} T(t)\, dt \quad \text{for } \operatorname{Re} z > \omega_0. \tag{14}$$

Let $\lambda \in \sigma(A)$ be an eigenvalue of A, the kernel $\mathcal{N}(\lambda I - A)$ is called the eigenspace at λ and its dimension d_λ the *geometric multiplicity*. The generalized eigenspace $\mathcal{M}_\lambda$ is the smallest closed subspace that contains all $\mathcal{N}((\lambda I - A)^j)$, $j = 1, 2, \ldots$ and its dimension is called the *algebraic multiplicity*. It is known that there is a close connection between the spectral properties of the infinitesimal generator A and the characteristic matrix $\Delta(z)$ associated with (10) (cf. [10] and [16]). In particular, the geometric multiplicity d_λ equals the dimension of the null space of $\Delta(\lambda)$ at λ and the algebraic multiplicity is equal to the multiplicity of $z = \lambda$ as a zero of $\det \Delta(z)$. Furthermore, the generalized eigenspace at λ is given by

$$\mathcal{M}_\lambda = \mathcal{N}\left((\lambda I - A)^{k_\lambda}\right),$$

where k_λ is the order of $z = \lambda$ as a pole of $\Delta(z)^{-1}$. The eigenvalue λ is called *simple* if $m_\lambda = 1$.

Let $\{\phi_1, \ldots, \phi_{m_\lambda}\}$ be a basis of eigenvectors and generalized eigenvectors of A at λ. Define the row m_λ-vector $\Phi = \{\phi_1, \ldots, \phi_{m_\lambda}\}$. Since $\mathcal{M}_\lambda$ is invariant under A, there exists a $m_\lambda \times m_\lambda$ matrix M such that $A\Phi = \Phi M$. The action of the semigroup $T(t)$ on Φ is given by $T(t)\Phi = \Phi e^{tM}$. In particular, it follows that $\Phi(\theta) = \Phi(0)e^{\theta M}$,

$-1 \le \theta \le 0$. So, in order to obtain a complete description of the subspace $\mathcal{M}_\lambda$, we have to compute $\Phi(0)$.

From standard spectral theory it follows that the spectral projection onto $\mathcal{M}_\lambda$ along $\mathcal{R}(\,(\lambda I - A)^{k_\lambda}\,)$ can be represented by a Dunford integral

$$P_\lambda = \frac{1}{2\pi i}\int_{\Gamma_\lambda} (zI - A)^{-1}\,dz, \tag{15}$$

where Γ_λ is a small circle such that λ is the only singularity of $(zI - A)^{-1}$ inside Γ_λ. A direct computation yields the following representation for the resolvent of A

$$(zI - A)^{-1}\varphi = \frac{1}{\det \Delta(z)} P(z)\varphi, \tag{16}$$

where

$$\big(P(z)\varphi\big)(\theta) = e^{z\theta}\Big\{\operatorname{adj}\Delta(z)[\varphi(0) + B_1\int_0^1 e^{-z\sigma}\varphi(\sigma-1)\,d\sigma] \\ + \det\Delta(z)\int_\theta^0 e^{-z\sigma}\varphi(\sigma)\,d\sigma\Big\}. \tag{17}$$

For example, if λ is a simple eigenvalue, this representation together with formula (15) can be used to obtain an explicit representation for the spectral projection onto the eigenspace corresponding to a simple eigenvalue

$$\begin{aligned} P_\lambda\varphi &= \operatorname*{Res}_{z=\lambda}\ (zI - A)^{-1}\varphi \\ &= e^{\lambda\cdot} H(\lambda)\big(\varphi(0) + B_1\int_0^1 e^{-\lambda\sigma}\varphi(\sigma-1)\,d\sigma\big). \end{aligned} \tag{18}$$

where $H(\lambda) = \lim_{z\to\lambda}(z-\lambda)\Delta(z)^{-1}$ is the residue of $\Delta(z)^{-1}$ at $z = \lambda$. Equivalently, we can write $P_\lambda\varphi = \langle\phi_\lambda^*, \varphi\rangle\phi_\lambda$, where ϕ_λ is an eigenvector at λ for A, ϕ_λ^* is an eigenvector at λ for A^* and $\langle\phi_\lambda^*, \phi_\lambda\rangle = 1$ (here $\langle\cdot,\cdot\rangle$ denotes the duality pairing between $\mathcal{C}$ and the dual space $\mathcal{C}^*$). For example, consider the system of delay equations (10) with $B_0 = 0$. The characteristic matrix is given by $\Delta(z) = zI - B_1 e^{-z}$ and, for every simple root of $\det\Delta(z)$, the spectral projection P_λ is given by

$$\big(P_\lambda\varphi\big)(\theta) = \Big[\frac{d}{dz}\det\Delta(\lambda)\Big]^{-1}\operatorname{adj}\Delta(\lambda)\Big(\varphi(0) + B_1\int_0^1 e^{-\lambda\tau}\varphi(\tau-1)\,d\tau\Big)e^{\lambda\theta}.$$

See [6] for details and for applications to the large time behaviour of solutions of differential delay equations.

Thus $\Phi(0)$ can be found by providing a basis for the range of P_λ. In [16] and Section IV.4 of [10] a systematic procedure has been developed to construct a canonical basis using Jordan chains. We shall here describe the underlying idea.

Definition 1. *A sequence of vectors* $\{\phi_0, \phi_1, \ldots, \phi_{r-1}\}$ *in* C *with* $\phi_0 \neq 0$ *is called a Jordan chain of* A *at* λ *if*

$$A\phi_0 = \lambda\phi_0, \qquad A\phi_j = \lambda\phi_j + \phi_{j-1},$$

for $j = 1, 2, \ldots, r-1$. The integer r is called the rank of the Jordan chain. Note that ϕ_0 is an eigenvector of A, the vectors $\phi_1, \ldots, \phi_{r-1}$ are called generalized eigenvectors at λ.

One can organize the Jordan chains according to the following procedure [9]. Choose an eigenvector, say $\phi_\lambda^{1,0}$, with maximal rank, say r_1. Next, choose a Jordan chain

$$\{\phi_\lambda^{1,0}, \ldots, \phi_\lambda^{1,r_1-1}\}$$

of length r_1 and let N_1 be the complement in $\mathcal{N}(\lambda I - A)$ of the subspace spanned by $\phi_\lambda^{1,0}$. In N_1 we choose an eigenvector $\phi_\lambda^{2,0}$ of maximal rank, say r_2, and let

$$\{\phi_\lambda^{2,0}, \ldots, \phi_\lambda^{2,r_2-1}\}$$

be a corresponding Jordan chain of length r_2. Next let N_2 be the complement in N_1 of the subspace spanned by $\phi_\lambda^{2,0}$ and replace N_1 by N_2 in the above-described procedure.

This construction yields a basis $\{\phi_\lambda^{1,0}, \ldots, \phi_\lambda^{d_\lambda,0}\}$ for $\mathcal{N}(\lambda I - A)$ and a corresponding canonical system of Jordan chains

$$\phi_\lambda^{1,0}, \ldots, \phi_\lambda^{1,r_1-1}, \ldots, \phi_\lambda^{d_\lambda,0}, \ldots, \phi_\lambda^{d_\lambda,r_{d_\lambda}-1}.$$

Note that $\max\{r_j \mid j = 1, \ldots, d_\lambda\} = k_\lambda$ and that

$$\sum_{j=1}^{d_\lambda} r_j = m_\lambda.$$

Because of the construction, the canonical system of Jordan chains is a basis for the generalized eigenspace $\mathcal{M}_\lambda$. With respect to this basis, the matrix M has Jordan normal form. We shall call such a basis a canonical basis of eigenvectors and generalized eigenvectors for A at λ.

To apply these ideas to delay equations, we shall first restrict ourselves to the case that $z = \lambda$ is a simple pole of $\Delta(z)^{-1}$, i.e., $\Delta(z)^{-1} = (z-\lambda)^{-1}H(z)$ with H analytic at $z = \lambda$.

Lemma 1. *If $z = \lambda$ is a simple pole of $\Delta(z)^{-1}$, then the generalized eigenspace at λ, $\mathcal{M}_\lambda = \mathcal{R}(P_\lambda)$, is given by*

$$\mathcal{M}_\lambda = \{\theta \mapsto e^{\lambda\theta}v \mid v \in \mathcal{N}(\Delta(\lambda))\}$$

and equals the eigenspace at λ (i.e., the geometric and the algebraic multiplicity of λ are equal).

Proof. First note that $\mathcal{R}\big(H(\lambda)\big) = \mathcal{N}\big(\Delta(\lambda)\big)$. Indeed, since

$$\Delta(z)H(z) = H(z)\Delta(z) = (z-\lambda)I.$$

It follows that $\Delta(\lambda)H(\lambda) = 0$. Assume that $v \in \mathbb{C}^n$ is such that $\Delta(\lambda)v = 0$ and define

$$w = \lim_{z\to\lambda} \frac{1}{z-\lambda}\Delta(z)v = \Delta'(\lambda)v.$$

Then

$$H(\lambda)w = \lim_{z\to\lambda} \frac{1}{z-\lambda}H(z)\Delta(z)v = \lim_{z\to\lambda} \frac{z-\lambda}{z-\lambda}v = v.$$

So $H(\lambda)w = v$. The proof of the lemma is now easily completed using the representation (18) for P_λ.

From the characterization of $\mathcal{M}_\lambda$ for simple poles of $\Delta(z)^{-1}$ we obtain $\Phi(0) = (v_1,\ldots,v_m)$, where $\{v_j\}_{j=1}^m$ is a basis for $\mathcal{N}\big(\Delta(\lambda)\big)$. To describe the procedure in general, one first has to extend the notion of a Jordan chain. An ordered set $(x_0, x_1, \ldots, x_{k-1})$ of vectors in $\mathbb{C}^n$ is called a Jordan chain for $\Delta(z)$ at $z = \lambda$ if $x_0 \neq 0$ and

$$\Delta(z)[x_0 + (z-\lambda)x_1 + \cdots + (z-\lambda)^{k-1}x_{k-1}] = O((z-\lambda)^k)$$

for $|z-\lambda| \to 0$. The number k is called the length of the chain and the maximal length of a chain starting with x_0 is called the rank of x_0.

We now have the following theorem (see [16] for a proof).

Theorem 2. *The spectrum of the generator A consists of eigenvalues of finite type only,*

$$\sigma(A) = \{\lambda \mid \det \Delta(\lambda) = 0\}.$$

For $\lambda \in \sigma(A)$, the geometric multiplicity of λ equals the dimension of the null space of $\Delta(\lambda)$, the algebraic multiplicity of λ equals the order of λ as a zero of $\det \Delta(z)$ *and the ascent of λ equals the order of λ as a pole of $\Delta^{-1}(z)$. Furthermore, a canonical basis of eigenvectors and generalized eigenvectors for A at λ can be obtained in the following way: If*

$$\{(\gamma_{i,0},\ldots,\gamma_{i,k_i-1}) \mid i = 1,\ldots,d_\lambda\}$$

is a canonical system of Jordan chains for $\Delta(z)$ at $z = \lambda \in \Omega$, then

$$\{\chi_{i,0},\ldots,\chi_{i,k_i-1} \mid i = 1,\ldots,d_\lambda\},$$

where

$$\chi_{i,\nu}(\theta) = e^{\lambda\theta}\sum_{l=0}^{\nu} \gamma_{i,\nu-l}\frac{\theta^l}{l!}$$

is a canonical basis for A at λ.

The generalized eigenspace $\mathcal{M}$ is the smallest subspace that contains all $\mathcal{M}_\lambda$, $\lambda \in \sigma(A)$. If $\mathcal{M}$ is dense in $\mathcal{C}$, we call the system of eigenvectors and generalized eigenvectors of A *complete*. A Jordan chain of A gives rise to a special solution of (10). If $\{\varphi_0, \varphi_1, \ldots, \varphi_{r-1}\}$ is a Jordan chain of A at λ, then

$$x(t) = e^{\lambda t} \sum_{\nu=0}^{r-1} \frac{1}{\nu!} t^\nu \varphi_{r-1-\nu}(0), \tag{19}$$

is a solution of (10) for all t. This solution corresponds to the special solutions in Theorem 1. So, if the linear space spanned by all eigenvectors and generalized eigenvectors of the infinitesimal generator is dense in $\mathcal{C}$, then each solution of (10) can be approximated by a linear combination of solutions of elementary solutions of the form (19).

4 Spectral Theory for Autonomous Equations

In this section we are interested in the question whether we can obtain a convergent series by letting $\gamma \to -\infty$ in (8). In Section 2 we have seen that the initial data corresponding to solutions $x(t) = p(t)e^{\lambda t}$ that arise in (8) are precisely the eigenvectors and generalized eigenvectors of the infinitesimal generator A defined in (12). Therefore, the question of whether we obtain a convergent series by letting $\gamma \to -\infty$ in (8) can be rephrased as a question concerning the convergence of the spectral projections of the infinitesimal generator A

$$T(t)\varphi = \sum_{\lambda \in \sigma(A)} T(t) P_\lambda \varphi, \tag{20}$$

where the convergence is in the state space $\mathcal{C}$. The initial question is contained in this problem. Indeed, every solution of (10) has a convergent series expansion if and only if (20) holds for every $t > 1$.

To analyse the behaviour of sums of spectral projections, we shall use the Riesz projection and the Cauchy theorem on residues. In order to do so, we need good estimates for the resolvent of A near infinity. The explicit representation for the resolvent of A given in (16) allows us to obtain these estimates using estimates for $\Delta(z)^{-1}$ near infinity.

4.1 Basic estimates

In this section we recall estimates for the inverse of the characteristic matrix in the complex plane outside small circles centered around the zeros of $\det \Delta(z)$. These estimates were first given in [27] and provide the basic tool to estimate the resolvent of the generator near infinity.

Consider the characteristic matrix (7)

$$\Delta(z) = zI - B_0 - B_1 e^{-z},$$

where B_0 and B_1 are constant $(n \times n)$-matrices. In order to estimate $|\Delta(z)^{-1}|$, we first analyse the behaviour of $|\Delta_1(z)^{-1}|$, where

$$\Delta_1(z) = zI - B_1 e^{-z}. \tag{21}$$

Since $e^z \Delta_1(z) = ze^z - B_1$, estimates for $|\Delta_1(z)^{-1}|$ follow from estimates for the resolvent $R(w) := (wI - B_1)^{-1}$ of B_1.

From the Neumann series, we have

$$|R(w)| \le (|w| - |B_1|)^{-1} \qquad \text{for} \quad |w| > |B_1|. \tag{22}$$

Therefore, for $z \in \mathbb{C}$ with $|ze^z| > |B_1|$,

$$|\Delta_1(z)^{-1}| = |e^z R(ze^z)| \le |e^z|(|ze^z| - |B_1|)^{-1}$$

and hence, for $z \in \mathbb{C}$ with $|ze^z| \ge C_0^{-1}|B_1|$ and $C_0 < 1$, we find

$$|\Delta_1(z)^{-1}| \le \frac{1}{|z|} \frac{|ze^z|}{|ze^z| - |B_1|} \le \frac{1}{|z|}(1 - C_0)^{-1}. \tag{23}$$

In particular, it follows that $\det \Delta_1(z) \ne 0$ for $z \in \mathbb{C}$ in the region with $|ze^z| \ge C_0^{-1}|B_1|$.

To estimate $|\Delta_1(z)^{-1}|$ for $z \in \mathbb{C}$ with $|ze^z|$ close to zero, we need further information about the behaviour $R(w)$, the resolvent of B_1, near the poles of $R(w)$. The poles of $R(w)$ are exactly the eigenvalues of B_1 and on the resolvent set $\rho(B_1) = \mathbb{C} \setminus \sigma(B_1)$, the function $w \mapsto R(w)$ is holomorphic. To extend the estimate (22) away from the poles of $R(w)$, it remains to estimate $|R(w)|$ on the compact set K, where

$$K = \{w \in \mathbb{C} \mid |w| \le |B_1| \text{ and } d(w, \sigma(B_1)) \ge \epsilon\}$$

and $d(w, H) = \inf\{d(w, h) \mid h \in H\}$ denotes the distance from a point $w \in \mathbb{C}$ to the set $H \subset \mathbb{C}$.

Since $w \mapsto R(w)$ is continuous on K, we conclude that given $\epsilon > 0$ and $w \in \mathbb{C}$ with $d(w, \sigma(B_1)) \ge \epsilon$, there exists a constant $M = M(\epsilon)$ such that $|R(w)| \le M$. Therefore, for $z \in \mathbb{C}$ with $d(ze^z, \sigma(B_1)) \ge \epsilon$, we find

$$|\Delta_1(z)^{-1}| \le |e^z| M. \tag{24}$$

Furthermore, to each nonzero eigenvalue μ_j, $j = 1, 2, \ldots, m$, of B_1, there corresponds a chain of zeros z_{jk}, $k = 1, 2, \ldots$, of $\det \Delta_1(z)$ defined by

$$z_{jk} e^{z_{jk}} = \mu_j.$$

Continuity of the complex function $z \mapsto ze^z$ implies that, given $\epsilon > 0$ and $z \in \mathbb{C}$ with $d(ze^z, \sigma(B_1)) \ge \epsilon$, there exists a $\delta > 0$ and a $C > 0$ such that, if $|z| > C$, then z is outside circles of radius δ centered at the zeros of $\det \Delta_1(z)$.

If B_1 is invertible, then $0 \notin \sigma(B_1)$ and we obtain that, given $\epsilon > 0$, there exists a positive constant C_0 such that, for $z \in \mathbb{C}$ with $|ze^z| \le C_0$, we have that

$d(ze^z, \sigma(B_1)) \geq \epsilon$. Hence, if B_1 is invertible, then for $z \in \mathbb{C}$ and $|ze^z| \leq C_0$, we have the following estimate

$$|\Delta_1(z)^{-1}| \leq |e^z|M. \tag{25}$$

In particular, it follows that $\det \Delta_1(z) \neq 0$ for $z \in \mathbb{C}$ with $|ze^z| \leq C_0$. Furthermore, it follows from (24) that, for $z \in \mathbb{C}$ and $C_0 < |ze^z| < C_0^{-1}$, the same estimate (25) holds as long as $d(ze^z, \sigma(B_1)) > \epsilon$. Or, equivalently, as long as $z \in \mathbb{C}$ is outside circles of radius δ centered around the zeros of $\det \Delta_1(z)$.

Summarizing we have proved the following lemma.

Lemma 2. *If* $\Delta_1(z) = zI - B_1e^{-z}$ *and* B_1 *is invertible, then, for every* $\delta > 0$, *there exists a constant* $M = M(\delta)$ *such that*

$$|\Delta_1(z)^{-1}| \leq M \min\{\frac{1}{|z|}, |e^z|\}, \tag{26}$$

for $z \in \mathbb{C}$ *outside circles of radius* δ *centered around the zeros of* $\det \Delta_1(z)$. *Furthermore, there exists a constant* C_0, $0 < C_0 < 1$, *such that the zeros of* $\det \Delta_1(z)$ *are inside the set* $V(C_0) = \{z \in \mathbb{C} \mid C_0 < |ze^z| < C_0^{-1}\}$.

Next we use estimate (26) for $|\Delta_1(z)^{-1}|$ to find an estimate for $|\Delta(z)^{-1}|$. Fix $\delta > 0$ and let $z \in \mathbb{C}$ outside circles of radius δ centered around the zeros of $\det \Delta_1(z)$. Since $\det \Delta_1(z) \neq 0$, we have

$$\Delta(z) = \Delta_1(z)\big(I - \Delta_1(z)^{-1}B_0\big) \tag{27}$$

and it follows that

$$\Delta(z)^{-1} = (I - \Delta_1(z)^{-1}B_0)^{-1}\Delta_1(z)^{-1}.$$

So, it remains to estimate $|(I - \Delta_1(z)^{-1}B_0)^{-1}|$. From the Neumann series, it follows that it suffices to estimate $|\Delta_1(z)^{-1}B_0|$ and, from Lemma 2, we obtain that, for $|z| > 2M|B_0|$,

$$\begin{aligned} \big|(I - \Delta_1(z)^{-1}B_0)^{-1}\big| &\leq \big(1 - |\Delta_1(z)^{-1}B_0|\big)^{-1} \\ &\leq 2. \end{aligned}$$

Thus, for $|z| > 2M|B_0|$, we have that $\Delta(z)$ is invertible and

$$|\Delta(z)^{-1}| \leq 2M \min\{\frac{1}{|z|}, |e^z|\}, \tag{28}$$

for $z \in \mathbb{C}$ outside circles of radius δ centered around the zeros of $\det \Delta_1(z)$. In particular, for $|z| > 2M|B_0|$, the zeros of $\det \Delta(z)$ are inside circles of radius δ centered around the zeros of $\det \Delta_1(z)$. In particular, inside the set $V(C_0)$ and (28) holds for $z \in \mathbb{C}$ outside circles of radius 2δ centered around the zeros of $\det \Delta(z)$.

We summarize the results of the discussion in a lemma.

Lemma 3. *If $\Delta(z) = zI - B_0 - B_1 e^{-z}$ and B_1 is invertible, then, for every $\delta > 0$, there exists a constants C and $M = M(\delta)$ such that*

$$|\Delta(z)^{-1}| \le M \min\{\frac{1}{|z|}, |e^z|\},$$

for $z \in \mathbb{C}$ with $|z| > C$ outside circles of radius δ centered around the zeros of $\det \Delta(z)$. *Furthermore, there exists a constant C_0, $0 < C_0 < 1$, such that the zeros of* $\det \Delta(z)$ *are inside the set $V(C_0) = \{z \in \mathbb{C} \mid C_0 < |ze^z| < C_0^{-1}\}$.*

Actually, we can use Lemma 2 to prove estimates for $|\Delta(z)^{-1}|$ for much more general characteristic matrices $\Delta(z)$. In fact, the same proof shows that Lemma 3 remains true for the characteristic matrix

$$\Delta(z) = \Delta_1(z) - \int_0^1 e^{-z\theta} d\zeta_0(\theta),$$

where ζ_0 is a matrix-valued function of bounded variation that is continuous at $\theta = 1$. See [25] for a proof of this fact using different arguments.

Next we consider the case that $B_1 \neq 0$ is not invertible. In this case we need more information about the singularity of the resolvent $R(w)$ of B_1 at $w = 0$. This information can be obtained from the Laurent series of $R(w)$ at $w = 0$

$$R(w) = \sum_{n=-k_0}^{\infty} w^n A_n.$$

The coefficients A_n are given by

$$A_n = \frac{1}{2\pi i} \int_\Gamma z^{-n-1} R(z)\, dz,$$

where Γ is a positively-oriented small circle of radius δ with center at $w = 0$ excluding any other eigenvalue of B_1. The coefficient A_{-1} is exactly the spectral projection onto the generalized eigenspace of B_1 at $\lambda = 0$ (see [17]).

Therefore, for $w \in \mathbb{C}$ with $0 < |w| \le \delta$, there exists a constant $M_1 = M_1(\delta)$ such that

$$|R(w)| \le M_1 |w|^{-k_0},$$

where k_0 is the order of the pole of $R(w)$ at $w = 0$. Therefore, for $z \in \mathbb{C}$ with $0 < |ze^z| \le \epsilon$, we find

$$|\Delta_1(z)^{-1}| \le M|z|^{-k_0}|e^{(1-k_0)z}|.$$

Thus, in the general case, Lemma 2 becomes

Lemma 4. *If $\Delta_1(z) = zI - B_1 e^{-z}$ and k_0 is the order of pole of the resolvent of $B_1 \neq 0$ at zero, then there is a constant C_0, $0 < C_0 < 1$, such that, for every $\delta > 0$, there exists a constant $M = M(\delta)$ and*

$$|\Delta_1(z)^{-1}| \leq M \min\{\frac{1}{|z|}, |e^z|\},$$

for $z \in \mathbb{C}$ *with* $|ze^z| > C_0$ *and* z *outside circles of radius* δ *centered around the zeros of* $\det \Delta_1(z)$. *If* $0 < |ze^z| \leq C_0$, *then* $\Delta_1(z)$ *is invertible and*

$$|\Delta_1(z)^{-1}| \leq M|z|^{-k_0}|e^{(1-k_0)z}|.$$

Furthermore, the nonzero zeros of $\det \Delta_1(z)$ *are inside the set* $V(C_0) = \{z \in \mathbb{C} \mid C_0 < |ze^z| < C_0^{-1}\}$.

As before, we use Lemma 4 and (27) to derive estimates for $|\Delta(z)^{-1}|$. In the general case, however, we have that $|\Delta_1(z)^{-1}|$ can grow in the region

$$W = \{z \in \mathbb{C} \mid |z| \geq C \text{ and } |ze^z| \leq C_0\}.$$

Therefore, we cannot use the Neumann series to estimate $|(I - \Delta_1(z)^{-1}B_0)^{-1}|$. To provide the estimate for $|\Delta(z)^{-1}|$ in the region W, we have to use a different approach. First we claim that $\Delta(z)$ is invertible for $z \in W$. To prove this claim, note that Lemma 4 implies that, for $z \in W$, the matrix $\Delta_1(z)$ is invertible. Furthermore, for C_0 sufficiently small, we have the inequality

$$\frac{1}{2}|B_1| \leq |ze^z I - B_1| \leq 2|B_1|$$

and hence

$$\frac{1}{2}|B_1| \leq |e^z \Delta_1(z)| \leq 2|B_1|.$$

Since $e^z\Delta(z) = e^z\Delta_1(z) - e^z B_0$ and $|z| > 2C_0$

$$|e^z B_0| \leq C_0|z|^{-1}|B_0| \leq \frac{1}{2}|B_1,$$

it follows that, for C large, C_0 small and $z \in W$, the matrix $\Delta(z)$ is invertible. Secondly, we claim that, for $z \in W$, the spectrum of $\Delta_1(z)^{-1}B_0$ is outside a circle of radius $r = r(C, C_0)$ centered at 1. To prove this claim, we assume that the statement is not true. Then there exists a $\mu \in \mathbb{C}$ with $|\mu - 1| < r$ and a vector $v \in \mathbb{C}^n$ with $v \neq 0$ such that $\Delta_1(z)^{-1}B_0 v = \mu v$. But then $B_0 v = \mu\Delta_1(z)v$ and

$$[\Delta(z) + (\mu - 1)\Delta_1(z)]v = 0.$$

Since $\Delta(z)$ is invertible, it suffices to show that, for $|\mu - 1|$ close to zero, we have $|\Delta(z)| > |\mu - 1||\Delta_1(z)|$. Indeed, this estimate implies that $\Delta(z) + (\mu - 1)\Delta_1(z)$ is invertible and hence $v = 0$ which contradicts the assumption. Note that if $z \in W$, then we can choose $C_0 > 0$ small and $C > 0$ large such that

$$|e^z\Delta(z)| = |ze^z I - e^z B_0 - B_1| > \frac{1}{2}|B_1|$$

and

$$|e^z \Delta_1(z)| = |ze^z I - B_1| \leq 2|B_1|.$$

Thus if we choose $r < 1/4$ and $\mu \in \mathbb{C}$ with $|\mu - 1| < r$, then $|\Delta(z)| > |\mu - 1||\Delta_1(z)|$ and this proves the claim.

This observation allows us to use the Laurent series to conclude that there exists a constant $M_2 = M_2(r)$ such that

$$\left|\left(I - \Delta_1(z)^{-1} B_0\right)^{-1}\right| \leq M_2.$$

So representation (27) together with Lemma 4 yields an estimate for $|\Delta(z)^{-1}|$ when $z \in W$.

Thus, in the general case, we arrive at the following estimates for $|\Delta(z)^{-1}|$.

Lemma 5. *If $\Delta(z) = zI - B_0 - B_1 e^{-z}$ and k_0 is the order of pole of the resolvent of $B_1 \neq 0$ at zero, then there is a constant C_0, $0 < C_0 < 1$, such that, for every $\delta > 0$, there exists a constant $M = M(\delta)$ and*

$$|\Delta(z)^{-1}| \leq M \min\{\frac{1}{|z|}, |e^z|\},$$

for $z \in \mathbb{C}$ with $|ze^z| > C_0$ and z outside circles of radius δ centered around the zeros of $\det \Delta(z)$. *If $0 < |ze^z| \leq C_0$ and $|z| > C$, then $\Delta(z)$ is invertible and*

$$|\Delta(z)^{-1}| \leq M|z|^{-k_0}|e^{(1-k_0)z}|.$$

Furthermore, there exist positive constants C_0 and C such that the zeros of $\det \Delta(z)$ *with $|z| > C$ are inside the set $V(C_0) = \{z \in \mathbb{C} \mid C_0 < |ze^z| < C_0^{-1}\}$.*

Remark that Lemma 5 does not immediately generalize to more general perturbations of $\Delta_1(z)$ and the precise estimates for $|\Delta(z)^{-1}|$ now do depend on the lower order terms in $\Delta(z)$.

To illustrate this fact consider the system of delay equations

$$\dot{x}(t) = B_0 x(t - \frac{1}{2}) + B_1 x(t - 1).$$

The characteristic matrix $\Delta(z)$ is given by

$$\Delta(z) = zI - B_0 e^{-z/2} - B_1 e^{-z}.$$

Therefore

$$e^{-z}\Delta(z)^{-1} = \left(I - \Delta_1(z) B_0 e^{z/2}\right)^{-1} \Delta_1(z)^{-1}.$$

Note that in the region where $\Delta_1(z)$ is invertible, we do not necessary have that $\Delta(z)$ is invertible. Indeed, $|\Delta_1(z) B_0 e^{z/2}|$ is close to 1 for $|ze^{z/2}|$ close to 1. This leads to chains of zeros of $\det \Delta(z)$ with different asymptotics then the zeros of $\det \Delta_1(z)$ and the behaviour of $\Delta(z)^{-1}$ is not completely controlled by the behaviour of $\Delta_1(z)^{-1}$. See [22] and Chapter V of [10] for a different approach based on an analysis of $\det \Delta(z)$ directly.

We end this subsection with the introduction of a standard sequence of contours that will be used to compute the complex line integral in the next section.

Using the estimates for $\Delta(z)^{-1}$ in Lemma 3 and Lemma 5, we can construct a sequence of real numbers ρ_l, such that $\rho_l \to \infty$, and a sequence of closed contours $\Gamma_l, l = 0, 1, \ldots$, such that for some positive constants k, ϵ and δ:

(i) Γ_l is contained in the interior of Γ_{l+1} and there are at most k zeros between Γ_l and Γ_{l+1};
(ii) the contours have at least distance $\epsilon > 0$ from the set of zeros of $\det \Delta(z)$;
(iii)the contour Γ_l lies along the circle $|z| = \rho_l$ outside $V(C_0)$; inside $V(C_0)$, the contour lies between the circle $|z| = \rho_l - \delta$ and the circle $|z| = \rho_l + \delta$;
(iv)the length of the portion of Γ_l within $V(C_0)$ is bounded for $l \to \infty$.

For any real γ we denote the part of the sequence of contours Γ_l contained in the left half-plane $\{z \mid \operatorname{Re} z \leq \gamma\}$ by $\Gamma_l^-(\gamma)$.

We end this section with an auxiliary result from complex analysis (see, for example, Lemma V.5.10 of [10]).

Lemma 6. *For any real number* γ

$$\lim_{l\to\infty} \Big| \int_{\Gamma_l^-(\gamma)} e^{zt} \min(\frac{|e^{-z}|}{|z|}, 1)\, dz \Big| = 0, \qquad \textit{for } t > 0.$$

The convergence is uniform for t in any interval $0 < t_0 < t < t_1 < \infty$.

4.2 Series expansions for autonomous equations

In this section we use the estimates for $|\Delta(z)^{-1}|$ to analyse the behaviour of the series of spectral projections. As explained before, the idea is to obtain an explicit representation for $T(t)\varphi$ itself, using the inverse of the Laplace transform. The starting point is the following inversion formula, see Theorem 11.6.1 of [15],

$$T(t)\varphi = \lim_{\omega\to\infty} \frac{1}{2\pi i} \int_{\gamma - i\omega}^{\gamma + i\omega} e^{zt}(zI - A)^{-1}\varphi\, dz, \qquad t > 0, \tag{29}$$

for $\gamma > \omega(A)$ and for every $\varphi \in \mathcal{D}(A)$.

In order to use the contours Γ_l introduced in the previous subsection to compute the integral in (29), we need the following observation. If $\lambda \in \sigma(A)$ is a pole of $z \mapsto e^{zt}(zI - A)^{-1}\varphi$, then

$$\operatorname*{Res}_{z=\lambda} e^{zt}(zI - A)^{-1}\varphi = T(t)P_\lambda\varphi,$$

where P_λ denotes the spectral projection onto $\mathcal{M}_\lambda$. This identity can be derived from (29), (15) and the resolvent equation

$$(zI - A)^{-1} - (\lambda I - A)^{-1} = (\lambda - z)(zI - A)^{-1}(\lambda I - A)^{-1}.$$

We are now ready to prove the following theorem.

Theorem 3. *Let $T(t)$ denote the semigroup associated with*

$$\dot{x}(t) = B_0 x(t) + B_1 x(t-1),$$

where B_1 is a nonsingular matrix. If $A(\mathcal{C} \to \mathcal{C})$ denotes the generator of $T(t)$, then, for every $\varphi \in \mathcal{D}(A)$, we have

$$\lim_{N\to\infty} \|T(t)\varphi - \sum_{j=0}^{N} T(t)P_{\lambda_j}\varphi\| = 0 \qquad \text{for } t > 0$$

uniformly on compact t-sets. Here λ_j, $j = 0, 1, \ldots$ denote the eigenvalues of A ordered by increasing modulus and P_{λ_j} denotes the spectral projection given by (15).

Proof. The Cauchy theorem implies that

$$T(t)\varphi = \lim_{l\to\infty} \Big\{ \sum_{j=0}^{m(l)} T(t)P_{\lambda_j}\varphi - \frac{1}{2\pi i}\int_{\Gamma_l^-} e^{zt}(zI - A)^{-1}\varphi \, dz \Big\},$$

where $\lambda_0, \ldots, \lambda_{m(l)}$ are the zeros of $\det \Delta$ inside the area enclosed by the line $\operatorname{Re} z = \gamma$ and the contour Γ_l^-. And it suffices to prove that, for every $\varphi \in \mathcal{D}(A)$,

$$\lim_{l\to\infty} \|\frac{1}{2\pi i}\int_{\Gamma_l^-} e^{zt}(zI - A)^{-1}\varphi \, dz\| = 0. \tag{30}$$

To analyse this limit, we shall use the representation for the resolvent of A given in (16) which we rewrite as follows

$$(zI - A)^{-1}\varphi = \Delta(z)^{-1}C(z)\varphi, \tag{31}$$

where

$$\begin{aligned}
\big(C(z)\varphi\big)(\theta) &= e^{z\theta}\{\varphi(0) + B_1 \int_0^1 e^{-z\sigma}\varphi(\sigma - 1)\, d\sigma + \Delta(z)\int_\theta^0 e^{-z\sigma}\varphi(\sigma)\, d\sigma\} \\
&= e^{z\theta}\varphi(0) + B_1 \int_0^{\theta+1} e^{-z\sigma}\varphi(\sigma - 1)\, d\sigma \\
&\qquad\qquad + (zI - B_0)\int_0^{-\theta} e^{-z\sigma}\varphi(\sigma + \theta)\, d\sigma.
\end{aligned} \tag{32}$$

Note that

$$\|C(z)\varphi\| = \sup_{-1\le\theta\le 0} \big|\big(C(z)\varphi\big)(\theta)\big| \le M|z|\,|e^{-z}|\,\|\varphi\|. \tag{33}$$

Together with the identity

$$(zI - A)^{-1}\varphi = \frac{1}{z}(zI - A)^{-1}A\varphi + \frac{1}{z}\varphi, \qquad \varphi \in \mathcal{D}(A),$$

it follows that, for $\varphi \in \mathcal{D}(A)$,

$$\|(zI - A)^{-1}\varphi\| \le C_1 \big|e^{-z}\Delta(z)^{-1}\big| \, \|A\varphi\| + C_2|z|^{-1}\|\varphi\|.$$

Because of the estimates for $|\Delta(z)^{-1}|$ and the definition of the contours Γ_l, there exists a constant l_0 such that, for $z \in \Gamma_l$, $l \ge l_0$, we have

$$\big|e^{-z}\Delta(z)^{-1}\big| \le K \min\{\frac{|e^{-z}|}{|z|}, 1\}. \tag{34}$$

Therefore, we obtain the following estimate for the resolvent of A

$$\|(zI - A)^{-1}\varphi\| \le C_1 K \min\{\frac{|e^{-z}|}{|z|}, 1\}\|A\varphi\| + C_2\frac{1}{|z|}\|\varphi\|$$

and an application of Lemma 6 shows (30).

Since $\mathcal{R}\big(T(t)\big) \subset \mathcal{D}(A)$ for $t > 1$, the next result follows immediately, see [24] for details and further results.

Corollary 2. *For every $\varphi \in C$, the solution $x(\,\cdot\,;\varphi)$ of the differential delay equation*

$$\dot{x}(t) = B_0 x(t) + B_1 x(t-1), \qquad x_0 = \varphi,$$

where B_1 is a nonsingular matrix, has a convergent series expansion

$$x(t) = \sum_{j=0}^{\infty} e^{\lambda_j t} p_j(t), \qquad t > 0.$$

Example 1. Consider the retarded equation

$$\dot{x}(t) = B_1 x(t-1), \quad t \ge 0, \qquad x_0 = \varphi, \tag{35}$$

where $B_1 \neq 0$ is an $n \times n$-matrix. The characteristic matrix is given by

$$\Delta(z) = zI - B_1 e^{-z}. \tag{36}$$

For every simple root of $\det \Delta(z)$, the spectral projection is given by

$$(P_\lambda\varphi)(\theta) = \Big[\frac{d}{dz}\det \Delta(\lambda)\Big]^{-1} \operatorname{adj} \Delta(\lambda)\Big(\varphi(0) + B_1 \int_0^1 e^{-\lambda\tau}\varphi(\tau - 1)d\tau\Big)e^{\lambda\theta}.$$

In the scalar case, a root λ of Δ is not simple if and only if

$$\begin{cases} \lambda - B_1 e^{-\lambda} = 0, \\ 1 + B_1 e^{-\lambda} = 0. \end{cases}$$

Therefore, if $B \neq -1/e$ or equivalently $\lambda = -1$ is not a root of Δ, then all roots of (36) are simple. So the spectral projections are given by

$$(P_\lambda\varphi)(\theta) = \frac{1}{1+\lambda}\Big(\varphi(0) + B_1\int_0^1 e^{-\lambda\tau}\varphi(\tau-1)d\tau\Big)e^{\lambda\theta}, \tag{37}$$

where λ satisfies $\lambda - B_1 e^{-\lambda} = 0$. Furthermore, it follows from Theorem 3 that

$$T(t)\varphi = \sum_{j=1}^{\infty} P_{\lambda_j}T(t)\varphi = \sum_{j=1}^{\infty} T(t)P_{\lambda_j}\varphi, \quad t > 0,$$

where λ_j, $j = 0, 1, \ldots$, denote the roots of $\lambda - Be^{-\lambda} = 0$, ordered according to decreasing real part. Using (37) and the fact that $T(t)e^{\lambda_j \cdot} = e^{\lambda_j(t+\cdot)}$, we can now explicitly compute the solution of (35) with initial condition $x_0 = \varphi$

$$x(t;\varphi) = \sum_{j=0}^{\infty} \frac{1}{1+\lambda_j}\Big(\varphi(0) + B_1\int_0^1 e^{-\lambda_j\tau}\varphi(\tau-1)d\tau\Big)e^{\lambda_j t}, \qquad t > 0.$$

If $B_1 = -1/e$, then all zeros of $\Delta(z)$ are simple except for $\lambda = -1$. For the simple zeros we can again use (37). For the double zero $\lambda = -1$, we have to work directly with (15) to compute the projection onto the two dimensional space $\mathcal{M}_{-1}$ and a residue computation yields that P_{-1} is given by

$$\begin{aligned}(P_{-1}\varphi)(\theta) = \Big(-\frac{2}{3}\varphi(0) + \frac{8}{3}\int_{-1}^0 e^{\tau}\varphi(\tau)d\tau + 2\int_{-1}^0 \tau e^{\tau}\varphi(\tau)d\tau\Big)e^{-\theta} \\ + 2\Big(\varphi(0) - \int_{-1}^0 e^{\tau}\varphi(\tau)d\tau\Big)\theta e^{-\theta}.\end{aligned}$$

Since $T(t)\phi = \phi(t+\cdot)$, where $\phi(\theta) = \theta e^{-\theta}$, we can again give the solution explicitly

$$\begin{aligned}x(t;\varphi) = {} & \Big(-\frac{2}{3}\varphi(0) + \frac{8}{3}\int_{-1}^0 e^{\tau}\varphi(\tau)d\tau + 2\int_{-1}^0 \tau e^{\tau}\varphi(\tau)d\tau\Big)e^{-\theta} \\ & + 2\Big(\varphi(0) - \int_{-1}^0 e^{\tau}\varphi(\tau)d\tau\Big)\theta e^{-\theta} \\ & + \sum_{j=1}^{\infty}\frac{1}{1+\lambda_j}\Big(\varphi(0) - \int_0^1 e^{-\lambda_j\tau-1}\varphi(\tau-1)d\tau\Big)e^{\lambda_j t}, \qquad t > 0,\end{aligned}$$

where λ_j, $j = 1, 2, \ldots$ are the zeros of Δ with real part less than -1 ordered according to decreasing real part. See [6] for further results.

The next corollary is a completeness result. See [22–24, 30] for much more general results.

Corollary 3. *Let $T(t)$ denote the semigroup associated with*

$$\dot{x}(t) = B_0 x(t) + B_1 x(t-1),$$

where B_1 is a nonsingular matrix. If $A(\mathcal{C} \to \mathcal{C})$ denotes the generator of $T(t)$, then the system of eigenvectors and generalized eigenvectors of A is complete in $\mathcal{C}$.

Proof. Let $\mathcal{M}$ denote the linear space spanned by the eigenvectors and generalized eigenvectors of A. To prove the statement we have to show that $\overline{\mathcal{M}} = \mathcal{C}$. Let $\varphi \in \mathcal{C}$. Since $\mathcal{D}(A)$ is dense in $\mathcal{C}$, we can choose a sequence $\varphi_j \in \mathcal{D}(A)$ such that $\varphi_j \to \varphi$ in $\mathcal{C}$. Then $T(t)\varphi_j \to T(t)\varphi$ uniformly in t on compact subsets of $[0, \infty)$. Because of Theorem 3, for every φ_j and $\epsilon > 0$, we have for $t \geq 0$

$$\lim_{N\to\infty} \|T(t+\epsilon)\varphi - \sum_{j=0}^{N} T(t+\epsilon)P_{\lambda_j}\varphi\| = 0.$$

Define $\varphi_{j,\epsilon} = T(\epsilon)\varphi_j$ and recall from the $\mathcal{C}_0$-semigroup property of $T(t)$ that

$$\|\varphi_{j,\epsilon} - \varphi_j\| \to 0 \qquad \text{as } \epsilon \downarrow 0.$$

Therefore, we can construct a subsequence $\{\hat{\varphi}_j\}$ of $\{\varphi_{j,\epsilon}\}$, such that $T(t)\hat{\varphi}_j$ has a convergent spectral projection series uniformly in t on compact subsets of $\mathbb{R}_+$ and $\hat{\varphi}_j$ converges to φ in $\mathcal{C}$ as $j \to \infty$. So, we have proved $\hat{\varphi}_j \in \overline{\mathcal{M}}$ and hence $\varphi \in \overline{\mathcal{M}}$. Thus $\overline{\mathcal{M}} = \mathcal{C}$ and this proves the corollary.

To illustrate Corollary 3, we consider the following example

$$\begin{cases} \dot{x}_1(t) = x_1(t) + x_2(t-1), \\ \dot{x}_2(t) = x_1(t-1). \end{cases} \tag{38}$$

Since B_1 is invertible, the system of eigenvectors and generalized eigenvectors of the generator of the semigroup associated with (38) is complete.

Next we consider the case that B_1 is not invertible. From an inspection of the proof of Theorem 3, we can make the following observation. If, for $\varphi \in \mathcal{D}(A)$, there exist positive constants τ, l_0 and K_4 such that, for $l \geq l_0$ and $z \in \Gamma_l^-(\gamma)$,

$$\|e^{z\tau}(zI - A)^{-1}\varphi\| \leq K_4 \min\{\frac{\max\{|e^{-z}|, 1\}}{|z|}, 1\}\left(\|\varphi\| + \|A\varphi\|\right), \tag{39}$$

then the conclusion of Theorem 3 holds. Therefore, for $t > \tau$,

$$T(t)\varphi = \lim_{N\to\infty} \sum_{j=0}^{N} T(t)P_{\lambda_j}\varphi.$$

To investigate condition (39) further, we use (31) and the basic estimates for $|\Delta(z)^{-1}|$. First note that, it follows from Lemma 5, that, for $z \in \mathbb{C}$ and $|ze^z| > C_0$ for some C_0, $0 < C_0 < 1$, the estimate for $|\Delta(z)^{-1}|$ is the same as in the case that B_1 is invertible. Thus to investigate (39), we can assume that $|ze^z| \leq C_0$. In this case, Lemma 5 yields the following estimate

$$\left|e^{-k_0 z}\Delta(z)^{-1}C(z)\varphi\right| \leq M_1,$$

where k_0 is the order of the pole of the resolvent of B_1 at zero.

This proves the following generalization of Theorem 3 and Corollary 2.

Theorem 4. *Let k_0 denote the order of the pole of the resolvent of B_1 at zero. For every initial condition $\varphi \in C$, the solution $x(\,\cdot\,;\varphi)$ of the differential delay equation*

$$\dot{x}(t) = B_0 x(t) + B_1 x(t-1), \qquad x_0 = \varphi,$$

has a convergent series expansion for $t > k_0$.

Definition 2. *A solution x of* (10) *is called a small solution if*

$$\lim_{t\to\infty} x(t)e^{kt} = 0 \qquad \text{for every } k \in \mathbb{R}.$$

Small solutions that are not identically zero are called nontrivial.

Let $\mathcal{S} = \{\varphi \in C \mid z \mapsto (zI - A)^{-1} \text{ is entire }\}$. If $\varphi \in C$ is such that $x(\,\cdot\,;\varphi)$ is a small solution, then

$$\lim_{t\to\infty} e^{kt}\|T(t)\varphi\| = 0 \qquad \text{for every } k \in \mathbb{R}$$

and hence

$$z \mapsto \int_0^\infty e^{-zt}T(t)\varphi\, dt$$

defines an entire function. But the Laplace transform of $T(t)\varphi$ equals the resolvent of the infinitesimal generator, see (14). Therefore, by analytic continuation, we have $\varphi \in \mathcal{S}$. Thus, the small solutions correspond to solutions with initial data belonging to $\mathcal{S}$. On the other hand if $\varphi \in \mathcal{S}$, then all the spectral projections of φ are identically zero. Thus, it follows from Theorem 4 that the corresponding solution $x = x(\,\cdot\,;\varphi)$ is identically zero on (k_0, ∞).

The following theorem gives a complete characterization of $\mathcal{S}$ and $\overline{\mathcal{M}}$ and we refer to [25,27] for a complete proof.

Theorem 5. *If k_0 denotes the order of the pole of the resolvent of B_1 at zero and if $T(t)$ denotes the solution semigroup generated by* (12)*, then for $t \geq k_0$*

$$\mathcal{S} = \mathcal{N}\big(T(t)\big) \quad \text{and} \quad \overline{\mathcal{M}} = \overline{\mathcal{R}\big(T(t)\big)}$$

and these relations do not hold for any t smaller than k_0.

So, in particular, it follows that, the small solutions of a linear autonomous delay equation are identically zero after finite time [14,21].

Corollary 4. *Let k_0 denote the order of the pole of the resolvent of B_1 at zero. If the solution $x = x(\,\cdot\,;\varphi)$ of the differential delay equation*

$$\dot{x}(t) = B_0 x(t) + B_1 x(t-1), \qquad x_0 = \varphi,$$

is a small solution, then x is identically zero on $[k_0 - 1, \infty)$.

To illustrate Theorem 4 and Theorem 5 we conclude with the following example.

Example 2. Consider the system of differential delay equations

$$\begin{cases} \dot{x}_1(t) &= x_2(t-1), \\ \dot{x}_2(t) &= x_1(t-1) + x_3(t-1), \\ \dot{x}_3(t) &= x_1(t-1) - x_2(t-1) + x_3(t-1). \end{cases} \tag{40}$$

So $B_0 = 0$ and

$$B_1 = \begin{pmatrix} 0 & 1 & 0 \\ 1 & 0 & 1 \\ 1 & -1 & 1 \end{pmatrix}.$$

A simple computation shows that $k_0 = 2$. So every solution has a convergent series expansion for $t > 2$. In order to investigate the small solutions, we provide an initial vector function φ on the interval $[-1, 0]$. Suppose that $\varphi = (\varphi_1, \varphi_2, -\varphi_1)$, where $\varphi_i : [-1, 0] \to \mathbb{R}$, $i = 1, 2$, are continuous functions. The solution to (40) on the interval $[0, 1]$ satisfies

$$\begin{cases} \dot{x}_1(t) &= \varphi_2(t-1), \\ \dot{x}_2(t) &= 0, \\ \dot{x}_3(t) &= -\varphi_2(t-1). \end{cases}$$

So, the solution to (40) on the interval $[0, 1]$ is given by

$$\begin{cases} x_1(t) &= \varphi_1(0) + \int_0^t \varphi_2(s-1)\, ds, \\ x_2(t) &= \varphi_2(0), \\ x_3(t) &= -\varphi_1(0) - \int_0^t \varphi_2(s-1)\, ds. \end{cases} \tag{41}$$

Suppose that $\varphi_2(0) = 0$. Given the solution on $[0, 1]$, we can now compute the solution to (40) on the interval $[1, 2]$. Indeed, from (41), it follows that the solution to (40) on the interval $[1, 2]$ satisfies

$$\begin{cases} \dot{x}_1(t) &= 0, \\ \dot{x}_2(t) &= 0, \\ \dot{x}_3(t) &= 0. \end{cases}$$

Therefore, the solution to (40) on the interval $[1, 2]$ is given by

$$\begin{cases} x_1(t) &= \varphi_1(0) + \int_0^1 \varphi_2(s-1)\, ds, \\ x_2(t) &= 0, \\ x_3(t) &= -\varphi_1(0) - \int_0^1 \varphi_2(s-1)\, ds. \end{cases} \tag{42}$$

Thus, if $\varphi_1(0) + \int_0^1 \varphi_2(s-1)\, ds = 0$, the solution is identically zero on the interval $[1, 2]$ and therefore identically zero on $[1, \infty)$. This shows that any solution of (40) with initial date $x_0 = \varphi$, where $\varphi = (\varphi_1, \varphi_2, -\varphi_1) \in \mathcal{C}$ such that $\varphi_2(0) = 0$ and

$$\varphi_1(0) + \int_0^1 \varphi_2(s-1)\, ds = 0$$

is identically zero on the interval $[1, \infty)$. Since we already know that small solutions are identically zero on $[k_0 - 1, \infty) = [1, \infty)$. This shows that the value k_0 in Theorem 5 is indeed the best possible.

References

1. Banks, H.T. and A. Manitius, Projection series for retarded functional differential equations with applications to optimal control problems, *J. Diff. Eqn.* **18** (1975), 296–332.
2. Bellman, R. and K.L. Cooke, *Differential-Difference Equations*, Academic Press, New York, 1963.
3. Delfour, M.C. and A. Manitius, The structural operator F and its role in the theory of retarded systems I, *J. Math. Anal. Appl.* **73** (1980), 466–490.
4. Delfour, M.C. and A. Manitius, The structural operator F and its role in the theory of retarded systems II, *J. Math. Anal. Appl.* **74** (1980), 359–381.
5. Diekmann, O., S.A. van Gils, S.M. Verduyn Lunel and H.O. Walther, *Delay Equations: Functional-, Complex-, and Nonlinear Analysis*, Springer-Verlag, New York, Applied Mathematical Sciences Vol. 110, 1995.
6. Frasson, M.V.S. and S.M. Verduyn Lunel, Large time behaviour of linear functional differential equations, *Integral Equations and Operator Theory* **47** (2003), 91-121.
7. Glass, L. and M.C. Mackey, Pathological conditions resulting from instabilities in physiological control systems, *Ann. N.Y. Acad. Sci.* **316** (1979), 214–235.
8. Gohberg, I., S. Goldberg and M.A. Kaashoek, *Classes of Linear Operators I*, Birkhäuser Verlag, Basel, 1990.
9. Gohberg, I.C. and E.I. Sigal, An operator generalization of the logarithmic residue theorem and the theorem of Rouché, *Mat. Sb.* **84** (1971), 609–629 (Russian)(Math. USSR Sb. **13** (1971), 603–625).
10. Hale, J.K. and S.M. Verduyn Lunel, *Introduction to Functional Differential Equations*, Springer-Verlag, New York, Applied Mathematical Sciences Vol. 99, 1993.
11. Hale, J.K. and S.M. Verduyn Lunel, Effects of small delays on stability and control, In: Operator Theory and Analysis, The M.A. Kaashoek Anniversary Volume (eds. H. Bart, I. Gohberg and A.C.M. Ran), Operator Theory: Advances and Applications, Vol. 122, Birkhäuser, 2001, pp. 275–301.
12. Hale, J.K. and S.M. Verduyn Lunel, Strong stabilization of neutral functional differential equations, *IMA Journal Math. Control Inform.* **19** (2002), 5-23.
13. Hale, J.K. and S.M. Verduyn Lunel, Stability and control of feedback systems with time delays, *Int J. Systems* to appear.
14. Henry, D., Small solutions of linear autonomous functional differential equations, *J. Differential Eqns.* **8** (1970), 494–501.
15. Hille, E. and R. Phillips, *Functional Analysis and Semigroups*, American Mathematical Society, Providence, RI, 1957.
16. Kaashoek, M.A. and S.M. Verduyn Lunel, Characteristic matrices and spectral properties of evolutionary systems, *Trans. Amer. Math. Soc.* **334** (1992), 479–517.
17. Kato, T., *Perturbation Theory for Linear Operators (2nd edn.)*, Springer-Verlag, Berlin, 1976.
18. Levinson, N. and C. McCalla, Completeness and independence of the exponential solutions of some functional differential equations, *Studies in Appl. Math.* **53** (1974), 1–15.
19. Manitius, A., Completeness and F-completeness of eigenfunctions associated with retarded functional differential equations, *J. Differential Eqns.* **35** (1980), 1–29.

20. Stépán, G., *Retarded dynamical systems: stability and characteristic functions*, Pitman, Boston, 1989.
21. Verduyn Lunel, S.M., A sharp version of Henry's theorem on small solutions, *J. Differential Eqns.* **62** (1986), 266–274.
22. Verduyn Lunel, S.M., Series expansions and small solutions for Volterra equations of convolution type, *J. Differential Eqns.* **85** (1990), 17–53.
23. Verduyn Lunel, S.M., The closure of the generalized eigenspace of a class of infinitesimal generators, *Proc. Roy. Soc. Edinburgh Sect.* **A 117** (1991), 171–192.
24. Verduyn Lunel, S.M., Series expansions for functional differential equations, *Integral Equations and Operator Theory* **22** (1995), 93–123.
25. Verduyn Lunel, S.M., About completeness for a class of unbounded operators, *J. Differential Eqns.* **120** (1995), 108–132.
26. Verduyn Lunel, S.M., Inverse problems for nonself-adjoint evolutionary systems, In: Topics in Functional Differential and Difference Equations (eds. T. Faria and P. Freitas), Fields Institute Communications, Vol. 29, American Mathematical Society, Providence, 2001, pp. 321–347.
27. Verduyn Lunel, S.M., Spectral theory for delay equations, In: Systems, Approximation, Singular Integral Operators, and Related Topics", International Workhop on Operator Theory and Applications, IWOTA 2000 (eds. A.A. Borichev and N.K. Nikolski), Operator Theory: Advances and Applications, Vol. 129, Birkhäuser, 2001, pp. 465-508.
28. Verduyn Lunel, S.M., Parameter identifiability of differential delay equations, *Int. J. Adapt. Control Signal Process.* **15** (2001), 655-678.
29. Verduyn Lunel, S.M., Spectral theory for neutral delay equations with applications to control and stabilization, In: Mathematical Systems Theory in Biology, Communications, Computation, and Finance (eds. J. Rosenthal and D.S. Gilliam), The IMA Volumes in Mathematics and its Applications, Vol. 134, Springer, 2003, pp. 415–468.
30. Verduyn Lunel, S.M. and D.V. Yakubovich, A functional model approach to linear neutral functional differential equations, *Integral Equations and Operator Theory* **27** (1997), 347–378.

Part II

Stability and Robust Stability

Complete Type Lyapunov-Krasovskii Functionals

Vladimir L. Kharitonov

Department of Automatic Control, CINVESTAV-IPN, A.P. 14-740, Mexico, D.F., Mexico

Summary. In this chapter we give a general description of the complete type quadratic Lyapunov-Krasovskii functionals. Special Lyapunov matrices associated with the functionals are also defined. Uniqueness conditions, as well as a numerical scheme for computation of the Lyapunov matrices, are discussed. Some robust stability conditions, based on the functional, close the chapter. All main results are illustrated with numerical examples.

1 Introduction

It is well known that Lyapunov-Krasovskii functionals play an important role for stability and robust stability analysis of time delay systems, see references in [8], [3].

Starting from the first publications on the topic, see [9], [11], the principal goal was to obtain general expressions for such functionals, at least for the case of linear time delay systems. This study was continued later in several of very important works, see [2], [4], [6]. But technical difficulties, arising in construction of such functionals, along with some open problems associated with the positivity check of the functionals, have prevented a wide spread application of the functionals in the engineering practice.

The main goal of this chapter is to demonstrate that some of these difficulties and problems can be overcome.

Section 2 is devoted to definitions of basic concepts for linear time delay systems.

Main results are given in Section 3. Here we introduce first a general description of the complete type quadratic Lyapunov-Krasovskii functionals. Special Lyapunov matrices, associated with these functionals, are also defined. The Lyapunov matrices satisfy a time delay matrix equation. Uniqueness conditions for these matrices are discussed in Theorems 1-2. Some elegant results concerning the uniqueness issue have been reported in [10]. The approach adopted there is based on analysis of a special two-point boundary problem associated with the Lyapunov matrices. Our approach to the uniqueness probles is more rough and direct. A numerical scheme for computation of the Lyapunov matrices is also proposed. It is shown that the functionals admit a positive quadratic low bound, see Theorem 3. Some robust stability conditions, based on the functional, close the Section. All main results of the chapter are illustrated with numerical examples.

2 Time Delay Systems

Given a time delay system of the form

$$\dot{x}(t) = A_0 x(t) + A_1 x(t-h), \tag{1}$$

where A_0 and A_1 are given $n \times n$ matrices, and $h \geq 0$.

2.1 Basic notations

In this chpater we will use some standard notations. For any piece-wise continuous initial function $\varphi : [-h, 0] \to \mathbb{R}^n$ there exists the unique solution, $x(t, \varphi)$, of (1) satisfying the initial condition

$$x(\theta, \varphi) = \varphi(\theta), \ \theta \in [-h, 0].$$

If $t \geq 0$ we denote by $x_t(\varphi)$ the trajectory segment

$$x_t(\varphi) : \theta \mapsto x(t+\theta, \varphi), \ \theta \in [-h, 0].$$

In this chapter we will use the Euclidean norm for vectors and the induced matrix norm for matrices. The space of piece-wise continuous initial functions is provided with the supremum norm $\|\varphi\|_h = \max_{\theta \in [-h,0]} \|\varphi(\theta)\|$. When it will not cause an ambiguity we write $x(t)$ and x_t instead of $x(t, \varphi)$ and $x_t(\varphi)$.

2.2 Fundamental matrix

The fundamental matrix of system (1), see [1], is $n \times n$ matrix $K(t)$ which satisfies the matrix equation

$$\frac{d}{dt} K(t) = A_0 K(t) + A_1 K(t-h), \text{ for } t \geq 0,$$

and the following initial conditions

$$K(\theta) = 0_{n \times n}, \text{ for } \theta \in [-h, 0), \text{ and } K(0) = E.$$

Here E denotes the identity matrix.

Remark 1. The fundamental matrix satisfies also the equation

$$\frac{d}{dt} K(t) = K(t) A_0 + K(t-h) A_1, \text{ for } t \geq 0.$$

2.3 Cauchy formula

Given an initial condition, φ, the corresponding solution $x(t,\varphi)$ can be written as

$$x(t,\varphi) = K(t)\varphi(0) + \int_{-h}^{0} K(t-\theta-h)A_1\varphi(\theta)d\theta, \tag{2}$$

see [1].

2.4 Stability concept

Definition 1. *[1] System (1) is said to be exponentially stable if there exist $\gamma \geq 1$ and $\sigma > 0$ such that any solution of the system satisfies the inequality*

$$\|x(t,\varphi)\| \leq \gamma \|\varphi\|_h e^{-\sigma t}, \textit{ for } t \geq 0. \tag{3}$$

Remark 2. System (1) is exponentially stable if and only if it is Lyapunov asymptotically stable.

3 Complete Type Lyapunov-Krasovskii Functionals

3.1 General formula

We start with the observation that given matrices W_1, W_2 then

$$\frac{d}{dt}\left(\int_{-h}^{0} x^\top(t+\theta)\left[W_1+(h+\theta)W_2\right]x(t+\theta)d\theta\right) = x^T(t)\left[W_1+hW_2\right]x(t)-$$

$$-x^T(t-h)W_1x(t-h) - \int_{-h}^{0} x^\top(t+\theta)W_2x(t+\theta)d\theta.$$

So, if there exists a functional $v_0(\cdot)$ such that

$$\frac{d}{dt}v_0(x_t) = -w_0(x_t) = -x^T(t)\left[W_0+W_1+hW_2\right]x(t),\ t \geq 0, \tag{4}$$

then the first time derivative of the functional

$$v(x_t) = v_0(x_t) + \int_{-h}^{0} x^\top(t+\theta)\left[W_1+(h+\theta)W_2\right]x(t+\theta)d\theta \tag{5}$$

is given by

$$\frac{d}{dt}v(x_t) = -x^T(t)W_0x(t) - x^T(t-h)W_1x(t-h)-$$

$$-\int_{-h}^{0} x^\top(t+\theta)W_2x(t+\theta)d\theta. \tag{6}$$

Let system (1) be exponentially stable. Then functional $v_0(\cdot)$ exists and can be written as

$$v_0(\varphi) = \int_0^\infty x^T(t,\varphi)\left[W_0 + W_1 + hW_2\right]x(t,\varphi)dt.$$

Substituting under the integral at the right hand side of the last equality $x(t,\varphi)$ by Cauchy formula (2), we arrive at the following expression

$$v_0(x_t) = x^\top(t)U(0)x(t) + 2x^\top(t)\int_{-h}^0 U(-h-\theta)A_1x(t+\theta)d\theta +$$

$$+\int_{-h}^0\int_{-h}^0 x^\top(t+\theta_2)A_1^\top U(\theta_2-\theta_1)A_1x(t+\theta_1)d\theta_1 d\theta_2, \tag{7}$$

where

$$U(\tau) = \int_0^\infty K^\top(t)\left[W_0 + W_1 + hW_2\right]K(t+\tau)dt. \tag{8}$$

is the *Lyapunov matrix for delay system (1).*

3.2 Lyapunov matrix

Let system (1) be exponentially stable, then for every symmetric matrix W, the corresponding Lyapunov matrix

$$U(\tau) = \int_0^\infty K^\top(t)WK(t+\tau)dt, \tag{9}$$

is well defined for all $\tau \in R$.

By direct calculations one can verify that matrix (9) satisfies the following properties.

- Dynamics

$$\frac{d}{d\tau}U(\tau) = U(\tau)A_0 + U(\tau-h)A_1, \text{ for } \tau \geq 0. \tag{10}$$

- Symmetry

$$U(-\tau) = U^T(\tau), \text{ for } \tau \geq 0. \tag{11}$$

- Algebraic

$$-W = U(0)A_0 + A_0^\top U(0) + U^\top(h)A_1 + A_1^\top U(h). \tag{12}$$

Remark 3. Properties (11)-(12) admit natural extensions to the case of systems with several time delays, and systems with distributed delay.

3.3 Uniqueness issue

One can compute Lyapunov matrix $U(\tau)$ using equation (10) and conditions (11)-(12), instead of the formal definition of the matrix as the improper integral (9). This alternative looks very attractive if equation (10) admits only one solution which satisfies conditions (11)-(12). In this case the solution automatically coincides with matrix (9).

Theorem 1. *Let system (1) be exponentially stable. Matrix (9) is the unique solution of equation (10) which satisfies conditions (11)-(12).*

Proof: The fact that matrix (9) satisfies equation (10) and conditions (11)-(12) has been demonstrated in [7].

Assume by contradiction that there are two solutions, $U_1(\tau)$ and $U_2(\tau)$, of equation (10) which satisfy conditions (11)-(12). Using these solutions one can define two functionals of the form (7), the first one with $U(\tau) = U_1(\tau)$, and the second one with $U(\tau) = U_2(\tau)$. Let us denote these functionals as $v_1(x_t)$ and $v_2(x_t)$, respectively. By direct calculations one can check that

$$\frac{d}{dt}v_j(x_t) = -x^T(t)Wx(t), \text{ for } t \geq 0, \ j = 1,2.$$

It means that functional $\Delta v(x_t) = v_2(x_t) - v_1(x_t)$ satisfies the equality

$$\frac{d}{dt}\Delta v(x_t) = 0, \text{ for } t \geq 0,$$

which implies that

$$\Delta v(x_t(\varphi)) = \Delta v(\varphi), \text{ for } t \geq 0.$$

By exponential stability of (1) $\Delta v(x_t(\varphi)) \to 0$, as $t \to \infty$, therefore

$$\Delta v(\varphi) = 0,$$

for every initial vector function φ.

In the explicit form the last condition looks as

$$0 = \varphi^{\mathrm{T}}(0)V(0)\varphi(0) + 2\varphi^{\mathrm{T}}(0)\int_{-h}^{0} V(-h-\theta)A_1\varphi(\theta)d\theta +$$

$$+\int_{-h}^{0} \varphi^{\mathrm{T}}(\theta_2)A_1^{\mathrm{T}}\left[\int_{-h}^{0} V(\theta_2-\theta_1)A_1\varphi(\theta_1)d\theta_1\right]d\theta_2, \tag{13}$$

where matrix $V(\tau) = U_2(\tau) - U_1(\tau)$. Matrix $V(\tau)$ satisfies the equation

$$V'(\tau) = V(\tau)A_0 + V(\tau-h)A_1, \tau \geq 0. \tag{14}$$

For the initial vector function

$$\varphi(\theta) = \begin{cases} \gamma, \text{ for } \theta = 0 \\ 0, \text{ for } \theta \in [-h, 0), \end{cases}$$

condition (13) takes the form $\gamma^T V(0)\gamma = 0$, and we can state that

$$V(0) = 0, \tag{15}$$

because $V(0)$ is a symmetric matrix and γ is an arbitrary vector.

Now, let $\theta_0 \in [-h, 0)$ and $\varepsilon > 0$ be such that $\theta_0 + \varepsilon < 0$. For any given vectors γ and ξ one can define the initial function

$$\varphi(\theta) = \begin{cases} \gamma, & \text{for } \theta = 0 \\ \xi, & \text{for } \theta \in [\theta_0, \theta_0 + \varepsilon] \\ 0, & \text{for all other points of } [-h, 0]. \end{cases} \tag{16}$$

For this initial function condition (13) looks as

$$0 = 2\gamma^{\top} \left[\int_{\theta_0}^{\theta_0+\varepsilon} V(-h-\theta) A_1 d\theta \right] \xi +$$

$$+\xi^{\top} A_1^T \left[\int_{\theta_0}^{\theta_0+\varepsilon} \int_{\theta_0}^{\theta_0+\varepsilon} V(\theta_1 - \theta_2) d\theta_1 d\theta_2 \right] A_1 \xi.$$

The last equality can be written as

$$0 = 2\varepsilon\gamma^{\top} V(-h-\theta_0) A_k \xi + o(\varepsilon),$$

where $\frac{o(\varepsilon)}{\varepsilon} \to 0$, as $\varepsilon \to +0$. The fact that γ and ξ are arbitrary vectors, and ε can be made arbitrary small, implies

$$V(\tau - h) A_1 = 0, \ \text{for } \tau \in [0, h]. \tag{17}$$

So, for $\tau \in [0, h]$ equation (14) is the form

$$V'(\tau) = V(\tau) A_0,$$

and condition (15) implies that

$$V(\tau) = 0, \ \text{for } \tau \in [0, h], \tag{18}$$

i.e.,

$$U_2(\tau) = U_1(\tau), \ \text{for } \tau \in [-h, h].$$

This ends the proof.

Now, we would like to study situations when equation (10) has no solutions satisfying conditions (11)-(12). Of course, such situations may occur only if our basic assumption on exponential stability of system (1) fails. Let us start with the following statement.

Lemma 1. *Given two nontrivial vectors γ and ξ. There exists a symmetric matrix W such that $\gamma^T W \xi \neq 0$.*

Proof: If there is an index j such that $\gamma_j \xi_j \neq 0$, then $W = e_j e_j^T$ satisfies the lemma. If for any j the product is zero, then there exist i and k $(i \neq k)$ such that $\gamma_i \neq 0$ and $\gamma_k = 0$, while $\xi_i = 0$ and $\xi_k \neq 0$. In this case matrix $W = e_i e_k^T + e_k e_i^T$ satisfies the lemma.

Theorem 2. *If system (1) has two eigenvalues, s_1 and s_2, such that $s_1 + s_2 = 0$, then there exists a symmetric matrix W for which equation (10) has no solution satisfying conditions (11)-(12).*

Proof: Assume by contradiction that for any symmetric matrix W equation (10) has a solution satisfying conditions (11)-(12).

We can associate with eigenvalues s_1 and s_2 two solutions of system (1) of the form

$$x^{(1)}(t) = e^{s_1 t}\gamma, \text{ and } x^{(2)}(t) = e^{s_2 t}\xi,$$

where γ and ξ are eigenvectors corresponding to the eigenvalues. By Lemma 1 there exists a symmetric matrix W_0 such that

$$\gamma^T W_0 \xi \neq 0.$$

In accordance with the previous assumption, equation (10) has a solution $U(\tau)$ satisfying conditions (11)-(12), where $W = W_0$. Let us define the bilinear functional

$$z(\varphi,\psi) = \varphi^{\top}(0)U(0)\psi(0) + \varphi^{\top}(0)\int_{-h}^{0} U(-h-\theta)A_1\psi(\theta)d\theta +$$

$$+\left[\int_{-h}^{0} \varphi^{\top}(\theta)A_1^T U(h+\theta)d\theta\right]\psi(0) +$$

$$+\int_{-h}^{0} \varphi^{\top}(\theta_2)A_1^{\top}\left[\int_{-h}^{0} U(\theta_2-\theta_1)A_1\psi(\theta_1)d\theta_1\right]d\theta_2.$$

Given two solutions of system (1), $x(t)$ and $y(t)$, one can verify by direct calculations that

$$\frac{d}{dt}z(x_t,y_t) = -x^T(t)W_0 y(t),$$

see Appendix for details. In particular, for solutions $x^{(1)}(t) = e^{s_1 t}\gamma$ and $x^{(2)}(t) = e^{s_2 t}\xi$, we obtain

$$\frac{d}{dt}z(x_t^{(1)},x_t^{(2)}) = -\left[x^{(1)}(t)\right]^T W_0 x^{(2)}(t) = -\gamma^T W_0 \xi \neq 0. \tag{19}$$

On the other hand, substituting these solutions directly into the bilinear functional we arrive at the following expression

$$z(x_t^{(1)}, x_t^{(2)}) = e^{(s_1+s_2)t}\gamma^{\top} \Big[U(0) + \int_{-h}^{0} U(-h-\theta)A_1 e^{s_2\theta} d\theta +$$

$$+ \int_{-h}^{0} A_1^T U(h+\theta) e^{s_1\theta} d\theta + \int_{-h}^{0}\int_{-h}^{0} e^{s_2\theta_1 + s_1\theta_2} A_1^{\top} U(\theta_2 - \theta_1) A_1 d\theta_1 d\theta_2 \Big] \xi.$$

Observe that matrix in the square brackets does not depend of t, condition $s_1+s_2=0$ implies that

$$\frac{d}{dt} z(x_t^{(1)}, x_t^{(2)}) = 0.$$

The latter equality contradicts with that in (19).

The contradiction ends the proof.

Theorem 2 can be interpreted in the terms of eigenvalues of the Lyapunov operator associated with system (1). In fact, this theorem states that any sum of the form $s_i + s_k$, where s_i and s_k are eigenvalues of system (1), is an eigenvalue of the operator. This result is a natural extension of the well known statement about eigenvalues of the Lyapunov operator $V \to A^T V + VA$ for delay free systems.

3.4 Computational issue

A very useful property of Lyapunov matrices for systems with one delay will be presented next. Matrix (9) is a solution of the second order differential matrix equation

$$U''(\tau) = U'(\tau)A_0 - A_0^T U'(\tau) + A_0^T U(\tau) A_0 - A_1^T U(\tau) A_1, \tag{20}$$

with the following boundary conditions

- $-W = U(0)A_0 + A_0^{\top} U(0) + U^{\top}(h) A_1 + A_1^{\top} U(h);$
- $U'(0) = U(0)A_0 + U^T(h)A_1.$

Equation (20) and the boundary conditions may be used for computation of $U(\tau)$. If there exist just one solution of (20) which satisfies the boundary conditions, then this solution defines the Lyapunov matrix (9). If there are several of such solutions one has to select among them the one which satisfies also to equation (10) and conditions (11)-(12).

Example 1. System

$$\dot{x}(t) = \begin{pmatrix} 0 & 1 \\ -1 & -2 \end{pmatrix} x(t) + \begin{pmatrix} 0 & 0 \\ -1 & 1 \end{pmatrix} x(t-1)$$

is exponentially stable. Let $W = 3E$, components of matrix $U(\tau)$ for $\tau \in [0,1]$ are plotted on Figure 1.

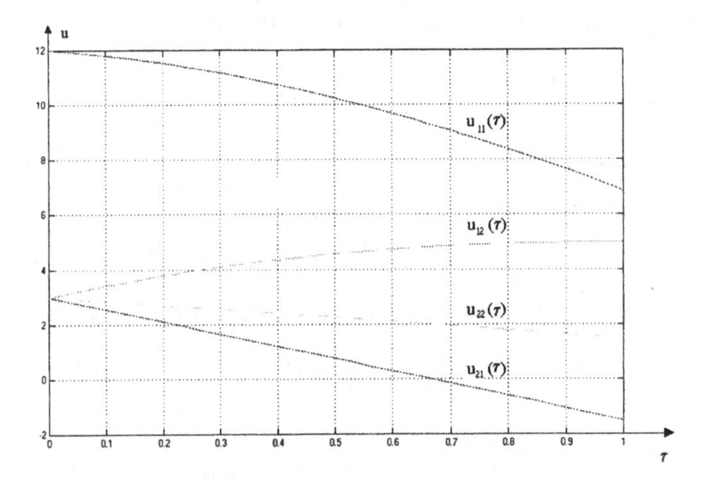

Fig. 1. Components of matrix $U(\tau)$

3.5 Quadratic low bound

One of the basic conditions for Lyapunov-Krasovskii functionals states that functional $v(\cdot)$ should admit a quadratic low bound of the form

$$\alpha_1 \|x(t)\|^2 \leq v(x_t),$$

where $\alpha_1 > 0$. Surprisingly enough, no such a bound has been demonstrated for functional $v_0(\cdot)$. The corresponding attempt, made in [6], resulted in a local cubic low bound only, see also [5]. The following theorem sheds light on the problem and indicates a modification of the functional needed to guarantee existence of a global quadratic low bound.

Theorem 3. *Let system (1) be exponentially stable. Given matrices $W_0 > 0$, $W_1 > 0$ and $W_2 \geq 0$, then there exists $\varepsilon > 0$ such that functional (5) admits the following quadratic low bound*

$$\varepsilon \|x(t)\|^2 \leq v(x_t).$$

Proof: Consider the modified functional

$$\widetilde{v}(x_t) = v(x_t) - \varepsilon \|x(t)\|^2 .$$

Then

$$\frac{d}{dt}\widetilde{v}(x_t) = -\widetilde{w}(x_t) \leq$$

$$\leq -(x^T(t), x^T(t-h)) \begin{pmatrix} W_0 + \varepsilon(A_0 + A_0^T) & \varepsilon A_1 \\ \varepsilon A_1^T & W_1 \end{pmatrix} \begin{pmatrix} x(t) \\ x(t-h) \end{pmatrix}.$$

For sufficiently small $\varepsilon > 0$ functional $\widetilde{w}(x_t) \geq 0$, therefore

$$\widetilde{v}(\varphi) = \int_0^\infty \widetilde{w}(x_t(\varphi))dt \geq 0, \text{ and } v(x_t) \geq \varepsilon \|x(t)\|^2 .$$

Example 2. For system

$$\dot{x}(t) = \begin{pmatrix} 0 & 1 \\ -1 & -2 \end{pmatrix} x(t) + \begin{pmatrix} 0 & 0 \\ -1 & 1 \end{pmatrix} x(t-1).$$

If we assume that $W_0 = W_1 = E$ then functional $v(x_t)$ satisfies inequality

$$v(x_t) \geq \varepsilon \|x(t)\|^2 ,$$

where ε is such that

$$\begin{pmatrix} W_0 + \varepsilon(A_0 + A_0^T) & \varepsilon A_1 \\ \varepsilon A_1^T & W_1 \end{pmatrix} \geq 0.$$

Direct calculations show that the last inequality holds for

$$\varepsilon \leq \sqrt{1.5} - 1 \simeq 0.2247.$$

3.6 Robust stability conditions

Consider a perturbed system of the form

$$\dot{y}(t) = (A_0 + \Delta_0)\, y(t) + (A_1 + \Delta_1)\, y(t-h). \tag{21}$$

Here matrices, Δ_0 and Δ_1, are unknown, but such that

$$\|\Delta_k\| \leq \rho_k, \; k = 0, 1. \tag{22}$$

Let system (1) be exponentially stable. We would like to find conditions on ρ_0 and ρ_1 under which system (21) remains stable for all Δ_0 and Δ_1 satisfying (22).

The first time derivative of (5) along solutions of system (21) is

$$\frac{d}{dt}v(y_t) = -w(y_t) + 2\left[\Delta_0 y(t) + \Delta_1 y(t-h)\right]^T \times$$
$$\times \left[U(0)y(t) + \int_{-h}^0 U(-h-\theta)A_1 y(t+\theta)d\theta\right].$$

Let $\nu = \max_{\theta\in[0,h]} \{\|U(\theta)\|\}$, then

- $2y^T(t)\Delta_0^T U(0)y(t) \leq 2\nu\rho_0 y^T(t)y(t)$;
- $2y^T(t-h)\Delta_1^T U(0)y(t) \leq \nu\rho_1 y^T(t)y(t) + \nu\rho_1 y^T(t-h)y(t-h)$;

- $2y^T(t)\Delta_0^T \int_{-h}^0 U(-h-\theta)A_1 y(t+\theta)d\theta \leq$

$$\leq \nu\rho_0 \left[hy^T(t)y(t) + \int_{-h}^0 y^T(t+\theta)A_1^T A_1 y(t+\theta)d\theta \right];$$

- $2y^T(t-h)\Delta_1^T \int_{-h}^0 U(-h-\theta)A_1 y(t+\theta)d\theta \leq$

$$\leq \nu\rho_1 \left[hy^T(t-h)y(t-h) + \int_{-h}^0 y^T(t+\theta)A_1^T A_1 y(t+\theta)d\theta \right].$$

¿From these inequalities we obtain that

$$\frac{d}{dt}v(y_t) \leq -w(y_t) + \nu\left[2\rho_0 + h\rho_0 + \rho_1\right] y^T(t)y(t) +$$

$$+\nu\rho_1\left[1+h\right] y^T(t-h)y(t-h) +$$

$$+\nu\left[\rho_0 + \rho_1\right] \int_{-h}^0 y^T(t+\theta)A_1^T A_1 y(t+\theta)d\theta.$$

¿From this inequality one can deduce the following statement.

Theorem 4. *Let system (1) be exponentially stable. Given positive definite matrices W_0, W_1, W_2. System (21) remains exponentially stable for all Δ_0 and Δ_1 satisfying (22) if*

i) $\lambda_{\min}(W_0) \geq \nu\left(2\rho_0 + h\rho_0 + \rho_1\right)$,
ii) $\lambda_{\min}(W_1) \geq \nu\rho_1\left(1+h\right)$,
iii) $W_2 \geq \nu\left(\rho_0 + \rho_1\right) A_1^T A_1$.

Example 3. Consider again the system

$$\dot{x}(t) = \begin{pmatrix} 0 & 1 \\ -1 & -2 \end{pmatrix} x(t) + \begin{pmatrix} 0 & 0 \\ -1 & 1 \end{pmatrix} x(t-1).$$

Given the perturbed system

$$\dot{y}(t) = \left(A_0 + \Delta_0\right) y(t) + \left(A_1 + \Delta_1\right) y(t-1),$$

where

$$\|\Delta_k\| \leq \rho.$$

Let $W_0 = W_1 = W_2 = E$, then from Figure 1 $\nu \leq 13$. Theorem 4 implies that perturbed system remains stable if

$$\rho \leq \frac{1}{56}.$$

4 Conclusions

In this chapter we present some results on construction of complete type Lyapunov-Krasovskii functionals for time delay systems.

5 Acknowledgment

This work was supported by the Université de Technologie de Compiègne, France.

References

1. Bellman R.E. and K.L. Cooke (1963) Differential-Difference Equations, Academic Press. New York
2. Datko R. (1972) An algorithm for computing Liapunov functionals for some differential-difference equations. In Ordinary Differential Equations. NRL-MRC Conference, Academic Press, New York, 387–398
3. Niculescu S.-I. E.I. Verriest Dugard L. and J.-M. Dion. (1997) Stability and Robust Stability of Time-Delay Systems: A Guided Tour, In Stability and Control of Time Delay Systems, Dugard L. and E.I. Verriest (eds). Lecture Notes in Control and Information Sciences, Springer Verlag, London, 228:1–71
4. Infante E.F. and W.B. Castelan (1978) A Liapunov functional for a matrix difference-differential equation. Journal of Differential Equations, 29:439–451
5. Hale J.K. and S. M. Verduyn Lunel (1993) Introduction to Functional Differential Equations. Springer-Verlag, New York
6. Huang W. (1989) Generalization of Liapunov´s theorem in a linear delay system. Journal of Mathematical Analysis and Applications, 142:83–94
7. Kharitonov V.L. and A.P. Zhabko (2003) Lyapunov-Krasovskii approach to the robust stability analysis of time-delay systems. Automatica, 39:15–20
8. Kharitonov V.L. (1999) Robust stability analysis of time delay systems: A survey. Annual Reviews in Control, 23:185–196
9. Krasovskii N. N. (1956) On the application of the second method of Lyapunov for equations with time delays. Prikl. Mat. Mech., 20:315–327 (in Russian)
10. Louisell. J. (1997) Numerics of the stability exponent and eigenvalue abscissas of a matrix delay system. In Stability and Control of Time-Delay Systems, Dugard L. and E.I. Verriest, (eds), Lecture Notes in Control and Information Sciences, Springer Verlag, London, 228:140–157
11. Repin Yu.M. (1965), Quadratic Liapunov functionals for systems with delay, Prikl. Mat. Mech., 29:564–566 (in Russian)

Robust Stability Conditions of Quasipolynomials by Frequency Sweeping

Jie Chen[1]* and Silviu-Iulian Niculescu[2]

[1] Department of Electrical Engineering, University of California, Riverside, CA 92521, USA. jchen@ee.ucr.edu

[2] HeuDiaSyC (UMR CNRS 6599), UT Compiègne, BP 20529, 60205, Compiègne France. silviu@hds.utc.fr

Summary. In this chapter we study the robust stability independent of delay of some class of uncertain quasipolynomials, whose coefficients may vary in a certain prescribed range. Our main contributions include *frequency-sweeping* conditions for interval, diamond and spherical quasipolynomial families. The correspoding results provide *necessary* and *sufficient conditions*, and are easy to check, requiring only the computation of two simple frequency-dependent functions. Various extensions (polytopic uncertainty, multivariate polynomials) are also presented.

1 Introduction

The stability of delay systems is a subject of recurring interest in the study of dynamical systems, and received considerable attention in the last decade, see, for instance, [12, 12, 16, 18], and the references therein. Two particular stability notions, *delay-dependent* and *delay-independent stability*, respectively, have been extensively treated in the literature, and both time and frequency domain stability tests have been developed. Here by delay-independent stability of a system we mean that the system is stable for all nonnegative values of delay, and otherwise the system's stability is delay-dependent. It is known that with only commensurate delays, the stability of a linear time-invariant system, whether delay-dependent or delay-independent, can be determined by solving a matrix eigenvalue problem [4, 5]. On the other hand, for systems with incommensurate delays, the stability problem has been found to be *NP-hard* in general [22].

This chapter focuses on the robust stability *independent* of *delay* of linear time-invariant delay systems. Specifically, we consider uncertain *quasipolynomials* whose coefficients depend on uncertain parameters in an affine manner. Unlike in the general situation, where one may have to resort to the computation of *structured singular values* [3], in this chapter we seek *computable necessary and sufficient conditions*

* Author to whom all correspondences should be addressed.

for robust stability. We shall focus particularly on polytopic uncertain quasipolynomials, and its sub-families of interval, diamond and spherical quasipolynomials. Note that the robust stability of quasipolynomials has been a well-studied topic (see, e.g., [2, 11, 12, 14, 15, 20] and the references therein), though the results seem less well-developed.

We adopt a frequency domain approach, one that is built upon the frequency-sweeping conditions obtained in [7,8], which give necessary and sufficient robust stability conditions for uncertain *polynomials*. We develop frequency-sweeping robust stability tests. These results require only the computation of some simple frequency-dependent functions and the computation can be done rather efficiently. For example, for interval, diamond and spherical polynomials, the robust stability can be determined by computing two frequency-dependent functions only, with one for the robust stability in the absence of delay, and another serving as a distance measure.

The remainder of this chapter is organized as follows. Some preliminary facts are included in Section 2. Section 3 presents frequency-sweeping conditions for interval, diamond and spherical quasipolynomials. The results will then be extended to more general problems in Section 4 (uncertain coefficients in ℓ_p-balls, polytopic uncertainty and multivariate polynomials). Illustrative examples are given in Section 5. The chapter concludes in Section 6 with a number of concluding remarks. Our development is guided by a geometrical interpretation which not only leads to the readily computable robust stability conditions, but also furnishes intuitively transparent insights guiding the derivations. This, apart from the technical results, appears to be another contribution of this chapter[3].

2 Preliminaries

We consider the class of quasipolynomials given by

$$p\left(s;\, e^{-\tau_1 s},\, \cdots,\, e^{-\tau_m s}\right) = a_0(s) + \sum_{k=1}^{m} a_k(s) e^{-\tau_k s}, \tag{1}$$

$\tau_k \geq 0$, $k = 1, \cdots, m$, where

$$a_0(s) = s^n + \sum_{i=0}^{n-1} a_{0i} s^i, \;\; a_k(s) = \sum_{i=0}^{n-1} a_{ki} s^i.$$

This quasipolynomial corresponds to the characteristic function of delay systems described by

$$y^{(n)}(t) + \sum_{i=0}^{n-1} \sum_{k=0}^{m} a_{ki} y^{(i)}(t - \tau_k) = 0, \qquad \tau_k \geq 0, \tag{2}$$

[3] For the brevity of the chapter, all the proofs are omitted. However, the main ideas and interpretations are sufficiently well detailed. For complete proofs, see, for instance, the paper [9].

or more generally, those given in the state-space description

$$\dot{x}(t) = A_0\, x(t) + \sum_{k=1}^{r} A_k\, x(t - T_k), \qquad T_k \geq 0. \tag{3}$$

We study the stability properties of the quasipolynomial (1), and accordingly, the stability of the time-delay systems (2) and (3). In particular, we are interested in the stability of (1) independent of the delay values τ_k, $k = 1, \cdots, m$. This stability notion is stated formally as follows.

Definition 1. *The quasipolynomial (1) is said to be stable if*[4]

$$p\left(s;\, e^{-\tau_1 s}, \;\cdots,\; e^{-\tau_m s}\right) \neq 0, \quad \forall s \in \overline{\mathbb{C}}_+. \tag{4}$$

It is said to be stable independent of delay if the condition (4) holds for all nonnegative delays τ_k.

We shall consider only quasipolynomials with incommensurate, independent delays, by which we mean that in (1) the delay parameters τ_k, $k = 1, \cdots, m$ are independent of each other. In this case, a necessary and sufficient stability condition is available from, e.g., [3,6,12].

Lemma 1. *Let* $\tau_k, k = 1, \cdots, m$ *be independent delays. Then the quasipolynomial (1) is stable independent of delay if and only if*

(i) $a_0(s)$ *is stable;*

(ii) $\sum_{k=0}^{m} a_{k0} \neq 0$, *and*

(iii)

$$\frac{\sum_{k=1}^{m} |a_k(j\omega)|}{|a_0(j\omega)|} < 1, \quad \forall \omega > 0. \tag{5}$$

Our primary purpose in this chapter is to study the stability of uncertain quasipolynomials with incommensurate, independent delays. Thus, we assume that the coefficients of the quasipolynomial (1) vary in a prescribed set. In general, such quasipolynomials can be described as

$$p\left(s; e^{-\tau_1 s}, \cdots, e^{-\tau_m s};\, \alpha\right) = a_0(s, \alpha_0) + \sum_{k=1}^{m} a_k(s,\, \alpha_k) e^{-\tau_k s}, \tag{6}$$

$\tau_k \geq 0$, $k = 1, \cdots, m$ where $\alpha_k \in \mathcal{Q}_k \subset \mathbb{R}^n$, $k = 0, \;\cdots,\; m$ represent the uncertain parameters.

---

[4] Here $\mathbb{C}_+ := \{s : \Re(s) > 0\}$ denotes the open right half plane, and $\overline{\mathbb{C}}_+ := \{s : Re(s) \geq 0\}$ the closed right half plane.

Definition 2. *The quasipolynomial family (6) is said to be robustly stable if for all $\alpha_k \in \mathcal{Q}_k$, $k = 0, \cdots, m$,*

$$p\left(s;\, e^{-\tau_1 s},\, \cdots,\, e^{-\tau_m s};\, \alpha\right) \neq 0, \quad \forall s \in \overline{\mathbb{C}}_+. \tag{7}$$

It is robustly stable independent of delay if (7) holds for all nonnegative delays τ_k.

We shall assume that each uncertain vector α_k varies independently. Additionally, we assume that each uncertain polynomial $a_k(s, \alpha_k)$, $k = 1, \cdots, m$ is perturbed in an affine manner, so that it can be written as

$$a_k(s,\, \alpha_k) = a_k(s,\, \alpha_k^*) + \sum_{i=1}^{n} \gamma_{ki} p_{ki}(s) \delta_{ki}, \tag{8}$$

$k = 0, 1, \cdots, m$, where γ_{ki} and $p_{ki}(s)$ are known constants and polynomials, and δ_{ki} are unknown perturbations within some prespecified ranges. Write $\delta_k := [\delta_{k1} \cdots \delta_{kn}]^T$. Then, a common characterization for the unknown perturbations is furnished by the ℓ_p-Hölder norm, that is,

$$\|\delta_k\|_p := \begin{cases} \left(\sum_{i=1}^{n} |\delta_{ki}|^p \right)^{1/p}, & 1 \leq p \leq \infty \\ \max_{1 \leq i \leq n} |\delta_{ki}|, & p = \infty. \end{cases}$$

For example, for $p_{ki} = s^i$ and $p = \infty,\ 1,\ 2$, the uncertain polynomial $a_k(s,\, \alpha_k)$ defines correspondingly the families of interval, diamond, and spherical polynomials, respectively. In the sequel, the uncertain quasipolynomial (6) will be termed interval, diamond, and spherical quasipolynomials if all its coefficient polynomials $a_k(s,\, \alpha_k)$ are interval, diamond, and spherical polynomials, respectively.

3 Frequency-Sweeping Conditions

We shall present several necessary and sufficient frequency-sweeping conditions for the uncertain quasipolynomial (6) when its coefficients are characterized by ℓ_p balls. In this case, the polynomial families $a_k(s,\, \alpha_k)$ assume the form

$$a_0(s,\, \alpha_0) = s^n + \sum_{i=0}^{n-1} \alpha_{0i} s^i, \ a_k(s,\, \alpha_k) = \sum_{i=0}^{n-1} \alpha_{ki} s^i, \tag{9}$$

where each α_{ki}, $k = 0,\ 1,\ \cdots,\ m$ is assumed to lie in a given interval $[\underline{\alpha}_{ki},\ \overline{\alpha}_{ki}]$, and the vector α_k is to vary in a weighted ℓ_p ball. More specifically, define

$$\alpha_{ki}^* := \frac{\overline{\alpha}_{ki} + \underline{\alpha}_{ki}}{2}, \quad \gamma_{ki} := \frac{\overline{\alpha}_{ki} - \underline{\alpha}_{ki}}{2}, \quad \Gamma_k := \mathrm{diag}\left(\gamma_{k1}, \cdots, \gamma_{kn}\right).$$

Then for any $p \in [1,\ \infty]$, the coefficient vector α_k is characterized by the weighted ℓ_p ball

$$\mathcal{Q}_k := \{\alpha_k : \alpha_k = \alpha_k^* + \Gamma_k \delta_k, \ \|\delta_k\|_p \le 1\}.$$

Note that by defining $p_{ki}(s) = s^i$, the polynomial family $a(s, \alpha_k)$ coincides with that given by (8), with $\|\delta_k\|_p \le 1$. Assume, without loss of generality, that n is an even integer; an analogous analysis applies when n is odd. Let $q \in [1, \infty]$ satisfy the relation $(1/p) + (1/q) = 1$. For each $\omega \in [0, \infty)$, define

$$\begin{aligned} X_{k,q}(\omega) &: = \left(\gamma_{k0}^q + (\gamma_{k2}\omega^2)^q + (\gamma_{k4}\omega^4)^q + \cdots\right)^{1/q}, \\ Y_{k,q}(\omega) &: = \left(\gamma_{k1}^q + (\gamma_{k3}\omega^2)^q + (\gamma_{k5}\omega^4)^q + \cdots\right)^{1/q}, \\ R_k(\omega) &: = \alpha_{k0}^* - \alpha_{k2}^*\omega^2 - \alpha_{k4}^*\omega^4 + \cdots, \\ I_k(\omega) &: = \alpha_{k1}^* - \alpha_{k3}^*\omega^2 + \alpha_{k5}^*\omega^4 + \cdots \end{aligned}$$

Note that for $q = \infty$,

$$\begin{aligned} X_{k,\infty}(\omega) &= \max\left\{\gamma_{k0}, \ \gamma_{k2}\omega^2, \ \gamma_{k4}\omega^4, \ \cdots\right\}, \\ Y_{k,\infty}(\omega) &= \max\left\{\gamma_{k1}, \ \gamma_{k3}\omega^2, \ \gamma_{k5}\omega^4, \ \cdots\right\}. \end{aligned}$$

It also follows that $a_k(j\omega, \alpha_k^*) = R_k(\omega) + j\omega I_k(\omega)$.

3.1 Interval quasipolynomials

In this case, as specified above, all the polynomials $a_k(s, \alpha_k)$ are each interval polynomials, that is, $\alpha_k \in [\underline{\alpha}_{ki}, \overline{\alpha}_{ki}]$, or alternatively,

$$\mathcal{Q}_k := \{\alpha_k : \alpha_k = \alpha_k^* + \Gamma_k \delta_k, \ \|\delta_k\|_\infty \le 1\}. \tag{10}$$

For each $k = 0, 1, \cdots, m$, define the four Kharitonov vertex polynomials

$$\begin{aligned} K_{k,1}(s) &: = \underline{\alpha}_{k0} + \underline{\alpha}_{k1}s + \overline{\alpha}_{k2}s^2 + \overline{\alpha}_{k3}s^3 + \underline{\alpha}_{k4}s^4 + \underline{\alpha}_{k5}s^5 + \cdots \\ K_{k,2}(s) &: = \overline{\alpha}_{k0} + \overline{\alpha}_{k1}s + \underline{\alpha}_{k2}s^2 + \underline{\alpha}_{k3}s^3 + \overline{\alpha}_{k4}s^4 + \overline{\alpha}_{k5}s^5 + \cdots \\ K_{k,3}(s) &: = \overline{\alpha}_{k0} + \underline{\alpha}_{k1}s + \underline{\alpha}_{k2}s^2 + \overline{\alpha}_{k3}s^3 + \overline{\alpha}_{k4}s^4 + \underline{\alpha}_{k5}s^5 + \cdots \\ K_{k,4}(s) &: = \underline{\alpha}_{k0} + \overline{\alpha}_{k1}s + \overline{\alpha}_{k2}s^2 + \underline{\alpha}_{k3}s^3 + \underline{\alpha}_{k4}s^4 + \overline{\alpha}_{k5}s^5 + \cdots \end{aligned}$$

It is well-known that the interval polynomial $a_k(s, \alpha_k)$ will be stable whenever $\alpha_{kn} \neq 0$ and the four vertex polynomials are stable. In the present setting, of interest is the stability of the interval polynomial $a_0(s, \alpha_0)$. The following alternative condition [8] provides a frequency-sweeping test and will be required.

Lemma 2. *Suppose that $a_0(s, \alpha_0^*)$ is stable. Then, the interval polynomial $a_0(s, \alpha_0)$ is stable if and only if $\underline{\alpha}_{00} > 0$ and*

$$\min\left\{\frac{X_{0,1}(\omega)}{|R_0(\omega)|}, \frac{Y_{0,1}(\omega)}{|I_0(\omega)|}\right\} < 1, \quad \forall \omega > 0. \tag{11}$$

Based upon Lemma 1 and Lemma 2, we are now ready to state a frequency-sweeping condition for the interval quasipolynomial (6).

Theorem 1. *Let τ_k, $k = 1, \cdots, m$ be independent delays. Define*

$$\overline{\rho}_k(\omega) := \sqrt{\left(|R_k(\omega)| + X_{k,1}(\omega)\right)^2 + \omega^2\left(|I_k(\omega)| + Y_{k,1}(\omega)\right)^2},$$

$$\underline{\rho}_0(\omega) := \sqrt{M_R^2(\omega) + \omega^2 M_I^2(\omega)}\,,$$

where

$$M_R(\omega) = \begin{cases} |R_0(\omega)| - X_{0,1}(\omega) & \text{if } |R_0(\omega)| > X_{0,1}(\omega) \\ 0 & \text{if } |R_0(\omega)| \leq X_{0,1}(\omega), \end{cases}$$

$$M_I(\omega) := \begin{cases} |I_0(\omega)| - Y_{0,1}(\omega) & \text{if } |I_0(\omega)| > Y_{0,1}(\omega) \\ 0 & \text{if } |I_0(\omega)| \leq Y_{0,1}(\omega). \end{cases}$$

Then the interval quasipolynomial (6), with $a_k(s, \alpha_k)$ given by (9-10), is robustly stable independent of delay if and only if

(i) The interval polynomial $a_0(s,\ \alpha_0)$ is stable;

(ii) $\sum_{k=0}^{m} \underline{a}_{k0} > 0$ *;*

(iii)

$$\frac{\sum_{k=1}^{m} \overline{\rho}_k(\omega)}{\underline{\rho}_0(\omega)} < 1, \quad \forall \omega > 0. \tag{12}$$

Theorem 1 makes it clear that the robust stability of the interval quasipolynomial independent of delay can be ascertained by checking the stability of one interval polynomial and additionally performing a frequency-sweeping test. Here the stability of the interval polynomial can be determined by either checking the four vertex polynomials, or by checking the frequency-sweeping condition (11). We note that both frequency-sweeping conditions require only simple algebraic computations and both admit rather intuitive interpretations. To illustrate, consider the value set of $a_k(s, \alpha_k)$, $V_k^I(\omega) := \{a_k(j\omega, \alpha_k) : \alpha_k \in \mathcal{Q}_k\}$, which is a frequency-dependent rectangle, known as the Kharitonov box depicted in Figures 1 and 2. Clearly, at any $\omega > 0$, the maximum of $|a_k(j\omega, \alpha_k)|$ is achieved at the vertex farthest from the origin, whose distance from the origin is:

$$\sqrt{\left(|R_k(\omega)| + X_{k,1}(\omega)\right)^2 + \omega^2\left(|I_k(\omega)| + Y_{k,1}(\omega)\right)^2}.$$

When at $\omega > 0$ so that the rectangle lies strictly in any one of the four quadrants, the minimum of $|a_0(j\omega,\ \alpha_0)|$ is also achieved at one of the four vertices, one which is nearest the origin and whose distance from the origin is:

$$\sqrt{\left(|R_0(\omega)| - X_{0,1}(\omega)\right)^2 + \omega^2\left(|I_0(\omega)| - Y_{0,1}(\omega)\right)^2}.$$

However, when the rectangle crosses either the real or imaginary axis, as shown in Figure 2, the smallest distance between the rectangle and the origin is that on the either axis[5], which is either $|R_0(\omega)| - X_{0,1}(\omega)$ or $\omega\,(|I_0(\omega)| - Y_{0,1}(\omega))$. This consequently gives a rather simple geometrical interpretation to the frequency-sweeping condition (12). A geometrical interpretation for Lemma 2 can be found in [8].

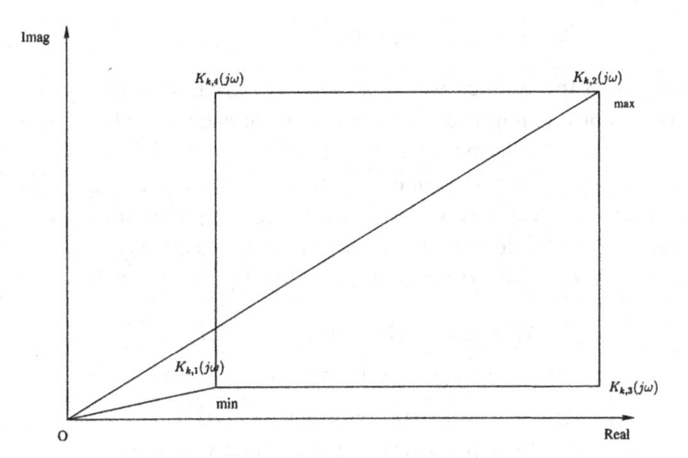

Fig. 1. Kharitonov boxes

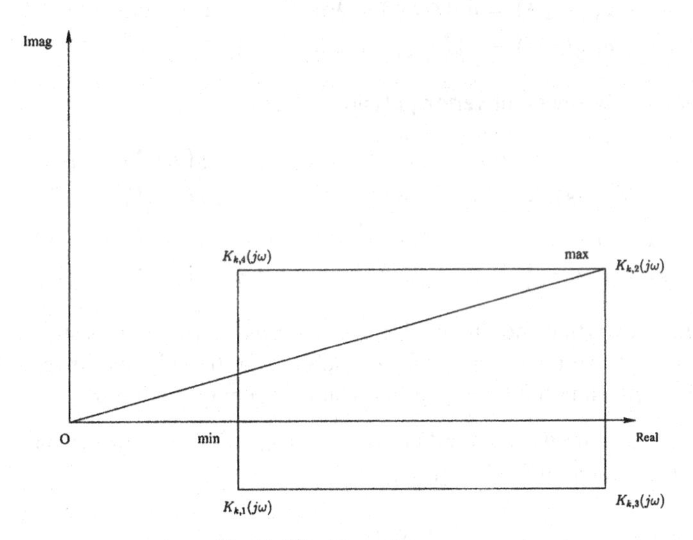

Fig. 2. Kharitonov boxes

[5] The former corresponds to $|I_0(\omega)| \leq Y_{0,1}(\omega)$, when the rectangle intersects the real axis, while the latter corresponds to $|R_0(\omega)| \leq X_{0,1}(\omega)$, when the rectangle intersects the imaginary axis.

3.2 Diamond quasipolynomials

For the diamond quasipolynomial family, each polynomial $a_k(s, \alpha_k)$ defines a diamond polynomial, whose coefficients vary in the parameter set

$$\mathcal{Q}_k := \{\alpha_k : \alpha_k = \alpha_k^* + \Gamma_k \delta_k, \ \|\delta_k\|_1 \leq 1\}. \tag{13}$$

For the brevity of the chapter, we shall consider *only* uniformly weighted diamond polynomials[6]. For this purpose we assume that for each $k = 0, 1, \cdots, m$, $\gamma_{k1} = \cdots = \gamma_{kn} = \gamma_k$. Denote the value set of $a_k(s, \alpha_k)$ by $V_k^D(\omega)$. It is known that $V_k^D(\omega)$ forms a frequency-dependent diamond with its center at $a_k(j\omega, \alpha_k^*)$. For $\omega \in (0, 1]$, the four vertices of the diamond are the polynomials $e_{k,1}(s)$, $e_{k,2}(s)$, $e_{k,3}(s)$, and $e_{k,4}(s)$, while for $\omega \in (1, \infty)$, the vertices are $e_{k,5}(s)$, $e_{k,6}(s)$, $e_{k,7}(s)$, and $e_{k,8}(s)$, which are the extremes of the eight edge polynomials

$$\begin{aligned}
e_{k,1}(s,\lambda) &= a_k(s, \alpha_k^*) - \lambda\gamma_k - (1-\lambda)\gamma_k s,\\
e_{k,2}(s,\lambda) &= a_k(s,\alpha_k^*) + \lambda\gamma_k - (1-\lambda)\gamma_k s,\\
e_{k,3}(s,\lambda) &= a_k(s,\alpha_k^*) + \lambda\gamma_k + (1-\lambda)\gamma_k s,\\
e_{k,4}(s,\lambda) &= a_k(s,\alpha_k^*) - \lambda\gamma_k + (1-\lambda)\gamma_k s,\\
e_{k,5}(s,\lambda) &= a_k(s,\alpha_k^*) - \lambda\gamma_k s^{n-1} - (1-\lambda)\gamma_k s^{n-2},\\
e_{k,6}(s,\lambda) &= a_k(s,\alpha_k^*) + \lambda\gamma_k s^{n-1} - (1-\lambda)\gamma_k s^{n-2},\\
e_{k,7}(s,\lambda) &= a_k(s,\alpha_k^*) + \lambda\gamma_k s^{n-1} + (1-\lambda)\gamma_k s^{n-2},\\
e_{k,8}(s,\lambda) &= a_k(s,\alpha_k^*) - \lambda\gamma_k s^{n-1} + (1-\lambda)\gamma_k s^{n-2}.
\end{aligned}$$

Correspondingly, the eight vertex polynomials are

$$\begin{aligned}
e_{k,1}(s) &= a_k(s, \alpha_k^*) - \gamma_k, & e_{k,2}(s) &= a_k(s,\alpha_k^*) + \gamma_k\\
e_{k,3}(s) &= a_k(s,\alpha_k^*) - \gamma_k s, & e_{k,4}(s) &= a_k(s,\alpha_k^*) + \gamma_k s,\\
e_{k,5}(s) &= a_k(s,\alpha_k^*) - \gamma_k s^{n-2}, & e_{k,6}(s) &= a_k(s,\alpha_k^*) + \gamma_k s^{n-2},\\
e_{k,7}(s) &= a_k(s,\alpha_k^*) - \gamma_k s^{n-1}, & e_{k,8}(s) &= a_k(s,\alpha_k^*) + \gamma_k s^{n-1}.
\end{aligned}$$

It is well-known that the diamond polynomial $a_0(s, \alpha_0)$ is robustly stable if and only if its eight vertex polynomials are stable. Additionally, its robust stability can be determined using the following frequency-sweeping condition[7].

Lemma 3. *Suppose that $a_0(s, \alpha_0^*)$ is stable. Then, the diamond polynomial $a_0(s, \alpha_0)$ is robustly stable if and only if $\underline{\alpha}_{00} > 0$ and*

$$\frac{X_{0,\infty}(\omega)}{|I_0(\omega)| + |R_0(\omega)|} < 1, \quad \forall \omega > 0. \tag{14}$$

---

[6] More generally, the diamond polynomials $a_k(s, \alpha_k)$ may not be uniformly weighted, and the analysis may become substantially more complex. Nevertheless, the geometrical argument employed remains useful and suggests a general, systematic approach. See, for instance, [9].

[7] Here $X_{k,\infty} = Y_{k,\infty} = \gamma_k \max\{1, \omega^{n-2}\}, \forall k = 0, 1, \cdots, m$.

We obtain a similar frequency-sweeping condition for the robust stability of the diamond quasipolynomial.

Theorem 2. *Let τ_k, $k = 1, \cdots, m$ be independent delays. Assume that for each $k = 0, 1, \cdots, m$, $\gamma_{k1} = \cdots = \gamma_{kn} = \gamma_k$. Define*

$$\overline{\rho}_k(\omega) := \max\left\{\sqrt{(|R_k(\omega)| + X_{k,\infty}(\omega))^2 + \omega^2 I_k^2(\omega)},\right.$$
$$\left.\sqrt{R_k^2(\omega) + \omega^2\left(|I_k(\omega)| + Y_{k,\infty}(\omega)\right)^2}\right\},$$
$$\underline{\rho}_k(\omega) := \min\left\{\sqrt{|R_k(\omega)|^2 + \omega^2(|I_k(\omega)| - Y_{k,\infty}(\omega))^2},\right.$$
$$\left.\sqrt{(|R_k(\omega)| - X_{k,\infty}(\omega))^2 + \omega^2|I_k(\omega)|^2}\right\}.$$

Furthermore, define

$$\Omega := \{\omega > 0 : |R_0(\omega)| - \omega^2|I_0(\omega)| + \omega^2 X_{0,\infty}(\omega) > 0,$$
$$|R_0(\omega)| - \omega^2|I_0(\omega)| - X_{0,\infty}(\omega) < 0\},$$

and

$$\rho_0(\omega) := \begin{cases} \frac{\omega}{\sqrt{1+\omega^2}}\left(|R_0(\omega)| + |I_0(\omega)| - X_{0,\infty}(\omega)\right) & \omega \in \Omega, \\ \underline{\rho}_0(\omega) & \omega \notin \Omega. \end{cases}$$

Then the diamond quasipolynomial (6), with $a_k(s, \alpha_k)$ given by (9), (13), is robustly stable independent of delay if and only if

(i) The diamond polynomial $a_0(s, \alpha_0)$ is stable;

(ii) $a_{00}^* > \sum_{k=1}^{m} \gamma_{k0}$;

(iii)

$$\frac{\sum_{k=1}^{m} \overline{\rho}_k(\omega)}{\rho_0(\omega)} < 1, \quad \forall\omega > 0. \tag{15}$$

The idea behind Theorem 2 is also rather simple. It is clear from Figure 3 that the *maximum* of $|a_k(j\omega, \alpha_k)|$ must be achieved on one of the four vertices. The *minimum* of $|a_0(j\omega, \alpha_0)|$, however, occurs on one of the four edges, but may not on the vertices. This is the case when a straight line from the origin is perpendicular to the edge closest to the origin, which occurs for $\omega \in \Omega$. Otherwise, the minimum will still be attained at one of the vertices. The calculation of $\rho_0(\omega)$ explains this intuition.

3.3 Spherical quasipolynomials

For the spherical quasipolynomial family, each polynomial $a_k(s, \alpha_k)$ is comprised of a spherical polynomial family, characterized by the parameter set

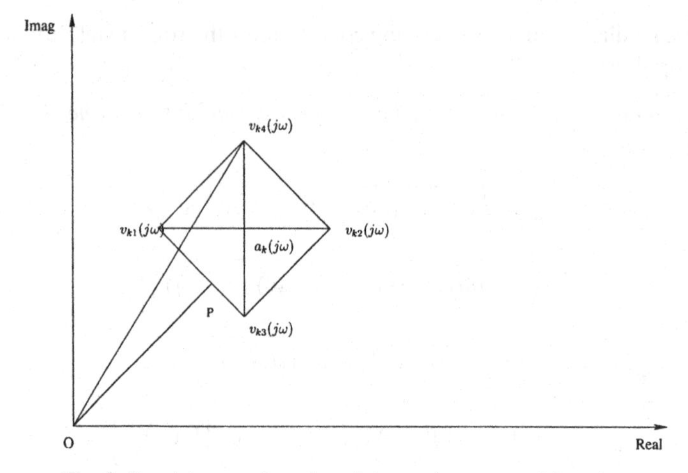

Fig. 3. Special case when the minimum is on one of the edges

$$\mathcal{Q}_k := \{\alpha_k : \alpha_k = \alpha_k^* + \Gamma_k \delta_k, \ \|\delta_k\|_2 \leq 1\}. \tag{16}$$

The robust stability of the spherical polynomial $a_0(s, \alpha_0)$ can be determined as well using a frequency-sweeping condition [8].

Lemma 4. *Suppose that $a_0(s, \alpha_0^*)$ is stable. Then, the spherical polynomial $a_0(s, \alpha_0)$ is robustly stable if and only if $\underline{\alpha}_{00} > 0$ and*

$$\frac{| R_0(\omega) |^2}{X_{0,2}^2(\omega)} + \frac{|I_0(\omega)|^2}{Y_{0,2}^2(\omega)} > 1, \quad \forall \omega > 0. \tag{17}$$

As in the case of interval and diamond quasipolynomials, we shall also calculate $\max_{\alpha_k \in \mathcal{Q}_k} |a_k(j\omega, \alpha_k)|$ and $\min_{\alpha_0 \in \mathcal{Q}_0} |a_0(j\omega, \alpha_0)|$. Define

$$h_k(\lambda) := \frac{|R_k(\omega)|^2 X_{k,2}^2(\omega)}{(\lambda + X_{k,2}^2(\omega))^2} + \frac{\omega^4 |I_k(\omega)|^2 Y_{k,2}^2(\omega)}{(\lambda + \omega^2 Y_{k,2}^2(\omega))^2}. \tag{18}$$

and

$$\overline{\lambda}_k := \begin{cases} X_{k,2}^2(\omega) & \text{if } I_k(\omega) = 0 \\ \omega^2 Y_{k,2}^2(\omega) & \text{if } R_k(\omega) = 0 \\ \max\left\{X_{k,2}^2(\omega), \ \omega^2 Y_{k,2}^2(\omega)\right\} & \text{otherwise,} \end{cases}$$

$$\underline{\lambda}_k := \begin{cases} X_{k,2}^2(\omega) & \text{if } I_k(\omega) = 0 \\ \omega^2 Y_{k,2}^2(\omega) & \text{if } R_k(\omega) = 0 \\ \min\left\{X_{k,2}^2(\omega), \ \omega^2 Y_{k,2}^2(\omega)\right\} & \text{otherwise.} \end{cases}$$

We have the following result:

Theorem 3. *Let τ_k, $k = 1, \cdots, m$ be independent delays. Define*

$$\overline{\rho}_k(\omega) := |\overline{\lambda}_k^*| \sqrt{\frac{|R_k(\omega)|^2}{(\overline{\lambda}_k^* + X_{k,2}^2(\omega))^2} + \frac{\omega^2 |I_0(\omega)|^2}{(\overline{\lambda}_k^* + \omega^2 Y_{k,2}^2(\omega))^2}},$$

$$\underline{\rho}_k(\omega) := |\underline{\lambda}_k^*| \sqrt{\frac{|R_k(\omega)|^2}{(\underline{\lambda}_k^* + X_{k,2}^2(\omega))^2} + \frac{\omega^2 |I_k(\omega)|^2}{(\underline{\lambda}_k^* + \omega^2 Y_{k,2}^2(\omega))^2}},$$

where for all $k = 0, 1, \cdots, m$, $\overline{\lambda}_k^* \in (-\infty, -\overline{\lambda}_k)$, $\underline{\lambda}_k^* \in (-\underline{\lambda}_k, \infty)$, and $h_k(\overline{\lambda}_k^*) = h_k(\underline{\lambda}_k^*) = 1$. Then the spherical quasipolynomial (6), with $a_k(s, \alpha_k)$ given by (9) and (16), is robustly stable independent of delay if and only if

(i) The spherical polynomial $a_0(s, \alpha_0)$ is stable;

(ii) $\sum_{k=0}^{m} \underline{\alpha}_{k0} > 0$;

(iii)

$$\frac{\sum_{k=1}^{m} \overline{\rho}_k(\omega)}{\underline{\rho}_0(\omega)} < 1, \quad \forall \omega > 0. \tag{19}$$

Theorem 3 makes it clear that the robust stability of the spherical quasipolynomial can be determined by solving essentially $\underline{\lambda}_0^*$ and $\overline{\lambda}_k^*$, $k = 1, \cdots, m$. With the properties of $h_k(\lambda)$ (increasing function on $(-\infty, -\overline{\lambda}_k)$, and decreasing function on $(-\underline{\lambda}_k, \infty)$, respectively), each of these can be solved readily using a line search method. They can also be found by solving the 4th order polynomial equation

$$\frac{|R_k(\omega)|^2 X_{k,2}^2(\omega)}{(\lambda + X_{k,2}^2(\omega))^2} + \frac{\omega^4 |I_k(\omega)|^2 Y_{k,2}^2(\omega)}{(\lambda + \omega^2 Y_{k,2}^2(\omega))^2} = 1,$$

at each $\omega \in (0, \infty)$. This can be easily accomplished as well.

4 Extensions

The preceding techniques and results can be extended in a number of ways, resulting in stability conditions for different and more general uncertain quasipolynomials. For example, it is easy to extend Theorems 1–3 to more general uncertainty descriptions characterized by the ℓ_p norm, for an arbitrary p with $1 \leq p \leq \infty$. Indeed, a direct generalization will yield necessary or sufficient conditions for robust stability independent of delay, by appropriately estimating the maximum of $|a_k(j\omega, \alpha_k)|$ and the minimum of $|a_0(j\omega, \alpha_0)|$; many bounds for these quantities can be obtained to this effect. One may also consider using different ℓ_p norms in the uncertainty characterizations, which will allow us to combine the formulas of $\overline{\rho}_k(\omega)$ and $\underline{\rho}_0(\omega)$ to obtain necessary and sufficient frequency-sweeping conditions in much the same spirit as in Theorem 1–3, despite that the uncertain coefficient polynomials are each described individually by any of the ℓ_1, ℓ_2, or ℓ_∞ norms.

This section presents a number of extensions beyond those alluded to above. We consider more general uncertainty descriptions and seek frequency-sweeping results. A generalization is also made to a special class of multivariate polynomials.

4.1 Uncertain coefficients in ℓ_p balls

Our first generalization is sought after for a case where uncertain coefficients are characterized by general ℓ_p norms. Consider the uncertain quasipolynomial (6) with the coefficients described by

$$\mathcal{Q}_k := \{\alpha_k : \alpha_k = \alpha_k^* + \Gamma_k \delta_k, \ \|\delta_k\| \le 1\}, \tag{20}$$

where

$$\|\delta_k\| := \max\{\|\delta_k^e\|_{p_1}, \ \|\delta_k^o\|_{p_2}\}. \tag{21}$$

In other words, we assume that the real and imaginary parts of the coefficient polynomials vary independently of each other. This enables us to generalize our preceding results immediately. Note that in this case a similar frequency-sweeping condition for the robust stability of $a_0(s, \alpha_0)$ can be obtained in much the same spirit as in [8]. Note also that uncertain polynomials with independent real and imaginary parts are studied in [21].

Lemma 5. *Suppose that $a_0(s, \alpha_0^*)$ is stable. Then, the uncertain polynomial $a_0(s, \alpha_0)$ with $\mathcal{Q}_0$ given by (20-21) is robustly stable if and only if $\underline{\alpha}_{00} > 0$ and*

$$\min\left\{\frac{X_{0,q_1}(\omega)}{|R_0(\omega)|}, \ \frac{Y_{0,q_2}(\omega)}{|I_0(\omega)|}\right\} < 1, \quad \forall \omega > 0. \tag{22}$$

The following result is a counterpart to Theorem 1.

Theorem 4. *Let τ_k, $k = 1, \cdots, m$ be independent delays. Define*

$$\overline{\rho}_k(\omega) := \sqrt{(|R_k(\omega)| + X_{k,q_1}(\omega))^2 + \omega^2 (|I_k(\omega)| + Y_{k,q_2}(\omega))^2},$$

$$\underline{\rho}_0(\omega) := \sqrt{M_R^2(\omega) + \omega^2 M_I^2(\omega)},$$

where

$$M_R(\omega) = \begin{cases} |R_0(\omega)| - X_{0,q_1}(\omega) & \text{if } |R_0(\omega)| > X_{0,q_1}(\omega) \\ 0 & \text{if } |R_0(\omega)| \le X_{0,q_1}(\omega) \end{cases}$$

$$M_I(\omega) := \begin{cases} |I_0(\omega)| - Y_{0,q_2}(\omega) & \text{if } |I_0(\omega)| > Y_{0,q_2}(\omega) \\ 0 & \text{if } |I_0(\omega)| \le Y_{0,q_2}(\omega) \end{cases}$$

Then the uncertain quasipolynomial (6), with $a_k(s, \alpha_k)$ given by (9) and (20-21), is robustly stable independent of delay if and only if

(i) The uncertain polynomial $a_0(s, \alpha_0)$ is stable;

(ii) $\sum_{k=0}^{m} \underline{\alpha}_{k0} > 0$;

(iii)

$$\frac{\sum_{k=1}^{m} \overline{\rho}_k(\omega)}{\underline{\rho}_0(\omega)} < 1, \quad \forall \omega > 0. \tag{23}$$

4.2 Polytopic uncertainty

Extensions may also be found to quasipolynomials with polytopic uncertainties. For this purpose, let the family of polynomials $a_k(s,\ \alpha_k)$ be the convex hull

$$\mathcal{P}_k := \text{conv}\left\{p_{k1}(s),\ \cdots,\ p_{k,l_k}(s)\right\}. \tag{24}$$

In other words, $a_k(s,\ \alpha_k)$ can be expressed as the convex combination of the *generating polynomials* $p_{kj}(s)$:

$$a_k(s,\ \alpha_k) = \sum_{j=1}^{l_k} \lambda_j p_{kj}(s), \quad \sum_{j=1}^{l_k} \lambda_j = 1, \ \ \lambda_j \geq 0, \ \ j = 1,\ \cdots,\ l_k, \tag{25}$$

where

$$p_{0j}(s) = s^n + \sum_{i=0}^{n-1} p_{0i}^{(j)} s^i, \tag{26}$$

$$p_{kj}(s) = \sum_{i=0}^{n-1} p_{ki}^{(j)} s^i, \quad k = 1,\ 2,\ \cdots,\ m. \tag{27}$$

We note that both the interval and diamond polynomials fall as special cases of this polytopic class. Note also that the stability of $a_0(s,\ \alpha_0)$ can be checked using the so-called *edge theorem* and other tools (see, e.g., [1,3]). We provide below frequency-sweeping results for the corresponding quasipolynomials.

Theorem 5. *Let $\tau_k \geq 0$, $k = 1,\ \cdots,\ m$ be independent delays. Define*

$$\rho_{ij}(\omega) := \begin{cases} \frac{|Im\{p_{0i}(-j\omega)p_{0j}(j\omega)\}|}{|p_{0i}(j\omega)-p_{0j}(j\omega)|} & \text{if} \\ \qquad Re\{p_{0i}(-j\omega)p_{0j}(j\omega)\} < \min\{|p_{0i}(j\omega)|^2,\ |p_{0j}(j\omega)|^2\} \\ \min\{|p_{0i}(j\omega)|,\ |p_{0j}(j\omega)|\} & \text{otherwise.} \end{cases}$$

Then the uncertain quasipolynomial (6), with $a_k(s,\ \alpha_k)$ given by (25-27), is robustly stable independent of delay if and only if

(i) The polytopic polynomial $a_0(s,\ \alpha_0)$ is robustly stable;

(ii) $\sum_{k=0}^{m} \min_{1\leq j\leq l_k} p_{k0}^{(j)} > 0$;

(iii)

$$\frac{\sum_{k=1}^{m} \max_{1\leq j\leq l_k} |p_{kj}(j\omega)|}{\min_{1\leq i<j\leq l_0} \rho_{ij}(\omega)} < 1, \qquad \forall\omega > 0. \tag{28}$$

We note that the numerator in (28) can be easily computed. The computation of the denominator can be more demanding, which requires $l_0(l_0-1)/2$ computations of $\rho_{ij}(\omega)$.

4.3 Multivariate polynomials

It is straightforward to extend the preceding results to a special class of multivariate polynomials which are also known as *disc polynomials* [3]. This class of multivariate polynomials are described as

$$p(s;\ z_1,\ \cdots,\ z_m) = a_0(s) + \sum_{k=1}^{m} a_k(s) z_k. \tag{29}$$

The multivariate polynomial $p(s;\ z_1,\ \cdots,\ z_m)$ is said to be stable if

$$p(s;\ z_1,\ \cdots,\ z_m) \neq 0, \qquad \forall s \in \overline{\mathbb{C}}_+,\ z_k \in \mathbb{D}^c,\ k = 1,\ \cdots,\ m.$$

A necessary and sufficient condition for the stability of $p(s;\ z_1,\ \cdots,\ z_m)$ is available from, e.g., [3, 17], which can also be seen rather trivially from [3].

Lemma 6. *The multivariate polynomial (29) is stable if and only if*

(i) $a_0(s)$ *is stable;*
(ii)

$$\frac{\sum_{k=1}^{m} |a_k(j\omega)|}{|a_0(j\omega)|} < 1, \qquad \forall\omega \geq 0. \tag{30}$$

Clearly, the sole difference between Lemma 6 and Lemma 1 lies at the frequency $\omega = 0$. It is thus unsurprising that the preceding results can all be extended readily to this class of multivariate polynomials with uncertain coefficients.

5 Illustrative Examples

We consider the uncertain quasipolynomial

$$\begin{aligned} p(s; e^{-s\tau_1}, e^{-s\tau_2}, e^{-s\tau_3}) &= s^4 + \alpha_{03}s^3 + \alpha_{02}s^2 + \alpha_{01}s + \alpha_{00} \\ &+(\alpha_{13}s^3 + \alpha_{12}s^2 + \alpha_{11}s + \alpha_{10})e^{-s\tau_1} + (\alpha_{23}s^3 + \alpha_{22}s^2 + \alpha_{21}s + \alpha_{20})e^{-s\tau_2} \\ &+(\alpha_{31}s + \alpha_{30})e^{-s\tau_3}, \end{aligned} \tag{31}$$

where $\alpha_{00} \in [4.85, 5.15], \alpha_{01} \in [7.85, 8.15], \alpha_{02} \in [9.85, 10.15], \alpha_{03} \in [5.85, 6.15]$ $\alpha_{10} \in [0.2, 0.4], \alpha_{11} \in [0.9, 1.1], \alpha_{12} \in [0.2, 0.4], \alpha_{13} \in [0.2, 0.4]$ $\alpha_{20} \in [0.1, 0.5]$, $\alpha_{21} \in [0.8, 1.2], \alpha_{22} \in [0.3, 0.7], \alpha_{23} \in [0, 0.4], \alpha_{30} \in [0, 0.6]$, and $\alpha_{31} \in [0.6, 1.2]$, respectively. Our purpose is to demonstrate how the preceding results can be effectively used to test robust stability independent of delay. For this purpose, we first obtain

$$\begin{aligned} &R_0(\omega) = 5 - 10\omega^2 + \omega^4, \quad I_0(\omega) = 8 - 6\omega^2, \\ &R_1(\omega) = 0.3(1 - \omega^2), \quad I_1(\omega) = 1 - 0.3\omega^2, \\ &R_2(\omega) = 0.3 - 0.5\omega^2, \quad I_2(\omega) = 1 - 0.2\omega^2, \\ &R_3(\omega) = 0.3, \quad I_3(\omega) = 0.9. \end{aligned}$$

Assume first that $p(s;\ e^{-s\tau_1},\ e^{-s\tau_2},\ e^{-s\tau_3})$ is an interval quasipolynomial. In this case,

$$\begin{aligned} &X_{0,1}(\omega) = 0.15(1 + \omega^2), \quad Y_{0,1}(\omega) = 0.15(1 + \omega^2), \\ &X_{1,1}(\omega) = 0.1(1 + \omega^2), \quad Y_{1,1}(\omega) = 0.1(1 + \omega^2), \\ &X_{2,1}(\omega) = 0.2(1 + \omega^2), \quad Y_{2,1}(\omega) = 0.2(1 + \omega^2), \\ &X_{3,1}(\omega) = 0.3, \quad Y_{3,1}(\omega) = 0.3. \end{aligned}$$

Figure 4 plots the frequency-dependent conditions in Lemma 2 and Theorem 1. From these plots, it is immediately clear that the interval quasipolynomial is robustly stable independent of delay. Next, suppose that $p(s;\ e^{-s\tau_1},\ e^{-s\tau_2},\ e^{-s\tau_3})$ is a diamond

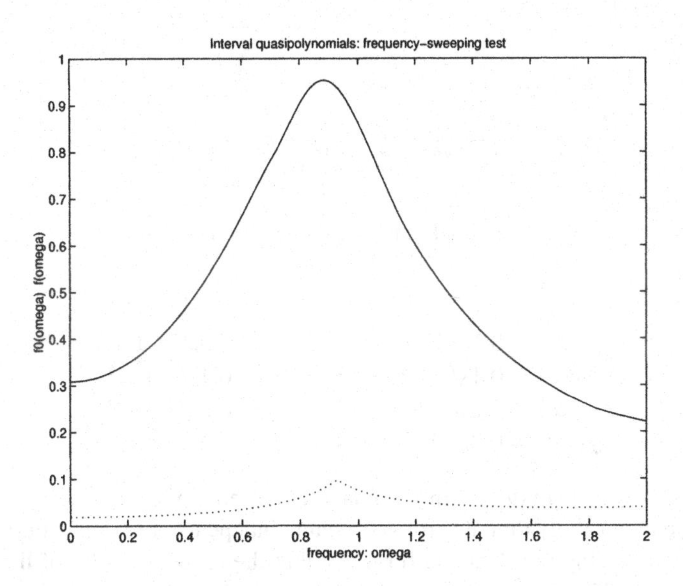

Fig. 4. Interval quasipolynomials

polynomial. One may immediately verify that it is a uniformly weighted diamond quasipolynomial with $\gamma_0 = 0.15, \gamma_1 = 0.1, \gamma_2 = 0.2$, and $\gamma_3 = 0.3$. Furthermore,

$$X_{0,\infty}(\omega) = 0.15\max\{1,\ \omega^2\},\quad Y_{0,\infty}(\omega) = 0.15\max\{1,\ \omega^2\},$$
$$X_{1,\infty}(\omega) = 0.1\max\{1,\ \omega^2\},\quad Y_{1,\infty}(\omega) = 0.1\max\{1,\ \omega^2\},$$
$$X_{2,\infty}(\omega) = 0.2\max\{1,\ \omega^2\},\quad Y_{2,\infty}(\omega) = 0.2\max\{1,\ \omega^2\},$$
$$X_{3,\infty}(\omega) = 0.3,\quad Y_{3,\infty}(\omega) = 0.3.$$

The conditions in Lemma 3 and Theorem 2 are plotted in Figure 5. Likewise, we conclude that the diamond quasipolynomial is robustly stable independent of delay. Finally, when $p(s;\ e^{-s\tau_1},\ e^{-s\tau_2},\ e^{-s\tau_3})$ is a spherical quasipolynomial, we find

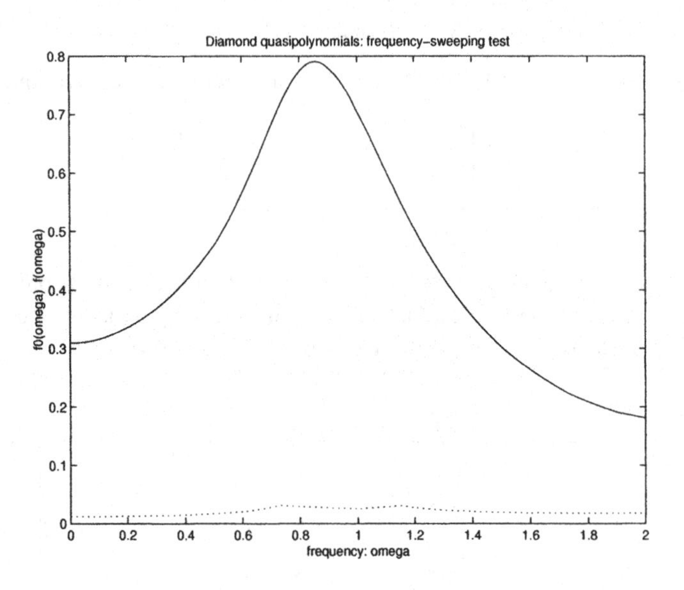

Fig. 5. Diamond quasipolynomials

$$X_{0,2}(\omega) = 0.15\sqrt{1+\omega^2}\,,\quad Y_{0,2}(\omega) = 0.15\sqrt{1+\omega^2},$$
$$X_{1,2}(\omega) = 0.1\sqrt{1+\omega^2},\quad Y_{1,2}(\omega) = 0.1\sqrt{1+\omega^2},$$
$$X_{2,2}(\omega) = 0.2\sqrt{1+\omega^2},\quad Y_{2,2}(\omega) = 0.2\sqrt{1+\omega^2},$$
$$X_{3,2}(\omega) = 0.3,\quad Y_{3,2}(\omega) = 0.3.$$

Figure 6 shows that conditions in Lemma 4 and Theorem 3, which also shows that the spherical quasipolynomial is robustly stable independent of delay. Furthermore, a comparison of Figures 4, 5 and 6 reveals that the robust stability of the interval quasipolynomial implies that of the spherical quasipolynomial, and the latter implies the robust stability of the diamond quasipolynomial. This is expected, as the ℓ_p norm $\|\cdot\|_p$ defines an increasing function of $p \in [1,\ \infty)$.

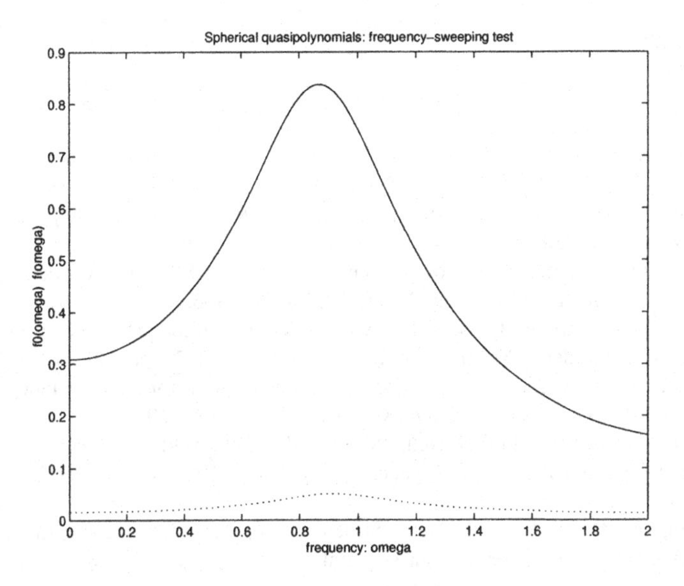

Fig. 6. Spherical quasipolynomials

6 Concluding Remarks

In this chapter we have studied the robust stability of uncertain quasipolynomials. We addressed specifically the notion of robust stability independent of delay, and considered interval, diamond and spherical quasipolynomials We derived necessary and sufficient conditions for these uncertain quasipolynomials, which can be checked by performing frequency-sweeping tests. Our technique is built upon the frequency domain approaches developed in [7, 8] for robust stability analysis of uncertain polynomials, and in [3] for stability of time-delay systems. A useful feature about this technique is that it provides a rather intuitive geometrical interpretation. This interpretation may shed light into related stability problems for other classes of uncertain quasipolynomials. The authors have also developed *vertex*-type *necessary* and *sufficient* conditions for stability independent of delay. These results are provided in [9].

Acknowledgements

The research of J. Chen was partially supported by NSF under the grant number ECS-9912533. The research of S.-I. Niculescu was partially supported by a PICS CNRS-NSF Grant: "Time-delay systems: Theory, computer aided design, and applications" (2002-2005).

References

1. B.R. Barmish, *New Tools for Robustness of Linear Systems*, New York, NY: Macmillan, 1994.
2. K.D. Kim and N.K. Bose, "Vertex implications of stability for a class of delay-differential interval systems," *IEEE Trans. Circ. Syst.*, vol. 37, no. 7, pp. 969-972, 1990.
3. S. P. Bhattacharyya, H. Chapellat, and L. H. Keel, *Robust control. The parametric approach*, Prentice Hall, 1995.
4. J. Chen, "On computing the maximal delay intervals for stability of linear delay systems," *IEEE Trans. Automat. Contr.*, vol. 40, pp. 1087-1093, 1995.
5. J. Chen, J., G. Gu, and C. N. Nett, "A new method for computing delay margins for stability of linear delay systems,'" *Syst. & Contr. Lett.*, vol. 26, pp. 101-117, 1995.
6. J. Chen, and H. A. Latchman, "Frequency sweeping tests for stability independent of delay," *IEEE Trans. Automat. Contr.*, vol. 40, pp. 1640-1645, 1995.
7. J. Chen, M.K.H. Fan, and C.N. Nett, "Structured singular values and stability analysis of uncertain polynomials, part 1: the generalized μ," *Syst. & Contr. Lett.*, vol. 23, pp. 53-65, 1994.
8. J. Chen, M.K.H. Fan, and C.N. Nett, "Structured singular values and stability analysis of uncertain polynomials, part 2: a missing link," *Syst. & Contr. Lett.*, vol. 23, pp. 97-109, 1994.
9. J. Chen and S.-I. Niculescu, "Robust stability of quasipolynomials: Frequency-sweeping conditions and vertex tests," *Internal Note HeuDiaSyC'03*, March 2003.
10. J. Chen and A.L. Tits, "Robust control analysis," in *Encyclopedia of Electrical and Electronics Engineering*, vol. 18, pp. 602-616, J.G. Webster, Ed., Wiley, 1998.
11. M. Fu, A. W. Olbrot, and M. P. Polis, "Robust stability for time-delay systems: The edge theorem and graphical tests," *IEEE Trans. Automat. Contr.*, vol. 34, pp. 813-820, 1989.
12. K. Gu, V.L. Kharitonov, and J. Chen, *Stability and robust stability of time-delay systems*, Birkhauser: Boston, 2003.
13. J.K. Hale, E.F. Infante, and F.S.P. Tsen, "Stability in linear delay equations," *J.Math. Anal. Appl.*, vol.105, pp. 533-555, 1985.
14. V. L. Kharitonov, and L. Atanassova, "On stability of wheighted diamond of real quasi-polynomials," *IEEE Trans. Automat. Contr.*, vol. 42, pp. 831-835, 1997.
15. V. L. Kharitonov, and A. Zhabko, "Robust stability of time delay systems," in *IEEE Trans. Automat. Contr.*, vol. 39, pp. 2388-2397, 1994.
16. J. Kogan, *Robust Stability and Convexity*, LNCIS., vol. 201, Springer: New York, 1995.
17. Y. Li, K. M. Nagpal, and E. Bruce Lee, "Stability analysis of polynomials with coefficients in disks," *IEEE Trans. Automat. Contr.*, vol. 37, pp. 509-513, 1992.
18. M.S. Mahmoud, *Robust Control and Filtering for Time-Delay Systems*, Mercel Dekker, 2000.
19. S. -I. Niculescu, *Delay effects on stability. A robust control approach*, Heidelberg, Germany: Springer-Verlag, LNCIS, vol. 269, 2001.
20. A.W. Olbrot and C.U.T. Igwe, "Necessary and sufficient conditions for robust stability independent of delay and coefficient perturbations," *Proc. 34th IEEE Conf. Decision Contr.*, New Orleans, LA, Dec. 1995, pp. 392-394.
21. E.R. Panier, M.K.H. Fan, and A.L. Tits, "On the robust stability of polynomials with no cross-coupling between the perturbations in the coefficients of even and odd parts," *Syst. & Contr. Lett.*, vol. 12, pp. 291-299, 1989.
22. O. Toker, and H. Ozbay, "Complexity issues in robust stability of linear delay-differential systems," *Math., Contr., Signals, Syst.*, vol. 9, pp. 386-400, 1996.

Improvements on the Cluster Treatment of Characteristic Roots and the Case Studies

Rifat Sipahi and Nejat Olgac*

191 Auditorium Rd., Engineering II Building, University of Connecticut, Storrs, CT 06269-3139

Summary. A recent methodology, the Direct Method (DM) , is considered for the stability analysis of the linear time-invariant time-delayed systems (LTI-TDS). Its fundamental strength is in the paradigm called the "cluster treatment of the characteristic roots" (CTCR) . Salient features of the CTCR and the steps of DM are described. An interesting extension to DM is its equally effective utilization for both the retarded and neutral TDS. A well studied necessary condition for NTDS is shown to be an inherent property imbedded within the steps of the DM. Example case studies are given to show the effectiveness of the procedure.

1 Introduction

This Quasipolynomial is formed in two parts. The first part presents several progressive steps on a recent methodology which deals with the stability of LTI-TDS (Linear Time-Invariant Time Delayed Systems). The treatment departs from the core concept of "cluster treatment of characteristic roots" (CTCR) [13]. And it addresses the improvements over it. The second part is on the applications of the CTCR paradigm on several practical problems. One of them is an active vibration control on a mutli-dimensional dynamics with MIMO structure which is influenced by a delay in the feedback control line. Another practical problem is a 2-D target tracking, again using a MIMO structure.

The treatment is on a relatively old problem of LTI-TDS in the form of

$$\dot{\mathbf{x}} = \mathbf{A}\mathbf{x}(t) + \mathbf{B}\mathbf{x}\,(t-\tau) \tag{1}$$

where $\mathbf{x}(n\times 1), \mathbf{A}, \mathbf{B} \in \Re^{(n\times n)}, \tau \in \Re^{+}$ [3,3,6,13,23]. We consider constant $\mathbf{A}$ and $\mathbf{B}$ matrices and try to assess the stability posture of the dynamics for the semi-infinite $\tau > 0$ domain. The objective is to determine all the stability intervals of τ exclusively. This process should be applicable independent of the initial stability nature of the system when non-delayed (i.e., $\tau = 0$). Such an objective is rather interesting in that, a dynamics which is unstable when there is no time delay may exhibit stable

* Author of correspondence: olgac@engr.uconn.edu

posture for larger τ values, or within several $\tau > 0$ intervals. Knowledge of these intervals can be of practical importance as demonstrated in the application example cases.

The characteristic equation of (1) is

$$\begin{aligned} CE(s,\tau) &= det(s\mathbf{I} - \mathbf{A} - \mathbf{B}e^{-\tau s}) \\ &= a_n(s)e^{-n\tau s} + a_{n-1}(s)e^{-(n-1)\tau s} + ... + a_0(s) \\ &= \textstyle\sum_{k=0}^{n} a_k(s)e^{-k\tau s}, \qquad \tau > 0 \end{aligned} \tag{2}$$

This equation is transcendental due to the exponential terms. Therefore it has infinite dimensional nature (i.e., infinitely many characteristic roots exist). The particular dynamics represented above is called "retarded" TDS with "commensurate" time delays, which implies that the highest order derivative term in (1) does not have delay in it (for "retarded" feature) and there are integer multiples of τ appearing in (2) yielding the "commensurate" property.

By definition, for the asymptotic stability of the dynamics all of the infinitely many characteristic roots have to be on the left half of the complex plane. To verify this is a very cumbersome task. Many earlier investigations [2, 3, 21–23, 25, 30] address this very question offering various paths to resolve the stability determination along the time delay axis. There are, however, limitations to these methodologies. Most common limitations can be listed as follows

- None of the existing techniques can offer a non-sequential procedure to identify all of the stable τ intervals, exclusively.
- These procedures depart with a stable system $\tau = 0$, and determine a "stability margin, τ_{max}" which assures stable operation for $0 \leq \tau \leq \tau_{max}$.
- These peer methodologies are all based on the detection of imaginary root crossings, the exclusivity of which cannot be achieved by the procedures suggested. Therefore the completeness of the methods are questionable especially when the dimension n increases. It is apparent from the literature that all of the peer methods can be implemented only for small dimensional dynamics ($n \leq 2$) and for relatively simple **A** and **B** matrices (mostly vacuous cases).

In response to these limitations we present a structured paradigm "the cluster treatment of characteristic roots" (CTCR) in the methodology presented in [13, 16, 25], the "Direct Method" (DM). We briefly recite, next, the underlying features of both the CTCR and the DM.

It is obvious that when a τ value causes a pair of characteristic roots $s = \mp\omega i$ (i.e., root crossing over the imaginary axis) this transition needs to be examined as to the nature of the crossing. Therefore there is a need of determining <u>all</u> such τ values yielding an imaginary crossing. Consider that such exhaustive analysis is done and a complete set of $\{\tau_k, \omega_k\}$, $k = 1...m$ is available. The following can be stated about these crossings (proofs of which are left to [13, 16, 25]):

i) $s = \omega_k i$ is generated not only by τ_k, but also infinitely many

$$\tau_{k,\ell} = \tau_k + \frac{2\pi}{\omega_k}\ell \quad k = 1...m\ , \ \ell = 0, \mp 1, \mp 2... \tag{3}$$

For simplicity and accounting for the positiveness of the time delay, τ we rename this set of delays as $\{\tau_{k\ell}\}$, $\ell = 0, 1, 2...$, where τ_{k0} is the smallest positive τ among the set in (3).

ii) For a given system in (1) there is a finite number of ω_k's, $k = 1...m$; and $m_{max} \leq n\times$ Fibonacci number$_{2n}$(1,1) [25], where the Fibonacci number(1,1) is a series with terms [24]:

$$\{1, 1, 2, 3, 5, 8, ... \ a_i = a_{i-1} + a_{i-2}\} \tag{4}$$

and the upper bound uses the $2n^{th}$ term of this series.

iii) At these *m* crossings the root tendency (RT) of *s* is invariant with respect to the generating value of τ. That is

$$RT|_{\tau_{k\ell}} = sgn[Re\left(\frac{\partial s}{\partial \tau}\right)\Bigg|_{\substack{s = \omega_k i \\ \tau = \tau_{k\ell}}}] \qquad \ell = 0, 1, 2... \tag{5}$$

is invariant with respect to ℓ. What this indicates is that at a given crossing $s = \omega_k i$ there can only be one directional passage of roots ($RT = +1$ from left to right-half-plane of *s*, and $RT = -1$ vice versa), no matter what the corresponding delay value is.

Utilizing the feature (i), we form *m* clusters of roots to be studied, which contain the *m* critical crossings and the respective delays,$\{\tau_{k\ell}\}$, $k = 1...m$, $\ell = 0, 1, 2...$ which create these *m* crossings as "stabilizing", i.e., with $RT = -1$ and "destabilizing",i.e., with $RT = +1$ [13,16,19,20,25]. These two clusters form the complete set of characteristic roots to be examined for stability, and again this set is exhaustive. This is the reason the method is called the "cluster treatment of characteristic roots" (CTCR). It enables the declaration of the complete picture of stability pockets .

The crucial point in this stability analysis is the determination of the complete set of imaginary crossings $\{\tau_{k0}, \omega_k\}$, $k = 1...m$. This problem is addressed by many researchers in the past and it is also a current research topic [2,3,11,17,23]. Throughout this study we deploy the methodology suggested by [17] in 1981, because of its simplicity and enabling properties. It starts with a substitution of

$$e^{-\tau s} = \frac{1 - Ts}{1 + Ts}, \quad \tau \in \Re^+, \ T \in \Re \tag{6}$$

which really represents an exact expression only when $s = \omega i$ and for the mapping condition between τ and T as:

$$\tau = \frac{2}{\omega}[tan^{-1}(\omega T) + \ell\pi] \quad \ell = 0, 1, 2, ...\infty \tag{7}$$

The substitution (6) converts the nth degree transcendental characteristic equation (2) into a $2n$th degree algebraic one as below.

$$\begin{aligned} CE(s,T) &= \sum_{k=0}^{n} a_k(s)(1-Ts)^k(1+Ts)^{n-k} \\ &= \sum_{k=0}^{2n} b_k(T)s^k = 0 \end{aligned} \tag{8}$$

Notice, however, that in (7) the mapping from T to τ is one on infinitely many. This really corresponds to the first clustering feature which is mentioned earlier, i.e., the finite number of T's (upper bounded by a Fibonacci number) which correspond to imaginary crossings will represent infinitely many time delays for the same crossing $s = \omega i$.

An important convenience that the Rekasius substitution offers is in numerical simplicity and exactness for determining the complete set of $\{T_k, \omega_k\}, k = 1...m$. As explained in [13] this procedure takes place as follows: The characteristic equation $CE(s,T)$, which is parameterized in T as in equation (8), is taken into account. Corresponding Routh's array (also parameterized in T) is formed. In order for $CE(s,T)$ to have a pair of imaginary roots two conditions must be satisfied:

i) the only term on the row corresponding to s^1 has to be zero for a real value of $\overline{T}$,
ii) For this $\overline{T}$ the two terms on the row of s^2 has to agree in sign. Notice that these 2 terms form the auxiliary equation which yields the ω_k, $k = 1...m$.

In other words, the s^1 term of the array, which is a polynomial of T can be solved for all real T roots, and the condition (ii) can be tested. These conditions, (i) and (ii), are the *numerical procedures* which result in the complete set of $\{T_k, \omega_k\}$, $k = 1...m$. This is probably the key uniqueness of the Rekasius substitution.

Once the critical 'clustering' step is finalized using $\{T_k, \omega_k, RT|_k\}$,$k = 1...m$ set, the structured framework of the Direct Method (DM) follows [13, 16, 25]:

i) Using the Rekasius substitution, determine **all** m imaginary roots of the system.
ii) Check the number of unstable roots of the non-delayed system,$NU(\tau = 0)$. A conventional Routh's array application suffices for this. If this number is zero, then the non-delayed system is asymptotically stable or at worst marginally stable. The DM does not require the stable non-delayed system.
iii) Form the Routh's array using equation (8). Study (NS v. T) variation at $T = 0$ where NS is the number of sign changes on the first column of the array. Check the following two conditions: a)$NS(T = 0^-) - NS(T = 0^+) = n$, and b)$NS(T = 0^+) = NU(\tau = 0)$. They correspond to the "τ-stabilizability" conditions of the system for the NTDS as explained further in the chapter as well as in [20].
iv) Form a table of $\{\tau_{k\ell}\}$, $k = 1...m$, $\ell = 0, 1, 2...$ and $RT|_{\tau_{k\ell}} = RT|_k$ in ascending order of $\tau_{k\ell}$. This table presents the complete picture of the **root clusters** with respective ω_k's and $RT|_k$'s.
v) Go to the smallest $\tau_{k\ell}$, determine the number of unstable roots, NU, using the $RT|_k$, as $\tau = \tau_{k\ell}^+$. If $RT = +1$, NU increases by 2, if $RT = -1$, it decreases by 2.

vi) Repeat the previous step (v) for the next $\tau_{k\ell}$ sequentially completing the analysis when the target value of τ is reached.
vii) Identify those regions in τ, where $NU(\tau) = 0$ as *stable* and others as *unstable*.

The Direct Method also renders a unique function of τ [13] avoiding the sequential treatment of these steps (from one $\tau_{k\ell}$ to the next). This function declares the number of unstable roots of the system in a *non-sequential form*:

$$NU(\tau) = NU(0^+) + \sum_{k=1}^{n} \Gamma(\frac{\tau - \tau_{k0}}{\Delta\tau_k})U(\tau, \tau_{k0})RT_k \tag{9}$$

where $NU(0^+)$ is the number of unstable roots when $\tau = 0^+$ and it is equal to $NU(0)$ for τ-stabilizable systems. $U(\tau, \tau_{k0})$ = Step function at $\tau = \tau_{k0}$.

$$U(\tau, \tau_{k0}) = \begin{cases} 0 & 0 < \tau < \tau_{k0} \\ 1 \; for & \tau \geq \tau_{k0}, \quad \omega_k = 0 \\ 2 & \tau \geq \tau_{k0}, \quad \omega_k \neq 0 \end{cases}$$

$\Gamma(x)$ =is the ceiling function of x. It returns the smallest integer greater than or equal to x. This expression $NU(\tau)$ requires the knowledge of four quantities:

i) $NU(0^+)$ is from Routh's array
ii) τ_{k0}, smallest τ corresponding to ω_k, $k = 1...m$
iii) $\Delta\tau_k = \tau_{k,\ell} - \tau_{k,\ell-1} = 2\pi/\omega_k$, $k = 1...m$
iv) RT(k), $k = 1...m$

Notice again that, NU is a nonsequentially evaluated function of τ. That is, in order to determine the $NU(\tau, \tau_k < \tau < \tau_{k+1})$ one does not need to know $NU(\tau, \tau_{k-1} < \tau < \tau_k)$. In this sense it is unique and useful when enumerating the stability outlook.

This completes the review of the DM. We present in Section 2 the further progress on the methodology and in Section 3 some practical applications.

2 Extension to the DM on Neutral TDS

The procedure described in Section 1 was first introduced for retarded TDS (RTDS). It is further elaborated for the neutral TDS (NTDS). Conventionally this class can be represented by

$$\dot{\mathbf{x}} = \mathbf{A}\mathbf{x}(t) + \mathbf{B}\mathbf{x}(t - \tau) + \mathbf{C}\dot{\mathbf{x}}(t - \tau) \tag{10}$$

and the respective characteristic equation

$$CE(s, \tau) = det(s\mathbf{I} - \mathbf{A} - \mathbf{B}e^{-\tau s} - \mathbf{C}se^{-\tau s}) \tag{11}$$

Again the transcendentality inducing infinitely many characteristic roots and the need for a clustering framework (CTCR) for assessing the stability are similar to the RTDS. The main difference between RTDS and NTDS appears in small delay,

$\tau = \epsilon$ region. It is proven by many investigators [1,5–8] that the "τ-stabilizability" of NTDS in (10) requires the discrete kernel operator

$$L(\mathbf{x}) = \mathbf{x}(t) - \mathbf{C}\mathbf{x}(t-\tau) \tag{12}$$

to be stable as a necessary condition. This is equivalent to say that the eigenvalues of $\mathbf{C}$ are all within the unit circle, or the spectral radius of $\mathbf{C}$, $\rho(\mathbf{C})$ is less than 1 [1,5–9, 30].If this condition is not satisfied an interesting property appears: regardless of the $NU(0)$ (i.e., the number of unstable roots when $\tau = 0$), $NU(0^+)$ becomes infinity. In other words the transition of $\tau = 0 \to 0^+$ brings a sudden appearance of infinitely many right-half plane roots (i.e., unstable roots). Furthermore all of these roots are at $+\infty$, injecting an incredibly strong instability. Other than $\tau = 0$, there is no point in $\tau \in \Re^+$ domain which exhibits this form of "root discontinuity". Therefore when the small delay instability appears with $NU(0^+) \to \infty$ no finite $\tau > 0$ region can exist bringing the system back to stability. That is primarily the reason we call $\rho(\mathbf{C}) < 1$ requirement, the "τ-stabilizability" condition.

It is analytically proven that the steps (iii)a and (iii)b in the steps of the DM (Section 2) are equivalent to saying $\rho(\mathbf{C}) < 1$ [16,25]. We avoid the repetition of the proof here in the interest of space and simply state that these conditions (iii)a and (iii)b of DM are satisfied automatically by all RTDS, and only by the "τ-stabilizable" NTDS (i.e., those with $\rho(\mathbf{C}) < 1$). Thus the "τ-stabilizability" condition, which is presented in the earlier studies as a necessary condition to be examined, happens to be an imbedded component within the DM. The "τ-stabilizability" does not need a seperate verification other than the verification of (iii)a and (iii)b. This statement also implies that the structured steps of the DM are identically utilized for both RTDS and NTDS. This feature gives an added strength to DM.

3 Application Case Studies

Two application cases are presented in this section, one is on active vibration suppression and the other is on target tracking.

3.1 Active vibration suppression with time delayed feedback:

This case study is fundamentally different procedure than what is reported in the literature as "active vibration absorber" [14]. Instead of bringing an absorber section to resonance for suppressing the vibration, we use a full state feedback control force in order to make the system asymptotically return to quiescence. This feedback, however, is influenced by a time delay. Take the system in Figure 1 from [18,19], with its dynamic model

$$\dot{\mathbf{x}} = \mathbf{A}\mathbf{x}(t) + \overline{\mathbf{B}}\mathbf{u} \tag{13}$$

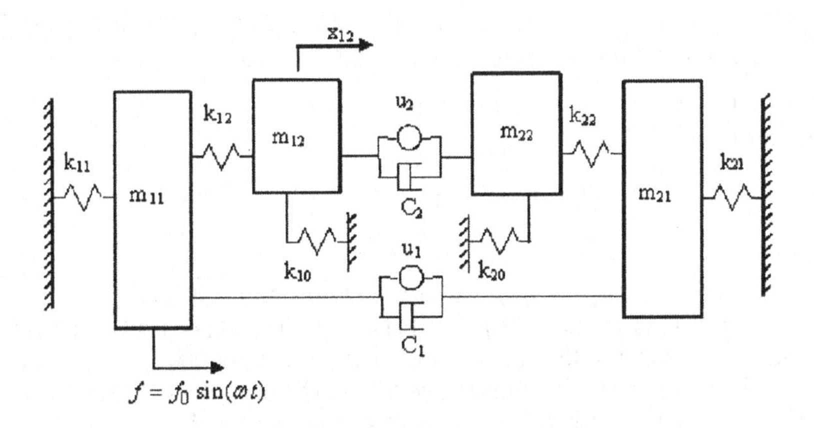

Fig. 1. 4 degree-of-freedom dynamic model

$$\mathbf{A} = \begin{pmatrix} 0 & 1 & 0 & 0 & 0 & 0 & 0 & 0 \\ -\frac{k_{11}+k_{12}}{m_{11}} & -\frac{c_1}{m_{11}} & \frac{k_{12}}{m_{11}} & 0 & 0 & 0 & 0 & \frac{c_1}{m_{11}} \\ 0 & 0 & 0 & 1 & 0 & 0 & 0 & 0 \\ \frac{k_{12}}{m_{12}} & 0 & -\frac{k_{10}+k_{12}}{m_{12}} & -\frac{c_2}{m_{12}} & 0 & \frac{c_2}{m_{12}} & 0 & 0 \\ 0 & 0 & 0 & 0 & 0 & 1 & 0 & 0 \\ 0 & 0 & 0 & \frac{c_2}{m_{22}} & -\frac{k_{22}+k_{20}}{m_{22}} & -\frac{c_2}{m_{22}} & \frac{k_{22}}{m_{22}} & 0 \\ 0 & 0 & 0 & 0 & 0 & 0 & 0 & 1 \\ 0 & \frac{c_1}{m_{21}} & 0 & 0 & \frac{k_{22}}{m_{21}} & 0 & -\frac{k_{22}+k_{21}}{m_{21}} & -\frac{c_1}{m_{21}} \end{pmatrix}$$

$$\overline{\mathbf{B}} = \begin{pmatrix} 0 & 0 \\ -\frac{1}{m_{11}} & 0 \\ 0 & 0 \\ 0 & -\frac{1}{m_{12}} \\ 0 & 0 \\ 0 & \frac{1}{m_{22}} \\ 0 & 0 \\ \frac{1}{m_{21}} & 0 \end{pmatrix}$$

where $\mathbf{x} = (x_{11}, \dot{x}_{11}, x_{12}, \dot{x}_{12}, x_{21}, \dot{x}_{21}, x_{22}, \dot{x}_{22},)'$ and the respective parameters taken as

$$\begin{aligned} &m_{11} = 0.2, \quad m_{12} = 0.15, \quad m_{21} = 0.2, \quad m_{22} = 0.15 \;\; kg \\ &c_1 = 2.2, \quad c_2 = 1.9 \;\; kg/s \\ &k_{10} = 2, \quad k_{11} = 4, \quad k_{12} = 2, \quad k_{20} = 4, \quad k_{21} = 2, \quad k_{22} = 2 \;\; N/m \end{aligned}$$

u_1 and u_2 in Figure 1 are the control forces created by hydraulic actuators and they are known for delayed reactions. Consider the feedback control law as

$$(u_1 \;\; u_2)^T = \mathbf{u}^T = \mathbf{K}\, \mathbf{x}(t - \tau) \quad with$$

$$\mathbf{K} = \begin{pmatrix} 3.5508\ 0.6220\ -2.0604\ -0.6741\ -2.8536\ -0.7806\ 3.2636\ 1.1393 \\ 2.0621\ 0.6296\ -2.5922\ -1.1130\ -2.5424\ -0.6635\ 1.3338\ 0.5483 \end{pmatrix}$$

Here $\mathbf{B}$ in Eequation (1) is indeed $\mathbf{B} = \overline{\mathbf{B}}\mathbf{K}$. For this 2-input, 4-output system the characteristic equation is

$$CE(s,\tau) = a_0(s) + a_1(s)e^{-\tau s} + a_2(s)e^{-2\tau s} = 0 \tag{14}$$

where

$$\begin{aligned} a_0(s) &= s^8 + 47.3333s^7 + 674s^6 + 4127.7777s^5 + 37244.4444s^4 \\ &\quad +111244.4444s^3 + 467911.1111s^2 + 911111.1111s + 444444.4444 \\ a_1(s) &= -5.5833s^7 - 130.3520s^6 - 568.2364s^5 - 10714.0002s^4 \\ &\quad -26942.9025s^3 - 162821.8942s^2 - 352026.9148s - 171542.7049 \\ a_2(s) &= 7.4630s^6 - 8.1770s^5 + 798.9722s^4 + 1065.1978s^3 \\ &\quad +13931.0291s^2 + 33937.2842s + 16607.8984 \end{aligned}$$

Notice that $rank(\mathbf{B}) = 2$ which causes the maximum commensurate delay term in (14) to appear as 2. When the DM and the root clustering procedures are deployed on (14) the following results appear: Rekasius substitution yields $CE(s,T)$, which is suppressed here for brevity. The Routh's array of this characteristic equation forms s^1 and s^2 rows which result in $T \in \Re$ solutions of interest. Symbolically:

$$Routh's\ \ array\ \begin{array}{c|cc} \cdots & \cdots & \cdots \\ s^2 & r_{21} & r_{22} \\ s^1 & r_{11} & \\ s^0 & r_{01} & \end{array} \quad \begin{array}{l} r_{11} = 0|_{T_i \in \Re} \\ verify\ \ sgn(r_{21}r_{22})|_{T_i \in \Re} > 0 \\ solve\ \ \omega = \sqrt{\frac{r_{22}}{r_{21}}} \end{array}$$

For the specific problem above, this procedure renders 16 real T roots, 4 of which are valid:

$$\begin{array}{lll} T_1 = -0.1105 & yields & \omega_1 = 3.9804 \\ T_2 = -0.0492 & yields & \omega_2 = 6.0076 \\ T_3 = 0.4614 & yields & \omega_3 = 6.7815 \\ T_4 = 1.8029 & yields & \omega_4 = 4.1906 \\ T_5 = -1.3954 & not\ \ valid & sign\ \ disagreement \\ \vdots & \vdots & \vdots \end{array}$$

And the results of the DM are shown on Table 1.

It is obvious that there are 3 stability pockets for this system and the vibration suppression can be performed effectively within these intervals of τ. A few example cases are taken from stable operation intervals (τ = 0.25,2.09,3.08 sec.) and their respective frequency response plots are superposed in Figure 2 for comparison. Notice that unless the preamble study for stability is completed these frequency response plots are meaningless. Thus, the importance of the DM in this procedure is clear. If one wishes to increase the suppression efficiency without changing the feedback gain matrix $\mathbf{K}$, the feedback delay τ can also be used as a parameter to achieve this. For instance when the excitation is at $\omega = 6\ rad/s$, $\tau = 0.25\ sec$ works better than no delay case ($\tau = 0$) as well as $\tau = 2.09\ sec$ and $\tau = 3.08\ sec$ cases.

Table 1. Stability table for the vibration suppression case study

τ [sec]	RT	Stable /Unstable NU	ω [rad/sec]
0			
		S, NU=0	
0.3720	+1		6.7815
		U, NU=2	
⋮	⋮	⋮	⋮
		U, NU=2	
1.9960	-1		6.0076
		S, NU=0	
2.1862	+1		4.1906
		U, NU=2	
⋮	⋮	⋮	⋮
		U, NU=2	
3.0418	-1		6.0076
		S, NU=0	
3.1516	+1		6.7815
		U, NU=2	
⋮	⋮	⋮	⋮

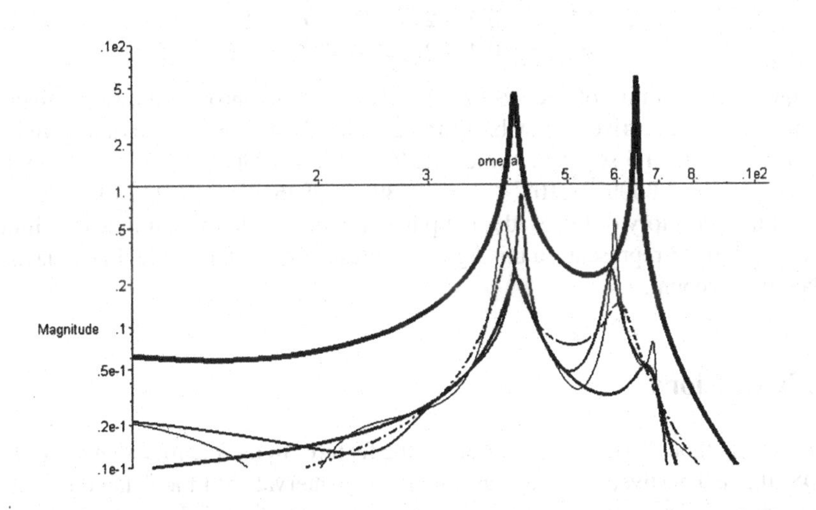

Fig. 2. Frequency response plots for various time delays(▬=uncontrolled, - -=controlled with no delay, ▬=controlled with $\tau = 0.25$ $s.$,-=controlled with $\tau = 2.08$ $s.$, —=controlled with $\tau = 3.09$ $s.$)

3.2 Target tracking with delayed feedback

In a 2-D platform a point mass (call it the pursuer) is expected to follow a target moving on a path unknown to the pursuer. A feedback control law which is formed over the error dynamics is supposed to enforce this, except the time delay which appears on the feedback line. In other words, the controller knows about the state being τ sec late. The immediate question is the stability posture of the system for various τ intervals. That brings us to the DM.

We start with the description of the error dynamics of

$$\dot{\mathbf{e}} = \mathbf{Ae} + \overline{\mathbf{B}}u = \mathbf{Ae} + \overline{\mathbf{B}}\mathbf{K}\mathbf{e}(t-\tau) \tag{15}$$

where $\mathbf{e} = (x - x_{target},\ \dot{x} - \dot{x}_{target},\ y - y_{target},\ \dot{y} - \dot{y}_{target})'$

$$\mathbf{A} = \begin{pmatrix} 0 & 1 & 1 & 1 \\ -k_x/m & -c_x/m & 0 & 0 \\ 2 & 1 & 0 & 1 \\ 0 & 0 & -k_y/m & -c_y/m \end{pmatrix}$$
$$\mathbf{B} = \begin{pmatrix} 0 & 0 & 0 & 0 \\ k1_x/m & k2_x/m & 0 & 0 \\ 0 & 0 & 0 & 0 \\ 0 & 0 & k1_y/m & k2_y/m \end{pmatrix} \tag{16}$$

As it is clear from (16) the e_1 and e_2 dynamics are coupled. For the system parameters:

$$\begin{array}{lll} m = 1 & k_x = 30.5 & c_x = 2.8 \\ k1_x = -5.5 & k2_x = 3 & c_y = 2 \\ k1_y = -0.4 & k2_y = -2.4 & k_y = 40 \end{array}$$

we query the stability of the system. The DM follows through the steps given in Section 2 and we arrive at a stability table, Table 2. It suggests that the given dynamics is stable in 5 separate τ intervals $0 \leq \tau < 0.2036$, $0.4630 < \tau < 0.9323$, $1.3368 < \tau < 1.6609$, $2.2107 < \tau < 2.3896$, $3.0845 < \tau < 3.1183$ *sec*. Regardless of the trajectory of the evader the pursuer could catch up with it as the simulations in Figure 3 represent. In this figure various stable and unstable time delays are taken into account.

4 Conclusions

The new method, DM, brings several unique features to the stability analysis of LTI-TDS. It is exhaustive, i.e., it reveals <u>all</u> stability intervals of time delay completely. It is exact in determining the bounds of these intervals and the numerical procedure suggested in a structured form is very efficient. This numerical procedure of DM is capable of handling large dimensional systems as well. Very importantly, the new method is equally applicable to the retarded and neutral TDS. It has the well known necessary condition of "τ-stabilizability" of NTDS imbedded within its steps. Example case studies on vibration and target tracking control problems are given just to display the strengths of the methodology.

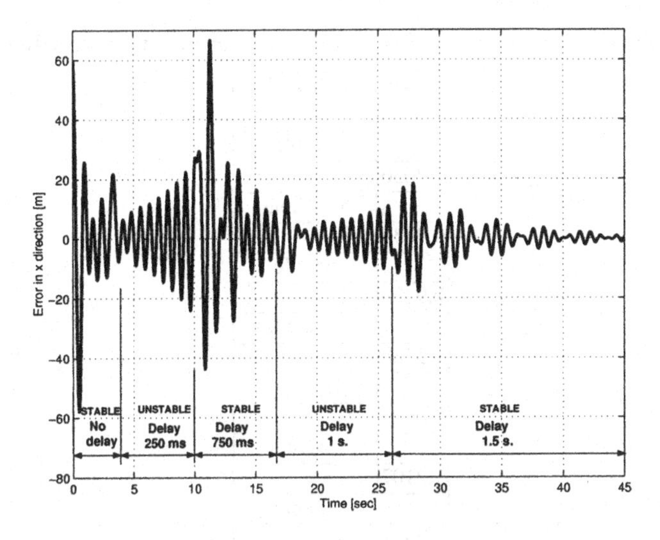

Fig. 3. Simulations of $e_x = x - x_{target}$ for various delay cases

5 Acknowledgements

The authors would like to thank Mr. Alper Ozturk for his help on transferring this chapter into LaTeX.

References

1. Bellman, R., Cooke, K., (1963), "Differential Difference Equations", Academic Press, New York
2. Chen, J., (1995), "On Computing the Maximal Delay Intervals for Stability of Linear Delay Systems", IEEE Transactions on Automatic Controls 40(6), 1087-1092.
3. Cooke, K.L., van den Driessche, P., (1986), "On Zeroes of Some Transcendental Equations", Funkcialaj Ekvacioj, 29, 77-90.
4. Datko, R., (1985), "Remarks Concerning the Asymptotic Stability and Stabilization of Linear Delay Differential Equations", Journal of Mathematical Analysis and Applications, 111, 571-584.
5. Halanay, A., (1966), "Differential Equations: Stability,Oscillations, Time Lags", Academic Press, New York
6. Hale, J.K., Verduyn Lunel, S.M., (2001), "Effects of Small Delays on Stability and Control", Operator Theory; Advances and Applications,122, 275-301.
7. Hale, J.K., Verduyn Lunel, S.M., (2001), "Strong Stabilization of Neutral Functional Differential Equations", IMA J. Math. Control Inform., 19, 1-19
8. Hale, J.K., Infante, E.F., Tsen, F.S.P, (1985), "Stability in Linear Delay Equations", Journal of Mathematical Analysis and Applications, 105, 533-555.
9. Hu,Guang-Di, Hu, Guang-Da, (1996), "Some Simple Criteria for Stability of Neutral Delay-Differential Sytems", Applied Mathematics and Computation, 80, 257-271.

Table 2. Abbreviated stability table for the target tracking case study

τ [sec]	Stability Chart
0	
	Stable
0.2036	
	Unstable
0.4630	
	Stable
0.9323	
	Unstable
1.3368	
	Stable
1.6609	
	Unstable
2.2107	
	Stable
2.3869	
	Unstable
3.0845	
	Stable
3.1183	
	Unstable
⋮	⋮

10. Kuo, B.C., (1991), "Automatic Control Systems", Prentice Hall.
11. Louisell, J., (2001), "A Matrix Method for Determining the Imaginary Axis Eigenvalues of a Delay System", IEEE Transactions on Automatic Control, 46 (12), 2008-2012.
12. Niculescu, S-I., (2001) "Delay Effects on Stability", Springer-Verlag.
13. Olgac,N., Sipahi, R., (2002), "An exact method for the stability analysis of time delayed LTI systems", IEEE Transactions on Automatic Control, 47 (5), 793-797.
14. Olgac, N., and Holm-Hansen, B., (1994), "A Novel Active Vibration Absorption Technique: Delayed Resonator", Journal of Sound and Vibration. Vol. 176, pp. 93-104.
15. Olgac,N, Sipahi, R., (2003), "Direct Method of Analyzing the Stability of Neutral Type LTI-Time Delayed Systems", to appear in Automatica.
16. Olgac,N., Sipahi, R., (2003), "Direct Method for Analyzing the Stability of Neutral Type LTI-Time Delayed Systems", to appear in IFAC03 Workshop on Time Delay Systems, France.
17. Rekasius, Z.V., (1980), "A Stability Test for Systems with Delays", in Proc. Joint Automatic Control Conf., Paper No. TP9-A.
18. Seto, K., (1995), "Structural Modeling and Vibration Control", Internal Report, Nihon University. Department of Mechanical Engineering, College of Science and Technology, Nihon University, 1-8-14 Kanda Surugadai Chiyoda-ku, Tokyo, 101-8308, Japan.
19. Sipahi, R., Olgac, N., (2003), "A New Perspective for Time Delayed Control Systems with Application to Vibration Suppression", in print, ASME Journal of Vibration and Acoustics

20. Sipahi, R., Olgac, N., (2003), "Degenerate Cases in Using the Direct Method", in print, ASME JDSMC, Special Issue on Time Delayed Systems.
21. Thowsen, A., (1981), "An Analytic Stability Test for a Class of Time-Delay Systems", IEEE Transactions on Automatic Controls 26(3), 735-736.
22. Thowsen, A., (1981), "The Routh-Hurwitz Method for Stability Determination of Linear Differential-Difference Systems", Int. J. Control 33(5), 991-995.
23. Walton, K.E., Marshall, J.E., (1987), "Direct Method for TDS Stability Analysis", IEE Proc. 29, 73-75.
24. Wells, D., (1986) "The Penguin Dictionary of Curious and Interesting Numbers", Middlesex, England: Penguin Books, pp.61-67.
25. Zhang, J., Knospe, C.R., Tsiotras, P., (2001), "Stability of time-delay systems: equivalence between Lyapunov and scaled small-gain conditions", IEEE Transactions on Automatic Control 46(3), 482-486.

From Lyapunov-Krasovskii Functionals for Delay-Independent Stability to LMI Conditions for μ-Analysis

Pierre-Alexandre Bliman

INRIA, Rocquencourt BP 105, 78153 Le Chesnay cedex, France
pierre-alexandre.bliman@inria.fr

Summary. Our scope in this note is to give a unified view on different approaches for studying stability of delay systems and parameter-dependent systems, and on estimation methods for some structured singular values. The classical approaches are exposed in Sections 1 to 3. A new result which links them together is given in Section 4, Elements of proof are gathered in Section 5. Comments are provided in Section 6. Finally, Section 7 proposes some open problems. For sake of space, exposure is kept to minimum, the reader is refered to the cited literature for more details.

Notations, Representation of Polynomials

By $\overline{\mathbb{C}^+}$ is meant the closed set of complex numbers with nonnegative real part. The closed unit ball (resp. circle) in $\mathbb{C}$ is denoted $\overline{\mathbb{D}}$ (resp. $\partial\mathbb{D}$). The symbol $\otimes$ denotes Kronecker product, the power of Kronecker products being used with the natural meaning: $M^{0\otimes} = 1$, $M^{p\otimes} \stackrel{\text{def}}{=} M^{(p-1)\otimes} \otimes M$. The transpose and transconjugate of a matrix are respectively denoted with a superscript ^{T} and ^{H}. We study here the stability of linear systems with m independent delays $h_1, \ldots, h_m$. In the whole note, we write $h \stackrel{\text{def}}{=} (h_1, \ldots, h_m)$, $\nabla \stackrel{\text{def}}{=} (\nabla_1, \ldots, \nabla_m)$, where ∇_i is the delay operator associated to delay h_i, acting on any convenient space of time functions. Also, $z \stackrel{\text{def}}{=} (z_1, \ldots, z_m)$ denotes a free variable in $\mathbb{C}^m$, and for simplicity, the notation $e^{-sh} \stackrel{\text{def}}{=} (e^{-sh_1}, \ldots, e^{-sh_m})$ is used in the transfers, where s is the Laplace variable.

For any integer n, let $\mathbb{R}^{n\times n}[z]$ (resp. $\mathbb{R}^{n\times n}[z,\overline{z}]$) be the ring of polynomials in $z \in \mathbb{C}^m$ (resp. in $z, \overline{z}$) with coefficients in $\mathbb{R}^{n\times n}$. The sets $\mathbb{C}^{n\times n}[z]$, $\mathbb{C}^{n\times n}[z,\overline{z}]$ are defined similarly. With $\mathcal{S}^n$ the subset of symmetric matrices in $\mathbb{R}^{n\times n}$, one defines analogously the set $\mathcal{S}^n[z,\overline{z}]$. An important subset of $\mathcal{S}^n[z,\overline{z}]$ is the set of those $M(z)$ such that $\forall z \in \mathbb{C}^m$, $M(z)^H = M(z)$; it is denoted $\mathcal{S}^n_H[z,\overline{z}]$.

To be able to represent and manipulate matrix-valued polynomials, define, for $l \in \mathbb{N}$, $i = 1, \ldots, m$ and for any $v \in \mathbb{C}$,

$$v^{[l]} \stackrel{\text{def}}{=} \begin{pmatrix} 1 \\ v \\ \vdots \\ v^{l-1} \end{pmatrix}, \quad \nabla_i^{[l]} \stackrel{\text{def}}{=} \begin{pmatrix} \text{Id} \\ \nabla_i \\ \vdots \\ \nabla_i^{l-1} \end{pmatrix}.$$

Notice that we denote in the same way the powers of complex numbers and the powers of delay operators (for the composition product), along the rule: $\nabla_i^2 \stackrel{\text{def}}{=} \nabla_i \circ \nabla_i \ldots$ This will permit in the sequel to apply polynomials in $\mathbb{C}^m$ to the operator ∇.

The expression $z_m^{[l]} \otimes \cdots \otimes z_1^{[l]}$ gathers all the monomials with degree at most $l-1$ in each of the components of z, so for any $M(z)$ in $\mathbb{C}^{n\times n}[z, \overline{z}]$, there exist $l \in \mathbb{N}$ and $M_l \in \mathbb{C}^{l^m n \times l^m n}$ such that, for all $z \in \mathbb{C}^m$,

$$M(z) = (z_m^{[l]} \otimes \cdots \otimes z_1^{[l]} \otimes I_n)^H M_l (z_m^{[l]} \otimes \cdots \otimes z_1^{[l]} \otimes I_n) .$$

The matrix M_l is just a concatenation, in prescribed order, of the matrices coefficients of $M(z)$. In this representation, which will be used as a central tool in the whole chapter, l and M_l are unique when taking minimal l. The matrix M_l is called the *coefficient matrix* of $M(z)$, $l-1$ the *degree* of the representation. Remark that, for M under the previous form, $M \in \mathcal{S}_H^n[z, \overline{z}]$ *iff* $M_l \in \mathcal{S}^{l^m n}$ for some $l > 0$.

The following identities are useful for calculations: for any l', $1 \leq l' \leq l$,

$$\begin{aligned} v^{l'-1} &= \left(0_{1\times(l'-1)}\ 1\ 0_{l-l'}\right) v^{[l]}, \\ \nabla_i^{l'-1} x &= \left(\left(0_{1\times(l'-1)}\ 1\ 0_{l-l'}\right) \otimes I_n\right) \nabla_i^{[l]} x , \end{aligned} \tag{1}$$

for any complex v and any time-function x taking values in $\mathbb{R}^n$. Last, let $\hat{J}_l, \check{J}_l \in \mathbb{R}^{l\times(l+1)}$ be defined by:

$$\hat{J}_l \stackrel{\text{def}}{=} \left(I_l\ 0_{l\times 1}\right),\ \check{J}_l \stackrel{\text{def}}{=} \left(0_{l\times 1}\ I_l\right) . \tag{2}$$

This corresponds to the matrix present in (1), for the values $l' = 1$ and $l' = l$.

1 Delay Systems and Associated Stability Properties

The delay system under study is denoted under the quite general form

$$\dot{x} = A(\nabla) x , \tag{3}$$

where $A(z) \in \mathbb{R}^{n\times n}[z]$ is a polynomial. By definition, we denote its degree $k-1$ (that is, the maximum of the m partial degrees with respect to $z_1, \ldots, z_m$). For example, for the affine map ($k-1=1$)

$$A(z) = A_0 + z_1 A_1 + \cdots + z_m A_m , \tag{4}$$

this yields: $\dot{x} = A_0 x + A_1 x(t-h_1) + \cdots + A_m x(t-h_m)$.

1.1 Basic properties

Let us first recall the following

Theorem 1 (Stability characterization). *System* (3) *is asymptotically stable* iff

$$\forall s \in \overline{\mathbb{C}^+},\ \det(sI_n - A(e^{-sh})) \neq 0\,.$$

As in [20, 21], we put:

Definition 1 (Delay-independent stability (DIS)). *System* (3) *is said to be delay-independently stable if it is stable for any $h \in [0, +\infty)^m$.*

The previous notion has been introduced in order to study the stability of systems with delays of imperfectly known values. The assumption that no information on the value of the delay is available may be coarse in practice, when bounds are already known. This has necessitated development of *delay-dependent* criteria too. This topic is not treated here.

Extension of results in [16, 17] permits the following claim.

Theorem 2 (Characterization of the delay-independent stability). *System* (3) *is DIS* iff

$$\forall (s, z) \in \overline{\mathbb{C}^+} \setminus \{0\} \times \overline{\mathbb{D}}^m \cup \{(0, 1, \ldots, 1)\},\ \det(sI_n - A(z)) \neq 0\,.$$

Extending [23] leads to introduce the slightly stronger property:

Definition 2 (Strong delay-independent stability (SDIS)). *System* (3) *is said to be strongly delay-independently stable if*

$$\forall (s, z) \in \overline{\mathbb{C}^+} \times \overline{\mathbb{D}}^m,\ \det(sI_n - A(z)) \neq 0\,. \tag{5}$$

Infinitely close (in terms of a metric on the coefficients of A) from any DIS system which is *not* SDIS, one may find systems which are *not* DIS. In other words, the set of SDIS systems is the *interior* of the set of DIS systems endowed with the topology whose neighborhoods are defined by the choice of a metric on the coefficient matrices [2].

1.2 The Lyapunov-Krasovskii functionals approach

For $P, Q_1, \ldots, Q_m \in \mathcal{S}^n$, define for any $\phi \in C([-(k-1)\max h_i, 0]; \mathbb{C}^n)$ the functional V by (see [6, 15, 22]):

$$V(\phi)(t) \stackrel{\text{def}}{=} \phi(0)^T P \phi(0) + \int_{-(k-1)h_1}^{0} \phi(\tau)^T Q_1 \phi(\tau)\, d\tau + \cdots + \int_{-(k-1)h_m}^{0} \phi(\tau)^T Q_m \phi(\tau)\, d\tau\,. \tag{6}$$

Denoting abusively $V(x|_{[t-(k-1)\max h_i,t]})$ by $V(x)(t)$, one has, along the trajectories of (3):

$$\frac{d[V(x)(t)]}{dt} = ((\nabla_m^{[k]} \otimes \cdots \otimes \nabla_1^{[k]} \otimes I_n)x)(t)^T R((\nabla_m^{[k]} \otimes \cdots \otimes \nabla_1^{[k]} \otimes I_n)x)(t) ,$$

where the exact value of the matrix $R = R(P, Q_1, \ldots, Q_m) \in \mathcal{S}^{(k+1)^m n}$ may be written using the formulas in (1). It is important to remark that R is *affine* in $P, Q_1, \ldots, Q_m$ and *independent* of the values of $h_1, \ldots, h_m$. Thus, searching for a Lyapunov-Krasovskii functional in the class (6) leads to the following.

Theorem 3 (Sufficient condition for SDIS). *If there exist $P, Q_1, \ldots, Q_m \in \mathcal{S}^n$ such that*

$$P > 0, Q_1 > 0, \ldots, Q_m > 0, R < 0 ,$$

then system (3) *is SDIS.*

The sufficient condition in Theorem 3 is a Linear Matrix Inequality (LMI), see [6].

2 Robust Stability of Parameter Dependent Systems

Associated to delay system (3) is the system with parameter $z \in \mathbb{C}^m$ given by

$$\dot{x} = A(z)x . \tag{7}$$

2.1 Basic properties

Definition 3 (Robust stability). *System* (7) *is said robustly stable if it is asymptotically stable for any $z \in \overline{\mathbb{D}}^m$.*

Trivially, this notion is linked with SDIS:

Theorem 4 (Link with SDIS). *System* (7) *is robustly stable* iff *system* (3) *is SDIS.*

2.2 Sufficient conditions for robust stability and the parameter-dependent Lyapunov functions approach

A number of published contributions have obtained robust stability criteria for systems similar to (7), by use of some prescribed class of parameter-dependent Lyapunov functions. The latter have been chosen *independent of the parameters, affine* [8, 11, 14, 24], *quadratic* [29, 30].

Existence of a Lyapunov function in these classes may be recast as solvability problem for certain LMI. Nevertheless, due to the fact that they assume a prespecified dependence of the Lyapunov function with respect to the parameters, they all lead to *sufficient conditions* for robust stability.

3 Structured Singular Values with Repeated Scalar Blocks

The notion of structured singular values is one of the basic tools of robust control [10].

3.1 Basic properties

Definition 4 (Structured singular values (ssv)). *For fixed $r_1, \dots, r_m \in \mathbb{N}$, let $\Delta \stackrel{\text{def}}{=} \{\text{diag}[\delta_1 I_{r_1}; \dots; \delta_m I_{r_m}] \ : \ (\delta_1, \dots, \delta_m) \in \mathbb{C}^m\}$. Then, for any $M \in \mathbb{C}^{r \times r}$, where $r \stackrel{\text{def}}{=} r_1 + \dots + r_m$, the structured singular value $\mu_\Delta(M)$ is null if no matrix $\varDelta \in \Delta$ makes $I_r - M\varDelta$ singular, and otherwise equal to*

$$(\min\{\overline{\sigma}(\varDelta) \ : \ \varDelta \in \Delta, \det(I_r - M\varDelta) = 0\})^{-1} .$$

Computing μ is generally a NP-hard task [28]. Using the change of variable $s = \frac{1+z_0}{1-z_0}$, which maps $\overline{\mathbb{C}^+}$ into $\overline{\mathbb{D}}$, one may exhibit, for any polynomial $A(z)$, a certain structure Δ_A and a real matrix M_A such that (5) holds *iff*

$$\mu_{\Delta_A}(M_A) < 1 . \tag{8}$$

Thus, checking SDIS of (3) or robust stability of (7) amounts to estimate a ssv with $m+1$ repeated scalar blocks. This more specific problem is also NP-hard [27].

Alternatively, this is equivalent [7] to check whether

$$\forall s \in j\overline{\mathbb{R}},\ \mu_{\tilde{\Delta}_A}(\tilde{M}_A(s)) < 1 , \tag{8'}$$

for certain structure $\tilde{\Delta}_A$ (with m repeated scalar blocks) and transfer $\tilde{M}_A$.

Conversely, let M be a real square matrix and Δ a block structure of compatible size, having $m+1$ repeated complex scalar blocks. Is it possible to find a polynomial $A(z)$ such that (3) is SDIS *iff* $\mu_\Delta(M) < 1$? The answer is no in general, as structured singular values may describe not only polynomial dependences, but also rational ones, via Linear Fractional Transform. As a matter of fact, the whole generality is obtained when considering delay-differential equations of *neutral type*, and not only of *retarded type*. In terms of parameter-dependent systems, this corresponds to *parameter-dependent singular (descriptor) systems*.

3.2 Upper bounds for ssv and the multiplier approach

Various upper bounds for the structured singular values have been proposed. Their principle relies on the use of *multipliers* [12] or *scaling technique* [1]. Some results are based on mixed methods [9, 13].

Interestingly enough, it has been shown [32] that checking SDIS by means of Lyapunov-Krasovskii functionals of the class (6), amounts to use in the previous inequality the conservative evalution of μ provided by D-scalings (the classical "μ upper bound").

Connection between the scaling approach and the parameter-dependent methods has been established by Iwasaki *et al.* [18,19]. Both approaches may be interpreted as special cases of the *quadratic separator*, separating in an appropriate space a graph associated to the "system" from a graph associated to the "perturbation", here the parameters. Roughly speaking, the previous results are obtained when looking for such a separator with prespecified, "simple", dependence, either with respect to the frequency (frequency-dependent scaling matrix in μ-analysis), or to the parameters (parameter-dependent Lyapunov functions).

4 A Key Result

For any $l \in \mathbb{N}$, for any $P_l \in \mathcal{S}^{l^m n}$, define $R_l = R_l(P_l) \in \mathcal{S}^{(k+l-1)^m n}$ to be the coefficient matrix[1] of $R(z) \stackrel{\text{def}}{=} A(z)^H P(z) + P(z)A(z)$, where $P(z)$ is defined by its coefficient matrix P_l. As an example [3,5], for $A(z)$ defined in (4), one has

$$R_l = \left(\left(\hat{J}_l^{m\otimes} \otimes A_0 \right) + \sum_{i=1}^{m} \left(\hat{J}_l^{(m-i)\otimes} \otimes \check{J}_l \otimes \hat{J}_l^{(i-1)\otimes} \otimes A_i \right) \right)^H P_l \left(\hat{J}_l^{m\otimes} \otimes I_n \right)$$
$$+ \left(\hat{J}_l^{m\otimes} \otimes I_n \right)^T P_l \left(\left(\hat{J}_l^{m\otimes} \otimes A_0 \right) + \sum_{i=1}^{m} \left(\hat{J}_l^{(m-i)\otimes} \otimes \check{J}_l \otimes \hat{J}_l^{(i-1)\otimes} \otimes A_i \right) \right) .$$

The following result is an extension of [3,5] to the cases where $k > 2$.

Theorem 5. *The following properties are* equivalent.

1. *System* (3) *is SDIS (resp. system* (7) *is robustly stable, resp. condition* (8) *or* (8') *is fulfilled with adequate structure and matrix choice).*

2. *There exists* $P(z) \in \mathcal{S}_H^n[z, \bar{z}]$ *such that,*

$$\forall z \in \overline{\mathbb{D}}^m,\ P(z) > 0_n,\ A(z)^H P(z) + P(z)A(z) < 0_n .$$

3. *There exist* $l \in \mathbb{N}$, m *matrices* $Q_{l,i} \in \mathcal{S}^{(k+l-2)^{m-i+1}(k+l-1)^{i-1}n}$, $i = 1, \dots, m$, *and matrix* $P_l \in \mathcal{S}^{l^m n}$ *such that*

$$P_l > 0_{l^m n} \tag{9a}$$

and

[1] There may exist a representation of $R(z)$ with coefficient matrix of size smaller than $(k + l - 1)^m n$, this aspect has no incidence on the sequel.

$$S_l(P_l, Q_{l,1}, \dots, Q_{l,m}) \stackrel{\text{def}}{=} R_l(P_l)$$
$$+ \sum_{i=1}^{m} \left(\hat{J}_{k+l-2}^{(m-i+1)\otimes} \otimes I_{(k+l-1)^{i-1}n} \right)^T Q_{l,i} \left(\hat{J}_{k+l-2}^{(m-i+1)\otimes} \otimes I_{(k+l-1)^{i-1}n} \right)$$
$$- \sum_{i=1}^{m} \left(\hat{J}_{k+l-2}^{(m-i)\otimes} \otimes \check{J}_{k+l-2} \otimes I_{(k+l-1)^{i-1}n} \right)^T$$
$$Q_{l,i} \left(\hat{J}_{k+l-2}^{(m-i)\otimes} \otimes \check{J}_{k+l-2} \otimes I_{(k+l-1)^{i-1}n} \right)$$
$$< 0_{(k+l-1)^m n} , \quad \text{(9b)}$$

where $\hat{J}_k, \check{J}_k$ are defined in (2).

Moreover, if LMI (9) *is solvable for the index l, then it is also solvable for any larger index.*

Thus, the conditions expressed in (9) are more and more precise (less and less conservative) when l increases, and the feasibility of any of them is sufficient to have the properties depicted in **1.** An important point is that *necessity* also holds, in the precise sense that: if the stability properties hold, then the corresponding LMIs are fulfilled *from a certain rank l and beyond.*

5 Elements of Proof of Theorem 5

System (3) is SDIS *iff* for any $z \in \overline{\mathbb{D}}^m$, there exists $P(z) > 0$ such that $A(z)^H P(z) + P(z)A(z) < 0$. The previous problem is a *parameter-dependent LMI*, in which z is the parameter vector. The dependence upon the latter being polynomial, and thus continuous, one may apply the result given in [4], and concludes that if (3) is SDIS, then without loss of generality P may be chosen *polynomial in z and $\overline{z}$*. This establishes the implication **1.** $\Rightarrow$ **2.**

To prove that **3.** $\Rightarrow$ **1.**, right- and left- multiply (9a) (resp. (9b)) by $(z_m^{[l]} \otimes \cdots \otimes z_1^{[l]} \otimes I_n)$ (resp. $(z_m^{[k+l-1]} \otimes \cdots \otimes z_1^{[k+l-1]} \otimes I_n)$) and its transconjugate. This yields $P(z) > 0_n$ and

$$R(z) + \sum_{i=1}^{m} (1-|z_i|^2)(z_m^{[k+l-2]} \otimes \cdots \otimes z_i^{[k+l-2]} \otimes z_{i-1}^{[k+l-1]} \otimes \dots z_1^{[k+l-1]} \otimes I_n)^H Q_{l,i}$$
$$(z_m^{[k+l-2]} \otimes \cdots \otimes z_i^{[k+l-2]} \otimes z_{i-1}^{[k+l-1]} \otimes \dots z_1^{[k+l-1]} \otimes I_n) < 0_n , \quad (10)$$

where $R(z) \stackrel{\text{def}}{=} A(z)^H P(z) + P(z)A(z)$. Indeed, this is a direct consequence of (1). Thus, $R(z) < 0_n$ if $|z_1| = \cdots = |z_m| = 1$, so the matrix $A(z)$ is Hurwitz for all $z \in (\partial\mathbb{D})^m$. This observation may be extended to the whole $\overline{\mathbb{D}}^m$, basically by subanalyticity, as in [5]. This proves that solvability of (9) implies robust stability of (7). In other terms, **3.** implies **1.**.

The difficult part of the proof is the implication **2.** $\Rightarrow$ **3.**, whose proof is adapted from [5].

First, it may be shown (but this is a non-trivial result) that the coefficient matrix P_l of $P(z)$, which is symmetric as $P \in \mathcal{S}_H^n[z,\overline{z}]$, may be supposed *positive definite, without loss of generality.*

The next stage consists in removing one by one the free variables $z_1, \dots, z_m$ and introducing concomitantly the multipliers $Q_{l,1}, \dots, Q_{l,m}$. Basically, this operation is achieved by applying recursively D-scaling with respect to z_1, ..., z_m. This procedure is lossless for one complex parameter (this is just the discrete-time counterpart of Kalman-Yakubovich-Popov lemma, see [26, 31], and [25] for recent statement and proof). The argument is the same than for the results in [5], up to some technical details. At each step, a new matrix is introduced, which however depends upon the remaining free-variables. Applying again [4], one may assume that this dependence is indeed polynomial, and the coefficient matrix of the latter turns out to be one of the $Q_{l,i}$. Some special care has to be taken, as the degree of the polynomial previously introduced is unknown: indeed, increases of the "degree" l may occur when passing from **2.** to **3.**, this is explained in detail in [5].

6 Interpretation of Theorem 5 and Comments

6.1 Link with parameter-dependent systems

Based on a solution $(P_l, Q_{1,l}, \dots, Q_{m,l})$ of LMI (9), construct $P(z) \in \mathcal{S}_H^n[z,\overline{z}]$ with coefficient matrix P_l. Then, along the trajectories of (7),

$$\frac{d[x(t)^T P(z) x(t)]}{dt} = x(t)^T R(z) x(t) ,$$

where $R(z) = A(z)^H P(z) + P(z) A(z)$ is defined by its coefficient matrix $R_l(P_l)$.

Remark however that $A(z)$, being polynomial, is analytic, and Hurwitzness of $A(z)$ for $z \in (\partial\mathbb{D})^m$ implies the same property in $\overline{\mathbb{D}}^m$, see [5]. In order to obtain a simpler LMI in the stability criterion, the smallest set has been considered, and the corresponding parameter-dependent Lyapunov function based on a solution of (9) is guaranteed to decrease only for $|z_1| = \dots = |z_m| = 1$. Positivity of the matrices $Q_{l,i}$ would ensure the property for the whole set $\overline{\mathbb{D}}^m$, see inequality (25) above. We conjecture that the previous positivity condition may be added without supplementary conservatism. This assertion is true at least for $m = 1$, $k = 2$ [2].

6.2 Link with delay systems

For any $l_1, \dots, l_m \in \mathbb{N}$, define

$$x^{[l_1,\dots,l_m]} \overset{\text{def}}{=} (\nabla_m^{[l_m]} \otimes \dots \otimes \nabla_1^{[l_1]} \otimes I_n) x ,$$

which takes values in $\mathbb{R}^{l_1 \dots l_m n}$. Consider the following functional (compare with (6)), parametrized by $(m+1)$ hermitian matrices $P_l, Q_{l,i}$ having the same size than in Theorem 5:

$$
\begin{aligned}
V_l(x)(t) &\stackrel{\text{def}}{=} x^{[l,\dots,l]}(t)^T P_l x^{[l,\dots,l]}(t) \\
&\quad + \int_{t-h_1}^{t} x^{[l+k-2,\dots,l+k-2]}(\tau)^T Q_{l,1} x^{[l+k-2,\dots,+l+k-2]}(\tau)\, d\tau \\
\int_{t-h_2}^{t} & x^{[l+k-1,l+k-2,\dots,+l+k-2]}(\tau)^T Q_{l,2} x^{[l+k-1,l+k-2,\dots,l+k-2]}(\tau)\, d\tau + \dots \\
+ \int_{t-h_m}^{t} & x^{[l+k-1,\dots,l+k-1,l+k-2]}(\tau)^T Q_{l,m} x^{[l+k-1,\dots,l+k-1,l+k-2]}(\tau)\, d\tau\,. \quad (11)
\end{aligned}
$$

The value of $V_l(x)$ at time t depends only upon the values of x on $[t-(k+l-2)\sum h_i; t]$. It turns out that

$$
\frac{d[V_l(x)(t)]}{dt} = x^{[l+k-1,\dots,l+k-1]}(t)^T S_l(P_l, Q_{l,1}, \dots, Q_{l,m}) x^{[l+k-1,\dots,l+k-1]}(t)\,,
$$

where S_l is defined in (9a).

$$
\begin{aligned}
&\frac{d}{dt}\left[\int_{t-h_1}^{t} x^{[l+k-2,\dots,l+k-2]}(\tau)^T Q_{l,1} x^{[l+k-2,\dots,l+k-2]}(\tau)\, d\tau\right] \\
&= x^{[l+k-2,\dots,l+k-2]}(t)^T Q_{l,1} x^{[l+k-2,\dots,l+k-2]}(t) \\
&\quad - x^{[l+k-1,l+k-2,\dots,l+k-2]}(t)^T Q_{l,1} x^{[l+k-1,l+k-2,\dots,l+k-2]}(t) \\
&= x^{[l+k-1,\dots,l+k-1]}(t)^T \left[\left(\hat{J}_{k+l-2}^{m\otimes} \otimes I_n\right)^T Q_{l,i} \left(\hat{J}_{k+l-2}^{m\otimes} \otimes I_n\right) \right. \\
&\quad \left. - \left(\hat{J}_{k+l-2}^{(m-1)\otimes} \otimes \check{J}_{k+l-2} \otimes I_n\right)^T Q_{l,i} \left(\hat{J}_{k+l-2}^{(m-1)\otimes} \otimes \check{J}_{k+l-2} \otimes I_n\right)\right] \\
&\quad x^{[l+k-1,\dots,l+k-1]}(t)\,,
\end{aligned}
$$

due to (1). Therefore, the appearance of LMI (9) is also related to the search for a Lyapunov-Krasovskii functional of the form (11). However, no positivity assumption has to be made in (9), see also the remark made previously in Section 6.1 for parameter-dependent systems. In the eventuality where the positivity assumption may be added without loss of enerality (e.g. $k=2, m=1$), strong delay-independent stability is equivalent to the existence of a certain Lyapunov-Krasovskii functional in the class (11) ensuring stability of delay system (3) for any nonnegative value of $h_1, \dots, h_m$.

7 Open Problems on μ Computation

To conclude, we present two open questions, linked to application and extension of the ideas and methods previously presented.

• Is it possible to extend the method, in order to associate to any problem (8) or (8'), a family of LMIs similar to (9), constituting sufficient conditions with increasing precision?
• How to use practically the above results for numerical estimation of structured singular values? In particular, how to choose in (9) the degree $l-1$ of the underlying parameter-dependent Lyapunov function? In the case $m = 1$, $k = 2$, an answer has been given in [33], which seems extendable to non affine systems ($k > 2$), but the general case is still unsolved.

References

1. T. Asai, S. Hara, T. Iwasaki (1996). Simultaneous modeling and synthesis for robust control by LFT scaling, *Proc. IFAC World Congress* part G, 309–314
2. P.-A. Bliman (2002). Lyapunov equation for the stability of linear delay systems of retarded and neutral type, *IEEE Trans. Automat. Control* **47** no 2, 327–335
3. P.-A. Bliman (2002). Nonconservative LMI approach to robust stability for systems with uncertain scalar parameters, *Proc. of 41th IEEE CDC*, Las Vegas (Nevada), December 2002
4. P.-A. Bliman (2003). An existence result for polynomial solutions of parameter-dependent LMIs Report research no 4798, INRIA. Available online at `http://www.inria.fr/rrrt/rr-4798.html`
5. P.-A. Bliman (2003, to appear). A convex approach to robust stability for linear systems with uncertain scalar parameters, *SIAM J. on Control and Optimization*
6. S. Boyd, L. El Ghaoui, E. Feron, V. Balakrishnan (1994). *Linear matrix inequalities in system and control theory*, SIAM Studies in Applied Mathematics vol. 15, SIAM, Philadelphia
7. J. Chen, H.A. Latchman (1995). Frequency sweeping tests for stability independent of delay, *IEEE Trans. Automat. Control* **40** no 9, 1640–1645
8. M. Dettori, C.W. Scherer (1998). Robust stability analysis for parameter dependent systems using full block S-procedure, *Proc. of 37th IEEE CDC*, Tampa (Florida), 2798–2799
9. M. Dettori, C.W. Scherer (2000). New robust stability and performance conditions based on parameter dependent multipliers, *Proc. of 39th IEEE CDC*, Sydney (Australia)
10. J.C. Doyle (1982). Analysis of feedback systems with structured uncertainties, *IEE Proc. Part D* **129** no 6, 242–250
11. E. Feron, P. Apkarian, P. Gahinet (1996). Analysis and synthesis of robust control systems via parameter-dependent Lyapunov functions, *IEEE Trans. Automat. Control* **41** no 7, 1041–1046
12. M. Fu, N.E. Barabanov (1997). Improved upper bounds for the mixed structured singular value, *IEEE Trans. Automat. Control* **42** no 10, 1447–1452
13. M. Fu, S. Dasgupta (2000). Parametric Lyapunov functions for uncertain systems: the multiplier approach, *in Advances in Linear Matrix Inequality Methods in Control* (L. El Ghaoui, S.-I. Niculescu eds.), SIAM, Philadelphia, 95–108
14. P. Gahinet, P. Apkarian, M. Chilali (1996). Affine parameter-dependent Lyapunov functions and real parametric uncertainty, *IEEE Trans. Automat. Control* **41** no 3, 436–442
15. J.K. Hale (1977). *Theory of functional differential equations*, Applied Mathematical Sciences 3, Springer Verlag, New York

16. J.K. Hale, E.F. Infante, F.S.P. Tsen (1985). Stability in linear delay equations, *J. Math. Anal. Appl.* **115**, 533–555
17. D. Hertz, E.I. Jury, E. Zeheb (1984). Stability independent and dependent of delay for delay differential systems, *J. Franklin Institute* **318** no 3, 143–150
18. T. Iwasaki (1998). LPV system analysis with quadratic separator, *Proc. of 37th IEEE CDC*, Tampa (Florida)
19. T. Iwasaki, S. Hara (1998). Well-posedness of feedback systems: insights into exact robustness analysis and approximate computations, *IEEE Trans. Automat. Control* **43** no 5, 619–630
20. E.W. Kamen (1982). Linear systems with commensurate time delays: stability and stabilization independent of delay, *IEEE Trans. Automat. Control* **27** no 2, 367–375
21. E.W. Kamen (1983). Correction to "Linear systems with commensurate time delays: stability and stabilization independent of delay", *IEEE Trans. Automat. Control* **28** no 2, 248–249
22. N.N. Krasovskii (1963). *Stability of motion. Applications of Lyapunov's second method to differential systems and equations with delay*, Stanford University Press, Stanford
23. S.-I. Niculescu, J.-M. Dion, L. Dugard, H. Li (1996). Asymptotic stability sets for linear systems with commensurable delays: a matrix pencil approach, *IEEE/IMACS CESA'96*, Lille, France
24. D.C.W. Ramos, P.L.D. Peres (2001). An LMI approach to compute robust stability domains for uncertain linear systems, *Proc. American Contr. Conf.*, Arlington (Virginia), 4073–4078
25. A. Rantzer (1996). On the Kalman-Yakubovich-Popov lemma, *Syst. Contr. Lett.* **28** no 1, 7–10
26. G. Szegö, R.E. Kalman (1963). Sur la stabilité absolue d'un système d'équations aux différences finies, *Comp. Rend. Acad. Sci.* **257** no 2, 338–390
27. O. Toker, H. Özbay (1996). Complexity issues in robust stability of linear delay-differential systems, *Math. Control Signals Systems* **9** no 4, 386–400
28. O. Toker, H. Özbay (1998). On the complexity of purely complex μ computation and related problems in multidimensional systems, *IEEE Trans. Automat. Control* **43** no 3, 409–414
29. A. Trofino (1999). Parameter dependent Lyapunov functions for a class of uncertain linear systems: a LMI approach, *Proc. of 38th IEEE CDC*, Phoenix (Arizona), 2341–2346
30. A. Trofino, C.E. de Souza (1999). Bi-quadratic stability of uncertain linear systems, *Proc. of 38th IEEE CDC*, Phoenix (Arizona)
31. V.A. Yakubovich (1962). Solution of certain matrix inequalities in the stability theory of nonlinear control systems, *Dokl. Akad. Nauk. SSSR* **143**, 1304–1307 (English translation in *Soviet Math. Dokl.* **3**, 620–623 (1962))
32. J. Zhang, C.R. Knospe, P. Tsiotras (2001). Stability of time-delay systems: equivalence between Lyapunov and scaled small-gain conditions, *IEEE Trans. Automat. Control* **46** no 3 482–486
33. X. Zhang, P. Tsiotras, T. Iwasaki (2003, submitted). Stability analysis of linear parametrically-dependent systems, *42nd IEEE Confernce on Decision and Control*, Maui (Hawaii)

Part III

Control, Identification, and Observer Design

Finite Eigenstructure Assignment for Input Delay Systems

Sabine Mondié[1] and Jean Jacques Loiseau[2]

[1] Departamento de Control Automático CINVESTAV-IPN, Av. IPN 2508, A.P. 14-740, 07300 México, D.F., México smondie@ctrl.cinvestav.mx
[2] Institut de Recherche en Communications et Cybernétique de Nantes, UMR CNRS 6597, Ecole Centrale de Nantes, BP 92101, 44 321 Nantes Cedex 03, France loiseau@irccyn.ec-nantes.fr

Summary. The problem of invariant factors assignment of input delay systems with classical finite spectrum assigment control laws with distributed delays is adressed. The multiplicities of the invariant factors are shown to be restricted by specified Rosenbrock type inequalities. The results are proved with the help of an equivalent linear assignment problem with no delay, and within the Bezout domain $\mathcal{E}$. Two different algorithms are hence provided. A bidimensional illustrative example is given.

1 Introduction

Consider a linear system with delays described by

$$\dot{x}(t) = \sum_{i=0}^{K} A_i x(t-ih) + \sum_{i=0}^{K} B_i u(t-ih) , \tag{1}$$

with control input $u(t) \in \mathbb{R}^m$, instantaneous state $x(t) \in \mathbb{R}^n$ for $t \geq 0$, delay $h \in \mathbb{R}$, $0 < h$, and the family of control laws described by integral Volterra equations of the second kind,

$$u(t) = \int_0^{Kh} f(\tau)u(t-\tau)d\tau + \sum_{i=0}^{K} g_i x(t-ih) + \int_0^{Kh} g(\tau)x(t-\tau)d\tau . \tag{2}$$

Such a control law was introduced in [13], where the spectral controllability of system (1) is shown to be a necessary and sufficient condition to freely assign a closed loop finite spectrum using a control law of the form(2). However, not only the location of the roots, but also their multiplicities, or equivalently the invariant factors of the closed loop, are crucial for the closed loop dynamics. The question that arises is to characterize the freedom in assigning the invariant factors. The answer to this query was given in the recent result in [10]: the constraint is that the sum of the degrees of the invariant factors must be equal to n.

In this work, we restrict our attention to input delay systems. This subfamily of the systems introduced above includes the models of a wide class of applications where the delay is due to transport phenomenons, time consuming information processing, sensors design, among others. These systems are described by

$$\dot{x}(t) = Ax(t) + \sum_{i=0}^{N} B_i u(t - ih). \tag{3}$$

As shown in [11,13], these systems are n-assignable by control laws that are simpler that those described by (2), of the form

$$u(t) = K[x(t) + \sum_{i=1}^{N} \int_{t-h}^{t} \mathrm{e}^{(t-\sigma-ih)A} B_i u(\sigma) d\sigma] \,. \tag{4}$$

This is indeed a particular case of the finite spectrum assignment problem introduced above. The motivation for using such control laws is their simplicity and their natural interpretation in terms of predictors. As shown below, they permit not only to assign a finite spectrum, but also finite invariant factors. The motivation for assigning invariant factors with a finite number of roots is the same as the one for spectrum assignment: the invariant factors give and additional freedom in shaping the dynamics. If they have a finite number of roots, they can be readily analyzed. The question is then whether or not the freedom, that exists in the general case, is restricted when such simpler laws are used. The problem under consideration is then the following.

Problem 1. Consider a spectrally controllable input time delay system of the form (3). Under what conditions does a control law described by (4) exists, such that the closed loop has prescribed invariant factors with finite number of roots.

The solution to this problem is organized as follows. Some backgrounds are recalled in Section 2. Then, the main result is established in Section 3. Further insight, including the design procedures for assigning invariant factors with control laws (4) and (2) are described in Section 4. An illustrative example is presented in Section 5 and some concluding remarks end the chapter.

2 Backgrounds

2.1 Notation

$\mathbb{R}[s]$ denotes the ring of polynomials over $\mathbb{R}$, the field of reals. The degree of a polynomial $\alpha(s)$ is denoted $\deg \alpha(s)$. $\mathbb{R}(s)$ stands for the field of rational functions over $\mathbb{R}$, while the ring of proper rational functions is denoted by $\mathbb{R}_p(s)$. Further, $\mathbb{R}^{m\times n}$, $\mathbb{R}^{m\times n}[s]$, $\mathbb{R}_p^{m\times n}(s)$ denote the sets of $m \times n$ matrices having elements in $\mathbb{R}$, $\mathbb{R}[s]$, $\mathbb{R}_p(s)$, respectively. Units of the ring $\mathbb{R}^{m\times m}[s]$ are called *unimodular matrices* and those of the ring $\mathbb{R}_p^{m\times m}(s)$ *bipropermatrices*. The set $\mathcal{E}$ is a Bezout domain whose elements are fractions of the form $\alpha(s, e^{-hs}) = n(s, e^{-hs})/d(s)$, where all the zeros of $d(s) \in \mathbb{R}[s]$ are zeros of $n(s, e^{-hs}) \in \mathbb{R}[s, e^{-sh}]$ the set of quasipolynomials. Notice that $\mathbb{R}[s] \subset \mathcal{E}$. Units of the Bezout domain $\mathcal{E}$ are called unimodulars over $\mathcal{E}$. For detailed information on these sets and their properties, see [4,5,7,9,10].)

2.2 Theorem of Rosenbrock

Consider a linear system described by

$$\dot{x}(t) = Ax(t) + Bu(t) \tag{5}$$

where $A \in \mathbb{R}^{n\times n}$, $B \in \mathbb{R}^{n\times m}$. Its input-state transfer is the rational matrix $T(s) = (sI_n - A)^{-1}B$, which admits a right matrix fraction description $T(s) = N(s)D^{-1}(s)$, where $N(s) \in \mathbb{R}^{n\times n}[s]$ and $D(s) \in \mathbb{R}^{m\times m}[s]$ are right coprime polynomial matrices. The denominator matrix can be chosen to be column reduced, with ordered column degrees, which means that it can be writen as $D(s) = B(s)\text{diag}\{s^{c_1}, \ldots, s^{c_m}\}$, where $B(s) \in \mathbb{R}_p(s)$ is a biproper matrix, and $c_1 \geq c_2 \ldots \geq c_m > 0$. The column degrees c_i are called the controllability indices of system (11). The system is called controllable if $\text{rank}[sI_n - A, B] = n$, $\forall s \in \mathbb{C}$, and in that case we have $\sum_{i=1}^{m} c_i = n$. Further, it is well known [5] that there exist unimodular polynomial matrices $U(s) \in \mathbb{R}^{n\times n}[s]$ and $V(s) \in \mathbb{R}^{m\times m}[s]$ so that $D(s)$ can be factored as

$$D(s) = U(s)\begin{pmatrix} diag\{\alpha_i(s)\}_{i=1}^{r} & 0 \\ 0 & 0 \end{pmatrix} V(s). \tag{6}$$

where $\{\alpha_i(s)\}_{i=1}^{r}$ are unique monic polynomials such that $\alpha_i(s)$ divides $\alpha_{i-1}(s)$, called the invariant factors of $D(s)$. We recall here the so-called control structure theorem of Rosenbrock [8,15], which is a key motivation of the present work.

Lemma 1. *[15] Let the linear system*

$$\dot{x}(t) = Ax(t) + Bu(t)$$

be controllable, having the controllability indices $\{c_i\}_{i=1}^{m}$ *where* $c_i \leq c_{i-1}$, *for* $i = 2, ..., m$. *Then there exists a static state feedback*

$$u(t) = Kx(t)$$

such that the invariant factors of the closed loop system are $\{\alpha_i'(s)\}_{i=1}^{n}$, *where* $\alpha_i'(s)$ *divides* $\alpha_{i-1}'(s)$, *for* $i = 2, ..., m$, *and* $\alpha_i'(s) = 1$, *for* $i = m+1, ..., n$, *if and only if*

$$\sum_{i=1}^{j} c_i \leq \sum_{i=1}^{j} \deg(\alpha_i'),\ j = 1, ..., m,$$

with equality for $j = m$.

2.3 Time-delay systems and pseudopolynomial matrices

The input delay system (3) is called *spectrally controllable* if

$$\operatorname{rank}[sI_n - A, B(\mathrm{e}^{-hs})] = n \ , \forall s \in \mathbb{C} \ ,$$

where

$$B(\mathrm{e}^{-hs}) = \sum_{i=0}^{N} B_i \mathrm{e}^{-ihs} \ .$$

The Bézout domain $\mathcal{E}$, whose definition and basic properties are recalled in the notations, section 1, has proved to be a powerful tool for the study of commensurate time delay systems, by providing a solid algebraic framework that allows the generalization of many results established for linear systems without delays [3]. The following remarks enlightens some subtleties of the machinery in $\mathcal{E}$, in connection with our problem. First of all, remark that $\mathcal{E}$ is also a so-called invariant factor domain. Following [10], every matrix $D(s, \mathrm{e}^{-hs})$ over $\mathcal{E}$ can also be factored in the form (6), were now the matrices $U(s)$ and $V(s)$ and their inverses are over $\mathcal{E}$, and $\{\alpha_i(s)\}_{i=1}^{r}$ are elements of $\mathcal{E}$, which are also called the invariant factors of $D(s)$. We first establish the following clue facts.

Proposition 1. *The invariant factors in the Bezout domain $\mathcal{E}$ of a polynomial matrix coincide with its invariant factors over the ring $\mathbb{R}[s]$.*

Proof. Consider a polynomial matrix $D(s) \in \mathbb{R}^{n \times m}[s]$ of rank r. Now, since $\mathbb{R}[s] \subset \mathcal{E}$, it is clear that when $D(s)$ is polynomial, the polynomial factorization (6) is also a factorization over $\mathcal{E}$. $U(s)$ and $V(s)$ are indeed matrices over $\mathcal{E}$, which are unimodular over $\mathcal{E}$, the rank of $D(s)$ is r over $\mathcal{E}$, and, since such a factorization over $\mathcal{E}$ is unique, the $\{\alpha_i(s)\}_{i=1}^{r}$, are also the uniquely defined invariant factors of $D(s)$ over $\mathcal{E}$. □

The degree of an element $\alpha(s, \mathrm{e}^{-hs}) = \frac{n(s,\mathrm{e}^{-hs})}{d(s)} \in \mathcal{E}$, where $n(s, \mathrm{e}^{-hs}) \in \mathbb{R}[s, \mathrm{e}^{-hs}]$ and $d(s) \in \mathbb{R}[s]$, is the difference $\delta = \deg_s n(s, \mathrm{e}^{-hs}) - \deg d(s)$, which lies in $\mathbb{Z}$, and we write $\deg \alpha(s, \mathrm{e}^{-hs}) = \delta$. A matrix $D(s, \mathrm{e}^{-hs}) \in \mathcal{E}^{m \times m}$ being given, and denoting c_i the degree of its ith column, it appears that

$$D(s, \mathrm{e}^{-hs}) = \sum_{k=-\infty}^{0} D_k(\mathrm{e}^{-hs}) \operatorname{diag}\{s^{c_1}, s^{c_2}, \ldots, s^{-c_m}\} \ ,$$

where the coefficient matrices $D_k(\mathrm{e}^{-hs})$ are uniquely defined. By analogy with the polynomial case, $D(s, \mathrm{e}^{-hs})$ is called column reduced whenever the rank of $D_0(\mathrm{e}^{-hs})$ equals m. Every matrix over $\mathcal{E}$ can be brought in column reduced form under unimodular operations over $\mathcal{E}$. Notice that the degrees of the elements of $\mathcal{E}$ can be negative, for instance $\deg \frac{1-\mathrm{e}^{-hs}}{s} = -1$. This implies that the reduced comumn degrees of a matrix over $\mathcal{E}$ are not uniquely defined. Consider for instance

$$U(s, \mathrm{e}^{-s}) = \begin{pmatrix} s + \ln 2 & \frac{1-\mathrm{e}^{-s}}{s} \\ s & \frac{2-\mathrm{e}^{-s}}{s+ln2} \end{pmatrix} . \tag{7}$$

This matrix is unimodular over $\mathcal{E}$. It is also column reduced, since

$$U_0(\mathrm{e}^{-s}) = \begin{pmatrix} 1 \; 1 - \mathrm{e}^{-s} \\ 1 \; 2 - \mathrm{e}^{-s} \end{pmatrix} .$$

is full rank and even unimodular, and its column degrees are 1 and -1. One can see that $U(s, \mathrm{e}^{-s})U^{-1}(s, \mathrm{e}^{-s})$ is also column reduced, with column degrees equal to 0.

3 Finite Eigenstructure Assignment

We first give a new interpretation of the result established in [1, 11] that has been extensively used in the literature, in the light of n-assignability, namely the ability to assign a finite spectrum to the closed loop system, provided that the system is spectrally controllable. Indeed, there is more to say: it is also possible to assign prescribed invariant factors with a finite number of roots to the closed loop characteristic matrix.

Lemma 2. *Consider a spectrally controllable linear multivariable system with delay in the input described by*

$$\dot{x}(t) = Ax(t) + \sum_{i=0}^{N} B_i u(t - ih) , \tag{8}$$

where $A \in \mathbb{R}^{n \times n}$, $B_i \in \mathbb{R}^{n \times m}$, for $i = 1$ to N, and the delays in the input are commensurate to $h \geq 0$. Then the following problems are equivalent.
(i) The control law

$$u(t) = K[x(t) + \sum_{i=0}^{N} \int_{t-ih}^{t} \mathrm{e}^{(t-\sigma-ih)A} B_i u(\sigma) d\sigma] \tag{9}$$

assigns to the system (8) a closed loop with invariant factors $\{\alpha_i'(s)\}_{i=1}^{n}$.
(ii) The control law $u(t) = Ky(t)$ assigns to the controllable system

$$\dot{y}(t) = Ay(t) + \sum_{i=0}^{N} \mathrm{e}^{-ihA} B_i u(t) \tag{10}$$

a closed loop system with a set of invariant factors $\{\alpha_i'(s)\}_{i=1}^{n}$.

Proof. Consider the description of the system (8) and the closed loop (9) in the frequency domain. They are respectively

$$(sI_n - A)x(s) = \sum_{i=0}^{N} B_i e^{-ihs} u(s)$$

and

$$\left(I_m - K(\textstyle\sum_{i=0}^{N} e^{-ihA}(sI - A)^{-1}(I - e^{-ih(sI-A)})\right) B_i)u(s) = Kx(s)$$

Defining the matrices B and M as

$$B = \sum_{i=0}^{N} B_i\, e^{-ihs}$$

and

$$M = \sum_{i=0}^{N} e^{-ihA}(sI - A)^{-1}(I - e^{-ih(sI-A)})B_i\ ,$$

it appears that the closed loop characteristic matrix writes

$$\begin{pmatrix} sI_n - A & -B \\ -K & I_m - KM \end{pmatrix} .$$

Notice that the matrices B and M are over $\mathcal{E}$. Thus the matrices

$$\begin{pmatrix} I_n & B + (sI_n - A)M \\ K & I_m \end{pmatrix}$$

and

$$\begin{pmatrix} I_n & -M \\ 0 & I_m \end{pmatrix}$$

are unimodular over $\mathcal{E}$, and since

$$\begin{pmatrix} I_n & B + (sI_n - A)M \\ K & I_m \end{pmatrix} \begin{pmatrix} sI_n - A & -B \\ -K & I_m - KM \end{pmatrix} \begin{pmatrix} I_n & -M \\ 0 & I_m \end{pmatrix}$$
$$= \begin{pmatrix} (sI_n - A)(I_n - MK) + BK & 0 \\ -K & I_m \end{pmatrix} ,$$

the invariant factors of the closed loop are those of

$$\begin{aligned} &(sI_n - A)(I_n - MK) + BK \\ &= sI_n - A - \sum_{i=0}^{N} e^{-ihA}(I - e^{-ih(sI-A)})B_iK) - \sum_{i=0}^{N} B_i e^{-ihs} K \\ &= sI_n - A - \sum_{i=0}^{N} e^{-ihA} B_i K\ . \end{aligned}$$

One can finally observe that

$$B + (sI_n - A)M = \sum_{i=0}^{N} \mathrm{e}^{-ihA} B_i \ ,$$

which ends the proof. □

The two matrices

$$(sI_n - A, -B)$$

and

$$(sI_n - A, -B - (sI_n - A)M)$$

are also equivalent under unimodular transformations. Hence the above proof also permits to establish the following key result.

Lemma 3. *[13] The input delay system (8) is spectrally controllable if and only if the linear system (10) is controllable.*

We are now able to state our main result.

Theorem 1. *Consider a spectrally controllable linear multivariable system with delay in the input described by*

$$\dot{x}(t) = Ax(t) + \sum_{i=0}^{N} B_i u(t - ih), \tag{11}$$

where $A \in \mathbb{R}^{n\times n}, B_i \in \mathbb{R}^{n\times m}$ and $h \geq 0$ is the delay. Let $\{c_i\}_{i=1}^m$ be the controllability indices, in non decreasing order, of the pair $(A, \sum_{i=0}^N \mathrm{e}^{-ihA} B_i)$ and let $\{\alpha_i'(s)\}_{i=1}^n$ be a set of monic polynomials such that $\alpha_i'(s)$ divides $\alpha_{i-1}'(s)$, for $i = 2, ..., m$.
There exist a control law

$$u(t) = K[x(t) + \sum_{i=0}^{N} \int_{t-h}^{t} \mathrm{e}^{(t-\sigma-ih)A} B_i u(\sigma) d\sigma] \tag{12}$$

that assigns a closed loop with invariant factors $\{\alpha_i'(s)\}_{i=1}^n$ if and only if

$$\sum_{i=1}^{j} c_i \leq \sum_{i=1}^{j} \deg \alpha_i' \ , \ j = 1, ..., m \tag{13}$$

with equality for $j = m$, and, by convention, $\alpha_i'(s) = 1$, for $i = m + 1, ..., n$.

Proof. According to Lemma 3, the pair $(A, \sum_{i=0}^N \mathrm{e}^{-ihA} B_i)$ is controllable. Let $\{c_i\}_{i=1}^m$ denote its controllability indices. From the Rosenbrock Control Structure Theorem (1), there exists a control law (12) that assigns to the system (10) a closed loop with invariant factors $\{\alpha_i'(s)\}_{i=1}^n$, where $\alpha_i'(s)$ divides $\alpha_{i-1}'(s)$, $i = 2, ..., m$, if and only if conditions (13) hold. Finally, the result follows from Lemma 2. □

Remark 1. Clearly, the use of simpler control laws has a cost: the degrees of the invariant factors cannot be assigned arbitrarily.

Remark 2. Lemma 2 implies that the control law described by (12) permits to assign to the input delay system (11) a set of desired invariant factors that satisfies the inequalities (13). The parameter K of the control law can be calculated in a straightforward manner as the solution to the problem of a static state feedback invariant factors assignment for the linear system with no delay (10). Notice that toolboxes for the analysis and design of linear systems are available [14]. If a set of desired invariant factors does not satisfies the inequalities (13), the results obtained in [10] guarantee that there exists a control law of the more genaral form (2) that allows to assign them. This control law if fully determined by g_i, $i = 1, 2, ...,$ $f(\tau)$ and $g(\tau)$.

4 Further Comments and Design Algorithms

The design of a control law assigning the invariant factors of the closed-loop factors is based on the computation of a right coprime factorization of the open-loop transfer of the system. The following is a key for this aim.

Proposition 2. *If the input delay system (8) is spectrally controllable, then its transfer matrix can be factored in the form*

$$(sI_n - A)^{-1}B = N(s, \mathrm{e}^{-hs})D^{-1}(s) \ ,$$

where $D(s)$ *and* $N(s, \mathrm{e}^{-hs})$ *are right coprime matrices that are respectively polynomial in the variables* s *and* s, e^{-hs}. *Moreover, if we expand* $N(s, \mathrm{e}^{-hs})$ *as a polynomial in* e^{-hs}, *of the form*

$$N(s, \mathrm{e}^{-hs}) = \sum_{i=0}^{N} \mathrm{e}^{-ihs} N_i(s) \ ,$$

then a factorization of the equivalent system without delay (10) is obtained as

$$\sum_{i=0}^{N} (sI_n - A)^{-1} \mathrm{e}^{-ihA} B_i = \sum_{i=0}^{N} \mathrm{e}^{-ihA} N_i(s) D^{-1}(s). \tag{14}$$

wher the column degrees of the column reduced polynomial denominator $D(s)$ *are equal to the controllability indices of the pair* $(A, \Sigma_{i=0}^{N} \mathrm{e}^{-ihA} B_i)$ *and they are uniquely defined.*

Proof. Consider the controllable linear system without delay described by (10), and let the pairs $(N_i'(s), D_i(s))$ be coprime factorizations for the pairs (A, B_i), $i = 0, N$. Hence

$$(sI_n - A)^{-1}B_i = N_i'(s)D_i^{-1}(s).$$

Let $N_i(s) = N_i'(s)D_i^{-1}(s)D(s)$ where $D(s)$ is the lowest right common multiple of $D_i(s)$, $i = 0, \ldots, N$. We then have

$$\sum_{i=0}^{N}(sI_n - A)^{-1}\mathrm{e}^{-ihA}B_i = \sum_{i=0}^{N}\mathrm{e}^{-ihA}N_i(s)D^{-1}(s).$$

The matrices $\sum_{i=0}^{N}\mathrm{e}^{-ihA}N_i(s)$ and $D(s)$ are right coprime. Observe that

$$\sum_{i=0}^{N}(sI_n - A)^{-1}B_i\mathrm{e}^{-ihs} = \sum_{i=0}^{N}\mathrm{e}^{-ihs}N_i(s)D^{-1}(s) \ ,$$

and that $\Sigma_{i=0}^{N}\mathrm{e}^{-ihs}N_i(s)$ and $D(s)$ are right coprime. Both systems (8) and (10) have the same denominator, $D(s)$, which is polynomial in s. As it is usual for systems without delays, we can assume that $D(s)$ is column reduced. Its column degrees are the controllability indices of the pair $(A, \Sigma_{i=0}^{N}\mathrm{e}^{-ihA}B_i)$, namely, $\{c_i\}_{i=1}^{m}$.

Any other polynomial denominator for (8), say $D'(s)$ is such that $D(s)U(s,\mathrm{e}^{-hs}) = D'(s)$, for some matrix $U(s,\mathrm{e}^{-hs})$ which is unimodular over $\mathcal{E}$. We can see that $U(s,\mathrm{e}^{-hs}) = D^{-1}(s)D'(s)$ is rational in the variable s. Since it is also a matrix over $\mathcal{E}$, an analytic function without pole, we conclude that it is actually polynomial in s. Since in $\mathbb{R}^{m\times m}[s]$ column degrees are not modified by post-multiplication by a unimodular, those of $D'(s)$ are equal to those of $D(s)$. In the sequel, we shall call such a factorization with a polynomial denominator a *natural* factorization of the system. □

Any coprime factorization $(N'(s,e^{-s}), D'(s,e^{-s}))$ in $\mathcal{E}$ for the system (8) can be written in terms of a natural coprime factorization as

$$\begin{aligned} N'(s,e^{-hs}) &= N(s,e^{-hs})U(s,e^{-hs}) \\ D'(s,e^{-hs}) &= D(s)U(s,e^{-hs}) \end{aligned}$$

where $U(s,e^{-hs})$ is a unimodular matrix over $\mathcal{E}$. It follows from Proposition 2 that the latter satisfies

$$(sI_n - A)N(s,e^{-hs}) = \sum_{i=0}^{N} B_i e^{-ihs} D(s) \tag{15}$$

and that the column indices of $D(s)$ are $\{c_i\}_{i=1}^{m}$. The control law (4) is described in the frequency domain as

$$(I_m - K(\Sigma_{i=0}^{N}e^{-ihA}(sI - A)^{-1}[I - e^{-ih(sI-A)}]B_i)u(s) = Kx(s)$$

and the closed-loop denominator is given by $D_K(s,e^{-s})$:

$$\begin{aligned} &((I_m - K\Sigma_{i=1}^{N}e^{-ihA}(sI - A)^{-1}[I - e^{-ih(sI-A)}]B_i)D'(s,e^{-s}) - KN'(s,e^{-s}) \\ &= ((I_m - K(\Sigma_{i=1}^{N}e^{-ihA}(sI - A)^{-1}[I - e^{-ih(sI-A)}]B_i))D(s)U(s,e^{-s}) \\ &\quad -KN(s,e^{-s})U(s,e^{-s}) \ . \end{aligned}$$

Using (15) gives

$$D_K(s,e^{-s}) = (I_m - K(sI-A)^{-1}\Sigma_{i=1}^N e^{-ihA}B_i)D(s)U(s,e^{-s}), \tag{16}$$

or equivalently,

$$D_K(s,e^{-s})U(s,e^{-s})^{-1} = (I_m - K(sI-A)^{-1}\Sigma_{i=1}^N e^{-ihA}B_i)D(s).$$

On the one hand, the matrix $(I_m - K(sI-A)^{-1}\Sigma_{i=1}^N e^{-ihA}B_i)$ is a biproper matrix in $\mathbb{R}^{m\times m}(s)$. Hence $(I_m - K(sI-A)^{-1}\Sigma_{i=1}^N e^{-ihA}B_i)D(s)$ is a polynomial matrix and so is $D_K(s,e^{-s})U(s,e^{-s})^{-1}$. Moreover, the column degrees of $D_K(s,e^{-s})U(s,e^{-s})^{-1}$ are the same as those of $D(s)$, namely, $\{c_i\}_{i=1}^m$.
On the other hand, $U(s,e^{-s})^{-1}$ is a unimodular matrix, therefore the invariant factors of $D_K(s,e^{-s})U(s,e^{-s})^{-1}$are the same as those of $D_K(s,e^{-s})$, namely $\{\alpha_i'(s)\}_{i=1}^m$.
The polynomial matrix $D_K(s,e^{-s})U(s,e^{-s})^{-1}$has invariant factors $\{\alpha_i'(s)\}_{i=1}^m$ and column degrees $\{c_i\}_{i=1}^m$. Then, according to [8], the lists $\{\alpha_i'(s)\}_{i=1}^m$ and $\{c_i\}_{i=1}^m$ are necessarily related by Rosenbrock inequalities.

The natural coprime factorization described in Claim 2 hence gives an alternative sufficiency proof of Theorem 1, and bases a design algorithm for the finite spectrum assignment. Next, two design procedure for finite invariant factors assignment of input delay systems are described. The first one is appropriate when the degrees of the invariant factors satisfy the Rosenbrock inequalities while the second, which is computationally more complex, works in the general case.

Design procedure if the Rosenbrock inequalities (13) are satisfied:
(i) Using Proposition 2, calculate a natural factorization $(N(s,\mathrm{e}^{-sh}), D(s))$ of the transfer matrix of system (8) where the matrix $D(s)$ is column reduced with column degrees $\{c_i\}_{i=1}^m$.
(ii) Choose a set of monic polynomials $\{\alpha_i'(s)\}_{i=1}^n$ where $\alpha_i'(s)$ divides $\alpha_{i-1}'(s)$ for $i = 2, ..., m$, such that $\{\alpha_i'(s)\}_{i=1}^m$ and $\{c_i\}_{i=1}^m$ satisfy the inequalities (13), and take $\alpha_{m+1}'(s) = \ldots = \alpha_n'(s) = 1$.
(iii) Construct a polynomial matrix $D_K(s)$, that is column reduced, with column degrees $\{c_i\}_{i=1}^m$ and invariant factors $\{\alpha_i'(s)\}_{i=1}^m$, according to the procedure in [8].
(iv) Determine the constant matrices X and Y of convenient dimensions, with X invertible, that satisfy the Diophantine equation

$$XD(s) + YN(s,\mathrm{e}^{-hs}) = D_K(s).$$

The existence of these matrices is guaranteed [8], and a toolbox is available to perform the computations [14].
(v) Substitute $K = -X^{-1}Y$ in the expression for the control law (3).

The situation is quite different when considering factorizations over $\mathcal{E}$. As shown in [10], contrarily to what happens in $\mathbb{R}(s)$, post multiplication of elements of $\mathcal{E}$ by unimodular over $\mathcal{E}$ modify the column degrees. An immediate consequence of this fact is the non unicity of the column degrees of the denominator of a left coprime factorization in $\mathcal{E}$. Moreover, there is complete freedom in choosing the column degrees of such a factorization.

If a set of invariant factors $\alpha'_i(s)$ does not satisfy the inequalities (13), except for $j = m$, it is still possible to achieve the desired assignment by employing a wider class of control laws. The design is based on the fact that the column degrees of the column reduced denominator of a coprime factorization in $\mathcal{E}$ are not unique.

Design procedure if the inequalities (13) are not satisfied:
(i) Using proposition 2, calculate a natural factorization $(N(s,\mathrm{e}^{-sh}), D(s))$ of the transfer matrix of system (8).
(ii) Obtain a coprime factorization $(N'(s,\mathrm{e}^{-sh}), D'(s,\mathrm{e}^{-sh}))$ as

$$\begin{aligned} D'(s,\mathrm{e}^{-s}) &= D(s)U(s,\mathrm{e}^{-s}), \\ N'(s,\mathrm{e}^{-s}) &= N(s,\mathrm{e}^{-s})U(s,\mathrm{e}^{-s}), \end{aligned} \tag{17}$$

where $U(s,\mathrm{e}^{-s})$ is an unimodular matrix such that $D'(s,\mathrm{e}^{-sh})$ is column reduced, with column degrees $\deg \alpha'_i(s)$, $i = 1,\ldots,m$. Such a transformation is obtained recursively using elementary operations as in (7).
(iii) Solve in $\mathcal{E}$ the Diophantine equation

$$\left(I_m - F(s,\mathrm{e}^{-sh})\right) D'(s,\mathrm{e}^{-sh}) + G(s,\mathrm{e}^{-sh})N'(s,\mathrm{e}^{-sh}) = Z(\mathrm{e}^{-hs})\mathrm{diag}\{\alpha'_i(s)\}$$

where $Z(\mathrm{e}^{-hs})$ is a a unimodular matrix in e^{-hs} and matrices $F(s,\mathrm{e}^{-sh})$ and $G(s,\mathrm{e}^{-sh})$ belong to $\mathcal{E}$. The existence of such a solution is proved in [10].
(iv) Determine the control law in the time domain (2) by inverse Laplace transform of

$$\left(I_m - F(s,\mathrm{e}^{-sh})\right) u(s) = G(s,\mathrm{e}^{-sh})x(s),$$

according to the algorithm described in [3,4].

This shows that a control law of the form (2) permits to assign the invariant factors $\alpha'_i(s)$ to the closed loop system (3–2). It should be clear that, in that case, a simpler control law of the form (4) cannot get this assignment.

5 Illustrative Example

The above results and design procedures are illustrated with the two dimensional academic example

$$\dot{x}(t) = \begin{pmatrix} 0 & 0 \\ 1 & 0 \end{pmatrix} x(t) + \begin{pmatrix} 1 & 1 \\ 0 & 0 \end{pmatrix} u(t-h).$$

This system is spectrally controllable because the pair $(A, \exp(-hA)B)$ is controllable (See Lemma 3). Its controllability indices are $\{2,0\}$.

The natural coprime factorization for this system is

$$N(s,\mathrm{e}^{-s}) = \begin{pmatrix} s\mathrm{e}^{-s} & 0 \\ \mathrm{e}^{-s} & 0 \end{pmatrix}, \quad D(s) = \begin{pmatrix} s^2 & -1 \\ 0 & 1 \end{pmatrix}.$$

It is possible to design a control law that assigns prescribed invariant factors with finite roots, of degree $\{2,0\}$. A solution is given by

$$(I_2 - F(s,e^{-s})\,D(s,e^{-s}) - G(s,e^{-s})N(s,e^{-s}) = D_K(s,e^{-s})\,. \quad (18)$$

According to Theorem 3, a control law where $G(s,e^{-s}) = K$, and $F(s,e^{-s}) = K(sI-A)^{-1}(e^{-hA} - e^{-hs}I)B$ does the job, and substituting $D(s)$, $N(s,e^{-s})$, A, B, and

$$K = \begin{pmatrix} k_{11} & k_{12} \\ k_{21} & k_{22} \end{pmatrix}$$

into (16), gives

$$D_K(s,e^{-s}) = \begin{pmatrix} s^2 - k_{11}s + k_{12}s - k_{12} & -1 \\ -k_{21}s + k_{22}s - k_{22} & 1 \end{pmatrix}. \quad (19)$$

The invariant factors of degree $\{2,0\}$, namely $\{s^2 + (-k_{11} + k_{12} - k_{21} + k_{22}) - (k_{12} + k_{22}), 1\}$, can indeed be assigned to arbitrary finite locations by choosing K. We have in this simple case obtained a parametrization of all the invariant factors that can be obtained with the control laws under consideration.

Now, if we want to assign invariant factors of degrees $\{1,1\}$, for instance both invariant factors taking the value $s+1$, let us consider, according to the design procedure described in the previous section, a coprime factorization with column degrees $\{1,1\}$ for the denominator. Substituting the unimodular

$$U(s,e^{-s}) = \begin{pmatrix} \frac{2-e^{-s}}{s+\ln 2} & -\frac{1-e^{-s}}{s} \\ -s & s+\ln 2 \end{pmatrix}.$$

in (17) gives the new factorization

$$D'(s,e^{-s}) = D(s)U(s,e^{-s}) = \begin{pmatrix} s^2 & -1 \\ 0 & 1 \end{pmatrix} \begin{pmatrix} \frac{2-e^{-s}}{s+\ln 2} & -\frac{1-e^{-s}}{s} \\ -s & s+\ln 2 \end{pmatrix},$$

$$N'(s,e^{-s}) = N(s,e^{-s})U(s,e^{-s}) = \begin{pmatrix} se^{-s} & 0 \\ e^{-s} & 0 \end{pmatrix} \begin{pmatrix} \frac{2-e^{-s}}{s+\ln 2} & -\frac{1-e^{-s}}{s} \\ -s & s+\ln 2 \end{pmatrix},$$

and a solution to the Diophantine equation in $\mathcal{E}$

$$\left(Z(e^{-s}) - F(s,e-s)\right) D'(s,e^{-s}) + G(s,e^{-s})N'(s,e^{-s}) = \operatorname{diag}\{s+1, s+1\},$$

is given by

$$G(s,e^{-s}) = \begin{pmatrix} (1+\ln 2)(2-e^{-s}) & \ln 2(2-e^{-s}) \\ 2-e^{-s} & 0 \end{pmatrix},$$

$$Z(e^{-s}) = \begin{pmatrix} 1 & 2-e^{-s} \\ 1 & 3-e^{-s} \end{pmatrix},$$

and

$$\begin{array}{ll} F_{11}(s,e^{-s}) & (s(1+\ln 2)+\ln 2)\left(\frac{1-e^{-s}}{s}\right)^2, \\ F_{12}(s,e^{-s}) & (1+\ln 2)\frac{(1-e^{-s})^2}{s} + \frac{1-e^{-s}}{s}, \\ F_{21}(s,e^{-s}) & (1-e^{-s})^2, \\ F_{22}(s,e^{-s}) & \frac{(1-e^{-s})^2}{s} + (1-\ln 2)\frac{2-e^{-s}}{s+n\ln 2}. \end{array}$$

Of course this calculation is uneasy, and so is the obtention of the corresponding control law in the time domain [3, 4], since there is no specialized toolbox to deal with $\mathcal{E}$.

6 Conclusion

It is shown that the control laws (4) for input delay systems that allows the assignment of a finite spectrum, can be used to assign as well invariant factors with a finite number of roots. This feature is indeed of interest in shaping the dynamic of the closed loop system. It is shown that the multiplicities of the invariant factors cannot be assigned freely: they are restricted by Rosenbrock inequalities by a set of well defined integers. The result is explained in the framework of the Bezout domain $\mathcal{E}$, where more general control laws allow a complete freedom in assigning the multiplicities of the invariant factors. Summarizing, finite invariant factors assigning control laws for input delay systems can be designed in a straightforward manner, but at a cost: the freedom in the assignment of the multiplicities of the invariant factors is restricted.

References

1. Artstein Z (1982) Linear systems with delayed controls: a reduction, IEEE Trans Autom Contr AC-27:869-879
2. Bellman R, Cooke KL (1963) Differential difference equations. Academic Press, London.
3. Brethé D (1997) Contribution à l'étude de la stabilisation des systèmes linéaires à retards. PhD Thesis, Université de Nantes, France
4. Brethé D, Loiseau JJ (1998) An effective algorithm for finite spectrum assignment of single-input systems with delays, Mathematics in Computers and Simulation 45:339-348
5. Gantmacher FR (1959) The theory of matrices, Vol 1. AMS Chelsea Publishing, New York
6. Kamen EW, Khargonekar PP, Tannenbaum A (1986) Proper stable bezout factorizations and feedback control of linear time delay systems, Int J Contr 43:837-857
7. Kailath T (1980) Linear systems. Prentice Hall, Englewood Cliffs, N.J.
8. Kučera V (1991) Analysis and design of discrete linear control systems. Prentice-Hall, London, and Academia, Prague
9. Loiseau JJ (2000) Algebraic tools for the control and stabilization of time-delay systems, Annual Reviews in Control 24:135-149
10. Loiseau JJ (2001) Invariant factors assignment for a class of time-delay systems, Kybernetika 37:265-275
11. Manitius AZ, Olbrot AW (1979) Finite spectrum assignment problem for systems with delays, IEEE Trans Autom Contr AC-24:541-553
12. Niculescu S-I (2001) Delays effects on stability, A robust control approach. Springer, Berlin Heidelberg New York
13. Olbrot A (1978) Stabilizability, detectability, and spectrum assignment for linear autonomous systems with general time delays, IEEE Trans Autom Contr AC-23:887-890
14. (1998) The polynomial toolbox, Polynomial methods for systems, signal and control. Polyx, Prague
15. Rosenbrock H.H. (1970) State space and multivariable theory. Wiley, New York

Control of Systems with Input Delay—An Elementary Approach

Vladimir Răsvan[1] and Dan Popescu[2]

[1] Department of Automatic Control, University of Craiova, A.I.Cuza Str. No. 13, RO 1100 Craiova, ROMANIA vrasvan@automation.ucv.ro
[2] dpopescu@automation.ucv.ro

Summary. The stabilization by feedback control of systems with input delays may be considered in various frameworks; a very popular is the abstract one, based on the inclusion of such systems in the Pritchard-Salamon class. In this chapter we consider the elementary approach based on variants of the Smith predictor, make a system theoretic analysis of the compensator and suggest a computer control implementation.This implementation is based on piecewise constant control which associates a discrete–time finite dimensional control system; it is this system which is stabilized, thus avoiding unpleasant phenomena induced by the essential spectrum of other implementations

1 Motivation and the State of Art

It has been pointed out that systems with input delays are of interest to control theorists and practitioners for various reasons. They originate from the simplest model of process control which assigns to the controlled plant a transfer function of the form $H(s)e^{-\tau s}$ with $H(s)$ a strictly proper rational function.

To this transfer function one may associate one of the following state representations

$$\begin{aligned} \dot{x} &= Ax + bu(t-\tau) \\ y &= c^* x(t) \end{aligned} \tag{1}$$

or

$$\begin{aligned} \dot{x} &= Ax + bu(t) \\ y &= c^* x(t-\tau) \end{aligned} \tag{2}$$

where in both cases $c^*(sI-A)^{-1}b \equiv H(s)$. Systems arising from such transfer functions are of special type: their state space is finite dimensional but either the input or the output operators are defined on infinite dimensional extensions and are unbounded. Throughout several decades it has been established that such systems have "a finite dimensional flavor". Indeed, there exists a remark of V.M. Popov [17] which tells that problems as stability and optimality in the framework of hyperstability (or dissipativity/passivity as it is called now) may be solved for these systems like for systems without delay; nevertheless Popov did not follow this line in his research. We may add to this the results on feedback stabilization due to Olbrot [11], Manitius and Olbrot [9], Watanabe and Ito [23]. The techniques of these papers are essentially finite dimensional; the opinion of Pandolfi [12–15] was that those results were not obtained in a standard way since they were not deduced from an abstract theory. In order to sustain his ideas, Pandolfi re-introduced in the model the *propagation effects* taken into account by the introduction of the delay and make use of the theory of the singular control. On the other hand, the finite dimensional results seem legitimate if the transform introduced by Artstein [1] is used. The papers of Tadmor (e.g. [21,22]) also support the idea that finite dimension is the most adequate framework for systems with input delays (see also [10]).

In the same line of research we may cite the papers of A. E. Pearson and his co-workers [3,4,8,16]. Worth mentioning that research on various subjects which is underlined by the above mentioned ideas is in progress within various groups (see e.g. the chapters in this book).

2 Artstein Transform and Stabilization Results

We shall consider here the case of [8]

$$\dot{x}(t) = Ax(t) + B_0u(t) + B_1u(t-\tau) \tag{3}$$

We might have chosen a more general structure of the input operator but this one helps a better understanding; at the same time it is not the simplest case and illustrates the advantages of the taken approach as well as its drawbacks.

Obviously the solution is defined for $t > 0$ if there are given the initial conditions $(x_0, u_0(\cdot))$ and the control $u(t)$ for $t > 0$; here $u_0(\theta)$ is some initial function defined for $\theta \in [-\tau, 0)$. The Artstein transform (Artstein, 1982) in this case is given by

$$z(t) = x(t) + \int_{-\tau}^{0} e^{-A(\theta+\tau)}B_1u(t+\theta)d\theta \tag{4}$$

and the result of Artstein takes the form of the following equivalence.

Proposition 1. *Let $(x(t), u(t); t > 0)$ be a solution (admissible pair) for (3), defined by some initial condition $(x_0, u_0(\cdot))$. Then $(z(t), u(t); t > 0)$ with $z(t)$ defined by (4) is a solution (admissible pair) for the system*

$$\dot{z}(t) = Az(t) + \left(B_0 + e^{-A\tau}B_1\right)u(t) \tag{5}$$

with the initial condition $z_0 = z(0)$. *Conversely, let* $(z(t), u(t); t > 0)$ *be a solution of (5) defined by some initial condition* z_0. *Then, given some* $u_0(\cdot)$ *defined on* $(-\tau, 0)$ *and taking*

$$x_0 = z_0 - \int_{-\tau}^{0} e^{-A(\theta+\tau)} B_1 u_0(\theta) d\theta \tag{6}$$

the solution of (3) defined by these initial conditions and by $u(t), t > 0$ *is given by*

$$x(t) = z(t) - \int_{-\tau}^{0} e^{-A(\theta+\tau)} B_1 u(t+\theta) d\theta \tag{7}$$

The proof of this result is straightforward. The simplest result to be obtained within this framework is the structure of the stabilizing feedback. The result on stabilization appears as such in the paper of Kwon and Pearson [8] but it may be found included in e.g. [7,9,23] also in [1] as well as [19,20]. The result reads as follows

Proposition 2. *Let* $u = Fz$ *be a feedback stabilizing scheme for (5). Then the control*

$$u(t) = F\left[x(t) + \int_{-\tau}^{0} e^{-A(\theta+\tau)} B_1 u(t+\theta) d\theta\right] \tag{8}$$

is stabilizing for (3).

The proof is straightforward and relies entirely on Proposition 1.

The structure defined by (8) may be used as a stabilizing compensator since the solution of the closed loop system

$$\begin{aligned} \dot{x}(t) &= Ax(t) + B_0 u(t) + B_1 u(t-\tau) \\ u(t) - F \int_{-\tau}^{0} & e^{-A(\theta+\tau)} B_1 u(t+\theta) d\theta = Fx(t) \end{aligned} \tag{9}$$

may be constructed by steps. Let $(x_0,\ u_0(\theta),\ -\tau \le \theta \le 0)$ be some initial condition with u_0 of appropriate smoothness; using it we may construct $x(t)$ for $0 \le t \le \tau$ by integrating the first equation of (9). Now if $0 \le t \le \tau$ and $-\tau \le \theta \le 0$ then $t - \tau \le t + \theta \le t$ and $u(t)$ may be constructed on $(0, \tau)$ using values that precede it with one step at most. If $u(t)$ is known on $(0, \tau)$ then $x(t)$ may be constructed on $(\tau,\ 2\tau)$ and the process will continue.

It is interesting to comment on the second equation of (9) : it describes the compensator which has some dynamics and may be viewed also as a linear functional on system's state space; at the same time this equation may be viewed as

$$(\mathcal{D}u)(t) = Fx(t)$$

with $\mathcal{D}$ a difference operator in the sence used e.g. in the book by Hale and Verduyn Lunel [6]. Coupled delay–differential and difference equations have been considered by the first author of the present chapter since 1973 (see [18] as a more recent survey) and they account for propagation phenomena thus sending to the already mentioned papers of Pandolfi. generally speaking such systems belong to the class of neutral functional differential equations; it is not the case here and the best argument is that discontinuity propagation does not occur. Indeed we have

$$u(0_+) = F\left(x_0 + \int_{-\tau}^{0} e^{-A(\theta+\tau)} B_1 u_0(\theta) d\theta\right) \neq u(0_-) := \lim_{t \to 0, t<0} u_0(t) \quad (10)$$

hence $u(t)$ is, generally speaking, discontinuous at 0. But for $t > 0$ $u(t)$ is continuous - due to the specific structure of the difference operator; moreover $u(t)$ is absolutely continuous and we may differentiate it to obtain the system

$$\begin{aligned} \dot{x}(t) &= Ax(t) + B_0 u(t) + B_1 u(t-\tau) \\ \dot{u}(t) &= F\left(Ax(t) + (B_0 + e^{-A\tau} B_1) u(t) + A \int_{-\tau}^{0} e^{-A(\theta+\tau)} B_1 u(t+\theta) d\theta\right) \end{aligned} \quad (11)$$

which is of delayed type.

Even more interesting is to check the characteristic equations of the two systems of Functional Differential Equations namely (9) and (11). Considering the Euler solutions we obtain for the first system

$$\det(sI - A - (B_0 + e^{-A\tau} B_1) F) = 0 \quad (12)$$

hence the spectrum is finite and may be assigned from the stabilization problem solved by Proposition 2 . For the second system we obtain

$$s^m \cdot \det(sI - A - (B_0 + e^{-A\tau} B_1) F) = 0 \quad (13)$$

where $m = \dim u$ and the factor s^m is a consequence of differentiation showing that the system evoluates confined to an invariant manifold.

3 The Discrete-Time Implementation. Spectral Properties of the Closed Loop

The implementation of the designed compensator requires memorizing of a trajectory segment i.e. a set of data that has infinite size. The practical implementation is finite and based on a suitable discretization. Following the line of the paper of Halanay and Răsvan [5] and of the book of Drăgan and Halanay [2] we shall use *piecewise constant control signals*, defined as follows

$$u(t) = u_k, \quad k\delta \leq t < (k+1)\delta, \quad k = 0, 1, 2, \ldots \quad (14)$$

where $\delta = \tau/N$. For the system (3) we associate the discrete time system

$$x_{k+1} = \mathbf{A}(\delta) x_k + \mathbf{B}_0(\delta) u_k + \mathbf{B}_1(\delta) u_{k-N} \quad (15)$$

where

$$\mathbf{A}(\delta) = e^{A\delta}, \quad \mathbf{B}_i(\delta) = \left(\int_0^\delta e^{A\theta} d\theta\right) B_i, \quad i = 0, 1 \tag{16}$$

Let $(x_0, u_0(\cdot))$ be the initial condition associated with (3). Since the discretized system is satisfied by $x_k = x(k\delta)$, $x(\cdot)$ being the solution of (3) with piecewise constant control, it is only natural to choose the discretized initial condition $(x_0; u_{-i}^0 = u_0(-i\delta), i = \overline{0, N})$. We may define

$$z_k = x_k + \sum_{-N}^{-1} \mathbf{A}(\delta)^{-(N+j+1)} \mathbf{B}_1(\delta) u_{k+j} \tag{17}$$

which is the discrete analogue of Artstein transform and find the associate system

$$z_{k+1} = \mathbf{A}(\delta) z_k + \left(\mathbf{B}_0(\delta) + \mathbf{A}(\delta)^{-N} \mathbf{B}_1(\delta)\right) u_k \tag{18}$$

It is worth mentioning that (17) might be obtained by writing (4) at $t = k\delta$ and computing the integral for piecewise constant control signals.

Let F be a stabilizing feedback for (18), i.e. is such that

$$\mathbf{A}(\delta) + \left(\mathbf{B}_0(\delta) + \mathbf{A}(\delta)^{-N} \mathbf{B}_1(\delta)\right) F$$

has its eigenvalues inside the unit disk. We deduce that the compensator

$$u_k = F x_k + \sum_{-N}^{-1} F \mathbf{A}(\delta)^{-(N+j+1)} \mathbf{B}_1(\delta) u_{k+j} \tag{19}$$

is stabilizing for (15). On the other hand, if we consider the closed loop system

$$\begin{aligned} x_{k+1} &= \mathbf{A}(\delta) x_k + \mathbf{B}_0(\delta) u_k + \mathbf{B}_1(\delta) u_{k-N} \\ u_k &= F x_k + \sum_{-N}^{-1} F \mathbf{A}(\delta)^{-(N+j+1)} \mathbf{B}_1(\delta) u_{k+j} \end{aligned} \tag{20}$$

one may see that this is a feedback system with an augmented dynamics

$$\begin{aligned} x_{k+1} &= \mathbf{A}(\delta) x_k + \mathbf{B}_1(\delta) v_k + \mathbf{B}_0(\delta) u_k \\ v_{k+1} &= w_k^1 \\ &\cdots \\ w_{k+1}^{N-1} &= u_k \\ u_k &= F[x_k + \mathbf{A}(\delta)^{-1} \mathbf{B}_1(\delta) v_k + \ldots \\ &\quad + \mathbf{A}(\delta)^{-(N-1)} \mathbf{B}_1(\delta) w_k^{N-2} + \mathbf{A}(\delta)^{-N} \mathbf{B}_1(\delta) w_k^{N-1}] \end{aligned} \tag{21}$$

Since $w_k^{N-1} = u_{k-1}$ the corresponding initial condition is $w_0^{N-1} = u_{-1} = u_0(-\delta)$; further, $w_0^{N-2} = u_{-2} = u_0(-2\delta)$, ..., $w_0^1 = u_0(-(N-1)\delta)$, $v_0 = u_0(-N\delta)$.

Obviously (21) is exponentially stable. This follows from the fact that $u = Fz$ is exponentially stabilizing system (18) and making use of (17).

To end the analysis we have to show how stabilization of the associated discrete time system ensures stabilization for the initial continuous time system. This problem will be tackled on the transformed system (5) with the stabilizing feedback

$$u(t) = Fz(k\delta) = Fz_k, \quad k\delta \le t < (k+1)\delta, \quad k = 0, 1, 2, \ldots \tag{22}$$

The system is of the type considered in [5]; a straightforward application of the results from the above paper will ensure the exponential stability of the closed loop hybrid system. Further, if $z(t)$ satisfies an exponential estimate then using (4) the exponential estimate for $x(t)$ is obtained what ends the proof of the following result

Proposition 3. *Consider the system (3) under the assumption that* $(A\,, B_0{+}e^{-A\tau}B_1)$ *is a stabilizable pair. Then the pair* $(\mathbf{A}(\delta)\,,\quad \mathbf{B}_0(\delta) +$ $\mathbf{A}(\delta)^{-N}\mathbf{B}_1(\delta))$ *is stabilizable and a stabilizing feedback for this couple is stabilizing for (3) provided* $\delta > 0$ *is small enough. Here* $\mathbf{A}(\delta)$, $\mathbf{B}_0(\delta)$, $\mathbf{B}_1(\delta)$ *are defined by (16) and* $\delta = \tau/N$. *Moreover, a stabilizing feedback for (5) is stabilizing also if the implementation is performed using samples i.e. state values measured at* $k\delta$, $k = 0, 1, 2, \ldots$

4 Concluding Remarks

We would like to point out a single but most important feature of our approach, confirmed also by simulations (nevertheless the proofs are rigorous and on a sound basis – see again [2, 5]). Most implementation approaches are based on the discretization of the integral what leads to continuous-time compensators described by difference equations hence to systems of neutral type with an essential spectrum. Stability of such systems require this spectrum to be inside the unit disk which *is not automatically ensured by a refined (with the step small enough) discretization*; consequently such systems often destabilize being either non-robust or fragile. The introduction of a Low Pass filter changes the system into one of delayed type and may re-stabilize, the paid price being another dimension augmentation.

The method of this chapter makes a difference in the sense that a specific control is used – the piece-wise constant control. In this way a discrete-time system is associated and it is *this* system that is stabilized; its augmented dynamics replaces the discretized integral term. Under these circumstances, the closed loop system (which is hybrid since it contains a continuous time controlled plant and a sampled-data compensator) is always stable provided the sampling stepδ is small enough (*op cit*).

The small sampling step is helpful in stabilization from another point of view also [5]. Let $\mathbf{F}(\delta)$ be the stabilizing feedback for the discretized system. Following the way from [5], of the asymptotic expansions, it is easily found that

$$\mathbf{F}(\delta) = F + F_1\delta + o(\delta)$$

where F is a stabilizing feedback for the continuous time system; one may use for implementation with piece-wise constant control F instead of $\mathbf{F}(\delta)$ and the stability is preserved provided δ is small enough.

References

1. Artstein Z (1982) Linear Systems with Delayed Controls: a Reduction. IEEE Trans Aut Control 27:869–879
2. Drăgan V, Halanay A (1999) Stabilization of Linear Systems. Birkhäuser Verlag, Boston
3. Fiagbedzi Y A, Pearson A E (1986) Feedback Stabilization of autonomous Time-lag Systems. IEEE Trans Aut Control 31:847–855
4. Fiagbedzi Y A, Pearson A E (1987) A Multistage Reduction Technique for Feedback Stabilizing Distributed Time-lag Systems. Automatica 23:311–326
5. Halanay A, Răsvan Vl (1977) General Theory of Linear Hybrid Control. Int J Control 20:621–634
6. Hale J K, Verduyn Lunel S M (1993) Introduction to Functional Differential Equations. Springer, Berlin Heidelberg New York
7. Ichikawa A (1982) Quadratic control of evolution equations with delays in control. SIAM J. Control 27:12–22
8. Kwon W H, Pearson A E (1980) Feedback stabilization of linear systems with delayed control. IEEE Trans Aut Control 25:266–269
9. Manitius A, Olbrot A W (1979) Finite spectrum assignment for systems with delays. IEEE Trans Aut Control 24:541–553
10. Mirkin L, Tadmor G (2002) H_∞ control of systems with I/O delay: a review of some problem–oriented methods. IMA Journ Math Contr Inf 19:185–199
11. Olbrot A W (1978) Stabilizability, detectability and spectrum assignment for linear systems with general time delays. IEEE Trans Aut Control 23:887–890
12. Pandolfi L (1981) A state space approach to control systems with delayed controls. In: Kisielewicz M (ed) Proceedings of Conference on Functional Differential Equations and Related Topics Zielona Gora, Poland
13. Pandolfi L (1989) Dynamic stabilization of systems with input delays. In: System Structure and Control: State Space and Polynomial Methods. Prague
14. Pandolfi L (1990) Generalized control systems, boundary control systems and delayed control systems. Math Contr Sign Syst 3:165–181
15. Pandolfi L (1991) Dynamic stabilization of systems with input delays. Automatica 27:1047–1050
16. Pearson A E (2003) Parallel System Modeling for Observer based Control of Input Delayed Plants. 4th Int Conference on Dynamic Syst Appl, Atlanta, 21–24 May, 2003
17. Popov V M (1960) Stability criteria for systems containing non-univocal elements (in Romanian). In: Probleme de Automatizare III:143–151, Ed.Academiei, Bucharest
18. Răsvan Vl (1998) Dynamical systems with lossless propagation and neutral functional differential equations. In: Mathem Theory of Networks and Systems MTNS98. Il Poligrafo, Padova
19. Răsvan Vl, Popescu D (2001) Feedback Stabilization of Systems with Delays in Control. Control Engineering and Applied Informatics 3:62-66
20. Răsvan Vl, Popescu D (2001) Control of systems with input delay by piecewise constant signals. 9th Medit. Conf. on Control and Automation, Paper WM1-B/122, Dubrovnik, Croatia

21. Tadmor G (1995) The Nehari problem in systems with distributed input delays is inherently finite dimensional. Syst and Contr Let 26:11-16
22. Tadmor G (1998) Robust control of systems with a single input lag In: Stability and Control of Time Lag Systems.LNCIS–228. Springer , Berlin New York London
23. Watanabe K, Ito M,(1981) An observer for linear feedback control laws of multivariable systems with multiple delays in controls and output. Syst and Contr Let 1:54–59

On the Stabilization of Systems with Bounded and Delayed Input

Frédéric Mazenc[1], Sabine Mondié[2], and Silviu-Iulian Niculescu[3]

[1] INRIA Lorraine, Projet CONGE, ISGMP Bât A, Ile du Saulcy, 57045 Metz Cedex 01, France. mazenc@loria.fr
[2] Depto. de Control Automático, CINVESTAV-IPN. A.P. 14-740. 07360, México D.F. smondie@ctrl.cinvestav.mx
[3] Heudiasyc, UTC, UMR CNRS 6599, BP 20529, 60205 Compiègne, France. silviu@hds.utc.fr

Summary. The problem of globally asymptotically stabilizing by bounded feedback an oscillator with an arbitrary large delay in the input is solved. A first solution follows from a general result on the global stabilization of null controllable linear systems with delay in the input by bounded control laws with a distributed term. Next, it is shown through a Lyapunov analysis that the stabilization can be achieved as well when neglecting the distributed terms. It turns out that this main result is intimately related to the output feedback stabilization problem.

1 Introduction

The family of the linear systems described by

$$\dot{x}(t) = Ax(t) + Bu(t - \tau) \tag{1}$$

where $A \in R^{n \times n}, B \in R^{n \times m}$, τ is a delay, u is the input and the initial condition is

$$x(\theta) = \phi(\theta), \theta \in [-\tau, 0]$$

is one of the simplest class of models with delay. Nevertheless, they describe properly a number of phenomena commonly present in controlled processes such as transport of information or products, lengthy computations, information processing, delays inherent to sensors, etc... Well known approaches for the control of these systems include in an explicit or implicit manner a predictor of the state at time $t + \tau$. Some of the more widely used are the Smith predictor [34], [13], Process-Model Control schemes [38], and finite spectrum assignment techniques [21], [4]. A common drawback, linked to the internal instability of the prediction, is that they fail to stabilize unstable systems [10]. As shown in [11], it is possible to overcome this problem by introducing a periodic resetting of the predictor.

A common concern in practical problems is the use of bounded control laws. It is well known that for linear systems with poles in the open right half plane, only locally stabilizing control laws can be obtained if there is a bound on the input, and that global asymptotic stability can be achieved only for systems with poles in the closed left half plane, named null-controllable systems. As shown in [18], the problem can be decomposed into two fundamental subproblems, namely, the control of chains of integrators of arbitrary finite length, and that of oscillators of arbitrary finite multiplicity. Solutions to this problem in the framework of systems with no delay are given in [19], [18], [7].

From the above, it follows that the study of linear systems whose inputs are both delayed and bounded is useful in many applications.

In this chapter, we focus our attention on the stabilization of a simple oscillator with arbitrarily large delay in the input described by

$$\begin{cases} \dot{x}_1(t) = x_2(t), \\ \dot{x}_2(t) = -x_1(t) + u(t-\tau). \end{cases} \tag{2}$$

We restrict our analysis to the case where the frequency is one for the sake of simplicity, but all the results that are presented can be adapted easily to the system

$$\begin{cases} \dot{x}_1(t) = \alpha x_2(t), \\ \dot{x}_2(t) = -\beta x_1(t) + u(t-\tau), \end{cases} \tag{3}$$

where α and β are strictly positive or strictly negative real numbers because this system can be transformed into the system (2) by a time rescaling and a change of coordinates.

To the best authors' knowledge, the issue of the stabilization of the oscillator was first discussed in [9] in the 1940s for a second-order (delayed) friction equation. Further comments and remarks on delayed oscillatory systems can also be found in [2,3].

The nature of oscillators is significantly different from that of chains of integrators: for example, when the input is zero, the solutions of oscillators are trigonometric functions of the time while those of integrators are polynomial functions of the time. Not surprisingly, the technique of proof we use is significantly different from the approach of [8] for chains of integrators with bounded delayed inputs based on the use of saturated control laws introduced in [19] that only require the knowledge of an upper bound on the delay. In the case of oscillators, we take advantage of the properties of the explicit solutions to zero inputs to determine expressions of the control laws which depend explicitly on the exact value of the delay. The proof of the result relies on the celebrated Lyapunov-Razumikhin theorem. The key feature of our control design are that for arbitrarily large delays it allows us to determine a family of globally asymptotically stabilizing state feedbacks which contains elements arbitrarily small in norm. The reinterpretation of this result in the context of output feedback stabilization leads to the following result : for *any* linear output one can find an arbitrarily large delay for which the problem of globally asymptotically stabilizing the oscillator by bounded feedback is solvable.

The chapter is organized as follows. In Section 2 a general result on the stabilization of null-controllable systems is established via distributed control laws. The procedure is applied to the oscillator. Next, a stabilizing bounded state feedback is proposed. In Section 3 this last result is embedded in the output feedback context. Some concluding remarks are presented in Section 4.

Preliminaries:

- For a real valued C^1 function $k(\cdot)$, we denote by $k'(\cdot)$ its first derivative.
- A function $\alpha : [0, +\infty) \longrightarrow [0, +\infty)$ is said to be of class $\mathcal{K}_\infty$ if it is zero at zero, strictly increasing and unbounded.
- We denote by $\sigma(\cdot)$ an odd nondecreasing smooth saturation such that
  - $\sigma(s) = s$ when $s \in [0, 1]$,
  - $\sigma(s) = \frac{3}{2}$ when $s \geq 2$,
  - $-s \leq \sigma(s) \leq s, 0 \leq \sigma'(s) \leq 1$.

2 Stabilization by Bounded Delayed Control Laws

We establish first a general result on the stabilization of null-controllable linear systems with delayed bounded inputs. by control law that contain distributed elements. Next we particularize this result to the cases of the oscillator. Finally, we exploit this last result in order to determine state feedbacks which do not contain distributed elements.

2.1 Distributed control laws for null-controllable systems

Lemma 1. *Consider a spectrally controllable linear multivariable system with delay in the input described by*

$$\dot{x}(t) = Ax(t) + Bu(t - \tau) \tag{4}$$

where $A \in R^{n \times n}, B \in R^{n \times m}$ is such that (A, B) is null-controllable, and the delays in the input is $\tau \geq 0$, with initial conditions

$$x(\theta) = \phi(\theta), \theta \in [-\tau, 0]$$

where $\phi(\cdot)$ is a continuous function. Then the following problems are equivalent.
(i) The bounded control law

$$u(t) = \xi(x) \tag{5}$$

globally asymptotically stabilizes the system (1) when $\tau = 0$.
(ii) The bounded distributed control law

$$u(t - \tau) = \xi(e^{A\tau}x(t - \tau) + \int_{t-\tau}^{t} e^{(t-s)A}Bu(s - \tau)ds) \tag{6}$$

globally asymptotically stabilizes the system (1).

Proof. The result follows in a straightforward manner from the fact that

$$x(t) = e^{A\tau}x(t - \tau) + e^{At}\int_{t-\tau}^{t} e^{-As}Bu(s - \tau)ds.$$

2.2 Distributed control laws for the oscillator

As a preliminary step for the main part of the work, we particularize this general result to the case of an oscillator. Recall first a very simple result whose proof is omitted.

Lemma 2. *Consider the system*

$$\begin{cases} \dot{x}_1(t) = x_2(t), \\ \dot{x}_2(t) = -x_1(t) + u, \end{cases} \tag{7}$$

where u is the input. This system is globally asymptotically stabilized by the bounded control law

$$u(x_1(t), x_2(t)) = -\varepsilon\sigma(x_2(t)) \tag{8}$$

where ε is a strictly positive parameters and $\sigma(\cdot)$ is the bounded function defined in the preliminaries.

Using the Cauchy formula for the oscillator

$$x_1(t) = \cos(\tau)x_1(t-\tau) + \sin(\tau)x_2(t-\tau) - \int_{t-\tau}^{t} \sin(s-t)u(s-\tau)ds, \tag{9}$$

$$x_2(t) = -\sin(\tau)x_1(t-\tau) + \cos(\tau)x_2(t-\tau) \tag{10}$$

$$-\int_{t-\tau}^{t} \cos(s-t)u(s-\tau)ds, \tag{11}$$

and Lemma 1, it follows straightforwardly that the distributed delayed control law

$$\begin{aligned} u(t-\tau) = -\varepsilon\sigma(&-\sin(\tau)x_1(t-\tau) + \cos(\tau)x_2(t-\tau) \\ &- \textstyle\int_{t-\tau}^{t} \cos(s-t)u(s-\tau)ds) \end{aligned} \tag{12}$$

globally asymptotically stabilizes the system (2).

2.3 Delayed state feedbacks for the oscillator

The control law (6) involves distributed elements. The implantation of such terms using numerical approximation is time consuming and, in some cases, it leads to instability [10]. One can observe that in (11) the distributed term is of the order of ε whereas the other term does not depend on ε: so we conjecture that as ϵ decreases, the influence of the distributed term decreases as well. Then, the question that arises naturally, is wether or not stability is preserved if the integral terms are neglected, in other words, if instead of (6), the control law

$$u(t) = \xi(e^{A\tau}x(t))$$

is used. The answer to this query depends on various aspects of the problem, as the dynamic of the open loop system, and the bounded control law itself. These facts are now illustrated with the oscillator considered in this chapter.

We state now the main result of the work.

Theorem 1. *Let τ be a positive number. The origin of the system (2) is globally asymptotically stabilized by the control law*

$$u(x_1, x_2) = -\varepsilon\sigma(-\sin(\tau)x_1 + \cos(\tau)x_2) \tag{13}$$

where

$$\varepsilon \in \left]0, \min\left\{\frac{1}{2}, \frac{1}{324\tau^2}, \frac{1}{40\tau}\right\}\right] \tag{14}$$

and $\sigma(\cdot)$ is the bounded function defined in the preliminaries.

Proof. The system (2) in closed loop with the feedback (13) rewrites

$$\begin{cases} \dot{x}_1(t) = x_2(t), \\ \dot{x}_2(t) = -x_1(t) - \varepsilon\sigma\left(x_2(t) + B_2(t)\right), \end{cases} \tag{15}$$

with

$$\begin{aligned} B_2(t) &= -\int_{t-\tau}^{t} \cos(s-t)u_s(x_1(s-\tau), x_2(s-\tau))ds \\ &= \varepsilon \int_{t-\tau}^{t} \cos(s-t)\sigma(-\sin(\tau)x_1(s-\tau) + \cos(\tau)x_2(s-\tau))ds. \end{aligned} \tag{16}$$

One can check readily that

$$\begin{cases} \dot{x}_1(t) = x_2(t), \\ \dot{x}_2(t) = -x_1(t) - \varepsilon\sigma(x_2(t)) + \beta(t), \end{cases} \tag{17}$$

with

$$\beta(t) \;=\; \varepsilon B_2(t) \int_0^1 \sigma'\left(x_2(t) + lB_2(t)\right) dl. \tag{18}$$

The result is established through a Lyapunov-Razumikhin approach. Consider the function

$$U(x_1, x_2) \;=\; \lambda(x_1^2 + x_2^2) + x_1 x_2 \tag{19}$$

where $\lambda(\cdot)$ is a function of class $\mathcal{K}_\infty$ such that $\lambda(s) \geq s,\ s \geq 0$ to be specified later. With such a choice for $\lambda(\cdot)$, $U(x_1, x_2)$ is a positive definite and radially unbounded function.

Its time derivative along the trajectories of system (17) satisfies

$$\begin{aligned} \dot{U}(x_1, x_2) &= -2\varepsilon\lambda'(x_1^2 + x_2^2)x_2\sigma(x_2) + 2\lambda'(x_1^2 + x_2^2)x_2\beta(t) + x_2^2 - x_1^2 \\ &\quad -\varepsilon x_1\sigma(x_2) + x_1\beta(t) \\ &\leq -\varepsilon\lambda'(x_1^2 + x_2^2)x_2\sigma(x_2) + \tfrac{2\lambda'(x_1^2+x_2^2)x_2}{\varepsilon\sigma(x_2)}\beta(t)^2 + x_2^2 - \tfrac{1}{2}x_1^2 \\ &\quad +\varepsilon^2\sigma(x_2)^2 + \beta(t)^2. \end{aligned} \tag{20}$$

According to the properties of $\sigma(\cdot)$ and the fact that $\varepsilon \leq \frac{1}{2}$, it follows that

$$\dot{U}(x_1,x_2) \leq -\varepsilon\lambda'(x_1^2+x_2^2)x_2\sigma(x_2) + \left[\tfrac{2\lambda'(x_1^2+x_2^2)x_2}{\varepsilon\sigma(x_2)}+1\right]\beta(t)^2 + \tfrac{5}{4}x_2^2 - \tfrac{1}{2}x_1^2. \tag{21}$$

Let now choose

$$\lambda(s) = 2k\int_0^s \sqrt{\frac{l}{\sigma(l)}}dl \tag{22}$$

where $k \geq \frac{1}{2}$. The properties satisfied by $\sigma(\cdot)$ ensure that this function is well defined, of class $\mathcal{K}_\infty$, and such that $\lambda''(s) \geq 0$. Then,

$$\begin{aligned}\dot{U}(x_1,x_2) &\leq -\varepsilon k\sqrt{\tfrac{x_1^2+x_2^2}{\sigma(x_1^2+x_2^2)}}x_2\sigma(x_2) - \tfrac{\varepsilon k}{\sqrt{\sigma(x_2^2)}}x_2^2|\sigma(x_2)| \\ &\quad + \left[\frac{4k\sqrt{\frac{x_1^2+x_2^2}{\sigma(x_1^2+x_2^2)}}x_2}{\varepsilon\sigma(x_2)}+1\right]\beta(t)^2 + \tfrac{5}{4}x_2^2 - \tfrac{1}{2}x_1^2 \\ &\leq -\tfrac{1}{2}x_1^2 - \varepsilon k\sqrt{\tfrac{x_1^2+x_2^2}{\sigma(x_1^2+x_2^2)}}x_2\sigma(x_2) + \left[-\tfrac{2\varepsilon k}{3}+\tfrac{5}{4}\right]x_2^2 \\ &\quad + \left[\tfrac{4k}{\varepsilon\sigma(x_2)}\sqrt{\tfrac{x_1^2+x_2^2}{\sigma(x_1^2+x_2^2)}}x_2+1\right]\beta(t)^2.\end{aligned} \tag{23}$$

Choosing $k = \frac{4}{\varepsilon}$ (which, according to (14), is larger than $\frac{1}{2}$),

$$\begin{aligned}\dot{U}(x_1,x_2) &\leq -\tfrac{1}{2}x_1^2 - 4\sqrt{\tfrac{x_1^2+x_2^2}{\sigma(x_1^2+x_2^2)}}x_2\sigma(x_2) - \tfrac{4}{3}x_2^2 \\ &\quad + \left[\tfrac{16}{\varepsilon^2\sigma(x_2)}\sqrt{\tfrac{x_1^2+x_2^2}{\sigma(x_1^2+x_2^2)}}x_2+1\right]\beta(t)^2.\end{aligned} \tag{24}$$

On the other hand, the use of the inequalities $0 \leq \sigma'(s) \leq 1$ and (16) leads to

$$\begin{aligned}\beta(t)^2 &= \varepsilon^2B_2(t)^2\left[\int_0^1 \sigma'\left(x_2(t)+lB_2(t)\right)dl\right]^2 \\ &\leq \varepsilon^4\left[\int_{t-\tau}^t |\sigma(-\sin(\tau)x_1(s-\tau)+\cos(\tau)x_2(s-\tau))|ds\right]^2.\end{aligned} \tag{25}$$

We distinguish now between two cases.
<u>First case:</u> $x_1^2+x_2^2 \geq \frac{1}{4}$. Inequalities (24) and (25) imply that

$$\begin{aligned}\dot{U}(x_1,x_2) &\leq -\tfrac{1}{2}x_1^2 - 4\sqrt{\tfrac{x_1^2+x_2^2}{\sigma(x_1^2+x_2^2)}}x_2\sigma(x_2) - \tfrac{3}{4}x_2^2 \\ &\quad + \left(1+\tfrac{\varepsilon^2}{8}\right)\tfrac{16\varepsilon^2\tau^2}{\sigma(x_2)}\sqrt{\tfrac{x_1^2+x_2^2}{\sigma(x_1^2+x_2^2)}}x_2.\end{aligned} \tag{26}$$

One can check readily that $\frac{s}{\sigma(s)} \leq 1+s$ for all $s \geq 0$. It follows that

$$\begin{aligned}\dot{U}(x_1,x_2) &\leq -\tfrac{1}{2}x_1^2 - 4\sqrt{\tfrac{x_1^2+x_2^2}{\sigma(x_1^2+x_2^2)}}x_2\sigma(x_2) - \tfrac{3}{4}x_2^2 \\ &\quad + \left(1+\tfrac{\varepsilon^2}{8}\right)16\varepsilon^2\tau^2(1+|x_2|)\sqrt{1+x_1^2+x_2^2} \\ &\leq -\tfrac{1}{2}x_1^2 - 4\sqrt{\tfrac{x_1^2+x_2^2}{\sigma(x_1^2+x_2^2)}}x_2\sigma(x_2) - \tfrac{3}{4}x_2^2 \\ &\quad + \left(1+\tfrac{\varepsilon^2}{8}\right)16\varepsilon^2\tau^2 10(x_1^2+x_2^2).\end{aligned} \tag{27}$$

When $\varepsilon \leq \frac{1}{40\tau}$, the inequality

$$\left(1+\frac{\varepsilon^2}{8}\right) 16\varepsilon^2\tau^2 10 \leq \frac{1}{4} \tag{28}$$

holds, and

$$\dot{U}(x_1,x_2) \leq -\tfrac{1}{4}x_1^2 - 4\sqrt{\tfrac{x_1^2+x_2^2}{\sigma(x_1^2+x_2^2)}}\,x_2\sigma(x_2) - \tfrac{1}{4}x_2^2. \tag{29}$$

Second case: $x_1^2+x_2^2 \leq \frac{1}{4}$. In this case $|x_1| \leq \frac{1}{2}, |x_2| \leq \frac{1}{2}, \sigma(x_2) = x_2, \sigma(x_1^2+x_2^2) = x_1^2+x_2^2$. Then it follows from (24) that

$$\dot{U}(x_1,x_2) \leq -\tfrac{1}{2}x_1^2 - \tfrac{4}{3}x_2^2 + \left[\tfrac{16}{\varepsilon^2}+1\right]\beta(t)^2 \tag{30}$$

and

$$\begin{aligned}\beta(t)^2 &= \varepsilon^2 B_2(t)^2\left[\int_0^1 \sigma'\left(x_2(t)+lB_2(t)\right)dl\right]^2 \\ &\leq \varepsilon^4\left[\int_{t-\tau}^{t}\left[|x_1(s-\tau)|+|x_2(s-\tau)|\right]ds\right]^2.\end{aligned} \tag{31}$$

Combining (30) and (31) leads to

$$\dot{U}(x_1,x_2) \leq -\frac{1}{2}x_1^2 - \frac{4}{3}x_2^2 \tag{32}$$

$$+\left[16\varepsilon^2+\varepsilon^4\right]\left[\int_{t-\tau}^{t}\left[|x_1(s-\tau)|+|x_2(s-\tau)|\right]ds\right]^2 \tag{33}$$

$$\leq -\frac{1}{2}x_1^2 - \frac{4}{3}x_2^2 + 17\varepsilon^2 2\tau^2\left[\sup_{s\in[t-\tau,t]}\left[x_1(s-\tau)^2+x_2(s-\tau)^2\right]\right] \tag{34}$$

Next, we determine the values of ε for which the feedback (13) globally asymptotically stabilizes the system (2) with the help of Razumikhin Theorem (see Appendix A). To do so we prove by exploiting (29) and (34) that when the inequality

$$U(x_1(t+\theta),x_2(t+\theta)) < 2U(x_1(t),x_2(t)) \tag{35}$$

holds for all $\theta \in [-2\tau,0]$, there exists a continuous, nondecreasing functions $w(.)$ such that

$$\dot{U}(x_1,x_2) \leq -w(\|x\|)$$

with $x=(x_1,x_2)^T$. Recall that $U(\cdot)$ is positive definite and radially unbounded.
1. When $x_1(t)^2+x_2(t)^2 \geq \frac{1}{4}$, it follows from (29) that $\dot{U}(x_1(t),x_2(t)) \leq -\frac{1}{4}\|x(t)\|^2$.
2. When $x_1(t)^2+x_2(t)^2 \leq \frac{1}{4}$, the saturations act in their linear regions hence,

$$U(x_1(t),x_2(t)) = \frac{8}{\varepsilon}(x_1(t)^2+x_2(t)^2)+x_1(t)x_2(t). \tag{36}$$

One can check readily that

$$U(x_1(t), x_2(t)) \leq \frac{9}{\varepsilon}(x_1(t)^2 + x_2(t)^2). \tag{37}$$

On the other hand,

$$U(x_1, x_2) = \frac{8}{\varepsilon}\int_0^{{x_1}^2+{x_2}^2} \sqrt{\frac{l}{\sigma(l)}}dl + x_1x_2.$$

Since $\sigma(l) \leq l,\ l \geq 0$, it follows that

$$U(x_1, x_2) \geq \frac{8}{\varepsilon}({x_1}^2 + {x_2}^2) + x_1x_2 \geq (\frac{8}{\varepsilon} - \frac{1}{2})({x_1}^2 + {x_2}^2).$$

In particular, this inequality implies that

$$14(x_1(t+\theta)^2 + x_2(t+\theta)^2) \leq U(x_1(t+\theta), x_2(t+\theta)). \tag{38}$$

Combining (37), (35) and (38),

$$14(x_1(t+\theta)^2 + x_2(t+\theta)^2) \ \leq \ \frac{18}{\varepsilon}(x_1(t)^2 + x_2(t)^2). \tag{39}$$

It follows from (39) and (34) that

$$\begin{aligned} \dot{U}(x_1, x_2) &\leq -\tfrac{1}{2}x_1^2 - \tfrac{4}{3}x_2^2 + 34\varepsilon^2\tau^2\tfrac{9}{7\varepsilon}[x_1(t)^2 + x_2(t)^2] \\ &\leq -\tfrac{1}{2}x_1^2 - \tfrac{4}{3}x_2^2 + 162\varepsilon\tau^2[x_1(t)^2 + x_2(t)^2]. \end{aligned} \tag{40}$$

Choosing $\varepsilon \in]0, \min\left\{\frac{1}{2}, \frac{1}{40\tau}, \frac{1}{648\tau^2}\right\}]$, implies that

$$\dot{U}(x_1, x_2) \leq -\tfrac{1}{4}\left\|x\right\|^2. \tag{41}$$

This concludes the proof.

3 Output Feedback Stabilization

The above analysis shows that when the delay is τ, feedbacks depending only on

$$-\sin(\tau)x_1(t-\tau) + \cos(\tau)x_2(t-\tau)$$

stabilize the oscillator. This leads straightforwardly to the following interpretation of Theorem 1 in an output feedback framework.

Corollary 1. *Consider the system*

$$\begin{cases} \dot{x}_1(t) = x_2(t), \\ \dot{x}_2(t) = -x_1(t) + u(t), \end{cases} \tag{42}$$

with the output

$$y = ax_1 + bx_2 \tag{43}$$

where a and b are arbitrary real numbers such that $a^2 + b^2 > 0$. *Then the system is globally asymptotically stabilized by a delayed feedback*

$$u = -\varepsilon\sigma\left(\frac{y(t-\tau)}{\sqrt{a^2+b^2}}\right) \tag{44}$$

where $\sigma(.)$ *is the function defined in Theorem 1, and where* τ *is a strictly positive number such that*

$$\begin{cases} a = -\sqrt{a^2+b^2}\sin(\tau), \\ b = \sqrt{a^2+b^2}\cos(\tau), \end{cases}$$

and $\varepsilon \in]0, \min\left\{\frac{1}{2}, \frac{1}{40\tau}, \frac{1}{648\tau^2}\right\}]$.

An interesting and somehow surprising particular case is when the output is x_1. In this case, although the system is not stabilizable when the delay is zero, it is for suitably chosen delays, as shown in Corollary 1.

An analysis of the problem of output feedback stabilization by linear feedback, based on the study of the corresponding characteristic equation [12], or on the Nyquist criterion [1] can be carried out. It leads to the following result that provides additional informations on our problem.

Proposition 1. *[12], [1]The linear system*

$$\begin{cases} \dot{x}_1(t) = x_2(t), \\ \dot{x}_2(t) = -x_1(t) + u(t), \end{cases} \tag{45}$$

with the output

$$y(t) = x_1(t), \tag{46}$$

can be stabilized by delayed output feedback

$$u(t) = -ky(t-\tau) \tag{47}$$

for all the pairs (k, τ) *satisfying simultaneously:*

i) the gain $k \in (0, 1)$
ii) the delay $\tau \in (\underline{\tau}_i(k), \overline{\tau}_i(k))$ *where:*

$$\begin{cases} \underline{\tau}_i(k) = \dfrac{(2i-1)\pi}{\sqrt{1-k}} \\ \overline{\tau}_i(k) = \dfrac{2i\pi}{\sqrt{1+k}} \end{cases} \tag{48}$$

for $i = 1, 2, \ldots$.

Furthermore, if $\tau = \underline{\tau}_i(k)$ *or* $\tau = \overline{\tau}_i(k)$, *the corresponding characteristic equation in closed-loop has at least one eigenvalue on the imaginary axis.*

The regions of stabilizing k shrink as the delay τ gets larger, and furthermore for each k there exists a value $\tau^*(k)$, such that for any $\tau > \tau^*(k)$ the closed-loop system is unstable.

Remark 2 *Observe that in the case when $y = x_1$, the values of a and b in equation (2) in Corollary 1 are respectively 1 and 0. This implies that $\tau_i = -\frac{\pi}{2} + 2i\pi$, where i is an integer. Not surprisingly, these values are such that*

$$\tau_i = -\frac{\pi}{2} + 2i\pi \in]\underline{\tau}_i(k), \overline{\tau}_i(k)[$$

when k is sufficiently small: indeed, the feedback (44) is linear in a neighborhood of the origin which implies that the system (1) in closed-loop with $-\varepsilon\frac{y(t-\tau)}{\sqrt{a^2+b^2}}$ is globally asymptotically stable when $\tau = \tau_i$.

Remark 3 *This observation gives a new way to establish that the delayed output feedback (47) satisfying i) and ii) are stabilizing. Indeed, boundaries of the regions of the plane k-τ described by i) and ii) correspond to crossing of the imaginary axis of the roots of the closed loop quasipolynomial. It follows from a continuity argument that in each region, the quasipolynomial has a fixed number of roots in the right-half plane, hence it is either stable or unstable in the whole region. A consequence of Remark 2 is that we are able to exhibit a stabilizing element (k_i, τ_i) in each of the regions described by i) and ii), hence all the control laws all the pairs (k, τ) satisfying simultaneously i) and ii) are stabilizing as well.*

4 Concluding Remarks

The problem of globally asymptotically stabilizing by bounded feedback an oscillator with an arbitrary large delay in the input is solved. A first solution follows from a general result on the global stabilization of null-controllable linear systems with delay in the input by bounded control laws with a distributed term. Next, it is shown through a Lyapunov analysis that the stabilization can be achieved as well when neglecting the distributed terms. It turns out that this main result is intimately related to the output feedback stabilization problem. The robustness problems due to the need for the exact knowledge of the delay can be investigated with the help of the Lyapunov function we have constructed.

References

1. Abdallah C., Dorato P., Benitez-Read and Byrne R.: *Delayed positive feedback can stabilize oscillatory systems.* Proc. American Contr. Conf. (1993) 3106-3107.
2. Ansoff H.I.: *Stability of linear oscillating systems with constant time lag.* J. Appl. Mech. **16** (1949) 158-164.
3. Ansoff H.I. and Krumhansl J. A.: *A general stability criterion for linear oscillating systems with constant time lag.* Quart. Appl. Math. **6** (1948) 337-341.
4. Artstein S., *Linear systems with delayed controls: a reduction.* IEEE Trans. Autom. Contr., Vol. AC-27, No. 4, 869-879, 1982.
5. Hale J.K. and Verduyn Lunel S.M., *Introduction to Functional Differential Equations.* Applied Math, Sciences, 99, Springer-Verlag, New-York, 1993.

6. Manitius A.Z. and Olbrot A.W., *Finite Spectrum Assignment problem for Systems with Delays.* IEEE Trans. Autom. Contr., Vol. AC-24, No. 4, 541-553, 1979.
7. Mazenc F. and Praly L., *Adding an Integration and Global Asymptotic Stabilization of Feedforward Systems,* IEEE Trans. Aut. Contr., Vol. 41, no.11, pp.1559-1578, 1996.
8. Mazenc F., Mondié S. and Niculescu S., *global asymptotic stabilization for chains of integrators with a delay in the input.* Proc. 40th IEEE Conf. Dec. Contr., Orlando, Florida, 2001, and submitted to IEEE Trans. Autom. Contr., 2001.
9. Minorsky N.: *Self-excited oscillations in dynamical systems possessing retarded actions.* J. Applied Mech. **9** (1942) A65-A71.
10. Mondié S., Dambrine M. and Santos O., *Approximation of control laws with distributed delays: a necessary condition for stability.* IFAC Symposium on Systems, Structure and Control, Prague, Czek Republic, August 2001.
11. Mondié S., Lozano R. and Collado, J., *Resetting Process-Model Control for unstable systems with delay.* Submitted to IEEE Trans. Autom. Contr., 2001.
12. Mufti I.H.: *A note on the stability of an equation of third order with time lag.* IEEE Trans. Automat. Contr. AC-9 (1964) 190-191.
13. Neimark J. (1949). *$\mathcal{D}$-subdivisions and spaces of quasipolynomials. Prickl. Math. Mech.*, 13, 349-380.
14. Niculescu S.-I.: *Delay effects on stability. A robust control approach.* Springer-Verlag: Heidelberg, LNCIS, May 2001.
15. Niculescu S.-I. and Abdallah C.T.: *Delay effects on static output feedback stabilization.* Proc. 39th IEEE Conf. Dec. Contr., Sydney, Australia (December 2000).
16. Palmor Z.J., *Time delay Compensation- Smith predictor and its modifications*, in The Control Handbook, (W.S. Levine, Eds), CRSC Press, 224-237, 1996.
17. Smith O.J.M., *Closer Control of loops with dead time*, Chem. Eng. Prog.,53, 217-219, 1959.
18. Sussmann H.J., Sontag E.D. and Yang Y., *A general result on the stabilization of linear systems using bounded controls.* IEEE Trans. Autom. Contr., Vol. AC-39, 2411-2425, 1994.
19. Teel A., *Global stabilization and restricted tracking for multiple integrators with bounded controls.* Systems and Control Letters, 18, 165-171, 1992.
20. Watanabe K. and Ito M., *A process model control for linear systems with delay.* IEEE Trans. Autom. Contr., Vol. AC-26, No. 6, 1261-1268, 1981.

A Razumikhin's Theorem

Theorem 4. *[8] Consider the functional differential equation*

$$\dot{x}(t) = f(t, x_t), t \geq 0, \tag{49}$$

$$x_{t_0}(\theta) = \phi(\theta), \forall\theta \in [-\tau, 0] \tag{50}$$

(Note: $x_t(\theta) = x(t+\theta), \forall\theta \in [-\tau, 0]$). The function f:$R \times C_{n,\tau}$ is such that the image by f of $R\times$(a bounded subset of $C_{n,\tau}$) is a bounded subset of R^n and the functions u, v, w:$R^+ \longrightarrow R^+$ are continuous, nondecreasing, $u(s), v(s)$ positive for all $s > 0, u(0) = v(0) = 0$ and v is strictly increasing.
If there exists a function V: $R \times R^n \longrightarrow R$ such that
(a) $u(\|x\|) \leq V(t,x) \leq v(\|x\|), t \in R, x \in R^n$

(b) $\dot{V}(t,x) \leq -w(\|x\|)$ *if* $V(t+\theta, x(t+\theta)) \leq V(t, x(t)), \forall\theta \in [-\tau, 0]$
then the trivial solution of (49, 50) is uniformly stable.

Moreover, if $w(s) > 0$ *when* $s > 0$*, and there exists a function* $p : R^+ \to R^+$,
$p(s) > s$ *when* $s > 0$ *such that:*
(a) $u(\|x\|) \leq V(t,x) \leq v(\|x\|), t \in R, x \in R^n$
(b) $\dot{V}(t,x) \leq -w(\|x\|)$ *if*

$$V(t+\theta, x(t+\theta)) < p(V(t, x(t))), \forall\theta \in [-\tau, 0] \tag{51}$$

then the trivial solution of (49, 50) is uniformly asymptotically stable.

Identifiability and Identification of Linear Systems with Delays

Lotfi Belkoura[1], Michel Dambrine[2], Yuri Orlov[3], and Jean-Pierre Richard[2]

[1] LAIL, Universite des Sciences et Technologies de Lille, France
lotfi.belkoura@univ-lille1.fr,
[2] LAIL, Ecole Centrale de Lille, France
Michel.Dambrine@ec-lille.fr, Jean-Pierre.Richard@ec-lille.fr
[3] CICESE Research Center, Electronics and Telecom Dpt., San Diego, USA
yorlov@cicese.mx

Summary. Parameter identifiability and identification are studied for linear differential delay equations of neutral type and with distributed delays. It is shown how the identifiability property can be formulated in terms of controllability conditions, namely approximate controllability for the general case, and weak controllability for the retarded case with finitely many lumped delays in the state vector and control input. The notion of sufficiently rich input, which enforces identifiability, is also addressed, and the results are obtained assuming knowledge of the solution on a bounded time interval. Once the parameter identifiability is guaranteed, synthesis of an adaptive parameter identifier is developed for systems with finitely many lumped delays in the state vector and control input. Theoretical results are supported by numerical simulations.

1 Introduction

Numerous researches involve time-delay systems and their applications to modelling and control of concrete systems. To name a few, the two monographs [20, 28] give examples in biology, chemistry, economics, mechanics, viscoelasticity, physics, physiology, population dynamics, as well as in engineering sciences. In addition, actuators, sensors, field networks and wireless communications that are involved in feedback loops usually introduce such delays. As it was noted in the recent survey [34], delays are strongly involved in challenging areas of communication and information technologies: stability of networked controlled systems, quality of service in MPEG video transmission or high-speed communication networks, teleoperated systems, parallel computation, computing times in robotics... Finally, besides actual delays, time lags are frequently used to simplify very high order models.

Works on identification of FDEs have shown the complexity of the question [37]. Identifying the delay is not an easy task for systems with both input and state delays, or when the delay is varying enough to require an adaptive identifier. Several authors use the relay-based approach initiated by Astrom and Hagglund [26,35], which, however, is not a real-time procedure since it needs to close some switching feedback loop during a preliminary identification phase. The adaptive control of delay systems is not so much developed either [4,15,38] and the delay is generally assumed to be known. And yet, as noted in [11], the on-line delay estimation has a longstanding issue in signal processing.

The present work will focus on the question of identifiability (section 2) and, then, on some algorithms for identification (section 3).

2 Identifiability Analysis

In this section, the question of identifiability of time invariant systems described by linear delay differential equations is addressed. We consider here the general case of equation of neutral type and with distributed delay of the form:

$$\begin{aligned} \frac{d}{dt}x(t) &= \sum_{i=0}^{N} A_i x(t-h_i) + \sum_{i=1}^{N} A_{-i}\dot{x}(t-h_i) \\ &+ \int_0^h A_c(\theta)x(t-\theta)d\theta \\ &+ \sum_{i=0}^{N} B_i u(t-h_i) + \int_0^h B_c(\theta)u(t-\theta)d\theta, \end{aligned} \tag{1}$$

where $0 \leq h_0 < h_1 < ... < h_N = h$. We shall not consider the problem of identifiability of the initial conditions, and the reader interested in this subject can refer to the work of [25] for homogeneous (without control) and retarded systems. Therefore, and without loss of generality, we shall consider zero initial state. According to [43], equation (1) admits an input/output representation in terms of convolution equation:

$$P * x = Q * u \tag{2}$$

where P and Q are respectively $n \times n$ and $n \times m$ matrices with entries in the space $\mathcal{E}'$ of distributions with compact support and are given by:

$$P = \delta' I + \sum_{i=0}^{N} A_i\, \delta_{h_i} + \sum_{i=1}^{N} A_{-i}\, \delta'_{h_i} + A_c(\theta) \tag{3}$$

$$Q = \sum_{i=0}^{N} B_i\, \delta_{h_i} + B_c(\theta). \tag{4}$$

Note that a non zero initial state would result in an additional term in the right hand side of (2), consisting in an element of $\mathcal{E}'$, the support of which is included in $[0, h]$. In association with (1), let us consider a reference model governed by:

$$\hat{P} * \hat{x} = \hat{Q} * u \tag{5}$$

where $\hat{P}$ and $\hat{Q}$ share the same expressions as (3,19) with coefficients $\hat{N}$, $\hat{h}_i$, $\hat{A}_i, \hat{A}_{-i}, \hat{A}_c(\theta), \hat{B}_i$ and $\hat{B}_c(\theta)$. We shall further assume $0 \leq \hat{h}_0 < \hat{h}_1 < ... < \hat{h}_{\hat{N}} = \hat{h}$, with $\hat{h} \leq h$. However, we shall present in section 2.3 some particular cases for which the latter assumption can be removed. The problem of identifiability can be stated as follows:

Definition 1. *The linear system described by (2) is called identifiable by (5) if one can find a input u and a bounded time interval $\mathcal{I}$ such that the equalities $\hat{P} = P$ and $\hat{Q} = Q$ result from $\hat{x} = x$ on $\mathcal{I}$.*

We first introduce some notations and results we shall use in the sequel: $\mathcal{D}'_+$ (resp. $\mathcal{E}', \mathcal{E}'(\mathbb{R}_-)$) is the space of distributions with support bounded on the left (resp. compact support, compact support contained in $(-\infty, 0]$). It is an algebra with respect to convolution with identity δ, the Dirac distribution. When T is a matrix valued distribution, $supp\, T$ is the union of the support of its entries. With no danger of confusion, if $T \in (\mathcal{E}')^{n\times m}$, we denote $T(s)$, $s \in \mathbb{C}$, the $n \times m$ matrix of entire functions obtained by Laplace transform of T. A distribution $u \in \mathcal{E}'$ is called invertible for $\mathcal{D}'$ (in the sense of Ehrenpreis) if the map $\mathcal{D}' \ni v \rightarrow u * v \in \mathcal{D}'$ is onto. We let $\mathcal{O}$ denote the space of all entire functions and $E' \subset \mathcal{O}$ the Laplace transform of $\mathcal{E}'$. Denote also (in this paragraph) $\tilde{u} \in E'$ the Laplace transform of $u \in \mathcal{E}'$. The next result can be found in [12] where the general problem of division in various spaces is studied.

Theorem 1. *[12] The following statements are equivalent:*

i) $u \in \mathcal{E}'$ is invertible.
ii) For any $g \in \mathcal{O}$, if $\tilde{u}\, g \in E'$, then $g \in E'$.

Another property of the convolution product we shall use is the well known relation $supp\, \alpha * \beta \subset supp\, \alpha + supp\, \beta$, for $\alpha, \beta \in \mathcal{D}'_+$, which easily extends to the matrix valued case. Moreover, for a square matrix $T \in (\mathcal{E}')^{n\times n}$ with $supp\, T \subset [0, h]$, and in the convolution sense, one can easily show that the determinant and the adjoint of T have their support contained in $[0, nh]$ and $[0, (n-1)h]$ respectively.

2.1 Time interval and sufficiently rich input

This paragraph is devoted to the design of an input u and to the characterization of $\mathcal{I}$ such that the equality of the solutions on $\mathcal{I}$ results in that of the impulse response, *i.e.*

$$\hat{x} = x \quad on \quad \mathcal{I} \Rightarrow P^{-1} * Q = \hat{P}^{-1} * \hat{Q}. \tag{6}$$

In order to restrict the identifiability study to a bounded time interval of observation, we assume the input to have a compact support contained in some segment $\mathcal{U}$. The "richness" of u will be defined within this time interval. The proof of the next propositions is given in the appendix.

Proposition 1. *If $\mathcal{I} := [r_1, r_2]$ and $\mathcal{U} := [t_1, t_2]$ are such that*

$$r_1 + (n+1)h < t_1 < t_2 < r_2 - (n+1)h \tag{7}$$

then $x = \hat{x}$ on $\mathcal{I}$ implies $x = \hat{x}$ on $\mathbb{R}$.

Now, define u as a piecewise $\mathbb{R}^m$-valued polynomial function, and let the set $\Lambda_u = \{s_0, s_1, ..., s_L\}$ be its singular support (*i.e.* the set of points in $\mathbb{R}$ having no open neighborhood $\mathcal{N}$ such that the restriction of u to $\mathcal{N}$ is a C^∞ function). Besides, let us denote:

$$\sigma_u^k(s_l) = u^{(k)}(s_l + 0) - u^{(k)}(s_l - 0), \tag{8}$$

$$U_l(D) = \sum_{i=0}^{N} \sigma_u^{N-i}(s_l)\, D^i, \quad D = d/dt. \tag{9}$$

One of the fundamental properties of distributions is that, by differentiation, we do not miss something essential such as a discontinuity. In our case, if all the polynomials defining u are of order $\leq N$ for some $N \in \mathbb{N}$, differentiation at the order $N+1$ results in the singular distribution

$$u^{(N+1)} = \sum_{l=0}^{L} U_l(D)\,[\delta] * \delta_{s_l}. \tag{10}$$

Using a Smith form factorization together with the invertibility of differential operators in $\mathcal{D}'_+$, we have the following statement

Proposition 2. *Under assumption (7), relation (6) holds with a piecewise polynomial u such that*

i) $rank\ [U_0(D), ..., U_L(D)] = m$,
ii) $s_l - s_{l-1} > (n+1)h, \quad l = 1, .., L$.

The simplest example consists in a piecewise constant $\mathbb{R}^m$-valued function with discontinuity points s_l as in condition (2) and jumps $\sigma_u^0(s_l)$ forming a matrix of full rank. More generally, we can design an input u of class C^r for an arbitrary finite integer r, but the main property required in this approach remains the non-smoothness of the input.

2.2 Identifiability of the plant

Using the kernel representation with $R := [P, -Q]$, the equality of the impulse response can be equivalently written as:

$$\hat{R} = \alpha * R, \tag{11}$$

$$\alpha = \hat{P} * P^{-1}. \tag{12}$$

Using the notion of order of a distribution (see [43]), one can easily show that α is a measure with support contained in $[0, \infty)$. On the other hand, it is clear from (11) that $\hat{R}$ is compact supported if α is, too, but the converse is not true in general.

Proposition 3. *If $rank\ R(s) = n$, $s \in \mathbb{C}$, then α has its entries in $\mathcal{E}'$.*

Proof. Since $\hat{R}$ and R are compact supported distributions, $\hat{R}(s)$ and $R(s)$ are with entries in $E' \subset \mathcal{O}$. Hence, taking the Laplace transform of (11), we deduce that, if $rank\ R(s) = n$, $s \in \mathbb{C}$, then $\alpha(s)$ must be in $\mathcal{O}^{n\times n}$. It remains to show that $\alpha(s) \in E'^{n\times n}$. Since $(\det P)(s).\alpha(s) = \hat{P}(s).Adj(P)(s) \in E'^{n\times n}$, and $\det P$ is invertible, theorem 1 applies componentwise and thus $\alpha(s) \in E'^{n\times n}$.

The condition $rank\ R(s) = n$, $s \in \mathbb{C}$, is usually referred to as spectral controllability. Clearly, this condition is not sufficient to guarantee the uniqueness of the impulse response representation in terms of the kernel R. Additional information will be obtained using the properties of the support of a convolution product; For $A \in (\mathcal{E}')^{n\times m}$, we define

$$E(A) := sup\{t \in supp\, A\}\,. \tag{13}$$

Note that by assumption, $\hat{h} = E(\hat{R}) \le h = E(R)$, and clearly $E(\delta_\tau * A) = \tau + E(A)$, $\tau \in \mathbb{R}$. Now we recall here some algebraic results one can find in [43]; Let $\mathcal{F}$ be the quotient field of the quotient ring $\mathcal{A} := \mathcal{E}'(\mathbb{R}_-)/J$, where

$$J = \{\varphi \in \mathcal{E}'(\mathbb{R}_-); E(\varphi) < 0\}\,, \tag{14}$$

and denote $\theta : \mathcal{E}'(\mathbb{R}_-) \to \mathcal{F}$ the composition of the canonical projection $\mathcal{E}'(\mathbb{R}_-) \to \mathcal{A}$ with the inclusion $\mathcal{A} \to \mathcal{F}$. In this setting, an element $w \in (\mathcal{E}'(\mathbb{R}_-))^n$ is nonzero when considered over $\mathcal{F}$ if and only if $E(w) = 0$. The convention we shall use here slightly differs from [43] since, in what follows, when we speak of a rank of a matrix $W \in (\mathcal{E}')^{n\times m}$ over $\mathcal{F}$, we shall always mean the rank of the matrix $\theta(\delta_{-E(W)} * W)$ considered over $\mathcal{F}$. We can now state our main result.

Theorem 2. *Let $R \in (\mathcal{E}')^{n\times(n+m)}$ be such that:*

i) $rank\ R(s) = n$, $\quad s \in \mathbb{C}$
ii) $rank_{\mathcal{F}}(R) = n$.

Then, the system (2) is identifiable by (5).

Proof. From condition (1) and proposition 3, α has a compact support $\subset [0, E(\alpha)]$. Assume $E(\alpha) > 0$. From [43, Thm 3.9], if $rank_{\mathcal{F}}(R) = n$, then for any such α, $E(\delta_{-h} * \alpha * R) > 0$, that is from (11), $E(\delta_{-h} * \hat{R}) > 0$ which contradicts the assumption $\hat{h} \le h$. Hence $E(\alpha) \le 0$ so $supp\alpha = \{0\}$, which means that α is a linear combination of Dirac distributions and its derivatives. Since α is a measure, $\alpha = \alpha_0\delta$ for some invertible $n \times n$ real matrix α_0. The normalization of P and $\hat{P}$ (due to the term δ') in (3) yields $\alpha_0 = I$.

Note that condition (2) is automatically satisfied if $n = 1$. For distributions with punctual support, this condition takes also a very simple form if one observes that $rank_{\mathcal{F}}(R)$ uniquely depends on the behavior of R in the vicinity of $E(R)$. The following example illustrates the link between identifiability and controllability results. Let P as in (3), $Q = B_0\delta$, and consider the system described by:

$$P * x = Q * u, \quad y(t) = x(t - h_N). \tag{15}$$

The kernel is given here by $R = [\delta_{-h_N} * P, -Q]$, and the conditions of Theorem 2 are nothing but the necessary and sufficient conditions for the system (15) to be *approximately controllable* (in the sense that the reachable space is dense in the state space [43]).

2.3 The particular case of systems with discrete delays

We end this section with a result in a particular case for which the identifiability condition can be expressed in terms of the *weak controllability* rather than the approximate controllability. Let us assume the plant and the associated model to reduce to retarded systems with discrete delays (*i.e.* $\hat{A}_{-i} = A_{-i} = 0$, $\hat{A}_c(\theta) = A_c(\theta) = 0$ and $\hat{B}_c(\theta) = B_c(\theta) = 0$ (a.e.)), and consider the two following matrices of entire functions:

$$A(s) = \sum_{i=0}^{N} A_i \, e^{-h_i \, s}, \quad B(s) = \sum_{i=0}^{N} B_i \, e^{-h_i \, s}. \tag{16}$$

The next result [3, 29], which no longer requires the assumption $\hat{h} \leq h$, is easily derived from the (non) regularity of $P^{-1} * Q$ and its derivatives.

Theorem 3. *If, for some $s \in \mathbb{C}$,*

$$rank \, \left[B(s), A(s)B(s), ...A^{n-1}(s)B(s)\right] = n, \tag{17}$$

then system (2) is identifiable by (5).

In the next section, efficient algorithms are developed in order to provide in line parameter identification for those time delay systems.

3 Adaptive Parameter Estimation for Time-Delay Systems

Recursive identification theory is nowadays well developed for linear time-invariant systems, but it is not the case for time-delay systems besides the fact that the presence of a not well identified time lag in a closed-loop may limit seriously the system performances (see, for instance, [11]). There exist few papers on adaptive identification of linear time-delay system but most of them focussed on dead-time systems. A great number of these papers concern discrete-time models, which appears at first sight the easy case, since a pure time-delay can be exactly represented by introduction of zeros and zero poles. But, when the delay is unknown, the on-line identification is not so straightforward, most methods are based on the identification of an extended models, the selection of the delay is then computed using ad-hoc criteria (number of negligible coefficients [5], minimization of an error function [22],...). In continuous time, most of the identification schemes are obtained in replacing the delay term by a finite-dimensional approximation (for instance, Padé approximant [1] [2] [32],

Laguerre approximation [14], ...), other schemes are obtained using extension of standard methods such as least square method [36], or gradient-based algorithms ([24], [16]). Papers dealing with identification of linear systems with time-delay in the state variables are not so numerous, let us mention [31] which concerns parameter identification of time-delay systems with *commensurate delays*, and [30].

The following algorithm, first presented in [30], allows on-line identification of linear dynamic systems with finitely many lumped delays in the state vector and control input. These systems are governed by linear functional differential equations with uncertain time-invariant parameters. The state variables of the system are assumed to be available for measurements.

Let us consider a system of the form

$$\dot{x}(t) = \sum_{i=0}^{r} \left[A_i x(t-\tau_i) + B_i u(t-\tau_i)\right], \tag{18}$$

where $x(t) \in \mathbb{R}^n$ is the state vector, $u(t) \in \mathbb{R}^p$ is a piecewise continuous control input, $0 = \tau_0 < \tau_1 < \ldots < \tau_r$ are time delays whose values are assumed to be known, and A_i, B_i are matrices of adequate dimensions with unknow coefficients. Now, consider the following identifier system:

$$\dot{\hat{x}}(t) = \sum_{i=0}^{r} [\hat{A}_i(t) x(t-\tau_i) + \hat{B}_i(t) u(t-\tau_i)] - G\Delta x(t), \tag{19}$$

where $\Delta x(t) = x(t) - \hat{x}(t)$ is the state error, $G \in \mathbb{R}^{n\times n}$ is a Hurwitz matrix, and time-varying matrices $\hat{A}_i(t)$, $\hat{B}_i(t)$ satisfying

$$\begin{aligned} \dot{\hat{A}}_i(t) &= F_i P \Delta x(t) x^T(t-\tau_i),\ \hat{A}_i(0) = \hat{A}_i^0, \\ \dot{\hat{B}}_i(t) &= \Phi_i P \Delta x(t) u^T(t-\tau_i),\ \hat{B}_i(0) = \hat{B}_i^0 \end{aligned} \tag{20}$$

with adaptation gain matrices F_i, Φ_i being positive definite and of appropriate dimensions, and P the (positive definite) solution of the Lyapunov equation

$$G^T P + PG = -Q$$

for a given positive definite matrix Q. Using Lyapunov redesign technique arguments [30], we obtain the following result:

Theorem 4. *Assume that system (1) is asymptotically stable and identifiable. Then, if $u(t)$ is a periodic, sufficiently rich input signal, the state error $\Delta x(t)$ converges asymptotically to 0, and the time-varying matrices $\hat{A}_i(t)$, $\hat{B}_i(t)$ converge towards the plant parameter matrices A_i, B_i.*

Proof. Let us represent the over-all system (1), (19), (20) in terms of the output-parameter errors:

$$\begin{aligned}
\dot{e}(t) &= \Sigma_{i=0}^{r} A_i e(t-\tau_i), \\
\Delta\dot{x}(t) &= \Sigma_{i=0}^{r}\{\Delta A_i[e(t-\tau_i)+z(t-\tau_i)] + \Delta B_i u(t-\tau_i)\} + G\Delta x(t), \\
\Delta\dot{A}_i(t) &= -F_i P\Delta x(t)[e(t-\tau_i)+z(t-\tau_i)]^T, \\
\Delta\dot{B}_i &= -\Phi_i P\Delta x(t) u^T(t-\tau_i),\ i=0,\dots,r
\end{aligned} \tag{21}$$

where $e(t) = x(t) - z(t)$ is the state deviation with respect to the steady-state solution $z(t)$ of (1) with periodic input $u(t)$. Now let us prove that (21) is globally asymptotically stable. For this purpose, we introduce the Lyapunov functional

$$\begin{aligned}
V(e, \Delta x, \Delta A_0, \dots, \Delta B_l) =& < We, e > + \Delta x^T P \Delta x \\
&+ \Sigma_{i=0}^{r}[tr(\Delta A_i^T F_i^{-1} \Delta A_i) + tr(\Delta B_i^T \Phi_i^{-1} \Delta B_i)]
\end{aligned} \tag{22}$$

where tr denotes the trace of the matrix, $W = \int_0^\infty S^*(t)S(t)dt$ ($S(t)$ being the exponentially stable semigroup of the former equation of (21). The computation of the time-derivative of the Lyapunov functional V along the trajectories of (21) yields

$$\dot{V}(t) = - < e_t, e_t > - \Delta x^T(t) Q \Delta x(t).$$

Although this time-derivative is only negative semidefinite, global asymptotic stability of (21), the right-hand side of which is apparently time-periodic, is established by invoking the extension of the LaSalle's invariance principle to periodic delay systems. According to the invariance principle, there must be a convergence of the trajectories of (21) to the largest invariant subset of the set of solutions of (21) for which $\dot{V}(t) \equiv 0$, or equivalently

$$e(t) \equiv 0,\ \Delta x(t) \equiv 0. \tag{23}$$

The manifold of (23), however, does not contain nontrivial trajectories of (21). Indeed, if (23) is in force, then taking into account (20), it follows that

$$\dot{\hat{A}}_i(t) = 0,\ \dot{\hat{B}}_i(t) = 0,\ i = 0, \dots, r. \tag{24}$$

In turn, coupled to (21), relations (23), (24) result in

$$\Sigma_{i=0}^{r}[\Delta A_i\, x(t-\tau_i) + \Delta B_i u(t-\tau_i)] \equiv 0$$

which by identifiability of (1) implies that

$$\Delta A_i = 0,\ \Delta B_i = 0,\ i = 0, \dots, r.$$

Thus, by applying the invariance principle system (21) is globally asymptotically stable and the required identifier convergence is concluded.

Performance issues of the adaptive identifier were studied by simulation of a scalar system

$$\dot{x}(t) = a_0 x(t) + a_1 x(t-\tau_1) + b_0 u(t),\ x, u \in \mathbb{R} \tag{25}$$

with a single delay in the state. Apparently system (25) is weakly controllable if and only if $b_0 \neq 0$. Thus system (25) with $b_0 \neq 0$ turns out to be identifiable and its identifiability is enforced by a sufficiently nonsmooth control input. If, in addition, the unforced system (25) subject to $u = 0$ is asymptotically stable ($a_0 < 0,\ a_0 + |a_1| < 0$) then the parameters of the system can be identified via the proper adaptive identifier design. The equations of the identifier of equation (25) are:

$$\dot{\hat{x}}(t) = \hat{a}_0 x(t) + \hat{a}_1 x(t - \hat{\tau}_1) + \hat{b}_0 u(t) + a\Delta x(t),\ t \geq 0, \tag{26}$$

$$\dot{\hat{a}}_0 = \beta_0 \Delta x(t) x^T(t),\ \dot{\hat{a}}_1 = \beta_1 \Delta x(t) x^T(t - \hat{\tau}_1),$$

$$\dot{\hat{b}}_0 = \gamma_0 \Delta x(t) u^T(t), \tag{27}$$

where $\hat{\tau}_1$ is the a priori estimated value of the delay τ_1. The following parameters of the system were selected:

$$a_0 = -1.5,\ a_1 = 0.5,\ b_0 = 1,\ \tau_1 = 1.$$

The initial distribution of the system and that of the identifier as well as the initial conditions of the tunable parameters were set to zero, *i.e.* $x(t) = \hat{x}(t) = 0\ \ t \leq 0$, and $\hat{a}_i(0) = 0,\ \hat{b}_i(0) = 0,\ i = 0, \ldots, \hat{r}$. All of the adaptation parameters were set to eight and the input $u(t)$ was the sum of two square waves of amplitude 5 and frequencies 0.5 and 0.7 Hz.

In the first simulation, the delay of the identifier was matched to the plant's one: $\hat{\tau}_1 = \tau_1 = 1$. Figure 2 shows the convergence of tunable parameters $\hat{a}_0$, $\hat{a}_1$, and $\hat{b}_0$, to their nominal values.

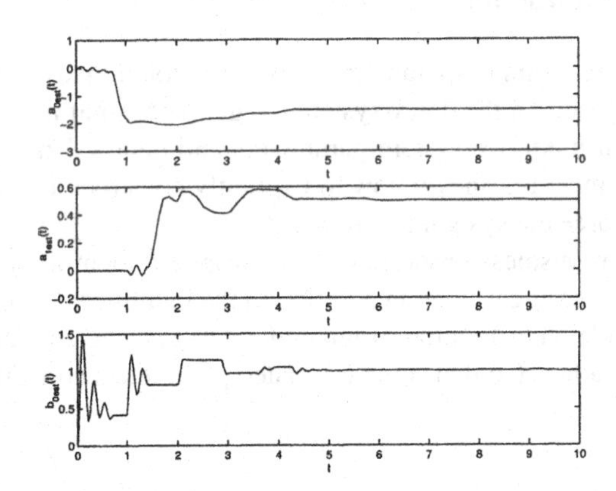

Fig. 1. Adaptive identifier with a priori known delay

The second simulation underline some robustness property of the identifier, allowing an estimation of an a priori unknown delay; We test the following identifier including two delays $\hat{\tau}_1 = 1.01$ and $\hat{\tau}_2 = 1.1$, both mismatching to that of the system:

$$\dot{\hat{x}}(t) = \hat{a}_0 x(t) + \hat{a}_1 x(t - \hat{\tau}_1) + \hat{a}_2 x(t - \hat{\tau}_2) + \hat{b}_0 u(t) + a\Delta x(t),$$

As shown in Figure 2, the tunable parameters converge to their nominal values. It is concluded from this figure that the only parameter $\hat{a}_2(t)$ becomes neglectable as $t \to \infty$ thereby establishing a fictitious delay $\hat{\tau}_2 = 1.1$ and approximately identifying another delay value $\hat{\tau}_2 = 1.01$.

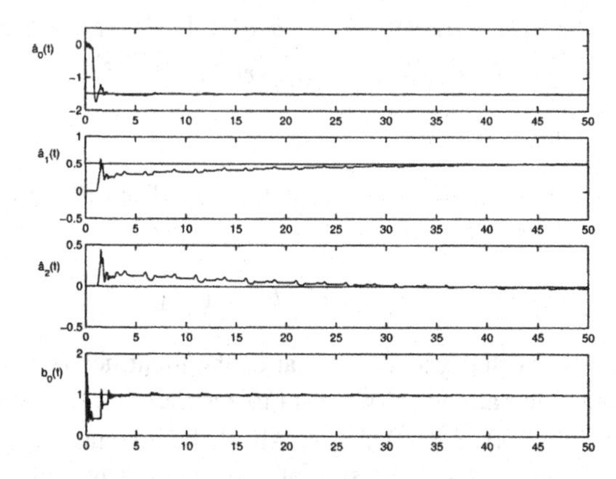

Fig. 2. Adaptive identifier with mismatched delays

4 Concluding Remarks

In analogy to linear lumped-parameters systems, counterparts of controllability conditions are imposed on the delay systems to guarantee that it is in principle possible to identify all unknown system parameters and delays. The approach used is also constructive since it is shown how to explicitly construct a sufficiently rich input in order to enforce the system identifiability.

The favorable robustness properties of the proposed identifier against small deviations of the delays suggest that, in addition to the identification of the parameters, we may also provide an estimation of unknown delays using an identifier involving a large number of delays (so that some estimated parameters correspond to fictitious delays).

Appendix

Proof. (of proposition 1) Let $\overline{x} := x - \hat{x} = \overline{x}_1 + \overline{x}_2$ with $supp\, \overline{x}_1 \subset [0, r_1]$ and $supp\, \overline{x}_2 \subset [r_2, \infty[$. Denote also $\Delta := \det P$, and consider $\Delta * \hat{P} * \overline{x}$. We get after some simple manipulations

$$S(Q) * u = \Delta * \hat{P} * \overline{x}_1 + \Delta * \hat{P} * \overline{x}_2. \tag{28}$$

where we have denoted formally

$$S(.) = \Delta * \hat{P} * (P^{-1} * (.) - \hat{P}^{-1} * \widehat{(.)}) \qquad (29)$$

$$= \hat{P} * Adj(P) * (.) - \Delta * \widehat{(.)}. \qquad (30)$$

Clearly, $S(Q)$ and $\Delta * \hat{P}$ have their support included in $[0,(n+1)h]$, so the respective supports in (28) are contained in $[t_1, t_2 + (n+1)h]$, $[0, r_1 + (n+1)h]$, and $[r_2, \infty[$. If (7) holds, these supports are disjoint, so each corresponding distribution must identically vanish. Invertibility of $\Delta * \hat{P}$ in (28) results in $\overline{x}_1 = \overline{x}_2 = 0$.

Proof. (of proposition 2) In (29), $\Delta * \hat{P}$ is invertible so it suffices to show that $S(Q) * u = 0 \Rightarrow S(Q) = 0$. Using (10), differentiation of $S(Q) * u = 0$ yields

$$\sum_{l=0}^{L} S(Q) * U_l(D)\,[\delta] * \delta_{s_l} = 0, \qquad (31)$$

with $supp\ S(Q) \subset [0,(n+1)h]$, while $supp\ U_l(D)\,[\delta] = \{0\}$. If condition (2) holds, all the terms of the sum in (31) have disjoint supports and then must identically vanish. Hence,

$$S(Q) * [U_0(D), ..., U_L(D)]\,[\delta] = 0. \qquad (32)$$

Now if $rank\ U(D) = m$, $U(D)$ admits a Smith form factorization, $U(D) = W(D)\,[\Lambda(D), 0]\,V(D)$ where $W(D), V(D)$ are unimodular matrices, and $\Lambda(D) = diag(\lambda_i(D))$, with no identically zero polynomial on its diagonal. Since every non zero polynomial $\lambda(D)\,[\delta]$, $D = \partial/\partial t$ is invertible in $\mathcal{D}'_+$, the conclusion follows.

References

1. Agarwal, M. and Canudas, C.: On line estimation of time delay and continuous-time process parameters, Int. Journal of Control, 1987, 46, pp. 295-311.
2. Bai, E.-W. and Chyung, D.H. Improving delay estimates derived from least-squares algorithms and Pade approximations, International Journal of Systems Science,1993, 24, 745-756.
3. L. Belkoura and Y. Orlov, Identifiability analysis of linear delay-differential systems, IMA J. of Mathematical Control and Information, 19:73-81, 2002.
4. F. Blanchini and E.P. Ryan. A Razumikhin-type lemma for functional differential equations with application to adaptive control. Automatica, 35(5):809–818, 1999.
5. Biswas, K.K. and Singh, G. (1978). Identification of stochastic time delay systems, IEEE Transactions on Automatic Control, AC-23, 504-505.
6. H.H. Choi and M.J. Chung. Observer-based H_∞ controller design for state delayed linear systems. Automatica, 32(7):1073–1075, 1996.
7. H.H. Choi and M.J. Chung. Robust observer-based H_∞ controller design for linear uncertain time-delay systems. Automatica, 33(9):1749–1752, 1997.
8. M. Darouach. Linear functional observers for systems with delays in state variables. IEEE Trans. Aut. Control, 46(3):491–496, March 2001.

9. M. Darouach, P. Pierrot, and E. Richard. Design of reduced-order observers without internal delays. IEEE Trans. Aut. Cont., 44(9):1711–1713, Sept. 1999.
10. C.E. DeSouza, R.E. Palhares, and P.L.D. Peres. Robust H_∞ filtering for uncertain linear systems with multiple time-varying state delays: An LMI approach. In 38th IEEE CDC99 (Conf. on Dec. and Control), pages 2023–2028, Phoenix, AZ, Dec. 1999.
11. S. Diop, I. Kolmanovsky, P. Moraal, and M. vanNieuwstadt. Preserving stability/performance when facing an unknown time delay. Control Eng. Practice, 9:1319–1325, Dec. 2001.
12. L. Ehrenpreis. Solutions of some problems of division, part IV, Invertible and elliptic operators, Amer. Journal of Mathematics, 82:522-588, 1960
13. F.W. Fairmar and A. Kumar. Delay-less observers for systems with delay. IEEE Trans. Aut. Control, 31(3):258–259, March 1986.
14. Fernandes, J.M. and Ferriera, A.R. An all-pass approximation to time delay, UKACC International Conference on CONTROL 96, 1996 1208-1213.
15. S.G Foda and M.S. Mahmoud. Adaptive stabilization of delay differential systems with unknown uncertainty bounds. Int. J. Control, 71(2):259–275, 1998.
16. Gawthrop, P.J., Nihtila, M. and Rad, A.B.: Recursive parameter estimation of continuous-time systems with unknown time delay, Control - Theory and Advanced Technology, 1989, 5, pp. 227-248.
17. A. Germani, C. Manes, and P. Pepe. A state observer for nonlinear delay systems. In 37th IEEE CDC98 Conf. on Dec. and Control), pages 355–360, Tampa, FL, Dec. 1998.
18. J.K. Hale and S.M. Verduyn-Lunel. Introduction to Functional Differential Equations, volume 99 of Applied Math. Sciences. Springer, NY, 1993.
19. Y.P. Huang and K. Zhou. Robust stability of uncertain time delay systems. IEEE Trans. Aut. Control, 45(11):2169–2173, Nov. 2000.
20. V.B. Kolmanovskii and A. Myshkis. Introduction to the theory and applications of functional differential equations. Kluwer Acad., Dordrecht, 1999.
21. V.B. Kolmanovskii and V.R. Nosov. Stability of functional differential equations. Academic Press, London, 1986.
22. Kurz, H. and Goedecke, W. (1981). Digital parameter-adaptive control of processes with unknown dead time, Automatica, 17, 245-252.
23. J. Leyva-Ramos and A.E. Pearson. An asymptotic modal observer for linear autonomous time lag systems. IEEE Trans. Aut. Cont., 40:1291–1294, July 1995.
24. Liu, G. Adaptive predictor for slowly time-varying systems with variable time-delay, Advances in modelling and simulation, Vol. 20, pp. 9-21, 1990
25. S.M.V. Lunel. Parameter identifiability of differential delay equations, Int J. of Adapt. Control Signal Process, 15: 2001.
26. S. Majhi and Atherton D.P. A novel identifcation method for time delay processes. In ECC99, European Control Conf., Karslruhe, Germany, 1999.
27. L. Mirkin and G. Tadmor. H_∞ control of systems with I/O delay: A review of some problem-oriented methods. IMA J. Math. Control Information, 19(1), 2002.
28. S.I. Niculescu. Delay Effects on Stability, volume 269 of LNCIS. Springer, 2001.
29. Y. Orlov, L. Belkoura, M. Dambrine and J.P. Richard, On identifiability of linear time-delay systems, IEEE Trans. Aut. Control, 47(8):1319-1324, Aug. 2002.
30. Y. Orlov, L. Belkoura, J.P. Richard, and M. Dambrine, On-Line Parameter Identification of Linear Time-Delay Systems, to appear in Proc. IEEE CDC 02, Las Vegas.
31. Pourboghrat F., Chyung, D.H., Parameter identification of linear delay systems, Int. J. Control, vol. 49, no. 2, pp. 595–627, 1989
32. Rad, A.B.: Self-tuning control of systems with unknown time delay: a continuous-time approach, Control Theory and Advanced Technology, 1994, 10, pp. 479-497.

33. O. Sename. New trends in design of observers for time-delay systems. Kybernetica, 37(4):427–458, 2001.
34. J.P. Richard. Time delay systems: An overview of some recent advances and open problems. Automatica, 39(9), 2003 (to appear).
35. K.K. Tan, Q.K. Wang, and T.H. Lee. Finite spectrum assignment control of unstable time delay processes with relay tuning. Ind. Eng. Chem. Res., 37(4):1351–1357, 1998.
36. Tuch, J., Feuer, A. and Palmor Z.J. Time delay estimation in continuous linear time-invariant systems, IEEE Trans. Aut. Control, Vol. 39, no. 4, pp. 823–827, 1994
37. S.M. Verduyn-Lunel. Identification problems in functional differential equations. In 36th IEEE CDC97 (Conf. on Dec. and Control), pages 4409–4413, San Diego, CA, Dec. 1997.
38. E.I. Verriest. Robust stability and adaptive control of time-varying neutral systems. In 38th IEEE CDC99 (Conf. on Dec. and Control), pages 4690–4695, Phoenix, AZ, Dec. 1999.
39. Z. Wang, B. Huang, and H. Unbehausen. Robust H_∞ observer design for uncertain time-delay systems: (I) the continuous case. In IFAC 14th World Congress, pages 231–236, Beijing, China, 1999.
40. Z. Wang, B. Huang, and H. Unbehausen. Robust H_∞ observer design of linear state delayed systems with parametric uncertainty: The discrete-time case. Automatica, 35(6):1161–1167, 1999.
41. K. Watanabe, E. Nobuyama, and K. Kojima. Recent advances in control of time-delay systems a tutorial review. In 35th IEEE CDC96 (Conf. on Dec. and Control), pages 2083–2089, Kobe, Japan, Dec. 1996.
42. Y.X. Yao, Y.M. Zhang, and R. Kovacevic. Functional observer and state feedback for input time-delay systems.Int. J. Control, 66(4):603–617, 1997.
43. Y. Yamamoto. Reachability of a class of infinite-dimensional linear systems:An external approach with application to general neutral systems, SIAM J. of Control and Optimization, 27: 217-234, 1989

A Model Matching Solution of Robust Observer Design for Time-Delay Systems

Anas Fattouh[1]* and Olivier Sename[2]

[1] Automatic Laboratory of Aleppo, Faculty of Electrical and Electronic Engineering, University of Aleppo, Aleppo, Syria fattouh@scs-net.org
[2] LAG, ENSIEG-BP 46, 38402 Saint Martin d'Hères Cedex, France
Olivier.Sename@inpg.fr

Summary. Uncertainties are unavoidable in practical situations and they have to be taken into consideration in control system design. In this chapter, a method for designing a robust observer for linear time-delay systems is proposed. Under the assumption that the considered time-delay system is spectrally controllable and spectrally observable, a double Bézout factorization of its nominal transfer matrix is obtained. Next, based on this factorization, all stable observers for the nominal system are parameterized. By applying those observers on the real system, the parameterization transfer matrix has to be found such that the error between the real estimation and the nominal one is minimized. This problem is rewritten as an infinite dimensional model matching problem for different types of uncertainty. In order to solve this infinite dimensional model matching problem, it is transformed into a finite dimensional one, and therefore a suboptimal solution can be obtained using existing algorithms.

1 Introduction

Time-delay systems appear naturally in many engineering applications and, in fact, in any situation in which transmission delays cannot be ignored. The control of such systems using state feedback has been thoroughly studied (see [2, 13, 39] and the references therein). One of the major difficulties in implementing such control laws is that all the state variables of the system (at least some state variables) are required for the controller synthesis. However, in most practical situations, this condition is rarely satisfied and an observer has to be built up in order to estimate the state variables from the measured output and the controlled input.

Many observer schemes have been proposed in the literature under the assumption that there are no model uncertainties [7, 10, 16, 20]. However in practice, model uncertainties have to be taken into consideration when designing observers. These uncertainties have been considered in the literature as additive disturbances [3,4,21] or as an unknown input [5].

* This work has been done during the post doctoral stay of Dr Anas Fattouh at LAG, INPG, France and it is supported by AUF (Agence Universitaire de la Francophonie).

In this chapter, unstructured uncertainties (additive uncertainty and input multiplicative uncertainty) are considered when designing observers. A set of all stable observers for the nominal system is firstly parameterized based on a double Bézout factorization of its transfer matrix. Then, the problem of finding the parameterization transfer matrix such that the resulting observer satisfies a specified robustness property is written as an infinite dimensional model matching problem. A suboptimal solution of this infinite dimensional problem has been proposed by the authors in [6, 17] in terms of multiple finite dimensional model matching problems. However the solution of those multiple finite dimensional model matching problems is not easy to obtain. Therefore, another solution is proposed in this chapter by transforming the infinite dimensional model matching problem into only one modified finite dimensional model matching problem. Then by applying an existing optimization algorithm on the modified finite dimensional problem a solution can be obtained.

The chapter is organized as follows. Section 2 gives some basic definitions concerning time-delay systems. Section 3 shows how to get a double Bézout factorization of a given delayed transfer matrix and based on this factorization how to parameterize the set of all stable observers. Section 4 describes how to find the parameterization transfer matrix for different types of system uncertainties. Section 5 provides a suboptimal solution of the infinite dimensional problem described in the previous section. Illustrative examples are given in Section 6 and some concluding remarks are given in Section 7.

Throughout this chapter the following notations will be used:

$\mathbb{R}$	is the field of real numbers
$\mathbb{N}$	is the set of integer numbers
$\mathbb{F}^+$	$= \{a \in \mathbb{F} : 0 < a < +\infty\}$, $\mathbb{F}$ denotes $\mathbb{R}$ or $\mathbb{N}$
$\mathbb{R}[\bullet]$	is the ring of polynomials in $\bullet$ with coefficients in $\mathbb{R}$
s	denotes the Laplace variable, $z = e^{-sh}$ and $h \in \mathbb{R}^+$ fixed
$\mathbb{R}[z][s]$	$=\{\sum_{k=0}^{m} a_k(z)s^k : a_k(z) \in \mathbb{R}[z], m \in \mathbb{N}^+\}$
$\mathbb{R}(s,z)$	is the field of rational functions in s and z with coefficients in $\mathbb{R}$
Θ	$= \{p(s,z) = \frac{b(s,z)}{a(s)} \in \mathbb{R}(s,z) : b(s,z) \in \mathbb{R}[z][s],\ a(s) \in \mathbb{R}[s],$ $deg_s(a(s)) > deg_s(b(s,z))$ and $p(s,z)$ is entire$\}$
$\Theta[z]$	is the ring of polynomials in z with coefficients in Θ
$\mathbf{F}$	$= \{p(s,z) = \frac{b(s,z)}{a(s)} \in \Theta[z] : a(s)$ is monic and stable$\}$
$\mathcal{M}(\bullet)$	denotes a transfer matrix of appropriate dimension with entries in $\bullet$
I_n	denotes the $(n \times n)$ identity matrix
$0_{i \times j}$	denotes the $(i \times j)$ zero matrix
X^T	denotes the transpose of a matrix X
$\lVert\cdot\rVert_\infty$	is the H_∞-norm defined by: $\lVert X(s)\rVert_\infty = \sigma_{max}(X(j\omega))$; $\sigma_{max}(X)$ denotes the maximum singular value of X, j is the imaginary number, ω denotes the frequency
$\mathcal{C}[a,b]$	is the set of continuous functions $[a,b] \to \mathbb{R}^n$

The elements of Θ can be viewed as transfer matrices of distributed time delays while the elements of $\Theta[z]$ can be viewed as transfer matrices of finite interconnections of

point and distributed time delays [9].
In order to simplify the notation, X *will denote either* $X(s,z)$ *or* $X(s)$ *when there is no confusion.*

2 Preliminaries

In this section, some basic definitions concerning time-delay systems are recalled.

Consider the following model of a time-delay system

$$\begin{cases} \dot{x}(t) &= A(\nabla)x(t) + B(\nabla)u_o(t) \\ y_o(t) &= C(\nabla)x(t) \\ x(t) &= \phi(t); \qquad t \in [-mh, 0] \end{cases} \tag{1}$$

where
• $x(t) \in \mathbb{R}^n$, $u_o(t) \in \mathbb{R}^r$ and $y_o(t) \in \mathbb{R}^p$ are the state, the control input and the measured output vectors respectively.
• $A(\nabla) \in \mathbb{R}[\nabla]^{n\times n}$, $B(\nabla) \in \mathbb{R}[\nabla]^{n\times r}$, $C(\nabla) \in \mathbb{R}[\nabla]^{p\times n}$ and $\phi(t) \in C[-mh, 0]$ is the functional initial condition of (1).
• ∇ is the delay operator ($(\nabla x)(t) = x(t-h), (\nabla^2 x)(t) = x(t-2h), \ldots$), $h \in \mathbb{R}^+$ is the fixed known delay duration and m is a positive integer such that mh represents the maximal delay in the system.

Definition 1. *[9] Consider the time-delay system (1). Let* z *be a complex variable, the pair* $(A(z), B(z))$ *is*

i) $\mathbb{R}[z]$*-controllable if and only if* $rank[sI_n - A(z), \quad B(z)] = n \ \forall\ (s,z) \in \mathbb{C}^2$.
ii) spectrally controllable if and only if $rank[sI_n - A(e^{-sh}), B(e^{-sh})] = n\ \forall\ s \in \mathbb{C}$.
iii) $\mathbb{R}(z)$*-controllable if and only if* $rank[sI_n - A(z), \quad B(z)] = n$ *for all but finitely many pairs* $(s,z) \in \mathbb{C}^2$.

The pair $(C(z), A(z))$ *is (i)* $\mathbb{R}[z]$*-observable (resp. (ii) spectrally observable or (iii)* $\mathbb{R}(z)$*-observable) if and only if the pair* $(A^T(z), C^T(z))$ *is (i)* $\mathbb{R}[z]$*-controllable (resp. (ii) spectrally controllable or (iii)* $\mathbb{R}(z)$*-controllable).*

Moreover, the realization (1) is

i) canonical if and only if the pair $(A(z), B(z))$ *is* $\mathbb{R}[z]$*-controllable and the pair* $(C(z), A(z))$ *is* $\mathbb{R}(z)$*-observable.*
ii) co-canonical if and only if the pair $(A(z), B(z))$ *is* $\mathbb{R}(z)$*-controllable and the pair* $(C(z), A(z))$ *is* $\mathbb{R}[z]$*-observable.*
iii) spectrally canonical if and only if the pair $(A(z), B(z))$ *is spectrally controllable and the pair* $(C(z), A(z))$ *is spectrally observable.* □

It should be noted that if the pair $(A(z), B(z))$ is $\mathbb{R}[z]$-controllable then there exists a polynomial matrix $K(z) \in \mathcal{M}(\mathbb{R}[z])$ (i.e. $K(z)$ *is the Laplace transformation of point time delays*) such that $det[sI_n - A(z) - B(z)K(z)] = \delta_1(s)$ where $\delta_1(s) \in \mathbb{R}[s]$ is a stable polynomial. In return, if the pair $(A(z), B(z))$

is spectrally controllable then there exists a matrix $K(s,z) \in \mathcal{M}(\Theta[z])$ (*i.e.* $K(s,z)$ *is the Laplace transformation of point and distributed time delays*) such that $det[sI_n - A(z) - B(z)K(s,z)] = \delta_2(s)$ where $\delta_2(s) \in \mathbb{R}[s]$ is a stable polynomial.

Remark 1. When system (1) is spectrally canonical then there exist feedback matrices $F_1(s,z)$, $F_2(s,z) \in \mathcal{M}(\Theta[z])$ and an observer gain matrix $K(s,z) \in \mathcal{M}(\Theta[z])$ such that

$$\begin{aligned} det\begin{bmatrix} sI_n - A(z) & -B(z) \\ -F_1(s,z) & I_r - F_2(s,z) \end{bmatrix} &= \alpha(s) \\ det[sI_n - A(z) - K(s,z)C(z)] &= \beta(s) \end{aligned} \tag{2}$$

where $\alpha(s), \beta(s) \in \mathbb{R}[s]$ and are stable polynomials. □

The transfer matrix of system (1) is given by [18]

$$G(s,z) = C(z)(sI_n - A(z))^{-1}B(z) \tag{3}$$

Note that $G(s,z)$ is a $(p \times r)$ matrix with i,j entry $G_{ij}(s,z) = \frac{b_{ij}(s,z)}{a_{ij}(s,z)}$ where $b_{ij}(s,z), a_{ij}(s,z) \in \mathbb{R}[z][s]$, $a_{ij}(s,z)$ is monic and

$$deg_s(a_{ij}(s,z)) \geq deg_s(b_{ij}(s,z)).$$

It has been shown in [9] that any transfer matrix $G(s,z)$ as defined above has a canonical realization and a co-canonical realization. *In the sequel it is assumed that the co-canonical realization (1) of $G(s,z)$ is spectrally controllable which will be noted as spectrally co-canonical realization.* Note that this assumption is not restrictive as $G(s,z)$ has a spectrally canonical realization if and only if the canonical realization of $G(s,z)$ is spectrally observable which is equivalent to requiring that the co-canonical realization of $G(s,z)$ be spectrally controllable (see [9] for details).

3 Parametrization of All Stable Observers

In this section, the set of all stable observers for system (1) are parameterized based on a polynomial factorization of its transfer matrix (3).

The following lemma provides us a polynomial factorization of the transfer matrix (3) associated with a spectrally co-canonical realization (1).

Lemma 1. *[14] Consider the transfer matrix (3) associated with a spectrally co-canonical realization (1). The transfer matrix (3) can be factorized as follows*

$$G(s,z) = N(s,z)M^{-1}(s,z) = \overline{M}^{-1}(s,z)\overline{N}(s,z) \tag{4}$$

where $N, M, \overline{M}, \overline{N}$ satisfy the following double Bézout equation

$$\begin{bmatrix} Y & X \\ -\overline{N} & \overline{M} \end{bmatrix}\begin{bmatrix} M & -\overline{X} \\ N & \overline{Y} \end{bmatrix} = \begin{bmatrix} I_r & 0_{r\times p} \\ 0_{p\times r} & I_p \end{bmatrix} \tag{5}$$

The eight matrices in (5) belong to $\mathcal{M}(\mathbf{F})$ and they can be obtained using the following equations

$$\begin{array}{ll} M = I_r + F_e(sI_e - A_o)^{-1}B_e & \overline{M} = I_p + C_e(sI_e - \overline{A}_o)^{-1}K_e \\ N = \quad C_e(sI_e - A_o)^{-1}B_e & \overline{N} = \quad C_e(sI_e - \overline{A}_o)^{-1}B_e \\ Y = I_r - F_e(sI_e - \overline{A}_o)^{-1}B_e & \overline{Y} = I_p - C_e(sI_e - A_o)^{-1}K_e \\ X = \quad F_e(sI_e - \overline{A}_o)^{-1}K_e & \overline{X} = \quad F_e(sI_e - A_o)^{-1}K_e \end{array} \tag{6}$$

where

$$A_o(s,z) = A_e + B_eF_e, \quad \overline{A}_o(s,z) = A_e + K_eC_e, \quad A_e(z) = \begin{bmatrix} A(z) & B(z) \\ 0_{r\times n} & -I_r \end{bmatrix}$$

$$B_e(z) = \begin{bmatrix} 0_{n\times r} \\ I_r \end{bmatrix}, \quad C_e(z) = \begin{bmatrix} C(z) & 0_{p\times r} \end{bmatrix}, \quad F_e(s,z) = \begin{bmatrix} F_1(s,z) & F_2(s,z) \end{bmatrix}$$

$$K_e(z) = \begin{bmatrix} K(z) \\ 0_{r\times p} \end{bmatrix}, \quad I_e = \begin{bmatrix} I_n & 0_{n\times r} \\ 0_{r\times n} & 0_{r\times r} \end{bmatrix}$$

and $F_1(s,z)$, $F_2(s,z) \in \mathcal{M}(\Theta[z])$, $K(z) \in \mathcal{M}(\mathbb{R}[z])$ are chosen such that (2) is satisfied for some stable polynomials $\alpha(s), \beta(s) \in \mathbb{R}[s]$. □

Now, let $r(t) = E(\nabla)x(t) \in \mathbb{R}^{k\times 1}$ be the vector to be estimated. An asymptotic observer of $r(t)$ is defined as a dynamic system of the following form

$$\hat{r}(s) = U(s,z)u_o(s) + V(s,z)y_o(s) \tag{7}$$

where $U(s,z), V(s,z) \in \mathcal{M}(\Theta[z])$, such that for all control input $\lim_{t\to\infty}(r(t) - \hat{r}(t)) = 0$ is satisfied.

A parameterization of all stable observers (7) based on the factorization (4) and (5) is given in the following lemma.

Lemma 2. *[21] Consider systems (1) and (7). The set of $\mathcal{M}(\Theta[z])$-matrices $U(s,z)$, $V(s,z)$ such that system (7) is an observer of $r(s) = E(z)x(s)$ is given by*

$$\begin{cases} U(s,z) = P(s,z)Y(s,z) - Q(s,z)\overline{N}(s,z) \\ V(s,z) = P(s,z)X(s,z) + Q(s,z)\overline{M}(s,z) \end{cases} \tag{8}$$

where $P(s,z) \in \mathcal{M}(\Theta[z])$ is given by

$$\begin{aligned} P(s,z) &= E_e(z)(sI_e - A_o(s,z))^{-1}B_e(z) \\ E_e(z) &= [E(z) \quad 0_{k\times r}] \end{aligned} \tag{9}$$

$Y, \overline{N}, X, \overline{M}$ are given by the double Bézout factorization (6) and $Q(s,z)$ is any matrix belonging to $\mathcal{M}(\Theta[z])$. □

It should be noted that $Q(s,z)$ can be properly chosen in order to achieve certain performance specifications as it will be shown in the next sections.

4 Model Matching Problem for Different Types of Uncertainties

Let us consider $\tilde{G}(s,z)$ as a real time-delay system and its corresponding model is given by (3). Let (7)-(8) be the set of all stable observers for the nominal model (3) associated with a spectrally co-canonical realization (1). Applying the set of observers (7)-(8) on the real system, the parameterized transfer matrix $Q(s,z)$ has to be found such that the difference between the real estimation and the nominal one is minimized as shown in Fig. 1. This problem is formulated in the following proposition as an infinite dimensional model matching problem for different types of uncertainties.

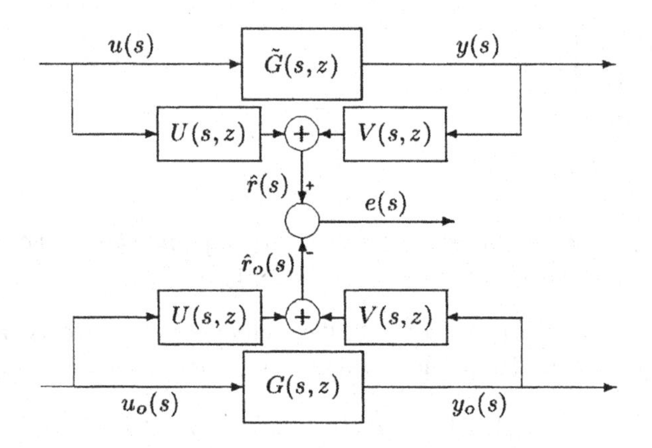

Fig. 1. Error due to system uncertainty.

Proposition 1. *Given a real time-delay system $\tilde{G}(s,z)$ with a nominal model $G(s,z)$ associated with a spectrally co-canonical realization (1). Let (7)-(8) be the set of all stable observers for nominal model. An observer of $\tilde{G}(s,z)$ is given by (7)-(8) where $Q(s,z)$ is the solution of the following model matching problem*

$$\min_{Q\in\mathcal{M}(\mathbf{F})} \|T_1(s,z) - Q(s,z)T_2(s,z)\|_\infty \tag{10}$$

where $T_1(s,z)$ and $T_2(s,z) \in \mathcal{M}([\Theta[z])$ are defined according to the considered model uncertainty.

Proof. Let $G(s,z) = N(s,z)M^{-1}(s,z) = \overline{M}^{-1}(s,z)\overline{N}(s,z)$ be the nominal model with a spectrally co-canonical state-space realization (1). Let (7)-(8) be a set of all stable observers of realization (1). Let $W(s)$ be a fixed stable transfer matrix and $\Delta(s,z)$ be variable stable transfer matrix with $\|\Delta(s,z)\|_\infty \leq 1$.

Applying (7)-(8) on the real system, the real estimation $\hat{r}(s)$ can be calculated for different model uncertainties as follows (see Fig. 1 and Fig. 2):

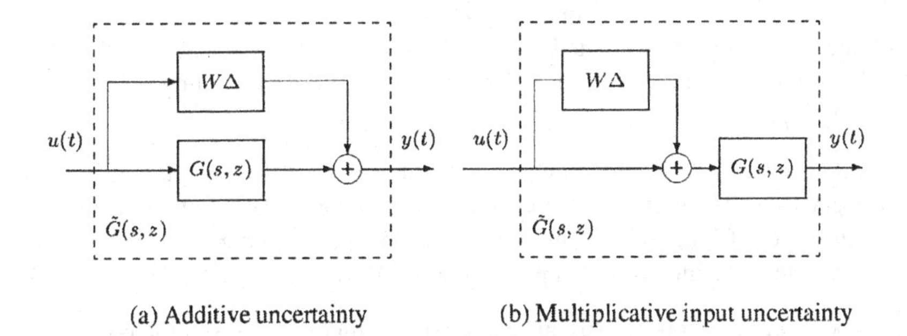

(a) Additive uncertainty (b) Multiplicative input uncertainty

Fig. 2. Different types of uncertainty.

1. Additive uncertainty:
In this case $\tilde{G} = G + W\Delta$ and $y(s) = y_o(s) + W\Delta u_o(s)$. The real estimation is given by

$$\hat{r}(s) = U(s,z)u_o(s) + V(s,z)y(s) = (U + VW\Delta)u_o(s) + Vy_o(s)$$

which implies that

$$e(s) = \hat{r}(s) - \hat{r}_o(s) = VW\Delta u_o(s) = (PXW + Q\overline{M}W)\Delta u_o(s)$$

Now, as W and Δ are stable, then $e(s)$ is stable. Moreover, as $u_0(s)$ is bounded then $e(s)$ is bounded. From the above equation, we can write

$$\|e(s)\|_2 \leq \|PXW + Q\overline{M}W\|_\infty \|\Delta(s,z)\|_\infty \|u_0(s)\|_2$$

As $\|\Delta(s,z)\|_\infty \leq 1$, then in order to minimize the effects of uncertainty on the real estimation, the optimization problem $min_Q \|T_1 - QT_2\|_\infty$ has to be solved for

$$\begin{cases} T_1(s,z) = P(s,z)X(s,z)W(s) \\ T_2(s,z) = -\overline{M}(s,z)W(s) \end{cases} \tag{11}$$

2. Multiplicative input uncertainty:
Following the same steps applied on the case of additive uncertainty, we get

$$e(s) = -UW\Delta u(s) = -(PYW - Q\overline{N}W)\Delta u(s)$$

Now, as W and Δ are stable, then $e(s)$ is stable. Moreover, as $u(s)$ is bounded, then $e(s)$ is bounded. In order to minimize the effects of uncertainty on the real estimation, the optimization problem $min_Q \|T_1 - QT_2\|_\infty$ has to be solved for

$$\begin{cases} T_1(s,z) = P(s,z)Y(s,z)W(s) \\ T_2(s,z) = \overline{N}(s,z)W(s) \end{cases} \tag{12}$$

From the above cases one can conclude that in order to minimize the effect of the model uncertainty on the estimated states one has to solve the model matching problem (10) for $T_1(s,z)$ and $T_2(s,z)$ are given by (11) or (12) according to the model uncertainty type. □

Remark 2. The result of Proposition 1 can be interpreted as follows: the estimation $\hat{r}(s)$ depends on the input and output information signals. When there are uncertainties in those information in some frequency intervals, this dependence has to be wasted away in these intervals by a proper choice of transfer matrix $Q(s,z)$. □

A suboptimal solution to (10) can be obtained by finding a transfer matrix $Q(s,z) \in \mathcal{M}(\mathbf{F})$ solution to

$$\|T_1(s,z) - Q(s,z)T_2(s,z)\|_\infty \leq \gamma \tag{13}$$

for some positive scalar γ. A solution for this optimization problem is given in the next section.

5 Suboptimal Solution

In this section, the infinite dimensional optimization problem (13) is transformed into a finite dimensional one.

Proposition 2. *Given two transfer matrices $T_1(s,z)$, $T_2(s,z) \in \mathcal{M}(\mathbf{F})$. Suppose that $deg_z(T_1(s,z)) \geq deg_z(T_2(s,z))$, then there exists a matrix $Q(s,z) \in \mathcal{M}(\mathbf{F})$ with $deg_z(Q(s,z)) = deg_z(T_1(s,z)) - deg_z(T_2(s,z))$ such that $\|T_1(s,z) - Q(s,z)$ $T_2(s,z)\|_\infty \leq \gamma$ where γ is some positive scalar.*

Proof. The proof provides a method to construct $Q(s,z) \in \mathcal{M}(\mathbf{F})$ solution to (13). First, note that

$$\|T_1(s,z) - Q(s,z)T_2(s,z)\|_\infty = \|T_1^T(s,z) - T_2^T(s,z)Q^T(s,z)\|_\infty \tag{14}$$

Note also that any matrix $T(s,z)$ belonging to $\mathcal{M}(\mathbf{F})$ can be written as $T(s,z) = \sum_{i=0}^{m} T_i(s)z^i$ where $T_i(s)$ are proper stable rational transfer matrices and $m \in \mathbb{N}^+$. Suppose that

$$T_1^T(s,z) = \sum_{i=0}^{m_1} T_{1i}(s)z^i \quad \text{and} \quad T_2^T(s,z) = \sum_{j=0}^{m_2} T_{2j}(s)z^j \tag{15}$$

where $T_{1i}(s)$ and $T_{2j}(s)$ are proper stable transfer matrices and $m_1, m_2 \in \mathbb{N}^+$. Since $deg_z(T_1(s,z)) \geq deg_z(T_2(s,z))$, then $m_1 \geq m_2$.

Let the parameterized matrix $Q^T(s,z)$ be of the form

$$Q^T(s,z) := \sum_{k=0}^{(m_1-m_2)} Q_k(s)z^k \tag{16}$$

Then, using (14), (15) and (16), the optimization problem (13) can be rewritten as follows

$$\begin{aligned}\|T_1(s,z) - Q(s,z)T_2(s,z)\|_\infty &= \| \left[1 \; z \; \dots \; z^{m_1} \right] (\overline{T_1}(s) - \overline{T_2}(s)\overline{Q}(s))\|_\infty \\ &\le \|\overline{T_1}(s) - \overline{T_2}(s)\overline{Q}(s)\|_\infty \le \gamma \end{aligned} \quad (17)$$

where

$$\overline{T_1}(s) = \begin{bmatrix} T_{10}(s) \\ T_{11}(s) \\ \vdots \\ T_{1m_1}(s) \end{bmatrix}, \quad \overline{T_2}(s) = \underbrace{\begin{bmatrix} T_{20}(s) & 0 & \dots & 0 \\ T_{21}(s) & T_{20}(s) & \dots & \vdots \\ \vdots & \vdots & & 0 \\ & & & T_{20}(s) \\ T_{2m_2}(s) & T_{2m_2-1}(s) & & \vdots \\ 0 & T_{2m_2}(s) & & \vdots \\ \vdots & 0 & & \vdots \\ 0 & \dots & 0 & T_{2m_2}(s) \end{bmatrix}}_{(m_1+1)\times(m_1-m_2+1)}, \quad \overline{Q}(s) = \begin{bmatrix} Q_0(s) \\ Q_1(s) \\ \vdots \\ Q_{m_1-m_2}(s) \end{bmatrix}$$

Now, using an algorithm from [8] or [1], a solution $\overline{Q}(s)$ to the finite dimensional optimization problem (17) can be found, which is a solution to the original infinite dimensional optimization problem (13). □

Remark 3. The problem is solved here as an optimization problem of the H_∞-norm of some transfer matrix. The analysis of the structure at infinity of $T_1(s,z)$ and $T_2(s,z)$ may allow to know whether there exists an exact solution for the model matching problem (13) or not (see [11, 15] for more details). □

6 Illustrative Examples

In this section the proposed method is applied on two examples. In the first one, the system has delay on its input but the value of this delay is uncertain. In the second one, the system has internal delay and the system parameters are uncertain.

6.1 Example 1

Consider the following transfer function [1]

$$G(s,z) = \frac{kz}{1+\tau s} \quad (18)$$

The parameters of the system are $k = 2.5$, $\tau = 2.5$ and the delay is presumed to be in the interval [2, 3], i.e. $2 \le h \le 3$ (note that $z = e^{-sh}$). The nominal value of the delay is $h = 2.5$, thus the nominal transfer function is

$$G_n(s,z) = \frac{z}{s+0.4} \quad (19)$$

A spectrally co-canonical state space representation of (19) is given by

$$\begin{cases} \dot{x}(t) = -0.4x(t) + \nabla u(t) \\ y(t) = x(t) \end{cases} \tag{20}$$

A double Bézout factorization for the transfer matrix (19) associated with the state space representation (20) can be obtained using (4) and (6) as follows

$$M = \overline{M} = \frac{s+0.4}{s+1.4} \qquad N = \overline{N} = \frac{z}{s+1.4}$$
$$Y = \overline{Y} = \frac{s^2+2.8s+1.96-e^{-1}z}{(s+0.4)(s+1.4)} \qquad X = \overline{X} = \frac{e^{-1}}{s+1.4}$$

Based on this factorization, the set of all stable observers for $r(t) = x(t)$ is given by (7) and (8) with

$$P(s) = \frac{z}{s+1.4}$$

and $Q(s,z)$ is any matrix belonging to $\mathcal{M}(\Theta[z])$.

Now, the changes in the delay h has to be specified in frequency domain by a weighting matrix $W(s)$ representing the uncertainties on the *nominal* transfer matrix (19). In this example, we have modelled these parameter variations as an additive uncertainty. Therefore, a transfer function $W(s)$ has to be found such that the following inequality is satisfied

$$\max_{2\leq h\leq 3}\left|G(s,e^{-sh}) - G_{nominal}(s,e^{-sh})\right| \leq |W(s)| \tag{21}$$

By plotting the left hand side of the above inequality, we can choose the following weighting function (see Fig. 3)

$$W(s) = \frac{0.1s^2+2.8s+0.15}{s^2+s+0.3} \tag{22}$$

Now the model matching problem (13) has to be solved for $T_1(s,z)$ and $T_2(s,z)$ given by (12). For this example they equal to

$$T_1(s,z) = \frac{(0.1s^2+2.8s+0.15)e^{-1}z}{(s+1.4)^2(s^2+s+0.3)} \qquad T_2(s,z) = -\frac{(0.1s^2+2.8s+0.15)(s+0.4)}{(s+1.4)^2(s^2+s+0.3)}$$

The model matching problem (13) can be rewritten in the form (17) with

$$\overline{T_1}(s) = \begin{bmatrix} 0 \\ \frac{0.1e^{-1}(s^2+28s+1.5)}{(s+1.4)^2(s^2+s+0.3)} \end{bmatrix}$$

$$\overline{T_2}(s) = \begin{bmatrix} -\frac{(0.1s^2+2.8s+0.15)(s+0.4)}{(s+1.4)^2(s^2+s+0.3)} & 0 \\ 0 & -\frac{(0.1s^2+2.8s+0.15)(s+0.4)}{(s+1.4)^2(s^2+s+0.3)} \end{bmatrix}$$

$$\overline{Q}(s) = \begin{bmatrix} Q_0(s) & Q_1(s) \end{bmatrix}^T$$

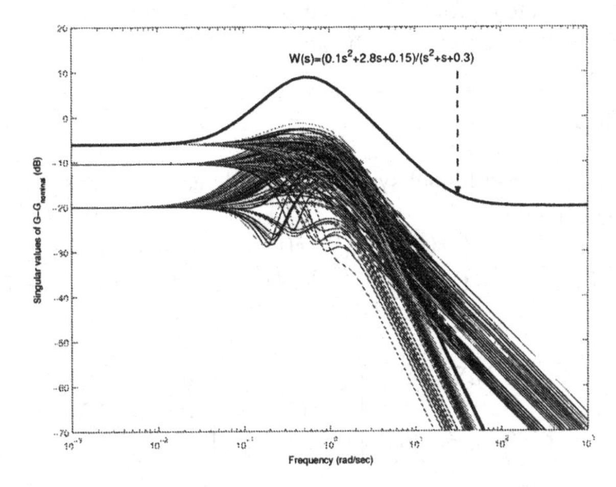

Fig. 3. Left hand side of (21) and the corresponding weighting function $W(s)$.

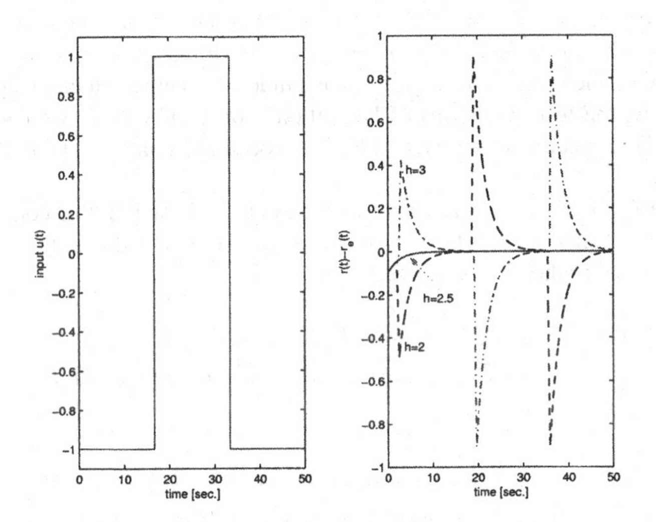

Fig. 4. Estimated error for for different values of h.

Applying the algorithm of Francis [8] on (17), we found that $Q_0(s) \equiv 0$ and $Q_1(s) = -0.3679/(s^2+1.8s+0.56)$ with $\gamma = 0$, i.e. the robust observer is an exact observer. The robust observer of $r(t) = x(t)$ is given by (7) with

$$U(s,z) = \frac{z}{s+0.4}, \qquad V(s,z) \equiv 0$$

Fig. 4 shows the estimated error for different values of h. From this figure, we can state that estimated error converges to zero for all admissible values of h but the convergence rate is different for each value.

6.2 Example 2

In this example, a wind tunnel system is considered which has the following state space model [12]

$$\begin{cases} \dot{x}(t) = A(\nabla)x(t) + B(\nabla)u(t) \\ y(t) = C(\nabla)x(t) \end{cases} \tag{23}$$

with

$$A(\nabla) = \begin{bmatrix} -a & ka\nabla & 0 \\ 0 & 0 & 1 \\ 0 & -\omega^2 & -2\zeta\omega \end{bmatrix}, \quad B(\nabla) = \begin{bmatrix} 0 \\ 0 \\ \omega^2 \end{bmatrix}, \quad C(\nabla) = \begin{bmatrix} 1 & 0 & 0 \end{bmatrix}$$

where $a = \frac{1}{\tau}$ and τ, k, ζ, ω are parameters depending on the operating point and they lie in the following intervals $0.739 \leq \tau \leq 2.58, \quad -0.0144 \leq k \leq -0.0029$. Their nominal values are $\tau = 1.964, \quad k = -0.0117$. Other parameters are presumed constant $\zeta = 0.8$, $\omega = 6$. The components of the state vector $x(t)$ are the change in Mach number, the guide vane angle and the guide vane angle velocity respectively.

Applying the test conditions of Definition 1 on system (23), we can find that system (23) is spectrally canonical, i.e., it is spectrally controllable and spectrally observable.

A state feedback control law is calculated in [12]. Based on this control law, an observer-based state feedback for system (23) can be obtained as follows (note that a constant gain for the observer is sufficient here)

$$\begin{cases} u(t) = F_1(\nabla)x(t) + F_2 \int_{-h}^{0} e^{a\theta} x(t+\theta)d\theta \\ \dot{\hat{x}}(t) = A(\nabla)\hat{x}(t) + B(\nabla)u(t) + K\left(y(t) - C(\nabla)\hat{x}(t)\right) \end{cases} \tag{24}$$

where

$$F_1(\nabla) = \begin{bmatrix} 661.5612 & 1.2399 + 0.0432\nabla & 0.1003 \end{bmatrix} \times 10^{-3}$$
$$F_2 = \begin{bmatrix} 0 & -4.6881 & 0.0510 \end{bmatrix} \times 10^{-3}$$
$$K = \begin{bmatrix} 0.5 & 0 & 0 \end{bmatrix}^T$$

The transfer matrix of system (23) is given by

$$G(s,z) = \frac{ka\omega^2 z}{(s+a)(s^2 + 2\zeta\omega s + \omega^2)} \tag{25}$$

Based on the above information, a double Bézout factorization for the transfer matrix (25) for nominal values of its parameters associated with the state space representation (23) can be obtained using (4) and (6) as follows

$$M = \frac{s^3+10.11s^2+40.89s+18.33}{s^3+10.11s^2+40.84s+18.48} \qquad \overline{M} = \frac{s+0.5092}{s+0.009165}$$

$$N = \frac{-0.2145}{s^3+10.11s^2+40.84s+18.48}z \qquad \overline{N} = \frac{-0.2145}{s^3+9.609s^2+36.09s+0.3299}z$$

$$Y = \frac{s^3+10.11s^2+40.84s+18.48}{s^3+10.11s^2+40.89s+18.33} \qquad \overline{Y} = \frac{s+0.009165}{s+0.5092}$$

$$+\frac{-3.168e-006s^2-0.001582s+0.07092}{s^4+10.12s^3+40.98s^2+18.7s+0.168}z \qquad +\frac{-3.168e-006s^2-0.001582s+0.07092}{s^4+10.61s^3+45.99s^2+39.27s+9.407}z$$

$$X = \frac{0.3308}{s+0.009165} \qquad \overline{X} = \frac{0.3308s^2+3.175s+11.91}{s^3+10.11s^2+40.84s+18.48}$$

The set of all stable observers for $r(t) = [0\ \ 1\ \ 0]x(t)$ is given by (7) and (8) with

$$P(s) = \frac{36s+18.33}{s^3+10.11s^2+40.84s+18.48}$$

and $Q(s,z)$ is any matrix belonging to $\mathcal{M}(\Theta[z])$.

The matrix $Q(s,z)$ has to be found such that the effects of the τ and k parameter changes on the estimation are minimized. To this end, changes in the parameters τ and k have to be specified in frequency domain by a weighting matrix $W(s)$ representing the uncertainties on the *nominal* transfer matrix (25). In this example, we have modelled these parameter variations as direct input multiplicative uncertainty. Therefore, a transfer function $W(s)$ has to be found such that the following inequality is satisfied

$$\max_{\substack{0.739\leq\tau\leq2.58\\-0.0144\leq k\leq-0.0029}} \left| \frac{G(s,e^{-sh})}{G_{nominal}(s,e^{-sh})} - 1 \right| \leq |W(s)| \qquad (26)$$

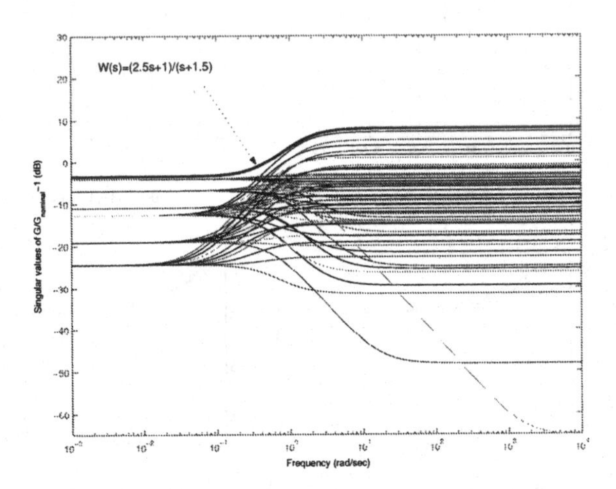

Fig. 5. Left hand side of (26) and the corresponding weighting function $W(s)$.}

By plotting the left hand side of the above inequality, we can choose the following weighting function (see Fig. 5)

$$W(s) = \frac{2.5s + 1}{s + 1.5} \tag{27}$$

Now the model matching problem (13) has to be solved for $T_1(s,z)$ and $T_2(s,z)$ given by (12). For this example, they equal to

$$\begin{aligned} T_1(s,z) &= \frac{90s+36}{s^3+11.1s^2+50.4s+54} \\ &+ \frac{-0.0002851s^3-0.1425s^2+6.326s+2.553}{s^7+21.21s^6+203.6s^5+1037s^4+2819s^3+3162s^2+1026s+9.144} z \\ T_2(s,z) &= \frac{-0.5362s-0.2145}{s^4+11.11s^3+50.5s^2+54.46s+0.4949} z \end{aligned}$$

The model matching problem (13) can be rewritten in the form (17) with

$$\overline{T_1}(s) = \begin{bmatrix} \frac{90s+36}{s^3+11.1s^2+50.4s+54} \\ \frac{-0.0002851s^3-0.1425s^2+6.326s+2.553}{s^7+21.21s^6+203.6s^5+1037s^4+2819s^3+3162s^2+1026s+9.144} \end{bmatrix}$$

$$\overline{T_2}(s) = \begin{bmatrix} 0 \\ \frac{-0.5362s-0.2145}{s^4+11.11s^3+50.5s^2+54.46s+0.4949} \end{bmatrix}$$

$$\overline{Q}(s) = Q_0(s)$$

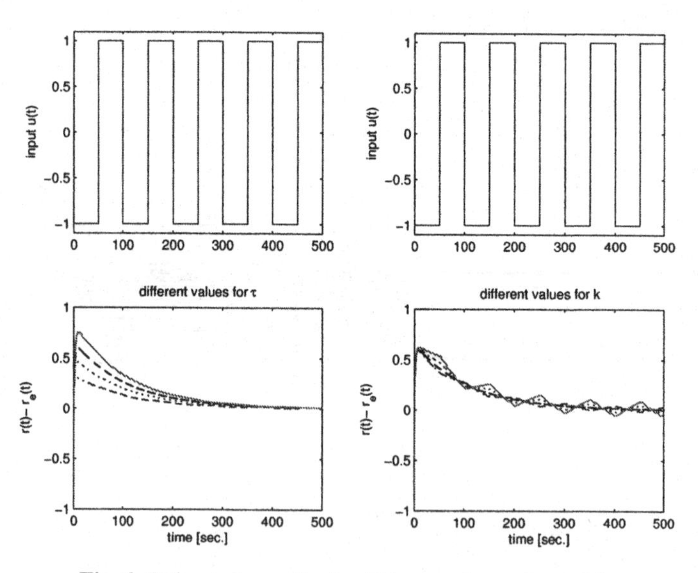

Fig. 6. Estimated error for for different values of τ and k.

Applying the algorithm of Francis [8] on (17), we found that $Q_0(s) \equiv 0$ with $\gamma = 1.26$ which means that the nominal (central) observer is robust. This observer is given by (7) with

$$U = \frac{36}{s^2 + 9.6s + 36} + \frac{-0.0001141s^2 - 0.05693s + 2.553}{s^6 + 19.71s^5 + 174s^4 + 775.9s^3 + 1655s^2 + 680.2s + 6.096} z$$

$$V = \frac{11.91s + 6.063}{s^4 + 10.11s^3 + 40.93s^2 + 18.85s + 0.1693}$$

Fig. 6 shows the estimated error for different values of τ and k. From those figures we can state that estimated error converges to zero for all admissible values of τ and k but the convergence rate is different for each value.

7 Concluding Remarks

A robust observer has been designed for time-delay systems. The time-delay system is assumed to be spectrally controllable and spectrally observable and the uncertainties are assumed to be defined by a weighting matrix in the frequency domain. The design method is based on a polynomial factorization of the nominal transfer matrix and on the parameterization of the set of all stable observers. The parameterization term is then found as a solution of an optimization problem depending on the uncertainty type. The assigned eigenvalues by the observer-based state feedback in the factorization stage influence the solution of the final optimization problem. This relationship between the closed-loop eigenvalues and the optimization problem will be considered in the future work.

References

1. Doyle, J. C., B. A. Francis and A. R. Tannenbaum (1992). *Feedback control theory*. Macmillan Publishing Company.
2. Dugard, L. and E. I. Verriest (1998). *Stability and control of time-delay systems*. Springer.
3. Fattouh, A., O. Sename and J.-M. Dion (1998). H_∞ observer design for time-delay systems. In 'Proc. 37th IEEE Confer. on Decision & Control'. Tampa, Florida, USA. pp. 4545–4546.
4. Fattouh, A., O. Sename and J.-M. Dion (1999*a*). 'Robust observer design for time-delay systems: A Riccati equation approach'. *Kybernetika* **35**(6), 753–764.
5. Fattouh, A., O. Sename and J.-M. Dion (1999*b*). An unknown input observer design for linear time-delay systems. In 'Proc. 38th IEEE Confer. on Decision & Control'. Phoenix, Arizona, USA. pp. 4222–4227.
6. Fattouh, A., O. Sename and J.-M. Dion (2000). Robust observer design for linear uncertain time-delay systems: A factorization approach. In '14th Int. Symp. on Mathematical Theory of Networks and Systems'. Perpignan, France, June, 19-23.
7. Fiagbedzi, Y. A. and A. E. Pearson (1990). 'Exponential state observer for time-lag systems'. *Int. Journal of Control* **51**(1), 189–204.

8. Francis, B. (1987). *A course in H_∞ control theory*. Springer-Verlag.
9. Kamen, E. W., P. P. Khargonekar and A. Tannenbaum (1986). 'Proper stable bezout factorizations and feedback control of linear time-delay systems'. *Int. Journal of Control* **43**(3), 837–857.
10. Lee, E. B. and A. Olbrot (1981). 'Observability and related structural results for linear hereditary systems'. *Int. Journal of Control* **34**(6), 1061–1078.
11. Malabre, M. and R. Rabah (1993). 'Structure at infinity, model matching and disturbance rejection for linear systems with delays'. *Kybernetika* **29**(5), 485–498.
12. Manitius, A. Z. (1984). 'Feedback controllers for a wind tunnel model involving a delay: Analytical design and numerical simulation'. *IEEE Trans. on Automatic Control* **29**(12), 1058–1068.
13. Niculescu, S.-I. (2001). *Delay effects on stability. A robust control approach.* Vol. 269. Springer-Verlag: Heidelbeg.
14. Nobuyama, E. and T. Kitamori (1990). Spectrum assignment and parameterization of all stabilizing compensators for time-delay systems. In 'Proc. 29th Confer. on Decision and Control'. Honolulu, Hawaii. pp. 3629–3634.
15. Picard, P., J. F. Lafay and V. Kučera (1998). 'Model matching for linear systems with delays and 2D systems'. *Automatica* **34**(2), 183–191.
16. Ramos, J. L. and A. E. Pearson (1995). 'An asymptotic modal observer for linear autonomous time lag systems'. *IEEE Trans. on Automatic Control* **40**(7), 1291–1294.
17. Sename, O. (2001). 'New trends in design of observers for time-delay systems'. *Kybernetika* **37**(4), 427–458.
18. Sontag, E. D. (1976). 'Linear systems over commutative rings: A survey'. *Ricerche di Automatica* **7**(1), 1–34.
19. Wang, Q.-G., T. H. Lee and K. K. Tan (1999). *Finite spectrum assignment for time-delay systems*. Springer-Verlag.
20. Watanabe, K. and T. Ouchi (1985). 'An observer of systems with delays in state variables'. *Int. Journal of Control* **41**(1), 217–229.
21. Yao, Y. X., Y. M. Zhang and R. Kovacevic (1996). 'Parameterization of observers for time delay systems and its application in observer design'. *IEE Proc.-Control Theory Appl.* **143**(3), 225–232.

Part IV

Computation, Software, and Implementation

Adaptive Integration of Delay Differential Equations

Alfredo Bellen and Marino Zennaro

Dipartimento di Scienze Matematiche, Università degli Studi di Trieste, 34100 Trieste, Italy
bellen@univ.trieste.it, zennaro@univ.trieste.it

Summary. We consider the numerical solution of nonstiff delay differential equations by means of a variable stepsize continuous Runge-Kutta method, the *advancing method*, of discrete order p and uniform order $q \leq p$. As in the well-known case of ordinary differential equations, the stepsize control mechanism is based on the use of a second method, the *error-estimating method*, of order $p' \neq p$, used to adapt the current stepsize in order that the local error fits a user-supplied tolerance TOL. We detect the minimal uniform order q needed for the correct performance of both the advancing and the error-estimating methods. After stating the relationships among global and local errors, we also discuss the effectiveness of the stepsize control mechanism in connection to the possible use of an additional continuous error-estimating method, which is used to monitor the uniform local error. A complete and detailed description of both the advancing method of uniform order 4 and the discrete error-estimating method of order 5 is given so as to enable the interested reader to implement his own code. A uniform error-estimating mechanism is also given which is recommendable for the reliability of the overall procedure. Finally, numerical experiments with a constructed equation are carried out aimed to test and check the results provided by the theory.

1 Introduction

We consider the numerical solution of nonstiff delay differential equations (DDEs) of the form

$$\begin{cases} y'(t) = f\Big(t, y(t), y\big(t - \tau(t, y(t))\big)\Big), & t_0 \leq t \leq t_f, \\ y(t) = \phi(t), \quad t \leq t_0, \end{cases}$$

by means of variable stepsize continuous Runge-Kutta (RK) methods of discrete (nodal) order p and uniform order $q \leq p$. Our aim is to ensure, as far as possible, the proportionality of the maximum global error to a given tolerance TOL and to minimize the computational cost.

It is well known that, for ordinary differential equations (ODEs), the stepsize control mechanism for a (discrete) *advancing method* of order p is mostly based on the use of a second method, the *error-estimating method*, used to adapt the current stepsize in order that the local error fits the tolerance TOL. The error-estimating method has order p', usually $p+1$ or $p-1$. When $p' = p+1$ the term TOL is used as a *tolerance per unit step* whereas, for $p' = p-1$, it is used as a *tolerance per step*. However, in both cases the global error turns out to be proportional to TOL, even though in the second case the proportionality can fail in presence of fortuitous and unpredictable superconvergence of the error-estimating method. Some imbedded pairs of advancing and error-estimating methods of order p and $p+1$ respectively has been proposed by Fehlberg for $p = 2$, $p = 4$ and $p = 7$ and are known as Runge-Kutta-Fehlberg pairs $RKF2(3)$, $RKF4(5)$ and $RKF7(8)$. The alternative approach with $p' = p-1$ has been considered much later by Dormand and Prince [2] who provided, for $p = 5$ and $p' = 4$ the pair DoPri5(4). For a comprehensive analysis of this topic and presentation of more general imbedded pairs, we refer the interested reader to the book by Hairer, Nørsett and Wanner [4].

Since the step-by-step numerical integration of DDEs is based on the use of continuous approximations with possibly different orders of discrete and continuous local errors, the extension of the error control mechanism is not straighforward, and its mathematical justification is far from being obvious.

For DDEs, the local problem to be solved by the advancing method is

$$\begin{cases} w'_{n+1}(t) = f\big(t, w_{n+1}(t), x(t - \tau(t, w_{n+1}(t)))\big), & t_n \le t \le t_{n+1}, \\ w_{n+1}(t_n) = y_n, \end{cases} \tag{1}$$

$$x(s) = \begin{cases} \phi(s) & \text{for } s \le t_0, \\ \eta(s) & \text{for } t_0 \le s \le t_n, \quad \text{(known)} \\ w_{n+1}(s) & \text{for } t_n \le s \le t_{n+1}, \quad \text{(to be computed)} \end{cases}$$

where, for $s < t_n$, the delayed approximate solution $\eta(s)$ has uniform global order p, that is

$$\max_{s \le t_n} \|y(s) - \eta(s)\| = O(h^p), \tag{2}$$

where $h = \max_n h_{n+1}$.

The continuous RK method for (1) takes the form

$$\eta(t_n + \theta h_{n+1}) = y_n + h_{n+1} \sum_{i=1}^{s} b_i(\theta) K_{n+1}^i, \quad 0 \le \theta \le 1, \tag{3}$$

where the system

$$\begin{aligned} K_{n+1}^i &= f\Big(t_{n+1}^j, Y_{n+1}^j, x\big(t_{n+1}^j - \tau(t_{n+1}^j, Y_{n+1}^j)\big)\Big), \\ Y_{n+1}^i &= y_n + h_{n+1} \sum_{j=1}^{s} a_{ij} K_{n+1}^j, \end{aligned} \qquad i = 1, \ldots, s, \tag{4}$$

has to be solved for the *stage* values K_{n+1}^i (and, hence, for the Y_{n+1}^i's, that are called stage values too). Here $t_{n+1}^i = t_{n+1} + c_i h_{n+1}$ and the c_i are the *abscissae* of the RK method.

In particular, for $\theta = 1$ we get the nodal value

$$y_{n+1} = \eta(t_{n+1}) = y_n + h_{n+1} \sum_{i=1}^{s} b_i K_{n+1}^i.$$

If overlapping occurs, i.e. if, for some index i, the argument $t_{n+1}^i - \tau(t_{n+1}^i, Y_{n+1}^i)$ of $x(s)$ lies in the current interval $[t_n, t_{n+1}]$, then the *spurious stage* value $\tilde{Y}_{n+1}^i = x(t_{n+1}^i - \tau(t_{n+1}^i, Y_{n+1}^i))$ arises which is given by the formula (3) itself for

$$\theta = \theta_{n+1}^i = c_i - \frac{\tau(t_{n+1}^i, Y_{n+1}^i)}{h_{n+1}},$$

that is

$$\tilde{Y}_{n+1}^i = y_n + h_{n+1} \sum_{j=1}^{s} b_j(\theta_{n+1}^i) K_{n+1}^j. \tag{5}$$

It is worth remarking that, in the case of overlapping, the overall RK method becomes implicit even if the underlying RK method is explicit, that is the coefficient matrix (a_{ij}) is strictly lower triangular. Remark also that overlapping may occur in equations with state-dependent delay, where the value of the delay τ is impredictable, as well as in equations with constant delay for large values of the stepsize.

It is known that, as long as the solution is smooth, in order for the advancing DDE method to perform to the discrete and uniform global order p it is sufficient (and necessary) to use a discrete RK method of order p (local error $O(h^{p+1})$) and a continuous extension of order q (uniform local error $O(h^{q+1})$) with $q = p - 1$ only. Remark that, in this case, the local and global uniform errors in the DDE method are both $O(h^p)$. In other words, the error of the continuous extension does not propagate as t increases.

On the other hand, if the initial function $\phi(t)$ and the solution $y(t)$ do not link smoothly (or even continuously) at t_0, a number of discontinuity points (called *breaking points*) propagate along the integration interval, where the solution is not regular enough and the desired order of the error is no longer guaranteed by the numerical method. In this case the required conditions are recovered by simply including the breaking points as mesh points in order to proceed across steps where the solution is piecewise as smooth as necessary. Whereas locating the breaking points is trivial for constant and time-dependent delays, it is not an easy task for state-dependent delays where their location depends on the solution itself. For an insight on possible strategies for going around these difficulties see Bellen and Zennaro [1, Chapter 4]. In particular, the recent code RADAR5 developed by Guglielmi and Hairer [3], gives up locating the breaking points and try to control the local error by simply suitably reducing the stepsize in their proximity.

However, the choice of the uniform order q in the advancing method has an impact on the accuracy of the error-estimating method. More precisely:

- If the (discrete) error-estimating method has order $p' = p + 1$ (local error $O(h^{p+2})$), then the uniform order of the continuous advancing method should be raised to $q = p$ (local and global uniform error $O(h^{p+1})$). On the contrary, if $p' = p - 1$, then $q = p - 1$ is sufficient. For a rigorous proof of this statement, see [1, Chapter 7].

2 The Stepsize Control Mechanism

In order to state the results on the stepsize control let us consider a mesh $\Delta = \{t_0, t_1, \ldots, t_N = t_f\}$ along the integration interval $[t_0, t_f]$ and denote the maximum discrete and uniform global errors by

$$e_n = \max_{i \leq n} \|y(t_i) - y_i\| \quad \text{and} \quad E_n = \max_{t \leq t_n} \|y(t) - \eta(t)\|.$$

Moreover, let

$$\sigma_{n+1} = \|w_{n+1}(t_{n+1}) - y_{n+1}\| / h_{n+1}$$

and

$$\Sigma_{n+1} = \max_{t_n \leq t \leq t_{n+1}} \|w_{n+1}(t) - \eta(t)\|$$

be the *discrete local error per unit step* and the *uniform local error* of the (local) problem (1) respectively. Finally, let us define

$$e_\Delta = \max_n e_n, \qquad \sigma_\Delta = \max_n \sigma_n \quad \text{and} \quad \Sigma_\Delta = \max_n \Sigma_n.$$

By a suitable modification of the standard propagation error analysis taking into account the error on the approximated delayed term x in (1), the conclusions we arrive at are synthetized in the following two points:

- The following relation among global and local errors holds independently of the presence of overlapping:

  $$e_\Delta \leq K(\sigma_\Delta + R\Sigma_\Delta). \tag{6}$$

  Therefore

  $$\sigma_n \leq TOL \quad \text{and} \quad \Sigma_n \leq TOL \implies e_\Delta \leq K' \cdot TOL.$$

  So we can say that both the discrete and the uniform local errors should be controlled, rather than the sole discrete local error as is done in ODE codes. More precisely, the given tolerance TOL should bound both the discrete local error per unit step and the uniform local error. For the latter, one can use a continuous error-estimating method of uniform order q' other than q, say $q' = q - 1$.
- Under the option $q = p$ in the uniform order of the advancing method, one might be satisfied with controlling the discrete local error only. In fact, we have

  $$p = q \implies \Sigma_\Delta = hO(\sigma_\Delta).$$

  Therefore, as TOL decreases the contribution of Σ_Δ in (6) becomes smaller and smaller and eventually it turns out to be negligible in comparison with σ_Δ.

A rigorous proof of the above statements is given in [1, Chapter 7] in the simplified case of variable non state-dependent delay.

The case $q = p - 1$ leads to local errors σ_Δ and Σ_Δ of the same order so that it is recommendable to submit both of them to the test of tolerance.

3 A Sample Method

In this section we illustrate the foregoing theory by means of a sample method. Our presentation shows some degrees of freedom, so that some parameters are not fixed a priori. They have to be chosen by the reader who wants to actually implement the method we propose.

We illustrate the strategy based on an advancing method of uniform order $q = p$ endowed with a discrete error-estimating method of order $p' = p + 1$.

In this setting, we base our procedure on explicit RK methods of order $p = 4$, namely the classical four-stage RK method

$$\begin{array}{c|cccc} 0 & & & & \\ \frac{1}{2} & \frac{1}{2} & & & \\ \frac{1}{2} & 0 & \frac{1}{2} & & \\ 1 & 0 & 0 & 1 & \\ \hline & \frac{1}{6} & \frac{1}{3} & \frac{1}{3} & \frac{1}{6} \end{array}$$

In order to maintain the order $p = 4$ of the discrete RK method, when implementing formulae (3), (4), (5), it is sufficient to use an interpolant of order $p-1 = 3$ in the current step $[t_n, t_{n+1}]$. It turns out that, as the interpolant of uniform order 3 in the current step $[t_n, t_{n+1}]$, it is convenient to consider the so-called *natural continuous extension* (NCE) of degree 3, introduced by Zennaro [5], defined as follows:

$$\hat{\eta}_{(0)}(t_n + \theta h_{n+1}) = y_n + h_{n+1} \sum_{i=1}^{4} b_i(\theta) K_{n+1}^i, \quad 0 \le \theta \le 1, \tag{7}$$

where

$$\begin{aligned} b_1(\theta) &= (\tfrac{2}{3}\theta^2 - \tfrac{3}{2}\theta + 1)\theta, \\ b_2(\theta) &= (-\tfrac{2}{3}\theta + 1)\theta^2, \\ b_3(\theta) &= (-\tfrac{2}{3}\theta + 1)\theta^2, \\ b_4(\theta) &= (\tfrac{2}{3}\theta - \tfrac{1}{2})\theta^2, \end{aligned}$$

and the stage values are

$$K_{n+1}^1 = f\big(t_n, y_n, \eta(t_n - \tau(t_n, y_n))\big);$$

$$K_{n+1}^2 = f\big(t_n + \tfrac{1}{2}h_{n+1}, y_n + \tfrac{1}{2}h_{n+1}K_{n+1}^1, \tilde{Y}_{n+1}^2\big),$$

with

$$\begin{aligned}\tilde{Y}_{n+1}^2 &= \hat{\eta}_{(0)}\big(t_n + \tfrac{1}{2}h_{n+1} - \tau(t_n + \tfrac{1}{2}h_{n+1}, y_n + \tfrac{1}{2}h_{n+1}K_{n+1}^1)\big) \\ &= y_n + h_{n+1}\sum_{i=1}^{4} b_i\Big(\frac{1}{2} - \frac{\tau(t_n + \frac{1}{2}h_{n+1}, y_n + \frac{1}{2}h_{n+1}K_{n+1}^1)}{h_{n+1}}\Big)K_{n+1}^i\end{aligned}$$

if $t_n + \frac{1}{2}h_{n+1} - \tau(t_n + \frac{1}{2}h_{n+1}, y_n + \frac{1}{2}h_{n+1}K_{n+1}^1) > t_n$, and

$$\tilde{Y}_{n+1}^2 = \eta\big(t_n + \tfrac{1}{2}h_{n+1} - \tau(t_n + \tfrac{1}{2}h_{n+1}, y_n + \tfrac{1}{2}h_{n+1}K_{n+1}^1)\big)$$

otherwise;

$$K_{n+1}^3 = f\big(t_n + \tfrac{1}{2}h_{n+1}, y_n + \tfrac{1}{2}h_{n+1}K_{n+1}^2, \tilde{Y}_{n+1}^3\big),$$

with

$$\begin{aligned}\tilde{Y}_{n+1}^3 &= \hat{\eta}_{(0)}\big(t_n + \tfrac{1}{2}h_{n+1} - \tau(t_n + \tfrac{1}{2}h_{n+1}, y_n + \tfrac{1}{2}h_{n+1}K_{n+1}^2)\big) \\ &= y_n + h_{n+1}\sum_{i=1}^{4} b_i\Big(\frac{1}{2} - \frac{\tau(t_n + \frac{1}{2}h_{n+1}, y_n + \frac{1}{2}h_{n+1}K_{n+1}^2)}{h_{n+1}}\Big)K_{n+1}^i\end{aligned}$$

if $t_n + \frac{1}{2}h_{n+1} - \tau(t_n + \frac{1}{2}h_{n+1}, y_n + \frac{1}{2}h_{n+1}K_{n+1}^2) > t_n$, and

$$\tilde{Y}_{n+1}^3 = \eta\big(t_n + \tfrac{1}{2}h_{n+1} - \tau(t_n + \tfrac{1}{2}h_{n+1}, y_n + \tfrac{1}{2}h_{n+1}K_{n+1}^2)\big)$$

otherwise;

$$K_{n+1}^4 = f\big(t_n + h_{n+1}, y_n + h_{n+1}K_{n+1}^3, \tilde{Y}_{n+1}^4\big),$$

with

$$\begin{aligned}\tilde{Y}_{n+1}^4 &= \hat{\eta}_{(0)}\big(t_n + h_{n+1} - \tau(t_n + h_{n+1}, y_n + h_{n+1}K_{n+1}^3)\big) \\ &= y_n + h_{n+1}\sum_{i=1}^{4} b_i\Big(1 - \frac{\tau(t_n + h_{n+1}, y_n + h_{n+1}K_{n+1}^3)}{h_{n+1}}\Big)K_{n+1}^i\end{aligned}$$

if $t_n + h_{n+1} - \tau(t_n + h_{n+1}, y_n + h_{n+1}K_{n+1}^3) > t_n$, and

$$\tilde{Y}_{n+1}^4 = \eta\big(t_n + h_{n+1} - \tau(t_n + h_{n+1}, y_n + h_{n+1}K_{n+1}^3)\big)$$

otherwise.

Once we have computed the discrete approximation $y_{n+1} = \hat{\eta}_{(0)}(t_{n+1})$ obtained from (7) for $\theta = 1$, we do not consider the interpolant $\hat{\eta}_{(0)}(t)$ any more, but we switch to the *cubic Hermite interpolant* $\eta_{(0)}(t)$ at the end-points t_n and t_{n+1}, which also makes use of the extra stage

$$K_{n+1}^5 = f\big(t_{n+1}, y_{n+1}, \tilde{Y}_{n+1}^5\big),$$

where

$$\begin{aligned}\tilde{Y}^5_{n+1} &= \hat{\eta}_{(0)}\big(t_{n+1} - \tau(t_{n+1}, y_{n+1})\big) \\ &= y_n + h_{n+1} \sum_{i=1}^{4} b_i \Big(1 - \frac{\tau(t_{n+1}, y_{n+1})}{h_{n+1}}\Big) K^i_{n+1}\end{aligned}$$

if $t_{n+1} - \tau(t_{n+1}, y_{n+1}) > t_n$, and

$$\tilde{Y}^5_{n+1} = \eta(t_{n+1} - \tau(t_{n+1}, y_{n+1}))$$

otherwise.

In other words, we consider the interpolant

$$\eta_{(0)}(t_n + \theta h_{n+1}) = d_1(\theta) y_n + d_2(\theta) y_{n+1} + d_3(\theta) h_{n+1} K^1_{n+1} + d_4(\theta) h_{n+1} K^5_{n+1},$$

where

$$\begin{aligned}d_1(\theta) &= (\theta - 1)^2 (2\theta + 1), \\ d_2(\theta) &= \theta^2 (3 - 2\theta), \\ d_3(\theta) &= \theta(\theta - 1)^2, \\ d_4(\theta) &= \theta^2 (\theta - 1).\end{aligned}$$

Note that, if overlapping does not occur in the current step, then K^5_{n+1} equals $K^1_{n+2} = f\big(t_{n+1}, y_{n+1}, \eta(t_{n+1} - \tau(t_{n+1}, y_{n+1}))\big)$, which is available for free from the next step $[t_{n+1}, t_{n+2}]$.

In order to construct the continuous approximation $\eta(t)$ of order $q = 4$ for the advancing method in the current step $[t_n, t_{n+1}]$, we perform one uniform correction of $\eta_{(0)}(t)$. Such a procedure is a particular case of a more general approach for raising the uniform order, based on successive uniform corrections starting from a lower order interpolant (see [1, Chapter 5]).

To this aim, consider the extra abscissa $\theta_1 \in (0, 1)$, $\theta_1 \neq \frac{1}{2}$, and the extra stage

$$K^6_{n+1} = f\big(t_n + \theta_1 h_{n+1}, \eta_{(0)}(t_n + \theta_1 h_{n+1}), \tilde{Y}^6_{n+1}\big),$$

where

$$\tilde{Y}^6_{n+1} = \eta_{(0)}\big(t_n + \theta_1 h_{n+1} - \tau(t_n + \theta_1 h_{n+1}, \eta_{(0)}(t_n + \theta_1 h_{n+1}))\big)$$

if $t_n + \theta_1 h_{n+1} - \tau(t_n + \theta_1 h_{n+1}, \eta_{(0)}(t_n + \theta_1 h_{n+1})) > t_n$, and

$$\tilde{Y}^6_{n+1} = \eta\big(t_n + \theta_1 h_{n+1} - \tau(t_n + \theta_1 h_{n+1}, \eta_{(0)}(t_n + \theta_1 h_{n+1}))\big)$$

otherwise.

Thus $\eta(t)$ is defined as the fourth-degree Hermite–Birkhoff interpolant relevant, for the generic polynomial $P(\theta)$ of degree four, to the interpolation conditions on $P(0)$, $P(1)$, $P'(0)$, $P'(1)$ and $P'(\theta_1)$. Note that $\theta_1 \neq \frac{1}{2}$ because the interpolation condition on $P'(\frac{1}{2})$ is linearly dependent on the other four conditions. We obtain

$$\eta(t_n + \theta h_{n+1}) = d_1(\theta)y_n + d_2(\theta)y_{n+1} + d_3(\theta)h_{n+1}K^1_{n+1}$$
$$+d_4(\theta)h_{n+1}K^5_{n+1} + d_5(\theta)h_{n+1}K^6_{n+1},$$

where

$$d_1(\theta) = \tfrac{1}{2\theta_1-1}(\theta-1)^2\big(-3\theta^2 + 2(2\theta_1-1)\theta + 2\theta_1 - 1\big),$$
$$d_2(\theta) = \tfrac{1}{2\theta_1-1}\theta^2\big(3\theta^2 - 4(\theta_1+1)\theta + 6\theta_1\big),$$
$$d_3(\theta) = \tfrac{1}{2\theta_1(2\theta_1-1)}\theta(\theta-1)^2\big((1-3\theta_1)\theta + 2\theta_1(2\theta_1-1)\big),$$
$$d_4(\theta) = \tfrac{1}{2(\theta_1-1)(2\theta_1-1)}\theta^2(\theta-1)\big((2-3\theta_1)\theta + \theta_1(4\theta_1-3)\big),$$
$$d_5(\theta) = \tfrac{1}{2\theta_1(2\theta_1-1)(\theta_1-1)}\theta^2(\theta-1)^2.$$

In order to estimate the discrete local error by means of an error-estimating method of higher order $p' = p + 1 = 5$, we construct an approximation $\tilde{y}_{n+1}$ by using a suitable quadrature formula. More precisely, we consider the Lobatto abscissae $\pi_1 = \frac{5-\sqrt{5}}{10}$ and $\pi_2 = \frac{5+\sqrt{5}}{10}$ and the additional stages

$$K^7_{n+1} = f\big(t_n + \pi_1 h_{n+1}, \eta(t_n + \pi_1 h_{n+1}), \tilde{Y}^7_{n+1}\big),$$

where

$$\tilde{Y}^7_{n+1} = \eta\big(t_n + \pi_1 h_{n+1} - \tau(t_n + \pi_1 h_{n+1}, \eta(t_n + \pi_1 h_{n+1}))\big)$$

if $t_n + \pi_1 h_{n+1} - \tau(t_n + \pi_1 h_{n+1}, \eta(t_n + \pi_1 h_{n+1})) > t_n$, and

$$\tilde{Y}^7_{n+1} = \eta\big(t_n + \pi_1 h_{n+1} - \tau(t_n + \pi_1 h_{n+1}, \eta(t_n + \pi_1 h_{n+1}))\big)$$

otherwise;

$$K^8_{n+1} = f\big(t_n + \pi_2 h_{n+1}, \eta(t_n + \pi_2 h_{n+1}), \tilde{Y}^8_{n+1}\big),$$

where

$$\tilde{Y}^8_{n+1} = \eta\big(t_n + \pi_2 h_{n+1} - \tau(t_n + \pi_2 h_{n+1}, \eta(t_n + \pi_2 h_{n+1}))\big)$$

if $t_n + \pi_2 h_{n+1} - \tau(t_n + \pi_2 h_{n+1}, \eta(t_n + \pi_2 h_{n+1})) > t_n$, and

$$\tilde{Y}^8_{n+1} = \eta\big(t_n + \pi_2 h_{n+1} - \tau(t_n + \pi_2 h_{n+1}, \eta(t_n + \pi_2 h_{n+1}))\big)$$

otherwise.

Then the quadrature formula is

$$\tilde{y}_{n+1} = y_n + h_{n+1}\big(\tfrac{1}{12}K^1_{n+1} + \tfrac{5}{12}K^7_{n+1} + \tfrac{5}{12}K^8_{n+1} + \tfrac{1}{12}\overline{K}^5_{n+1}\big),$$

where, as we have already observed,

$$\overline{K}^5_{n+1} = K^1_{n+2} = f\big(t_{n+1}, y_{n+1}, \eta(t_{n+1} - \tau(t_{n+1}, y_{n+1}))\big)$$

may or may not equal the already available stage K^5_{n+1}.

As is usually done, we estimate the discrete local error per unit step by

$$\tilde{\sigma}_{n+1} = \|\tilde{y}_{n+1} - y_{n+1}\|/h_{n+1}.$$

The current step is accepted if and only if

$$\tilde{\sigma}_{n+1} \leq TOL, \tag{8}$$

where TOL is the user-supplied tolerance.

The new stepsize h_{new}, to be used either as h_{n+2} in the next step $[t_{n+1}, t_{n+2}]$ if (8) is satisfied or, again, as h_{n+1} in the current step $[t_n, t_{n+1}]$ if (8) is violated, is computed by using the common formula

$$h_{\text{new}} = \max\left\{\omega_{\min}, \min\left\{\omega_{\max}, \rho\left(\frac{TOL}{\tilde{\sigma}_{n+1}}\right)^{1/4}\right\}\right\} \cdot h_{n+1}, \tag{9}$$

where $\omega_{\min} \in (0,1)$, $\omega_{\max} > 1$ and $\rho \in (0,1)$ are suitable *safety factors*.

A maximum number of rejections per step may be fixed by the user as well.

Since the uniform order is $q = p$, controlling the discrete local error is sufficient for the stepsize selection. Nevertheless, for more reliability, we also propose an additional optional control of the uniform local error Σ_{n+1} by a continuous error-estimating method of lower order $q' = q - 1 = 3$, by employing local extrapolation. More precisely, we consider the computable quantity

$$\hat{\Sigma}_{n+1} = \max_{0 \leq \theta \leq 1} \|\eta_{(0)}(t_n + \theta h_{n+1}) - \eta(t_n + \theta h_{n+1})\|. \tag{10}$$

It can be seen that

$$\Sigma_{\Delta} = O(TOL) \tag{11}$$

is assured if we impose

$$h_{n+1}\hat{\Sigma}_{n+1} \leq TOL. \tag{12}$$

Note that, by (6), relation (11) is necessary to ensure error-tolerance proportionality.

The test (12) is made whenever (8) holds. If (12) is not satisfied, then we reduce the stepsize by using the formula

$$h_{\text{new2}} = \max\left\{\omega_{\min}, \rho\left(\frac{TOL}{h_{n+1}\hat{\Sigma}_{n+1}}\right)^{1/5}\right\} \cdot h_{n+1} \tag{13}$$

and repeat the computations in the current step $[t_n, t_{n+1}]$ with $h_{n+1} = h_{\text{new2}}$.

In any case, even if (8) is satisfied, the stepsize h_{new2} suggested by (13) is used to bound the next step h_{n+2}, so that (9) is substituted by

$$h_{n+2} = \max\left\{\omega_{\min}, \min\left\{\omega_{\max}, \rho\left(\frac{TOL}{\tilde{\sigma}_{n+1}}\right)^{1/4}, \rho\left(\frac{TOL}{h_{n+1}\hat{\Sigma}_{n+1}}\right)^{1/5}\right\}\right\} \cdot h_{n+1}.$$

In order to compute $\hat{\Sigma}_{n+1}$ from (10), we set

$$\delta(\theta) = \eta_{(0)}(t_n + \theta h_{n+1}) - \eta(t_n + \theta h_{n+1}).$$

It turns out that

$$\delta(\theta) = h_{n+1}\frac{\theta^2(\theta-1)^2}{2(2\theta_1-1)}\Big(\frac{2\theta_1-1}{\theta_1}K^1_{n+1} - (2K^2_{n+1} + 2K^3_{n+1} + K^4_{n+1}) + \frac{3\theta_1-2}{\theta_1-1}K^5_{n+1} + \frac{1}{\theta_1(\theta_1-1)}K^6_{n+1}\Big),$$

so that

$$\begin{aligned}\hat{\Sigma}_{n+1} &= \|\delta(\tfrac{1}{2})\| \\ &= \frac{h_{n+1}}{32|2\theta_1-1|}\Big\|\frac{2\theta_1-1}{\theta_1}K^1_{n+1} - (2K^2_{n+1} + 2K^3_{n+1} + K^4_{n+1}) \\ &\quad + \frac{3\theta_1-2}{\theta_1-1}K^5_{n+1} + \frac{1}{\theta_1(\theta_1-1)}K^6_{n+1}\Big\|.\end{aligned}$$

4 A Numerical Experiment

We conclude by applying the method introduced in the previous section to the constructed equation

$$\begin{cases} y'(t) = -y(t) + y\left(\alpha(t)\right) + \frac{t}{20}\cos\left(\frac{t}{20}\right) + \sin\left(\frac{t}{20}\right) - \sin\left(\frac{\alpha(t)}{20}\right), \; 0 \le t \le 1000, \\ y(t) = \sin\left(\frac{t}{20}\right), \quad -20 \le t \le 0, \end{cases} \tag{14}$$

with variable (non state-dependent) deviated argument $\alpha(t) = t - 1 + \sin(t)$ corresponding to a delay $\tau(t) = 1 - \sin(t)$ which periodically vanishes at the points $\frac{\pi}{2} + 2k\pi$, k positive integer. The exact solution $y(t) = \sin\left(\frac{t}{20}\right)$ is smooth and hence no breaking points have to be included as mesh points for preserving the piecewise regularity of the solution and hence the accuracy of the methods. Remark that overlapping occurs in the proximity of every point where the delay vanishes, and possibly somewhere else. Being the exact solution available, the example is suitable to illustrate that, according to the theory, the local error estimation as well as the expected proportionality with the given tolerance are succesfully obtained also in presence of overlapping.

For the actual implementation of the method described in Section 3, we have chosen $\theta_1 = \frac{1}{3}$ for the computation of K^6_{n+1}, and $\omega_{\min} = 0.5$, $\omega_{\max} = 1.5$, $\rho = 0.9$ in (9) and (13).

Since $q = p$, the error-tolerance proportionality is guaranteed under the sole discrete error control $\sigma_n \le TOL$ (see Table 1 where $\sigma-$rejections stands for the number of steps rejected by the test on the discrete error). In Table 2 we provide the results obtained by performing also the continuous error control $\Sigma_n \le TOL$. The term Σ-rejections stands for the number of steps rejected by the test on the continuous error control. As expected from the theory, for sufficiently small tolerance

TOL, the continuous error control does not work any more and the global error is essentially determined by the discrete error controller. However, the estimation of the uniform error Σ_n is taken into account in the prediction of the next stepsize. This is why the results in the last three rows of Table 1 and Table 2 are not just identical.

TOL	$\sigma-$rejections	steps	overlappings	$\frac{e_N}{TOL}$
10^{-1}	5	17	17	32.17
10^{-2}	108	200	200	25.34
10^{-3}	210	429	429	26.72
10^{-4}	208	687	430	11.03
10^{-5}	253	1211	523	14.50
10^{-6}	280	2120	649	20.28
10^{-7}	265	3741	770	20.12

Table 1. Numerical results for equation (14) with the sole discrete test of tolerance.

TOL	$\sigma-$rejections	$\Sigma-$rejections	steps	overlappings	$\frac{e_N}{TOL}$
10^{-1}	2	158	276	276	1.47
10^{-2}	1	238	415	415	2.86
10^{-3}	16	238	573	462	3.61
10^{-4}	128	112	780	446	10.38
10^{-5}	209	0	1196	509	16.40
10^{-6}	235	0	2121	650	15.64
10^{-7}	253	0	3740	771	17.19

Table 2. Numerical results for equation (14) with the discrete and uniform test of tolerance.

References

1. A. Bellen, M. Zennaro, *Numerical Methods for Delay Differential Equations*, Numerical Mathematics and Scientific Computation Series, Oxford University Press, 2003.
2. J.R. Dormand and P.J. Prince, A family of embedded Runge–Kutta formulae, *J. Comput. Appl. Math.* **6** (1980), 19–26
3. N. Guglielmi, E. Hairer, Implementing Radau IIA methods for stiff delay differential equations, *Computing* **67** (2001), 1–12
4. E. Hairer, S.P. Nørsett and G. Wanner, *Solving Ordinary Differential Equations I, Nonstiff Problems*, Springer-Verlag, Berlin, 1993.
5. M. Zennaro, Natural Continuous Extensions of Runge-Kutta methods, *Math. Comput.* **46** (1986), 119–133

Software for Stability and Bifurcation Analysis of Delay Differential Equations and Applications to Stabilization

Dirk Roose, Tatyana Luzyanina, Koen Engelborghs, and Wim Michiels

K.U. Leuven, Department of Computer Science, Celestijnenlaan 200A, 3001 Heverlee, Belgium {firstname.lastname}@cs.kuleuven.ac.be

Summary. DDE-BIFTOOL is a Matlab software package for the stability and bifurcation analysis of parameter-dependent systems of delay differential equations. Using continuation, branches of steady state solutions and periodic solutions can be computed. The local stability of a solution is determined by computing relevant eigenvalues (steady state solutions) or Floquet Multipliers (periodic solutions). Along branches of solutions, bifurcations can be detected and branches of fold or Hopf bifurcation points can be computed. We outline the capabilities of the package and some of the numerical methods upon which it is based. We illustrate the usage of the package for the analysis of two model problems and we outline applications and extensions towards controller synthesis problems. We explain how its stability routines can be used for the implementation of the continuous pole placement method, which allows to solve stabilization problems where multiple parameters need to be tuned simultaneously.

1 Introduction

Mathematical modeling with delay differential equations (DDEs) is widely used in various application areas of engineering (e.g. semi-conductor lasers with delayed feedback) and in the life sciences (e.g. population dynamics, epidemiology, immunology, physiology, neural networks), see e.g. [8]. Most often, numerical methods are the only possible way to achieve a complete analysis, prediction and control of these models.

DDE-BIFTOOL is a collection of Matlab routines for the numerical stability and bifurcation analysis of systems of delay differential equations (DDEs) with multiple discrete delays,

$$\dot{x}(t) = f(x(t), x(t-\tau_1), \ldots, x(t-\tau_m), \eta), \tag{1}$$

where $x(t) \in \mathbb{R}^n$, $f : \mathbb{R}^{n(m+1)} \times \mathbb{R}^p \to \mathbb{R}^n$ is a nonlinear smooth function, $\eta \in \mathbb{R}^p$ are parameters and $\tau_i > 0$, $i = 1, \ldots, m$, denote the delays, which can be fixed or state-dependent, i.e. $\tau_i = g_i(x(t))$.

It can be used to compute branches of steady state solutions and branches of fold and Hopf bifurcation points using continuation. Given a steady state solution, it approximates the rightmost, stability determining roots of the characteristic equation using time-integration of the variational equation. Periodic solutions and their Floquet multipliers are computed using orthogonal collocation with adaptive mesh selection. Branches of periodic solutions can be continued starting from a previously computed Hopf point or from an initial guess of a periodic solution. The package does not perform a time integration of DDEs, but the Matlab routine dde23 [39] can be used in combination with DDE-BIFTOOL.

In this chapter we briefly outline the numerical methods upon which DDE-BIFTOOL is based and the structure of the package. We illustrate its capabilities by presenting results for two model problems and indicate how the package can be used to solve controller synthesis problems.

Detailed information on the numerical methods and the use of DDE-BIFTOOL can be found in [18–20, 23, 25] and the references therein. For example, DDE-BIFTOOL has been used to analyze mathematical models arising in physiology [21], immunology [30] and in analyzing the stability of semi-conductor lasers with delayed feedback and the associated nonlinear phenomena (low frequency fluctuations) [26], [37]. The package is freely available for scientific purposes.

2 Numerical Methods

Steady state solutions. A steady state solution $x^* \in \mathbb{R}^n$ of (1) is computed by solving the n-dimensional algebraic system

$$f(x^*, x^*, \ldots, x^*, \eta) = 0 \tag{2}$$

with η fixed. Note that x^* does not depend on the delays. The stability of a steady state solution is determined by the roots of the characteristic equation

$$\det\left(\lambda I - A_0 - \sum_{i=1}^{m} A_i e^{-\lambda \tau_i}\right) = 0, \tag{3}$$

where, using $f \equiv f(x^0, x^1, \ldots, x^m, \eta)$,

$$A_i \triangleq \left.\frac{\partial f}{\partial x^i}\right|_{(x^*, x^*, \ldots, x^*, \eta)}, \quad i = 0, \ldots, m. \tag{4}$$

The rightmost (stability determining) characteristics roots of (3) are approximated as follows. The solution operator of the linearization of (1),

$$\dot{y}(t) = A_0 y(t) + \sum_{i=1}^{m} A_i y(t - \tau_i), \tag{5}$$

is discretized over one time step h using a linear multistep method, combined with Lagrange interpolation to evaluate the delayed terms. In this way a linear map (a matrix) is constructed which approximates the linear solution operator. Eigenvalues of this matrix, μ_h, approximate a finite number of the eigenvalues μ of the solution operator which correspond, by the exponential transform $\mu = e^{\lambda h}$, to the characteristic roots λ of (3). Once μ_h are computed, the approximations λ_h to the roots λ are extracted using

$$\Re(\lambda_h) = \frac{1}{h}\ln(|\mu_h|), \quad \Im(\lambda_h) = \frac{1}{h}\arcsin(\frac{\Im(\mu_h)}{|\mu_h|}) \ (\mathrm{mod}\ \frac{\pi}{h}).$$

A steplength heuristic for h [18] is used to ensure accurate approximations of all the roots with real part greater than a given constant. The obtained approximations can be corrected using a Newton iteration on the characteristic matrix.

Periodic and homoclinic solutions. Periodic solutions are computed using piecewise polynomial collocation [23]. Adaptive mesh selection allows the efficient computation of solutions with steep gradients. Approximations to the Floquet multipliers, which determine the local asymptotic stability of a periodic solution, are computed as eigenvalues of the discretised Monodromy operator obtained from the collocation equations. Homoclinic orbits are computed in a similar way as periodic solutions by imposing projection boundary conditions on the profile, see [38].

Continuation and bifurcations. The dependence of a steady state solution (or periodic solution) on a physical parameter can be studied by computing a branch of steady state solutions (periodic solutions) using a continuation procedure. The branch is computed by a combination of secant predictions and Newton corrections. The steplength strategy is based on a combination of extrapolations and interpolations [19].

The stability of the steady state can change during continuation whenever characteristic roots cross the imaginary axis. Generically a fold bifurcation (or turning point) occurs when a real characteristic root passes through zero and a Hopf bifurcation occurs when a pair of complex conjugate characteristic roots crosses the imaginary axis. Once a fold point or Hopf point is detected it can be followed in a two parameter space using an appropriate determining system (Equation (2) extended with equations involving an eigenvalue and eigenvector of the characteristic matrix). In this way, one can compute, for instance, the stability region of a steady state solution in the two parameter space.

A branch of periodic solutions can be started from a Hopf point or from an initial guess (e.g. resulting from time integration). A bifurcations of periodic solutions occurs when Floquet multipliers move into or out of the unit circle. Generically this is a turning point when a real multiplier crosses through 1, a period doubling point when a real multiplier crosses through -1 and a torus bifurcation when a complex pair of multipliers crosses the unit circle.

Extra conditions. The package allows to add extra (algebraic) conditions and corresponding free parameters to a determining system. We mention two possible applications of such extra conditions.

(i) Extra conditions allow to continue branches in a higher-dimensional parameter space with (possibly nonlinear) dependence between some parameters. E.g., a branch of steady state solutions can be continued by varying two parameters η_1, η_2 under the condition $\eta_1 2 + \eta_2 2 = 1$.

(ii) Also specific properties of a given system may require the introduction of an extra (algebraic) condition to ensure a unique solution. This situation occurs e.g., with phase shifts in oscillators, because periodic solutions are invariant w.r.t. time-shifts.

Note that these situations differ from the case of a system of differential-algebraic equations. In this case, the algebraic equations are part of the system definition and determine the stability together with the differential equations. In our application extra conditions are used to select special solutions of the (given) differential system.

3 Structure of DDE-BIFTOOL

The package is structured into four layers, each with a different functionality.

Layer zero contains the system definition and is provided by the user. It consists of a routine to evaluate the right hand side f and a routine to evaluate the derivatives of f required by the different determining systems and their Jacobian matrices. A default file is available which implements finite difference approximations to the desired derivatives.

Layer one forms the numerical core of the package and is (normally) not directly accessed by the user. The functionality of this layer is hidden by and used through layers two and three.

Layer two contains routines to manipulate individual points. A point has one of the following four types. It can be a steady state point, steady state Hopf or fold bifurcation point, a periodic solution point or an homoclinic orbit point. Furthermore a point can contain additional information concerning its stability. Routines are provided to compute individual points, to compute and plot their stability and to convert points from one type to another. The latter allows to switch e.g. from a steady state solution to a steady state fold or Hopf bifurcation or from a Hopf bifurcation to the emanating branch of periodic solutions.

Layer three contains routines to manipulate branches. A branch is the combination of an array of points and three sets of method parameters. The array contains points of the same type ordered along the branch. Method parameters allow to guide the computation of individual points, the continuation strategy and the computation of stability. Default sets of method parameters are available which can easily be changed whenever required. Routines are provided to extend a given branch (that is, to compute extra points using continuation), to (re)compute stability along the branch and to visualize the branch and/or its stability.

4 Illustrative Examples

4.1 Model of coupled neurons

As a first example, we use the system of delay differential equations, taken from [40],

$$\begin{aligned}\dot{x}_1(t) &= \kappa x_1(t) + \beta \tanh(x_1(t-\tau_s)) + a_{12} \tanh(x_2(t-\tau_2)) \\ \dot{x}_2(t) &= \kappa x_2(t) + \beta \tanh(x_2(t-\tau_s)) + a_{21} \tanh(x_1(t-\tau_1)).\end{aligned} \tag{6}$$

This system models two coupled neurons with time delayed connections. We fix the parameters $\kappa = -0.5, \beta = -1, a_{12} = 1$ and $\tau_1 = \tau_2 = 0.2,\ \tau_s = 1.5$ and vary a_{21}. Continuation of branches of steady state and periodic solutions and their stability analysis results in the bifurcation diagram shown in Fig. 1.

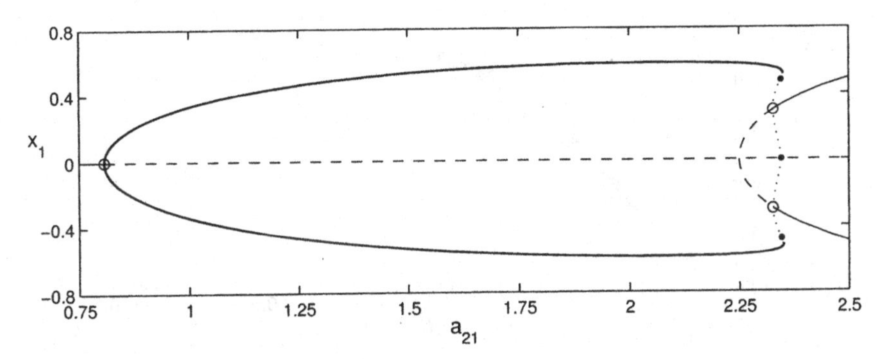

Fig. 1. Branches of stable (thin lines) and unstable (−−) steady state solutions. Hopf bifurcations (∘) and stable (thick lines) and unstable (· · ·) branches of periodic solutions emanating from these Hopf bifurcations (maximum and minimum values of $x_1(t)$ are shown). Homoclinic orbits are denoted by (•).

The branch of periodic solutions (thick lines) which emanates from the Hopf point of the zero steady state branch (thin lines) is initially stable, looses its stability in a turning point and approaches a double homoclinic loop (due to symmetry). Its solution profile is depicted in Fig. 4 (top). The symmetric branches of periodic solutions which emanate from the nonzero steady state branches are always unstable. As a_{21} grows both branches approach a (normal) homoclinic solution (see Fig. 4 (bottom)).

We now give some more details on the computation of this bifurcation diagram using DDE-BIFTOOL.

Stability of steady states. Equation (6) has a trivial steady state solution for all values of the parameters. We fix $a_{21} = 2.5$ and compute the rightmost roots of the characteristic equation at the zero steady state solution. By setting a system parameter, we indicate that we want to compute all roots λ with $\Re(\lambda) \geq -3.5$. The results are shown in Fig. 2. Approximations of these roots (obtained via time integration of the variational equations) are corrected via Newton iteration. For practical reasons, a lower bound for the steplength used in the time integration is imposed. If this bound is reached, a warning signals that approximated and corrected roots may diverge, possibly causing part of the wanted spectrum to be missed, (cf. left part of Fig. 2).

Periodic solutions with steep gradients. The possibility of adaptively refined meshes to compute periodic solutions in the collocation procedure is especially use-

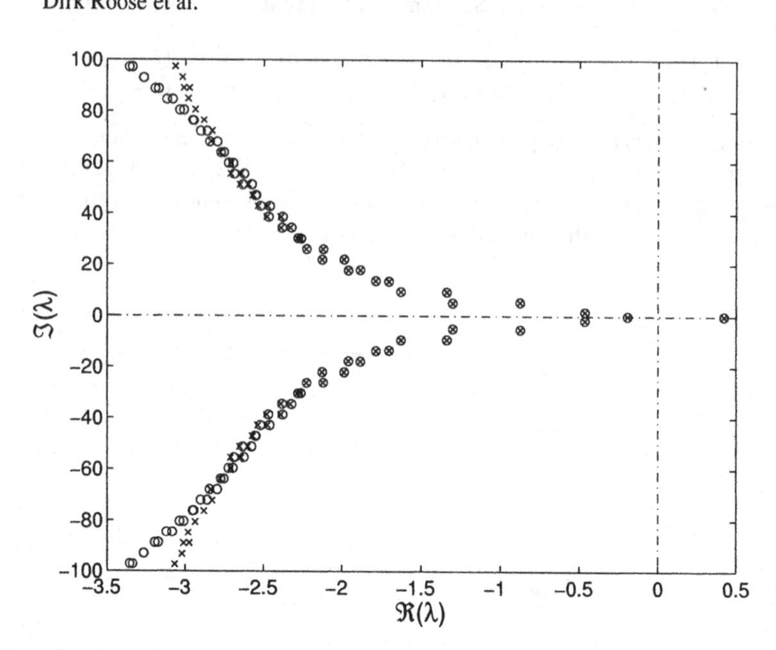

Fig. 2. Approximated (∘) and corrected (×) characteristic roots of (6) computed up to $\Re(\lambda) \geq -3.5$.

ful for solutions with steep gradients. This is illustrated by the results shown in Figs. 3 and 4. In this case, the periodic solution has a steep gradient since it is near a homoclinic orbit. When a coarse mesh is used, the solution is not computed accurately (leading also to wrong stability results), see Fig. 3. In Fig. 4 (top) we depict the same periodic solution computed using a refined mesh. The effectiveness of the mesh adaptation is apparent.

Continuation of homoclinic solutions. A branch of homoclinic solutions in two-parameter space can be approximated by imposing projection boundary conditions on the solution profile [38]. We freed a second parameter, τ_s, and take a periodic solution with a large period $T = 300$ as a starting point. In this way we computed a branch of homoclinic solutions in the (a_{21}, τ_s)-plane. This branch emanates from the Bogdanov-Takens point ($a_{21} = 2.25$, $\tau_s = 1.3$) and its continuation leads to the homoclinic solutions depicted in Fig. 1 for $\tau_s = 1.5$. The corresponding solution profiles are shown in Fig. 4.

4.2 Congestion control model

The following nonlinear delay equations, taken from [28, 31],

$$\begin{cases} \dot{W}(t) = \frac{1}{R} - \frac{1}{2}\frac{W(t)\,W(t-R)}{R}\,p(t-R) \\ \dot{q}(t) = N\frac{W(t)}{R} - C, \quad p(t) = K\,q(t) \end{cases}, \tag{7}$$

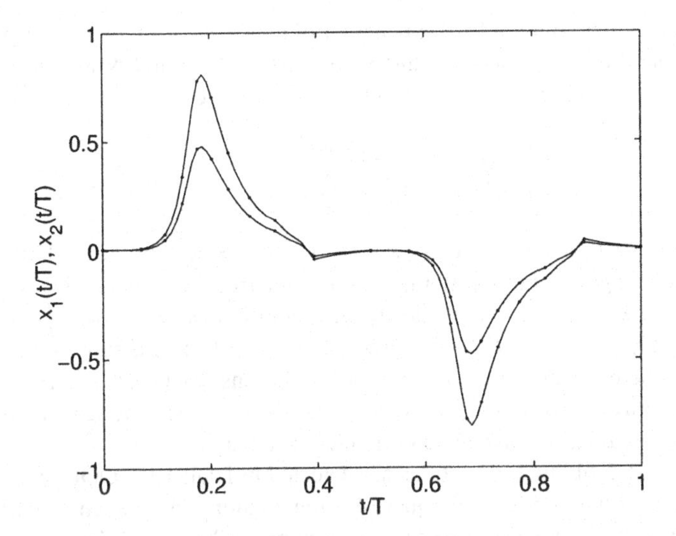

Fig. 3. Nonsmooth profile computed using a coarse mesh.

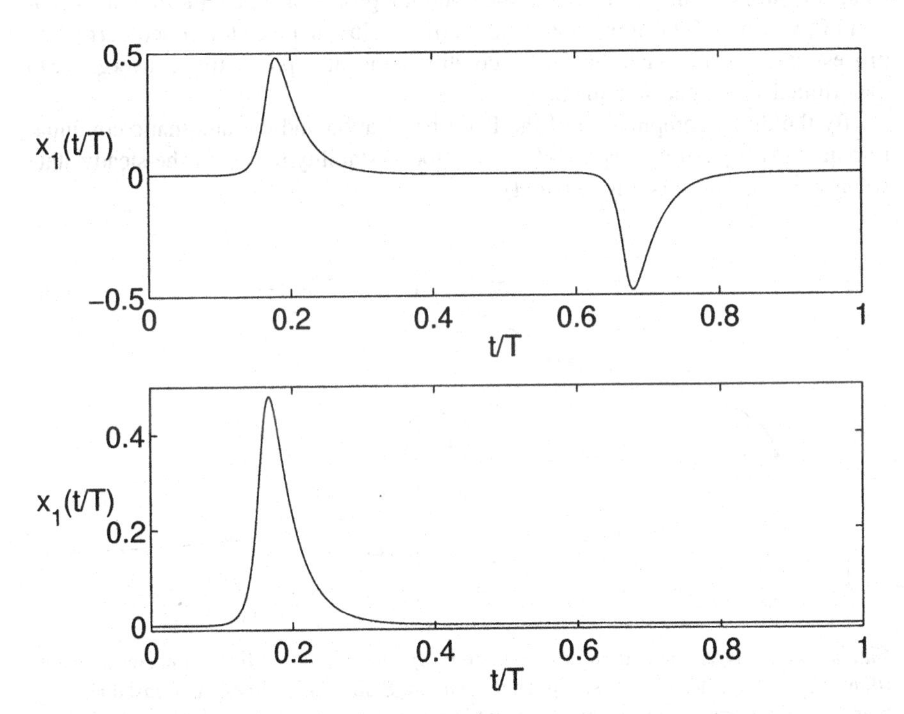

Fig. 4. Profiles of a double homoclinic (top) and a normal homoclinic (bottom) solutions. These solutions are approximated using an interval of length 300; $\tau_s = 1.5$, $a_{21} \approx 2.346$.

describe a simplified model for the behavior of congested routers in TCP-AQM networks. Under the assumption that the parameters K, C, R and N are all constant, a rescaling of state and time allows a transformation to the form

$$\begin{cases} \dot{w}(\bar{t}) = 1 - \frac{w(\bar{t})\ w(\bar{t}-1)}{2} kq(\bar{t}-1) \\ \dot{q}(\bar{t}) = w(\bar{t}) - c. \end{cases}, \qquad (8)$$

where $\bar{t} = t/R,\ w = W,\ q = QN,\ k = KN$ and $c = \frac{RC}{N}$. In Fig. 5 (left) we show a bifurcation diagram of this system, when k is the free parameter and $c = 1$ is fixed.

For small $k > 0$ the unique steady state solution $(w, q) = \left(c, \frac{2}{kc2}\right)$ is locally asymptotically stable. As k is increased, stability is lost in a subcritical Hopf bifurcation, where a branch of stable periodic solutions emanates. The latter become unstable after a period doubling bifurcation. As shown in [31], the sequence of period doubling bifurcations ultimately leads to chaotic behavior.

The period doubling bifurcations are detected in DDE-BIFTOOL by computing and monitoring the dominant Floquet Multipliers along the branch of periodic solutions. Near the bifurcation one can easily jump to the period doubled branch by using an automatically created periodic doubling profile on a new mesh (concatenation of two times the original mesh) as starting value in a Newton based correction process and an appropriate steplength condition, the latter preventing convergence to the original single period branch.

By the direct computation of the Hopf bifurcation and the automatic continuation in the two-parameter space (k, c) the (local) stability region of the steady state solution is obtained, see Fig. 5 (right).

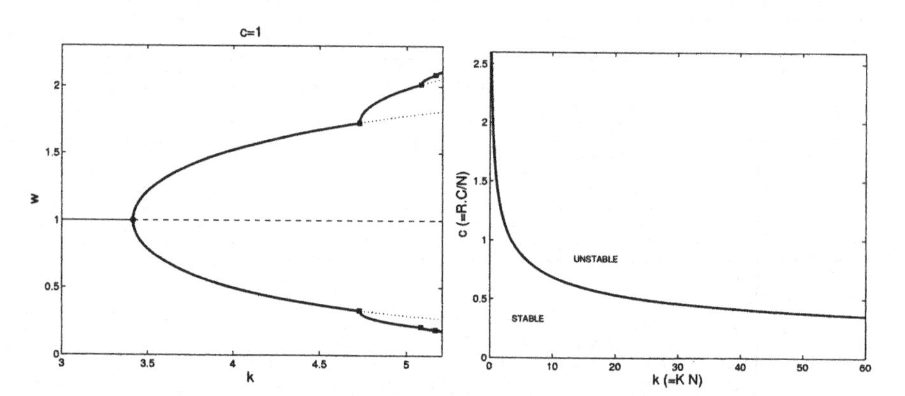

Fig. 5. (left) Bifurcation diagram of the system (8) with $c = 1$ and free parameter k. Stable (thin line) and unstable (−−) steady state solution. Stable (thick lines) and unstable $(\cdot\cdot\cdot)$ branches of periodic solutions (maximum and minimum values of $w(t)$ are shown). Connections are formed by a Hopf Bifurcation (∘) and by Period Doubling Bifurcations (□). (right) Stability region of the steady state solution in the (k, c)-plane, whose boundary is formed by a branch of Hopf bifurcations.

5 Applicability to Controller Synthesis Problems

We briefly outline the capabilities of DDE-BIFTOOL to solve controller *synthesis* problem, focusing on stabilization.

5.1 Systems with one parameter

When a stabilization problem consists of tuning *one parameter* the routines to compute the rightmost eigenvalues or dominant Floquet multipliers and the continuation facility of DDE-BIFTOOL are readily applicable. If a stabilizing solution is found, stability regions can be explored in a multi-parameter space. Here, note that in the continuation routines of the package constant time-delays are treated in the same way as other system's parameters.

The continuation approach is not necessarily limited to problems with only one physical parameter. Often theoretical considerations allow a *reduction* of multiple parameters to one parameter, as we now illustrate with an example.

Example. A chain of n integrators,

$$y^{(n)}(t) = u(t), \;\; n \geq 1, \tag{9}$$

is stabilizable with a delayed output feedback controller of the form

$$u(t) = \sum_{i=1}^{n} k_i \, y(t-\tau_i),$$

see [36]. Based on a derivative feedback approximation idea, a reduction of the $2n$ controller parameters (k_i, τ_i) to one parameter is possible, as expressed in the following result [36, Theorem 3.4]:

Theorem 1. *Assume that* $0 \leq \tau_1 < \ldots < \tau_n$ *and that the polynomial* $q(\lambda) = \lambda^n + \sum_{i=1}^{n} q_{n-i}\lambda^{n-i}$ *is Hurwitz. Then the control law*

$$u(t) = -\left[\varepsilon^n q_0 \;\; \frac{\varepsilon^{n-1}}{(-1)} q_1 \;\; \frac{2!\,\varepsilon^{n-2}}{(-1)^2} q_2 \;\; \cdots \;\; \frac{(n-1)!\,\varepsilon}{(-1)^{n-1}} q_{n-1}\right] \cdot \begin{bmatrix} 1 & \tau_1 & \cdots & \tau_1^{n-1} \\ \vdots & & & \vdots \\ 1 & \tau_n & \cdots & \tau_n^{n-1} \end{bmatrix}^{-1} \begin{bmatrix} y(t-\tau_1) \\ \vdots \\ y(t-\tau_n) \end{bmatrix} \tag{10}$$

achieves asymptotic stability of (9), for sufficiently small values of $\varepsilon > 0$.

An upper bound for parameter ε guaranteeing closed loop stability can be directly computed with DDE-BIFTOOL.

Analogous types of control laws, depending on one parameter, can be found in the context of low-gain control, see e.g. [35], and singular perturbations [14]. Note that numerical threshold computations are exact and often much better than analytically obtained bounds, which may be conservative. ◇

5.2 Systems with multiple parameters

The situation is different when one has to determine multiple controller parameters at the same time. For instance, consider an open loop unstable linear control system of the form

$$\dot{x}(t) = \sum_{i=1}^{n_x} A_i\, x(t-\tau_{x,i}) + \sum_{i=1}^{n_u} B_i\, u_i(t-\tau_{u,i}),\ x \in \mathbb{R}^n,\ u_i \in \mathbb{R}^{m_i}, \tag{11}$$

where all the time-delays in states and inputs are constant and a stabilizing feedback controller

$$u_i(t) = K_i^T\, x(t),\ \ i = 1 \ldots n_u, \tag{12}$$

is searched for. Tuning the elements of the feedback gains K_i one by one using continuation and eigenvalue computations, while keeping the others fixed, will generally not lead to a solution. Also a classical pole placement approach cannot be followed since the closed-loop system has infinitely many eigenvalues, while the number of controller parameters is finite.

In [25] we have developed a constructive numerical method for the stabilization of linear time-delay systems with multiple parameters, which combines an iterative *pole-shifting algorithm* with the computation the rightmost eigenvalues of the DDE. The main idea consists of moving the rightmost or unstable eigenvalues to the left half plane in a quasi-continuous way, by applying small changes to the controller parameters, and meanwhile monitoring the other eigenvalues with a large real part. Because the method heavily relies on the continuous dependence of the rightmost eigenvalues on the controller parameters, it is called *continuous pole placement*. Applied to the problem (11) and (12), the basic algorithm is as follows:

Algorithm 1 *The continuous pole placement method*

A. *Initialize $m = 1$ and the gains K_i.*
B. *Compute the rightmost eigenvalues of (11) and (12), using the stability routines of DDE-BIFTOOL.*
C. *Compute the sensitivity of the m rightmost eigenvalues w.r.t. changes in the feedback gains K_i.*
D. *Move the m rightmost eigenvalues in the direction of the left half plane by applying a small change to the feedback gains K_i, using the computed sensitivities.*
E. *Monitor the rightmost uncontrolled eigenvalues. If necessary, increase the number of controlled eigenvalues, m. Stop when stability is reached or when the available degrees of freedom in the controller do not allow to further reduce the real part of the rightmost eigenvalue. In the other case, go to step B.*

More details on the different steps can be found in [25]. A numerical example is presented below.

Example. We consider the system with distributed delay, taken from [25],

$$\begin{cases} \dot{x}(t) = A_1\, x(t) + A_2\, x(t-\tau_1) + \int_{t-\tau_1}^{t} A_3\, x(s)ds + Bu(t-\tau_2) \\ u(t) = K^T x(t) \end{cases}, \tag{13}$$

where the system matrices are given by

$$A_1 = \begin{bmatrix} 0.1 & 0 & 0 \\ 0.2 & 0 & -0.2 \\ 0.3 & 0.1 & -0.2 \end{bmatrix}, \quad A_2 = \begin{bmatrix} -0.2 & 0 & 0 \\ -0.4 & -0.2 & 0.4 \\ -0.4 & -0.1 & 0.2 \end{bmatrix},$$
$$A_3 = \begin{bmatrix} 0.1 & -0.2 & 0 \\ 0 & 0.1 & 0.1 \\ -0.1 & 0 & 0.1 \end{bmatrix}, \quad B = \begin{bmatrix} 0.1 \\ 0 \\ 0 \end{bmatrix}, \quad \tau_1 = 6, \ \ \tau_2 = 1. \tag{14}$$

and a feedback gain $K \in \mathbb{R}^{3\times 1}$ needs to be determined.

In order to apply the continuous pole placement method, we first remove the distributed delay by differentiating (13), which leads to the following closed-loop equation with only discrete delays:

$$\dot{z}(t) = \begin{bmatrix} 0 & I \\ A_3 & A_1 \end{bmatrix} z(t) + \begin{bmatrix} 0 & 0 \\ -A_3 & A_2 \end{bmatrix} z(t-\tau_1) + \begin{bmatrix} 0 \\ BK^T \end{bmatrix} z(t-\tau_2), \tag{15}$$

where $z(t) = \begin{bmatrix} x(t) \\ \dot{x}(t) \end{bmatrix}$. This *model transformation* introduces 3 additional zero eigenvalues in the spectrum, whatever the value of K. Therefore, the zero solution of (15) is never asymptotically stable. However, the continuous pole placement method can cope with this problem because the zero eigenvalues can be removed after applying step B of Algorithm 1 to (15).

To illustrate the capability of DDE-BIFTOOL to analyze *delay sensitivity*, we show in Fig. 6 the eigenvalues of the uncontrolled system (13) and (15) as a function of the delay τ_1. For the nominal delay $\tau_1 = 6$, there are 3 eigenvalues in the open right half plane, a real eigenvalue and a pair of complex conjugate eigenvalues.

Iterations of the continuous pole placement algorithm are shown in Fig. 7. First only the dominant real eigenvalue is controlled. This way control is lost over the other eigenvalues and at around 5 iterations an unstable pair of complex conjugate eigenvalues starts moving to the right in the complex plane. This is detected and after 10 iterations both the real eigenvalue and the real part of the complex conjugate pair of eigenvalues are controlled. After 50 iterations another pair of eigenvalues becomes almost dominant and from that moment on the three degrees of freedom in the controller are used to shift the real eigenvalue and the real parts of the two complex conjugate pairs of eigenvalues. Around 90 iterations the procedure terminates and an optimum is (almost) reached, characterized by 5 rightmost eigenvalues with the same real part: a complex conjugate pair of eigenvalues with multiplicity 2 and a real eigenvalue. In Fig. 8 we depict the rightmost eigenvalues of (13) for $K = 0$ and for the final, stabilizing value of the feedback gain. ◇

The continuous pole placement method emphasizes *stabilization*. Iteration steps can be taken until the stability exponent, the real part of the rightmost eigenvalue, is minimized over the controller parameters. Therefore, the method introduces no conservatism, i.e. in principle a stabilizing solution can be found if it exists. However,

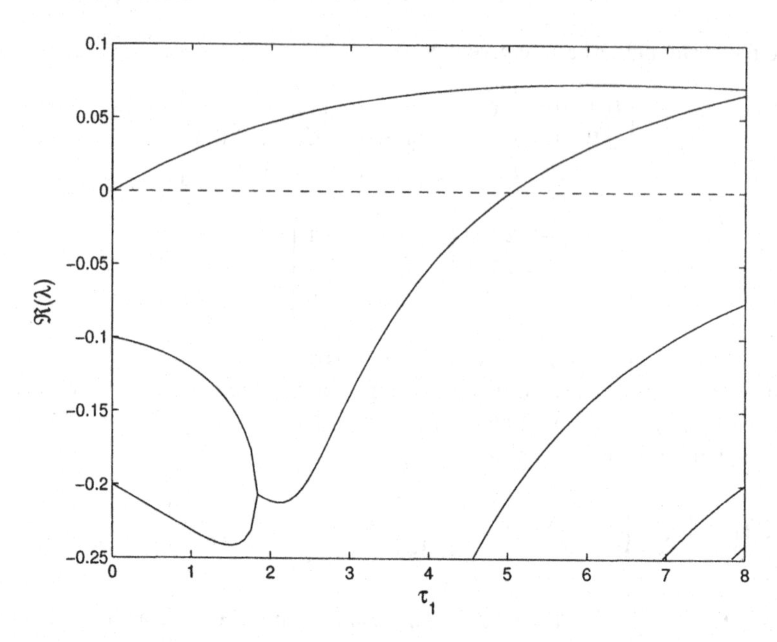

Fig. 6. Real parts of the rightmost eigenvalues of the system (13) and (15) for $K = 0$ as a function of the delay τ_1. The solid lines correspond to the eigenvalues of (13). Equation (15) has in addition a triple eigenvalue at zero (dashed line).

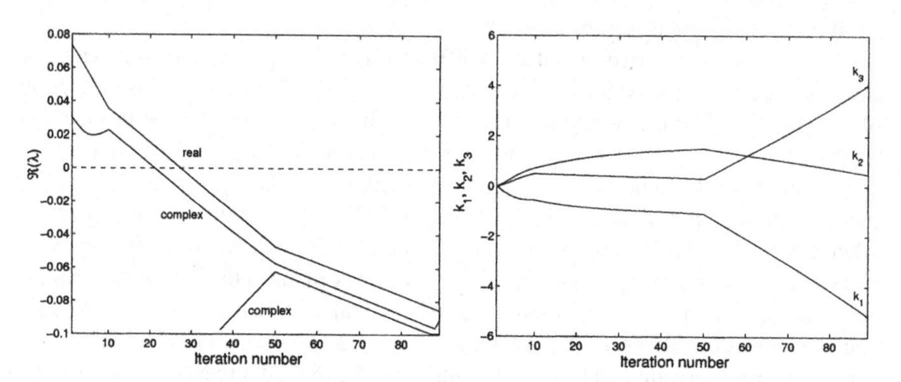

Fig. 7. Real parts of the rightmost eigenvalues of (13) and (15) (left) and components of the feedback gains $K = [k_1\ k_2\ k_3]^T$ (right) as a function of the iterations of the continuous pole placement algorithm. Starting from $K = 0$ a stabilizing feedback gain is computed.

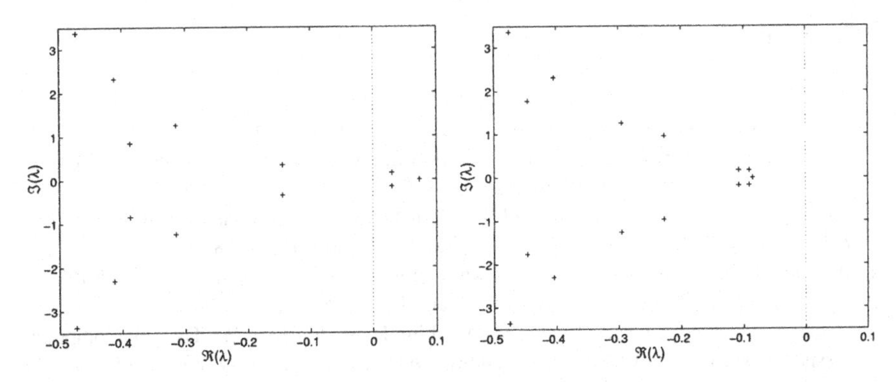

Fig. 8. Rightmost eigenvalues of (13) for $K = 0$ (left) and for the final value of the feedback gain (Iteration 89, $K = [-5.19\ 0.491\ 4.06]^T$) (right).

no attention is paid to the *robustness* of the achieved stability against perturbations of the system's parameters. Notice that precisely the minima of the stability exponent may be very sensitive to perturbations, because they are usually characterized by rightmost eigenvalues with a multiplicity larger than one, as seen in e.g. [33,34] and confirmed by the numerical example above. These observation are the motivation for the numerical procedure described in [34], which -staring from a stabilizing solution-improves the robustness of stability. In that reference static perturbations are assumed on the system matrices and the robustness of stability is expressed in terms of complex stability radii. In the numerical procedure, these stability radii are maximized as a function of the controller parameters. This corresponds to a $\mathcal{H}_\infty$ synthesis problem, which is solved by a quasi-continuous shaping of some frequency response plots. The structure of the algorithm is analogous to Algorithm 1.

6 Conclusions

In this chapter we described the software package DDE-BIFTOOL for the stability and bifurcation analysis of parameter-dependent delay differential equations. We described the usage of the package for the analysis of some model problems (model for coupled neurons, behavior of congested routers). We also indicated how the package can be used for controller synthesis problems, with the emphasis on the implementation of the continuous pole placement method.

Acknowledgments

This research has been supported by the Belgian Programme on Interuniversity Attraction Poles, initiated by the Belgian State, Prime Minister's Office for Science, Technology and Culture (IUAP P5/22) and the Research Council of the K.U.Leuven (OT/98/16). Wim Michiels is a postdoctoral fellow of the Fund for Scientific Research -Flanders(Belgium).

References

1. Argyris,J., Faust, G. and Haase, M.: *An Exploration of Chaos – An Introduction for Natural Scientists and Engineers*, North Holland Amsterdam: 1994.
2. Alsing, P.M., Kovanis, V., Gavrielides, A. and Erneux, T.: "Lang and Kobayashi phase equation," *Phys. Rev. A*, **53** (1996) 4429-4434.
3. Ascher, U.M., Mattheij, R.M.M., and Russel, R.D.: *Numerical Solution of Boundary Value Problems for Ordinary Differential Equations*, Prentice Hall: 1988.
4. Azbelev, N.V., Maksimov, V.P. and Rakhmatullina, L.F.: *Introduction to the Theory of Functional Differential Equations* (in Russian), Nauka: Moscow, 1991.
5. Back, A., Guckenheimer, J., Myers, M., Wicklin, F. and Worfolk, P.: "DsTool: Computer Assisted Exploration of Dynamical Systems," *AMS Notices*, **39** (1992) 303-309.
6. Bellman, R. and Cooke, K.L.: *Differential-Difference Equations*, Mathematics in Science and Engineering, Academic Press, 1963.
7. Bocharov, G.A.: "Modelling the dynamics of LCMV infection in mice: conventional and exhaustive CTL responses," *J.theor. Biol.* bf 192 (1998) 283-308.
8. Bocharov, G.A. and Rihan, F.A.: "Numerical modelling in biosciences using delay differential equations," *J. Comput. Appl. Math.*, **125** (2000) 183-199.
9. Chow, S.-N., and Hale, J.K.: *Methods of Bifurcation Theory*, Springer: New York, 1982.
10. Diekmann, O., van Gils, S.A., Verduyn Lunel, S.M. and Walther, H.-O.: *Delay Equations: Functional-, Complex-, and Nonlinear Analysis,* Springer: New York, 1995.
11. Doedel, E.J., Keller, H.B. and Kernevez, J.P.: "Numerical analysis and control of bifurcation problems (I): bifurcation in finite dimensions," *Internat. J. of Bifur. Chaos*, **1** (1991) 493-520.
12. Doedel, E.J., Keller, H.B. and Kernevez, J.P.: "Numerical analysis and control of bifurcation problems (II) bifurcation in infinite dimensions," *Internat. J. Bifur. Chaos*, **1** (1991) 745-772.
13. Doedel, E.J., Champneys, A.R., Fairgrieve, T.F., Kuznetsov, Y.A., Sandstede, B. and Wang, X.:, *AUTO97: Continuation and bifurcation software for ordinary differential equations*; available via `ftp.cs.concordia.ca` in directory `pub/doedel/auto`, 1997.
14. Dragan, V. and Ionita, A.: "Stabilization of singularly perturbed linear systems by state feedback with delays," *Proceedings of the Fourteenth International Symposium on Mathematical Theory of Networks and Systems*, Perpignan, France, 2000.
15. Driver, R.D.: *Ordinary and Delay Differential Equations,* Springer-Verlag: 1977.
16. El'sgol'ts, L.E. and Norkin, S.B.: *Introduction to the Theory and Application of Differential Equations with Deviating Arguments,* (Academic Press: 1973)
17. Engelborghs, K.: *Numerical Bifurcation Analysis of Delay Differential Equations* (Dept. of Computer Science, K.U.Leuven, May 2000, Leuven, Belgium).
18. Engelborghs, K. and Roose, D.: "On stability of LMS-methods and characteristic roots of delay differential equations," *SIAM J. Num. Analysis*, **40** (2002) 629-650.
19. Engelborghs, K., Luzyanina, T. and Samaey, G.: "DDE-BIFTOOL v. 2.00: a Matlab package for bifurcation analysis of delay differential equations," *Internal Report TW-330*, 2001, Department of Computer Science, K.U.Leuven, Belgium, Available from `www.cs.kuleuven.ac.be/~koen/delay/ddebiftool.shtml`.
20. Engelborghs, K. and Doedel, E.: "Stability of piecewise polynomial collocation for computing periodic solutions of delay differential equations," in *Numer. Math.*, **91** (2002) 627-648.

21. Engelborghs, K., Lemaire, V., Bélair, J. and Roose, D.: "Numerical bifurcation analysis of delay differential equations arising from physiology modeling," *J. Math. Biol.*, **42** (2001) 361-385.
22. Engelborghs, K., and Roose, D.: "Numerical computation of stability and detection of Hopf bifurcations of steady state solutions of delay differential equations," *Adv. Comput. Math.* **10** (1999) 271-289.
23. Engelborghs, K., Luzyanina, T., in 't Hout, K. J. and Roose, D.: "Collocation methods for the computation of periodic solutions of delay differential equations," *SIAM J. Sci. Comput.*, **22** (2000) 1593-1609.
24. Engelborghs, K., Luzyanina, T. and Roose, D.: "On the bifurcation analysis of a delay differential equation using DDE-BIFTOOL," *16th IMACS World Congress 2000 Proceedings*, 2000, 1-6.
25. Engelborghs, K., Luzyanina, K. and Roose, D.: "Numerical bifurcation analysis of delay differential equations using DDE-BIFTOOL," *ACM Transactions on Mathematical Software*, **28** (2002) 1-21.
26. Haegeman, B., Engelborghs, K., Roose, D., Pieroux, D. and Erneux, T.: "Stability and rupture of bifurcation bridges in semiconductor lasers subject to optical feedback," (2002) *Phys.Rev.E.* (accepted 2002).
27. Hohl, A. and Gavrielides, A.: "Bifurcation cascade in a semiconductor laser subject to optical feedback," *Phys. Rev. Lett.* **82** (1999) 1148-1151.
28. Hollot, C.V. and Chait, Y.: "Nonlinear Stability Analysis for a class of TCP/AQM Network," *Proc. of the 40th IEEE Conf. Dec. Contr.*, Orlando, FL, USA, 2001.
29. Lang, R. and Kobayashi, K.: "External optical feedback effects on semiconductor injection laser properties," *IEEE J. Quantum Electron.*, **QE-16** (1980) 347–355.
30. Luzyanina, T., Engelborghs, K., Ehl, S., Klenerman, P. and Bocharov, G.: "Low level viral persistence after infection with LCMV: a quantitative insight through numerical bifurcation analysis," *Internal Report TW-321* (2001), Department of Computer Science, K.U.Leuven, Belgium.
31. Melchor-Aguilar, D., Michiels, W. and Niculescu, S.-I.: "Remarks on Nonlinear Stability Analysis for a class of TCP/AQM Networks," (2003) (in preparation).
32. Michiels, W., Engelborghs, K., Vansevenant, P. and Roose, D.: "The continuous pole placement method for delay equations," *Automatica*, **38** (2002) 747-761.
33. Michiels, W. and Roose, D.: 'Stabilization with delayed state feedback: a numerical study," *International Journal of Bifurcation and Chaos*, **12**: 6 (2002) 1309-1320.
34. Michiels, W., and Roose, D.: "An eigenvalue based approach for the robust stabilization of linear time-delay systems," *International Journal of Control*, **76**:7 (2003) 678-686.
35. Michiels, W. and Roose, D.: "Global stabilization of multiple integrators with time-delay and input constraints," *Proc. 3th IFAC Workshop on Time-Delay Systems*, Santa Fe, NM, 266-271, 2001.
36. Niculescu, S.-I. and Michiels, W.: "Stabilizing a chain of integrators using multiple delays," *IEEE Transactions on Automatic Control* (2003) (accepted).
37. Pieroux, D., Erneux, T., Luzyanina, T. and Engelborghs, K.: "Interacting pairs of periodic solutions lead to tori in lasers subject to delayed feedback," *Physical Review E* **63** (2001).
38. Samaey, G., Engelborghs, K. and Roose, D.: "Numerical computation of homoclinic orbits in delay differential equations," *Numerical algorithms*, **30** (2002) 335-352.
39. Shampine, L.F. and Thompson, S.: *Solving ddes in matlab*, Southern Methodist University and Radford University, Dallas, Radford, (http://www.runet.edu/~thompson/webddes/), 2000.
40. Shayer, L.P. and Campbell, S.A.: "Stability, bifurcation and multistability in a system of two coupled neurons with multiple time delays," *SIAM J. Appl. Math.*, **61** (2000) 673-700.

Empirical Methods for Determining the Stability of Certain Linear Delay Systems

Richard Datko

Georgetown University, Department of Mathematics, Washington, DC 200257-1233

Summary. In this chapter we present two methods for examining the stability of certain delay systems. These methods make use of well-known computing packages such as MATLAB and are variations on both the Lyapunov and analytic function approaches. The first method is to consider systems where the delays are treated as parameters and the undelayed system is uniformly exponentially stable (u.e.s.). To find the "closest" parameters for which the parametric family loses u.e.s. we give a necessary and sufficient condition for a corresponding family of quadratic matrices to satisfy a Riccati matrix equation. This condition is relatively easy to verify and depends on the zeros of a quasi-polynomial defined on a compact Cartesian product. In particular if the delays are commeasureable this procedure either determines the smallest delays for which the system is u.e.s. or shows that it is u.e.s. for all delays.

The second method is more general and is based on a variation of the Poisson Integral representation for H-infinity mappings. This method is more useful in determining instability than stability, but is easily implemented by standard numerical packages since it involves the computation of a parametric family of scalar functions over a fixed compact interval.

1 Introduction

In this chapter we give two numerical procedures for determining the stability or instability of certain linear dynamical processes. The first method is applicable to a large class of linear autonomous delay-differential equations. It is based on a simple variant of the Lyapunov method used to determine the stability or instability of ordinary differential equations with constant coefficients and the observation that the $n \times n$ matrix equation

$$A^*X + XA = -C \tag{1}$$

(* denotes adjoint and X is unknown) has no unique solution if A has an eigenvalue on the imaginary axis.

The second method is more general and encompasses a wide variety of linear dynamical systems. It is more effective in determining instability than stability. It is based on the observations that a mapping from the complex plane into a complex Banach space is in H^∞ if and only if its Poisson Integral over the imaginary axis is analytic in the right half plane and the Cauchy-Riemann condition for analyticity. These two criteria for analyticity lead to the condition that a function F which is analytic and uniformly bounded in an open vertical strip which contains the imaginary axis is in H^∞ if and only if the integral

$$\int_{-\pi/2}^{\pi/2} e^{-2i\sigma} F[i(x \tan\sigma + y)]d\sigma = 0 \tag{2}$$

for all $x > 0$.

A note on the computations used in this report. The numerical calculations for the first method (presented in Sections 2 and 3) were performed on a TI-86 calculator using Newton's Method where differentiation was approximated by a difference scheme using the increment 1×10^{-6}. The numerical calculations for the second method (presented in Section 4) were also performed on a TI-86 calculator using Simpson's Rule and the interval $[-\frac{\pi}{2} + .0001, \frac{\pi}{2} - .0001]$. The number of divisions varied from 100 to 200. Any numerical values related to specific mathematical systems will be denoted by Eq. $\doteq$ value.

2 Preliminaries

Notation

i) R_+^m will denote the sets of vectors in the Euclidean space, R^m, which satisfy

$$h = \{(h_1, \ldots, h_m) : 0 \leq h_j, 1 \leq j \leq m\}$$

ii) K will denote the set of vectors in R^m which satisfy the condition

$$\alpha = \{(\alpha_1, \ldots, \alpha_m) : 0 \leq \alpha_j \leq 2\pi, 0 \leq j \leq m\}.$$

iii) $M(n)$ will denote the set of $n \times n$ matrices with complex entries. The identity matrix in $M(n)$ is denoted by I. If $A \in M(n)$, A^* will denote its conjugate transpose, $\overline{A}$ its complex conjugate and A' its transpose.

iv) If $B \in M(n)$ is Hermitian and positive definite, it will be denoted by $B > 0$. If it is Hermitian and negative definite, it will be denoted by $B < 0$.

v) $\det A$ denotes the determinant of $A \in M(n)$.

vi) C^n denotes the complex n-vectors. The norm in C^n is the usual l_2-norm denoted by $|\cdot|$. The matrix norm in $M(n)$ is given by

$$|A| = \sup\{|Ax| : |x| = 1\}.$$

vii) Let $A \in M(n)$. Then

$$\sigma(A) = \{Re\lambda : \lambda \text{ is an eigenvalue of } A\}.$$

viii) If $A \in M(n)$ and $\sigma(A)$ lies in the open left half complex plane, then A is said to be Hurwitzian. This is signified by

$$A \in H.$$

Some Properties of $M(n)$

Property 1. $A \in M(n)$ has no eigenvalues on the imaginary axis iff there is a unique Hermitian matrix, X, which satisfies the equation

$$XA + A^*X = -I. \tag{3}$$

Property 2. The solution of (3), if it exists, is given by the n^2-linear equation

$$[A^* \otimes I + I \otimes \overline{A}]\hat{x} = -\hat{I}, \tag{4}$$

where $\otimes$ represents the Kronecker Product of two matrices and

$$\hat{x} = \begin{pmatrix} x_{11} \\ \vdots \\ x_{nn} \end{pmatrix}, \quad \hat{I} = \begin{pmatrix} 1 \\ 0 \\ \vdots \\ 1 \end{pmatrix}$$

(see, e.g., [7], Chapter 8).

A more general form of (3), which will be used in the sequel is the following.

Property 3. Let A, B, and C be in $M(n)$. Then the solution of the matrix equation

$$AXB^* + BXA^* = -C, \tag{5}$$

if it exists, is

$$[A \otimes \overline{B} + B \otimes \overline{A}]x = -\hat{C}, \tag{6}$$

where

$$\hat{x} = \begin{pmatrix} x_{11} \\ \vdots \\ x_{nn} \end{pmatrix} \text{ and } \hat{C} = \begin{pmatrix} c_{11} \\ \vdots \\ c_{nn} \end{pmatrix}$$

(again see [7], Chapter 8).

Property 4. Let A and B be in $M(n)$ with $\det B \neq 0$. Then $B^{-1}A$ has no eigenvalues on the imaginary axis iff

$$\det[A \otimes \overline{B} + B \times \overline{A}] \neq 0. \tag{7}$$

Proof. From Property 1

$$B^{-1}AX + XA^*(B^{-1})^* = -I \tag{8}$$

has a unique solution iff $B^{-1}A$ has no eigenvalues on the imaginary axis. But (13) is equivalent to

$$AXB^* + BXA^* = -BB^*, \tag{9}$$

whose unique solution is, by Property 3, dependent on the condition that (10) be satisfied.

Corollary 1. *Assume $B^{-1} \in M(n)$. Then $B^{-1}A$ has an eigenvalue on the imaginary axis iff*

$$\det[A \otimes \overline{B} + B \otimes \overline{A}] = 0. \tag{10}$$

3 The Main Stability Results

Let $A_0, A_1, \ldots, A_m$ and $B_1, \ldots, B_m$ be matrices in $M(n)$ with real entries and $h \in R^m_+$. Consider the differential delay equation

$$\frac{d}{dt}\left[x(t) - \sum_{j=1}^{m} x(t-h_j)\right] = A_0 x(t) + \sum_{j=1}^{m} A_j x(t-h_j). \tag{11}$$

The following are standing assumptions in this section, with the exception of Example 4.

Assumption 1 *The matrices $\{B_j\}$ satisfy the condition*

$$\sum_{j=1}^{m} |B_j| < 1. \tag{12}$$

Assumption 2 *When $h = 0$, the system (11) is uniformly exponentially stable (u.e.s.) i.e., all eigenvalues of*

$$\left(I - \sum_{j=1}^{m} B_j\right)^{-1} \left(Ao + \sum_{J=1}^{m} A_j\right)$$

lie in $Re\lambda < 0$.

Definition 1. *Let $\alpha \in K$. Then*

$$B(\sigma) = I \sum_{j=1}^{m} B_j e^{-i\sigma j} \tag{13}$$

and

$$A(\sigma) = A_0 + \sum_{J=1}^{m} A_j e^{-i\sigma j}. \tag{14}$$

Property 5. If Assumptions 1 and 12 are satified, then either: (i) the system (11) is uniformly exponentially stable (u.e.s.) For all $h \in R^m$ or (ii) there exists and $h^0 \neq 0$ in R^m such that the system

$$\frac{d}{dt}\left[x(t) - \sum_{j=1}^{m} B_j x(t-h_j)\right] = A_0 x(t) = \sum_{j=1}^{m} A_j x(t-h_j) \tag{15}$$

has a nontrivial periodic solution.

Proof. The proof is a consequence of the observation that, because of Assumptions 1 and 2, for a fixed $h \in R^m$ the family of matrices

$$C(\delta h, \lambda) = \lambda \left(I - \sum_{j=1}^{m} B_j e^{-\delta\lambda h_j} \right) - A_0 - \sum_{j=1}^{m} A_j e^{-\delta\lambda h_j} \tag{16}$$

has the property that

$$\Upsilon(\delta, h) = \sup\{Re\lambda : \det C(\delta h, \lambda) = 0\} \tag{17}$$

is continuous on $[0, \infty)$ (see, e.g., [1]). Thus, if for all $h \in R_+^m \Upsilon(\delta, h) < 0$, then (i) must be satisfied. On the other hand if for some $h \in R^m, h \neq 0$, and $\delta > 0$ $\Upsilon(\delta, h) = 0$ then there exists $w > 0$ such that

$$\det \left[iw \left(I - \sum_{j=1}^{m} Bje^{iw\,\delta hj} \right) - Ao - \sum_{j=1}^{m} A_j e^{iw\,\delta hj} \right] = 0.$$

Hence (ii) must hold for $h_0 = \delta h$.

Property 6. The system (11) has a nontrivial periodic solution for some $h \in R_+^m$ iff for some $\sigma \in K$, $\sigma \neq 0$, the equations

$$\det[A(\sigma) \otimes \overline{B}(\sigma) + B(\sigma)x\overline{A}(\sigma)] = 0 \tag{18}$$

and

$$\det[iw(B(\sigma) - A(\sigma)\} = 0,\ w > 0, \tag{19}$$

are satisfied. In which case

$$C(h, iw) = 0 \quad \text{(i.e., (16)), where}$$

$$h = \frac{\sigma}{w}. \tag{20}$$

Proof. (i) If (11) has a nontrivial periodic solution for some $h \in R_+^m$, $h \neq 0$, then there exists $w > 0$ such that $C(h, iw) = 0$ (w cannot be zero because of Assumption 2). Let $\alpha = wh$, then (18) and (19) are both satisfied for the pair α and w. (ii) Assume (18) and (19) are satisfied. Then if $h = \frac{\alpha}{w}$, $C(h, iw) = 0$.

Recipe for Application of Property 6

Let h_0 be in R_+^m. Consider the system

$$d\left[x(t) - \sum_{j=1}^{m} B_j x(t - \delta h_j^0) \right] = A_0 x(t) + \sum_{j=1}^{m} Ajx(t - \delta h_j^0), \tag{21}$$

where $\delta > 0$ and Assumptions 1 and 2 hold. Thus for system (21) $\Upsilon(0, h^0) < 0$ and, by the continuity of $\{(\delta, h^0)$, there are five possibilities.

(i) $\Upsilon(\delta, h^0) < 0$ for all $\delta \geq 0$. Then (11) is u.e.s. for $h = h^0$.
(ii) The smallest δ for which $\Upsilon(\delta, h^0) = 0$ is $\delta > 1$. Then (11) is u.e.s. for $h = h_0$
(iii) $\Upsilon(1, h^0) = 0$ and (19) is satisfied for $h = h_0$ and some $w > 0$, where $\alpha = wh_0$. Then the system (11) has a nontrivial solution for $h = h_0$.
(iv) $\Upsilon(\delta, h^0) = 0$, but (19) is satisfied only when $w = 0$. Then the system is u.e.s. at $h = h_0$.
(v) $\Upsilon(\delta, h^0) = 0$ for some $\delta < 1$ and (19) has a solution $w > 0$ where $\alpha = w\delta h_0$. Then the stability of the system is indeterminant.

Remark 1. To find where $\Upsilon(\delta, h^0) = 0$ it is only necessary to solve the equation

$$\det[A(\tau h_0) \otimes \overline{B}(\tau h_0) + B(\tau h_0) \otimes \overline{A}(\tau h_0)] = 0 \tag{22}$$

where

$$\tau > 0.$$

Examples

Example 1.

$$\begin{aligned}\dot{x}(t) &= \begin{pmatrix} 0 & 0 \\ 0 & -1 \end{pmatrix} x(t) + \begin{pmatrix} -1 & 1 \\ 0 & 0 \end{pmatrix} x\left(t - \frac{1}{2}\right) + \begin{pmatrix} 0 & 0 \\ -1 & 0 \end{pmatrix} x\left(t - \frac{\pi}{2}\right) \\ &= A_0 x(t) + A_1 x(t - h_1) + A_2 x\left(t - \frac{\pi}{2}\right).\end{aligned} \tag{23}$$

Here $h_1 = \frac{1}{2}$ and $h_2 = \frac{\pi}{2}$, Assumptions 1 and 2 are satisfied and

$$B(\tau) = I, \quad A(\tau) = \begin{pmatrix} -e^{-i\frac{\tau}{2}}, & e^{-i\frac{\tau}{2}} \\ -e^{-i\frac{\pi}{2}\tau}, & -1 \end{pmatrix}.$$

The smallest value of τ for which (22) is satisfied is

$$\tau \doteq 1.180333.$$

At this point

$$w \doteq 1.8028 > 1.$$

Hence the system is u.e.s.

Example 2. Let A_0, A_1, and A_2 be the matrices in Example 36. But $h_1 = 1, h_2 = \pi$. Then

$$B(\tau) = I, \quad A(\tau) = A(\tau) = \begin{pmatrix} -e^{-i\tau}, & e^{-i\tau} \\ e^{-i\pi\tau}, & -1 \end{pmatrix}.$$

Here the solution of (22) yields

$$\begin{aligned} \tau &\doteq .590167, \\ w &\doteq .65472211 \text{ and} \\ \delta &\doteq .901401. \end{aligned}$$

Hence the test is indeterminant. However using the method described in Section 4 we shall show the system in this example is unstable.

Example 3. Let

$$B_1 = \begin{pmatrix} 0 & \frac{1}{2} \\ -\frac{1}{2} & 0 \end{pmatrix}, \quad A_0 = \begin{pmatrix} -1 & 0 \\ 0 & -1 \end{pmatrix}, \quad A_1 = \begin{pmatrix} 0 & 1 \\ -1 & 0 \end{pmatrix}. \tag{24}$$

Then

$$B(\alpha) = \begin{pmatrix} 1, & -\frac{1}{2}e^{-i\alpha} \\ \frac{1}{2}e^{-i\alpha}, & 1 \end{pmatrix}, \quad A(\alpha) = \begin{pmatrix} -1, & e^{-i\alpha} \\ -e^{-i\alpha}, & -1 \end{pmatrix}. \tag{25}$$

The corresponding delay system satisfies Assumptions 1 and 2 and has a zero when $\alpha = \pi/2$. However, the only point on the imaginary axis which satisfies (19) is $w = 0$. Thus the corresponding system (11) is u.e.s. for all delays.

The next example does not conform to the context of this section in that Assumption 2 is not satisfied. However, the general methodology of this section may still be applied to study its stability.

Example 4.

$$\frac{d}{dt}\left[x(t) + \frac{1}{4}x(t-h)\right) = y(t) \tag{26a}$$

$$\frac{dy}{dt}(t) = -x(t). \tag{26b}$$

Here

$$B(\alpha) = \begin{pmatrix} 1 + \frac{1}{4}e^{-i\alpha}, & 0 \\ 0, & 1 \end{pmatrix}, \quad A(\alpha) = \begin{pmatrix} 0 & 1 \\ -1 & 0 \end{pmatrix}. \tag{27}$$

When $h = 0$ this system has two independent periodic solutions. It is easy to show that for h "small" (26) is u.e.s. The object is to find the maximum interval $(0, h^0)$ for which it is u.e.s. Equation (18) and (19) are satisfied when

$$\alpha = \pi, \; w = \frac{2}{\sqrt{3}}.$$

Hence the system is u.e.s. when $h \in \left(0, \frac{\sqrt{3}}{2}\pi\right)$.

Example 5. Let

$$B_1 = \begin{pmatrix} 0 & \frac{1}{2} \\ -\frac{1}{2} & 0 \end{pmatrix}, \quad A_0 = \begin{pmatrix} 0, & 1 \\ -2, & -1 \end{pmatrix}. \tag{28}$$

Assumptions 1 and 2 are satisfied,

$$B(\alpha) = \begin{pmatrix} 1, & -\frac{1}{2}e^{-i\alpha} \\ \frac{1}{2}e^{-i\alpha}, & 1 \end{pmatrix}, \quad A(\alpha) = \begin{pmatrix} 0, & 1 \\ -2, & -1 \end{pmatrix} \tag{29}$$

and Equation (18) and (19) are satisfied for

$$\begin{aligned} w &\doteq 1.9463989, \\ \alpha &\doteq 2.635816 \text{ and} \\ h^0 &\doteq .73844. \end{aligned}$$

4 The Poisson Integral Method

In this section C will denote the complex plane, B is a complex Banach space with a zero vector denoted by 0.

Observation 1 *Let F be a vector-valued mapping from C into a Banach space B. Then F is analytic in a nonempty open set, $\triangle$, in C if and only if the Cauchy-Riemann condition*

$$\frac{\partial F}{\partial x} = \frac{1}{i}\frac{\partial F}{\partial y}, \qquad \lambda = x + iy,$$

is satisfied in $\triangle$.

Definition 2. *A mapping $F : C \to B$ is in H^∞ if F is analytic and uniformly bounded in $Re\lambda > 0$ (see, e.g., [6]).*

The Poisson Integral

Let $F : C \to B$ be analytic and uniformly bounded in a vertical strip

$$S(\delta) = \{\lambda : -\delta < Re\lambda < \delta\}.$$

Then the Poisson Integral of F

$$I(F,\lambda) = \frac{x}{\pi}\int_{-\infty}^{\infty} \frac{F(it)dt}{x^2 + (t-y)^2}, \qquad \lambda = x + iy, \quad x > 0, \tag{30}$$

is a uniformly bounded harmonic function on $Re\lambda > 0$. If F is also holomorphic and uniformly bounded on $Re\lambda > 0$, then

$$I(F,\lambda) = F(\lambda), \tag{31}$$

(see e.g. [6]).

Lemma 1. *The Poisson Integral also has the representation*

$$\frac{1}{\pi}\int_{-\frac{\pi}{2}}^{\frac{\pi}{2}} F[i(x\tan\sigma + y)]d\sigma = I(F,\lambda). \tag{32}$$

roof. Substitute $t = y + x\tan\sigma$ into (38).

Theorem 1. *Let F be analytic and uniformly bounded on a vertical open strip, $S(\delta)$, containing the imaginary axis, then the condition*

$$\int_{-\frac{\pi}{2}}^{\frac{\pi}{2}} e^{-2i\sigma} F[i(x\tan\sigma + y)]d\sigma = 0,\ x > 0, \tag{33}$$

is necessary and sufficient for F to be in H^∞.

Proof. The proof is a consequence of the Cauchy-Riemann condition applied to (40) (see e.g. [3]).

Remark 2. Since $I(F,\lambda)$ is harmonic in $Re\lambda > 0$ the mappings $\frac{\partial I}{\partial x}(F,\lambda)$ and $\frac{\partial I}{\partial y}(F,\lambda)$ are also. Consequently the function

$$\begin{aligned} G(F,\lambda) &= \frac{\partial I}{\partial x}(F,\lambda) - \frac{1}{i}\frac{\partial I}{\partial y}(F,\lambda) \\ &= \frac{1}{\pi x}\int_{-\frac{\pi}{2}}^{\frac{\pi}{2}} e^{-2i\sigma} F[i(x\tan\sigma + y)]d\sigma \end{aligned} \tag{34}$$

is also harmonic in $Re\lambda > 0$.

Remark 3. The functions $I(F,\lambda)$ and $G(F,\lambda)$ have harmonic conjugates which are unique up to constants. This implies that if either $I(F,\lambda)$ is holomorphic in a neighborhood, U, in $Re\lambda > 0$ or if $G(F,\lambda) \equiv 0$ in U, that $I(F,\lambda)$ is holomorphic in $Re\lambda > 0$. Consequently $F(\lambda) = I(F,\lambda)$ in $Re\lambda$ and $F \in H^\infty$. Conversely if either $I(F,\lambda_0) \neq F(\lambda_0)$ or $G(F,\lambda_0) \neq 0$ for some $\lambda_0 \in Re\lambda > 0$, then F is not in H^∞. This observation is encapsulated in the following theorem.

Theorem 2. *(i) If either $I(F,\lambda) = F(\lambda)$ or $G(F,\lambda) = 0$ is some neighborhood, U, in $Re\lambda > 0$, then $F \in H^\infty$.*

(ii) If either $F(\lambda_0) \neq I(F,\lambda_0)$ or $G(F,\lambda_0) \neq 0$ at some point λ_0 in $Re\lambda > 0$, then F is not in H^∞.

Examples

Example 6. Consider Example 4.

The stability of this system is determined by the analytic function

$$\left[\lambda^2\left(1 + \frac{e^{-\lambda h}}{4}\right) + 1\right]^{-1},\ h > 0. \tag{35}$$

In Example 4 it was shown that the system was stable for $0 < h < \frac{\sqrt{3}}{2}\pi$. Clearly $h = 2 < \frac{\sqrt{3}}{2}\pi$. Thus for $y = 0$ and $x = 1$

$$I(F,1) \doteq .491684093537 \tag{36}$$

(using Simpson's Rule, 200 partitions and the interval

$$I = \left[-\frac{\pi}{2} + .0001,\ \frac{\pi}{2} - .0001\right].$$

The actual value is

$$F(1) \doteq .491682255339.$$

On the other hand for $h = 3 > \frac{\sqrt{3}}{2}\pi$ and $\lambda = 1$, the value of (41) is approximately 2.137 which confirm's the instability of the system in Example 4 for this value of h.

Example 7. Consider the system in Example 37, where $h_1 = 1$ and $h_2 = \pi$. The methods of Part I led to an indeterminate conclusion concerning the stability of the system. However the value of (41) for the Laplace Transform of this system, i.e. the holomorphic function

$$[(\lambda + e^{-\lambda})(\lambda + 1) + e^{-\lambda(\pi+1)}]^{-1}, \tag{37}$$

at $\lambda = 1$ is approximately .933526. Hence the system is unstable (The integral (41) at $\lambda = 1$ was approximated using Simpson's Rule over $\left[-\frac{\pi}{2} + .0001,\ \frac{\pi}{2} - .0001\right]$ using 200 equispaced partitions).

References

1. R. Datko. A procedure for determination of the exponential stability of certain differential-difference equations. Quarterly Appl Math 36 (1978) 279-292
2. R. Datko. Lyapunov functions for certain linear delay differential equation in a Hilbert space. Jour Math Anal Appl 76 (1980) 37-57
3. R. Datko. Empirical methods for determining the stability of certain transfer functions. Jour of Math Anal and Appl 218 (1998) 409-420
4. J.K. Hale and S.M. Verduyn Lunel. "Introduction to Functional Differential Equation," Appl Math Sci 99, Springer-Verlag, New York, 1993
5. D. Henry. Linear neutral functional differential equations. J Diff Eq 15 (1974) 106-128
6. E. Hille. "Analytic Function Theory," II Ginn and Company, New York, 1962
7. P. Lancaster. "Theory of Matrices," Academic Press, New York, 1969

Stability Exponent and Eigenvalue Abscissas by Way of the Imaginary Axis Eigenvalues

James Louisell

Department of Mathematics, Colorado State University—Pueblo, Pueblo, CO 81001, USA

Summary. In this chapter we present a technique for accurate computation of the stability exponent and other eigenvalue abscissas of a matrix delay equation. Previously the author introduced a finite dimensional linear operator, determined by the delay equation coefficients, having spectrum containing all possible imaginary axis eigenvalues of the delay system. Using the eigenvalues of this operator, and introducing a translation in the equation's characteristic function, we can make an accurate numerical determination of the system stability or growth exponent and other eigenvalue abscissas. After giving the basic theorems for the method, we give an example in which we go over essentials of implementation. Then we explore the method in some special cases, beginning with second order scalar delay equations and an interesting example of positive delay feedback. We proceed to a rather detailed examination of the effect of the delay parameter in some simple first order delay equations, finding that accurate computation of the system abscissa leads us to some interesting and unconventional conclusions on its behavior with respect to this parameter. We then give an example in which the method is adapted to a certain distributed delay equation. We conclude with some comments on possible future research.

1 Introduction

In previous papers the author has investigated the computational determination of the stability exponent and other eigenvalue abscissas of a linear autonomous delay differential equation [5,6]. For an asymptotically stable system, the method involved first solving a boundary value problem having as its solution a system quadratic energy matrix of the type first explored by Repin [11] and Datko [3]. By varying a real parameter representing a translation of the characteristic function, and noting the behavior of the solution to the boundary value problem as this parameter varies, the stability exponent could be determined. In the more general case, where the stability or instability of the system is not known, a translation given in terms of the norms of the system coefficients allows us to use the same method.

In the time since publication the author, along with Niculescu [9], has been interested in the determination of the oscillatory eigenvalues of linear delay systems. Earlier contributions to this topic were made by Chen et al [2], and Marshal et al [8].

We now have techniques which work well for many commensurate delay systems. Since these eigenvalues play such an important part in determining system stability, it seemed natural to investigate whether one could develop an approach to determination of the stability exponent by way of the imaginary axis eigenvalues.

In Section 2 of this chapter we present such an approach, which will work for many commensurate delay systems. To keep the presentation manageable, we will give our theorems for matrix retarded single delay systems or systems that can be readily put in that form. We use a theorem previously given by the author on a finite dimensional linear operator which we associate with the delay system [7]. This operator has all possible oscillatory eigenvalues of the delay system included in its spectrum. We project this operator's eigenvalues horizontally onto the imaginary axis, and then evaluate the delay system's characteristic function at these pure imaginary values. Again varying a real parameter representing a translation of the characteristic function, we will find the abscissas of system eigenvalues. One advantage of the method is that the imaginary parts of the eigenvalues of the delay system can be readily recovered from the associated linear operator.

An example with MATLAB computation is given at the end of Section 2. In the sections following, rather than elaborating on details of the method, we give examples of the value of accurate computation of the stability exponent as a means of understanding some interesting questions that occur in stability analysis.

In Section 3 we first have a look at second order scalar single delay equations. After establishing an algebraic framework for the calculation of imaginary axis eigenvalues for these equations, we examine an example of Abdallah et al [1] on positive delay feedback for the oscillator. Here we give a computation of the stability exponent for a system of the type proposed by these authors. Readers interested in scalar single delay systems and the stability exponent can also see the references given in that section.

In Section 4 we show how accurate computation of the stability exponent can help us understand the stability behavior of a delay system as the delay parameter varies. Just using first order systems, we come to some surprising conclusions on the effect of the delay parameter. In Section 5 we give an example of computation of the stability exponent for a distributed delay equation which can be recast in terms of a matrix delay system or a scalar third order delay equation. We conclude with comments and suggestions for future research in Section 6.

There may be other ways to use the oscillatory eigenvalues in the determination of system eigenvalue abscissas. The method presented in Section 2 was chosen for simplicity of presentation and ease of implementation in MATLAB and similar programming environments. Considerations of computational size might dictate that many of the associated linear operator's eigenvalues be dismissed in advance, so that time not be spent evaluating the characteristic function at many more pure imaginary values than necessary. Other considerations, including interpretation and visualization, could lead to other modifications.

It is worth noting that Engelborghs and Roose [4] have also been developing a method for computational determination of delay system eigenvalues, particularly including the eigenvalues having the leading abscissas. This technique involves approximating the solution operator, and can be very fast. Since it involves the solution operator, it can of course at times have inconvenient size. The author's method, based on an associated linear operator determined by the system, involves only matrices of fixed size for any given system.

2 Imaginary Axis Eigenvalues

We begin with matrices $A_0, A_1 \in R^{nxn}$, and $h>0$, and we consider

$$(*)\ x'(t) = A_0 x(t) + A_1 x(t-h),$$

the delay differential equation, along with its characteristic matrix function $T(s) = sI - A_0 - A_1 e^{-sh}$ and the complex characteristic function $f(s) = |T(s)|$. We refer to any zero of $f(s)$ as an eigenvalue of the system $(*)$. Setting

$$T_d(s) = T(s+\ d) = sI - (A_0 -\ dI) - (e^{-hd} A_1) e^{-sh},$$

$f_d(s) = |T_d(s)|$ for real d, we note that $T_d(s)$ is the characteristic matrix function for the delay equation

$$(*d)\ x'(t) = (A_0 -\ dI)x(t) + (e^{-hd} A_1) x(t-h).$$

Noting the linear shift of s, we have the following simple lemma, which will be very useful in giving a numerical method for determining the stability exponent.

Lemma 1. *Let $A_0, A_1 \in R^{nxn}$, and $h>0$. Then the system $(*)$ $x'(t) = A_0 x(t) + A_1 x(t-h)$ has an eigenvalue with abscissa equal to d if, and only if, the system $(*d)$ $x'(t) = (A_0 - dI)x(t) + (e^{-hd} A_1)x(t-h)$ has an imaginary axis eigenvalue.*

The value of this lemma will be seen when we consider recent work on finite dimensional techniques for the computation of oscillatory eigenvalues of delay equations. For our purposes, we note that the author has shown [7] that the imaginary axis eigenvalues of the delay equation $(*)$ are also eigenvalues of the operator φ on $C^{nxn} \times C^{nxn}$ given by

$$\varphi \begin{pmatrix} X \\ Y \end{pmatrix} = \begin{pmatrix} A_0 X + A_1 Y \\ -X A_1^T - Y A_0^T \end{pmatrix}$$

for $X, Y \in C^{nxn}$. This operator was converted to the matrix J given in Lemma 4, operating on C^{2n^2}. We summarize in the lemma below, which follows directly from the author's previously derived theorem [7, Thm. 3.1] by setting the matrix B of the theorem in that paper equal to the zero matrix.

Lemma 2. *Let $A_0, A_1 \in R^{nxn}$, $h>0$, and let*

$$J = \begin{pmatrix} A_0 \otimes I & A_1 \otimes I \\ -I \otimes A_1 & -I \otimes A_0 \end{pmatrix},$$

where $\otimes$ denotes Kronecker product. Then all imaginary axis eigenvalues of the delay equation $(*)$ $x'(t) = A_0 x(t) + A_1 x(t-h)$ are eigenvalues of J.

For real d, we let J_d be the matrix associated with the system $(*d)$ just as J is associated with $(*)$. To be precise, we let

$$J_d = \begin{pmatrix} (A_0 - dI) \otimes I & e^{-hd} A_1 \otimes I \\ -I \otimes e^{-hd} A_1 & -I \otimes (A_0 - dI) \end{pmatrix}.$$

Now noting Lemma 3, we have it in mind to determine the x - coordinates of the eigenvalues of the delay equation $(*)$ by finding the values of d giving the system $(*d)$ pure imaginary eigenvalues. Noting Lemma 4, we know that imaginary axis eigenvalues of $(*d)$ are pure imaginary eigenvalues of J_d. We can evaluate these eigenvalues in $f_d(s) = |T_d(s)|$ to see if they are genuine eigenvalues of $(*d)$.

We would like to evaluate the characteristic function $f(s + d)$ only for those $s = s_d$ in $\mathrm{Eig}(J_d)$ which lie close to the imaginary axis, observing the convergence of $f(s_d + d)$ to zero as d converges to system eigenvalue abscissas. It might seem then that we would be making a numerical decision on which s_d lie close enough to the imaginary axis to consider. Since the eigenvalues we are looking for lie on the imaginary axis, we will instead, for clarity of presentation, simply project all values of s_d horizontally onto the imaginary axis. Thus, for $s_d = \delta_d + i\omega_d$, we will evaluate $f(i\omega_d + d)$ rather than $f(s_d + d)$. For this purpose, given any complex nxn matrix F, we define $\mathrm{Im\,Eig}(F) = \{\omega : s = \delta + i\omega \text{ is an eigenvalue of } F\}$.

Theorem 1. *Let $A_0, A_1 \in R^{nxn}$, $h>0$. For $d \in R$, define*

$$g(d) = \min_{\omega \in \mathrm{Im\,Eig}(J_d)} |f_d(i\omega)|.$$

If $a + ib$ is an eigenvalue of $(*)$ $x'(t) = A_0 x(t) + A_1 x(t-h)$, then $g(d) \to 0$ as $d \to a$. If $g(d) \to 0$ as $d \to a$, then a is an abscissa of an eigenvalue of $(*)$.

Proof. Noting that $\phi(d, s) = f_d(s)$ is continuous on C^2, and that the entries of J_d are continuous in d, we know that the function $g(d)$ is continuous over the reals.

Now if $a + ib$ is an eigenvalue of $(*)$, then ib is an eigenvalue of $(*a)$, so that ib is also an eigenvalue of J_a . This gives us $g(a) = 0$, and by continuity, we have $g(d) \to 0$ as $d \to a$. On the other hand, if $g(d) \to 0$ as $d \to a$, then $g(a) = 0$, i.e. $0 = \min_{\omega \in \mathrm{Im\,Eig}(J_a)} |f_a(i\omega)|$. We let b be that ω achieving the minimum, and we have $f(a + ib) = 0$.

The system stability exponent being the abscissa of most interest, we will frequently initialize our examination of system abscissas at d-values $d = d^+$ greater than the leading abscissa. For this we have the following lemma, proven in a few lines in the author's previous paper on the stability exponent [5].

Lemma 3. *For $\beta = ||A_0|| + ||A_1||$, there exist no complex zeros of $f(s) = |sI - A_0 - A_1 e^{-sh}|$ which lie in $\{\mathrm{Re}(s) > \beta\}$.*

From the above theorem and lemmas the ideas for our numerical determination of system eigenvalue abscissas are now clear. The above theorem gives us a numerical characterization of the eigenvalue abscissas which can be implemented by plotting the graph of the real function $y = g(d)$. To determine the leading abscissa, we take $d^+ > ||A_0|| + ||A_1||$, and begin by sweeping through decreasing d-values.

This new method for determining eigenvalue abscissas has some advantages over the method presented in [5]. Since there is no boundary value problem to solve, one of the most obvious advantages is simplicity of programming using software packages such as MATLAB. The items to calculate are the eigenvalues of the matrix J_d and the determinants of the matrices $T_d(i\omega)$. The matrix J_d has dimension $2n^2$, as does the boundary value problem, while the matrices $T_d(s)$ have dimension n.

Although the method's simplicity and directness are worthy of note, there are tradeoffs to mention. The method based on quadratic energy gives us the behavior of the system's quadratic energy as the exponent varies. For time delay perturbations of finite dimensional systems, the quadratic energy method allows us to simultaneously assess the effect of time delay on LQ energy, and determine the stability exponent of the time delay system resulting from the perturbation. There is also the advantage that after all the calculations involved in solving the boundary value problem, the item we plot in the end is just the maximum eigenvalue of a symmetric matrix. The method chosen will depend on the user's context.

To demonstrate the new technique, we give an example of abscissa computation previously carried out with the author's quadratic energy method [5].

Example 1. Consider the system $(*)\ x'(t) = A_0x(t) + A_1x(t-1.3)$, where

$$A_0 = \begin{pmatrix} -1.6 & .8 \\ 2.4 & 2.7 \end{pmatrix}, \qquad A_1 = \begin{pmatrix} 2.5 & 4.1 \\ 1.5 & -3.2 \end{pmatrix}.$$

Since $||A_0|| \leq 5.1$, $||A_1|| \leq 7.3$, we can set $d^+ = 12.5$. For $d \leq 12.5$, $f(s) = |T(s)|$, we then compute $g(d) = \min_{\omega \in \mathrm{Im\,Eig}(J_d)} |f_d(i\omega)|$ using MATLAB. Below we first graph $y = g(d)$ for $0 \leq d \leq 12.5$, and then focus on the leading zero of $g(d)$ in that interval. Then we do the same for $0 \leq d \leq 1$, and finally for $-1 \leq d \leq 0$. We use increments of size 0.01 in the larger intervals, and of size 1×10^{-5} in the smaller. We see here that the stability exponent d_0 lies in $(3.0982, 3.0983)$, while the eigenvalue abscissa d_1 closest to d_0 lies in $(0.2577, 0.2578)$, and the closest eigenvalue abscissa d_2 on the left of d_1 lies in $(-.1741, -.1740)$.

Figures 1,2 and 3 show the results of

$$(*)\ x'(t) = A_0x(t) + A_1x(t-1.3).$$

To find the imaginary part of the eigenvalue which accompanies the leading abscissa, we let $d = 3.09825$, and use MATLAB to compute $\mathrm{Eig}(J_d)$. We find that the only imaginary axis eigenvalues of J_d are $\pm 0.0002i$. Thinking now that the leading eigenvalue may be real, we consider the characteristic function $f(s)$ at $s = 3.0983$, $s = 3.0982$. Confirming, we find that $f(3.0983) \approx 2.6249x10^{-4}$, $f(3.0982) \approx -2.4263x10^{-4}$, so that $\lambda_0 \approx 3.09825$ is real. Next, we set $d = 0.25775$, and again compute $\mathrm{Eig}(J_d)$. This time we find imaginary axis eigenvalues $s = \pm 1.1369i$, $\pm 0.6739i$. We find $f(.25775 + 1.1369i) \approx -0.0007 + 0.0049i$, having absolute value approximately .00495, while $f(.25775 + .6739i) \approx -4.8836 + 6.3877i$. Thus

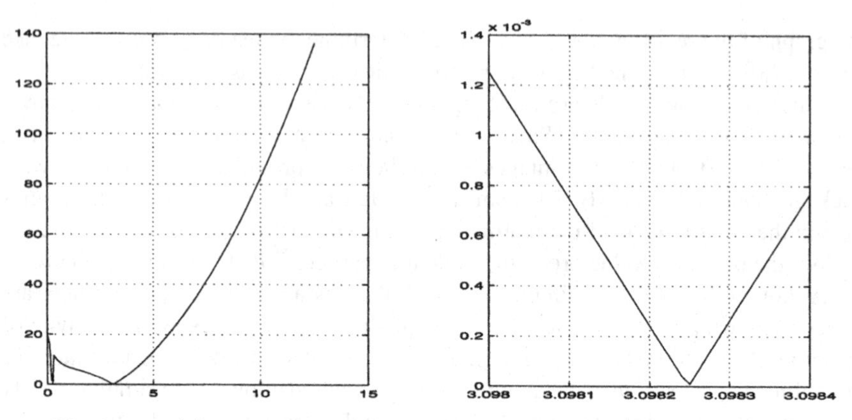

Fig. 1. $y = g(d)$. Shown on the left: $0 \leq d \leq 12.5$; right: $3.0980 \leq d \leq 3.0984$

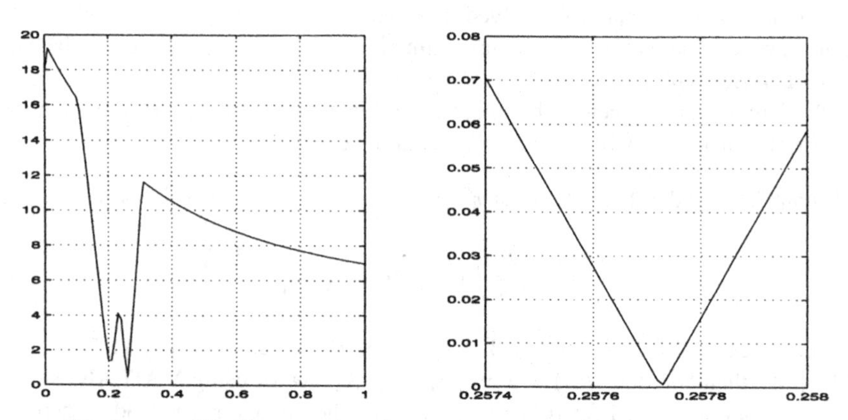

Fig. 2. $y = g(d)$. Shown on the left: $0 \leq d \leq 1$; right: $0.2574 \leq d \leq 0.2580$

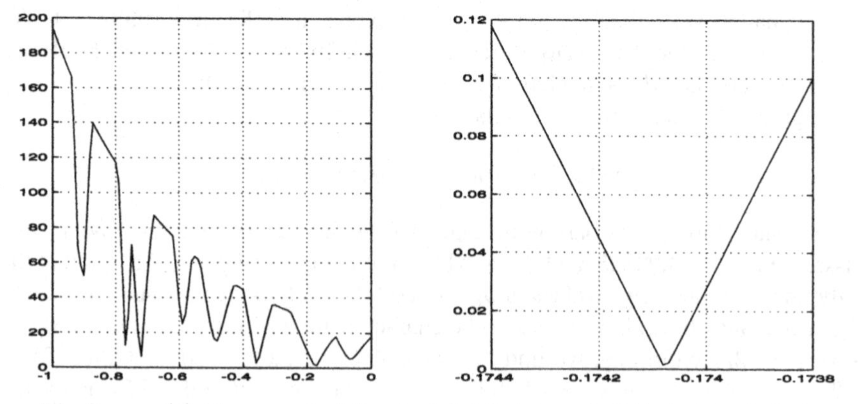

Fig. 3. $y = g(d)$. Shown on the left: $-1 \leq d \leq 0$; right: $-0.1744 \leq d \leq -0.1738$

the eigenvalues having abscissa d_1 are $\lambda_{1-}, \lambda_{1+} \approx 0.25775 \pm 1.1369i$. Doing the same at d_2, we will find that the eigenvalues with this abscissa are $\lambda_{2-}, \lambda_{2+} \approx -0.17405 \pm 3.5451i$.

3 Second Order Scalar Delay Equations

In this section we have a look at our new technique for determining eigenvalue abscissas in the well known case of a second order scalar delay equation having a single delay. Rather than apply Theorem 8 literally, we adhere to the ideas of Section 2, at the same time using special properties of low order systems for simplification. We first present algebraic details of the method which are relevant to our particular case here. Then we examine the idea of stabilization of the oscillator via positive delay feedback proposed by Abdallah et al [1], giving an accurate computation of the stability exponent for this interesting example. In Section 4 and Section 5 we will return to direct use of the matrix J_d and Theorem 8.

It is fair to mention that the idea of first solving an equation for pure imaginary eigenvalues, and then varying a real parameter to match the characteristic equation, was developed independently, just for the scalar single delay case, by Ozbay and Ulus [10,12]. It is interesting that these authors, motivated largely by engineering education, arrived at this notion which has such general value when viewed in the context of the appropriate linear operators as explained in Section 2.

We consider $(*)$ $x''(t) + a_1x'(t) + b_1x'(t-h) + a_2x(t) + b_2x(t-h) = 0$, the scalar second order delay equation with a single delay. We can write this in matrix form as $z'(t) = A_0z(t) + A_1z(t-h)$, with

$$A_0 = \begin{pmatrix} 0 & 1 \\ -a_2 & -a_1 \end{pmatrix}, \qquad A_1 = \begin{pmatrix} 0 & 0 \\ -b_2 & -b_1 \end{pmatrix}.$$

In either form we have scalar characteristic function $f(s) = s^2 + (a_1 + b_1e^{-hs})s + a_2 + b_2e^{-hs}$. For $f_d(s) = f(s+d)$, easy calculations lead to $f_d(s) = s^2 + (a_1 + 2d)s + (b_1e^{-hd})se^{-hs} + (d^2 + a_1d + a_2) + e^{-hd}(b_1d + b_2)e^{-hs}$, which we write as $f_d(s) = s^2 + \alpha_1 s + \beta_1 se^{-hs} + \alpha_2 + \beta_2 e^{-hs}$, noting the dependence of these coefficients on d.

The complex function f(s) has a zero with abscissa equal to d if, and only if, the function $f_d(s)$ has an imaginary axis zero. In this case we have $f_d(s) = 0 = f_d(-s)$. We can solve for e^{-hs} in $f(s+d) = 0$ and for e^{hs} in $f(-s+d) = 0$, obtaining $s^2+\alpha_1s+\alpha_2 = e^{-hs}(-\beta_2-\beta_1s)$, and $s^2-\alpha_1s+\alpha_2 = e^{hs}(-\beta_2+\beta_1s)$. Multiplying left and right sides of these equations and simplifying, we obtain a quadratic equation in s^2, i.e. $s^4 + Bs^2 + C = 0$, with $B = 2\alpha_2 + \beta_1^2 - \alpha_1^2$, $C = \alpha_2^2 - \beta_2^2$. Thus we have the four roots $s = \pm\sqrt{t}$, where $t = s^2$ is given by the quadratic formula $t = \frac{-B\pm\sqrt{B^2-4C}}{2}$.

Example 2. Consider $x''(t) + 4x(t) = 2x(t-1)$, an example of the type of Abdallah et al [1] showing that positive delay feedback can stabilize the oscillator. In matrix form we have $z'(t) = A_0z(t) + A_1z(t-1)$,

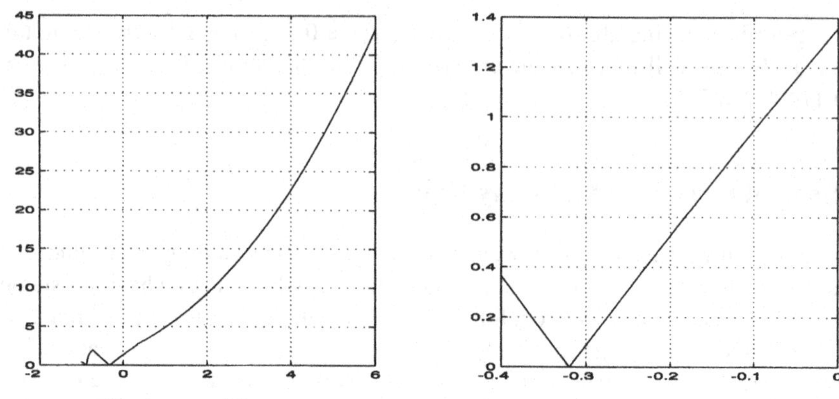

Fig. 4. $y = g(d)$. Shown on the left: $-1 \le d \le 6$; right: $-0.4 \le d \le 0$

$$A_0 = \begin{pmatrix} 0 & 1 \\ -4 & 0 \end{pmatrix}, \qquad A_1 = \begin{pmatrix} 0 & 0 \\ 2 & 0 \end{pmatrix}.$$

Since $||A_0|| = 4$, $||A_1|| = 2$, we set $d^+ = 6.001$, and examine $f_d(s)$ with $s = \pm\sqrt{t}$ determined by d as above, where $d < d^+$. With $h = 1$ and $f(s) = s^2 + 4 - 2e^{-hs}$, we have $f_d(s) = s^2 + 2ds + d^2 + 4 - 2e^{-hd}e^{-hs}$. We set $\alpha_1 = 2d$, $\alpha_2 = d^2 + 4$, and $\beta_2 = -2e^{-hd}$. From $f(s+d) = 0 = f(-s+d)$, we have $s^2 + \alpha_1 s + \alpha_2 = -\beta_2 e^{-hs}$, $s^2 - \alpha_1 s + \alpha_2 = -\beta_2 e^{hs}$. Proceeding as above, we obtain $s^4 + (2\alpha_2 - \alpha_1^2)s^2 + \alpha_2^2 - \beta_2^2 = 0$. This equation has roots $s = \pm\sqrt{t}$, with $t = \frac{-B \pm \sqrt{D}}{2}$, $B = 2\alpha_2 - \alpha_1^2$, $D = \alpha_1^4 + 4\beta_2^2 - 4\alpha_1^2\alpha_2$. Referring to these four root branches as $s_k(d) = \delta_k(d) + i\omega_k(d)$, $k = 1, ..., 4$, we can plot $g_k = f_d(i\omega_k)$ or $g = \min_{1 \le k \le 4} |g_k|$ as d varies. We have given the plot of g below.

Figures 4 and 5 show the results of

$$x''(t) + 4x(t) - 2x(t-1) = 0.$$

We see here that the stability exponent lies in the interval $[-.35, -.30]$. Below we get a closer look, finding that the stability exponent lies in $[-.32, -.3175]$.

4 The Effect of the Delay Parameter

In this section we use the previously developed techniques to examine some commonly held notions about the significance of the delay parameter for the stability of delay differential equations. In giving this presentation our goal is not merely to encourage new thinking on this question, but also to display the general value of accurate computational techniques in assessing the merits of conventional wisdom. To make comparisons between conventional wisdom and scientific understanding, we will only need first order delay equations. We are careful not to make very general

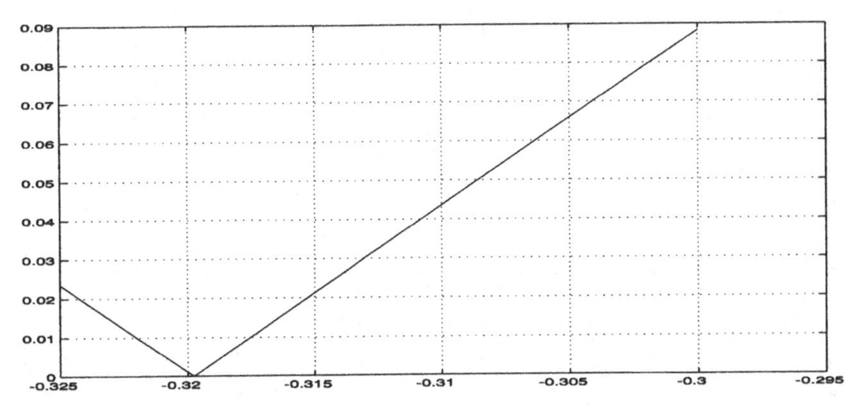

Fig. 5. $y = g(d)$, $-0.325 \leq d \leq -0.30$

claims about stability theory here. Our stability criterion will just be the behavior of the stability exponent as the delay parameter varies.

Consider now the idea that if a delay system is asymptotically stable, then increases in the delay time bring the system closer to instability. This idea is expressed at many conferences and in all kinds of engineering and mathematics journals. We use the method introduced in Section 2 to test this idea in the case of the system $(*)$ $x'(t) = 2x(t) - 3x(t-h)$. With $f(s) = s - 2 + 3e^{-hs}$, we have $f(s+d) = s - (2-d) + (3e^{-hd})e^{-hs}$, and this gives us

$$J_d = \begin{pmatrix} 2-d & -3e^{-hd} \\ 3e^{-hd} & -(2-d) \end{pmatrix},$$

having eigenvalues $\pm s_d$, with $s_d = ((2-d)^2 - 9e^{-2hd})^{1/2}$. For real ω we know $f_d(i\omega)$, $f_d(-i\omega)$ are conjugates. We write $g(\, d) = \min_{\omega \in \mathrm{Im\,Eig}(J_d)} |f_d(i\omega)|$, as in Section 2. With $s_d = \delta_d + i\omega_d$, we then have $g(d) = |f_d(i\omega_d)|$. Below we graph this function for $h = .05$, $h = .20$, $h = .25$, $h = .30$.

Figure 6 shows the results of

$$(*)\; x'(t) = 2x(t) - 3x(t-h).$$

When $h = 0$, the leading eigenvalue of the system $(*)$ is at $x = -1$. As one can see from the first three graphs above, the abscissa of the leading eigenvalue is at first to the *left* of $x = -1$ with increased delay time, and the system is enjoying enhanced decay. Only after the parameter h increases to values in the interval $(.25, .30)$ do we have the destabilizing effect of increased delay. The availability of this means of determining the stability abscissa has allowed us a careful look at a phenomenon which is considerably at odds with the folklore of the area.

It is certainly interesting that for $h = .20$, the stability exponent is more than twice the distance left of zero as the stability exponent for h = 0. If we desire to determine the h - intervals of stability or instability for the system $(*)$, we note that with $d = 0$, the matrix J_d has eigenvalues $s = \pm i\sqrt{5}$. Solving $f(i\sqrt{5}) = 0$ for h, we

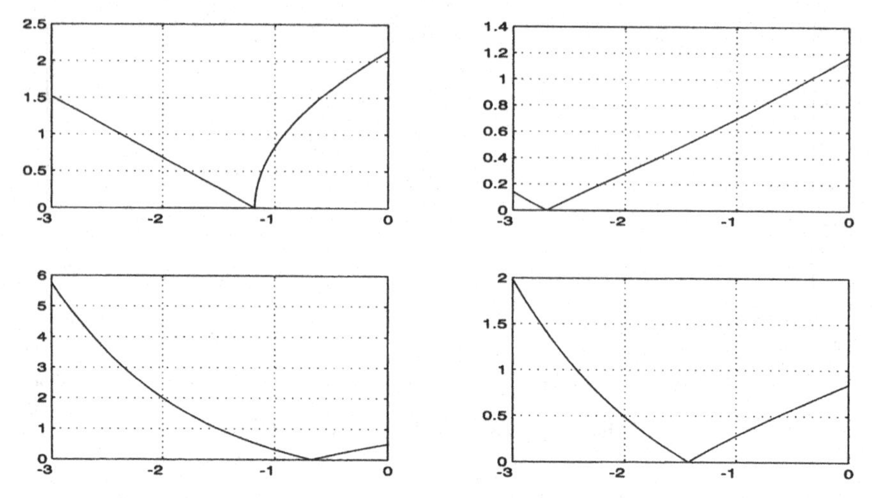

Fig. 6. $y = g(d)$. Clockwise from upper left: $h = .05$, $h = .20$, $h = .25$, $h = .30$

find that the smallest possible nonnegative value of h is $h_0 = \frac{1}{\sqrt{5}} \arctan(\frac{\sqrt{5}}{2}) \approx .376$, so that $(*)$ is exponentially stable for $0 \leq h < h_0$. With further routine analysis we will find this system exponentially unstable for $h_0 < h$.

It may seem surprising that the stability exponent can behave this way in a first order delay system with such simple stability structure. There is no multiple stability switching and the imaginary axis eigenvalues are simple. The conclusion is that the folklore is often wrong to regard zero delay as the optimal condition. If we use the stability exponent as our criterion here, it is more appropriate to regard $h = 0$ as the left endpoint of an interval containing the optimal h in the interior and having the value of h giving exponent zero at the right endpoint.

It is also interesting to examine the effect, on an unstable system, of changes in the delay. Here the conventional wisdom holds that the system is brought closer to stability as the delay decreases. We try this idea out on the system $(*)$ $x'(t) = 3x(t) + 2x(t-h)$, having leading abscissa $x = 5$ with $h = 0$. Here the matrix J_d will have eigenvalues $\pm s_d$, with $s_d = ((3 - d)^2 - 4e^{-2hd})^{1/2}$. This time we graph the function $g(d) = |f_d(i\omega_d)|$ for $h = .10$, $h = .01$.

Figure 7 shows the results of

$$(*)\ x'(t) = 3x(t) + 2x(t-h).$$

We see here that decreases in the delay from $h = .1$ to $h = .01$ to $h = 0$ bring increases in the leading abscissa, again contrary to conventional wisdom.

If we had used the system $x'(t) = 3x(t) - 2x(t-h)$, then we would have observed the leading abscissa decreasing with decreasing delay, as conventional wisdom would expect. Alert readers may by now have noticed what is happening with these simple first order systems. As h first increases from the value zero, the leading abscissa is moving in the direction of the dominant coefficient, e.g. would move

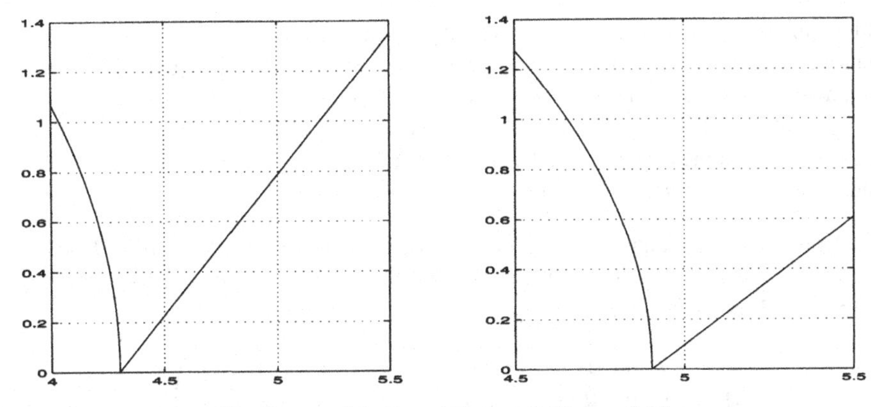

Fig. 7. $y = g(d)$. From left: $h = 0.10$, $h = 0.01$

in the direction of -3 in the system $(*)$ $x'(t) = 2x(t) - 3x(t-h)$. As the delay increases to very large values, the leading abscissa moves in the direction of the coefficient of present position, e.g. would move in the direction of 2 in the system $(*)$ $x'(t) = 2x(t) - 3x(t-h)$.

Again we emphasize that having a means of direct computational determination of the system abscissa has allowed us to have a careful look at the area's folklore, arriving at a better understanding of the effect of the delay in these simple systems. The author has small doubt that there are many cases in which our understanding of higher order delay equations would be enhanced by similar studies.

5 An Equation with Distributed Delay

In this section we show how the author's new technique can be adapted for use in determining the stability exponent of a delay differential equation having distributed delay. As in Section 4, we only give a rather simple first order example. We use a trigonometrically distributed delay function, anticipating that with advances in computing power, it may soon be practical to use the author's technique in analyzing the stability of distributed delay systems with delays approximated by finite Fourier series with many harmonics. We proceed entirely by example.

Example 3. Consider the delay equation $(*)$ $x'(t) = k\int_{-h}^{0} \sin(u)x(t+u)du$. Looking for exponential solutions of the form $x(t) = e^{st}$, we arrive at $\frac{se^{st}}{k} = \int_{-h}^{0} e^{st}e^{su}\sin(u)du$, i.e. $\frac{s}{k} = \int_{-h}^{0} e^{su}\sin(u)du$. Integrating by parts, we have $\frac{s}{k} = \frac{-1+(\sin h)se^{-hs}+(\cos h)e^{-hs}}{s^2+1}$. Note that $s = i$, $s = -i$ are zeros common to both numerator and denominator on the right. We have $f(s) = \frac{s}{k} - \frac{-1+(\sin h)se^{-hs}+(\cos h)e^{-hs}}{s^2+1}$ for characteristic function, and by L'Hopital's Rule we have $\lim_{s \to i} f(s) = \frac{i}{k} + \frac{i}{2}(e^{-ih}\sin h - h)$. With $h \geq 0$, it is interesting that this limit is equal to zero if, and only if, we have $h = n\pi$, $k = \frac{2}{n\pi}$ for some natural number n. In this case, and only in this case, will the equation $(*)$ have eigenvalues $s = \pm i$. In all other cases we multiply $f(s)$ by $k(s^2+1)$, obtaining $F(s) = s^3 + s + k - (k\sin h)se^{-hs} - (k\cos h)e^{-hs}$. For $h{>}0$ the set of zeros of $F(s)$ is equal to the union of $\{\pm i\}$ with the set of zeros of $f(s)$. Note that with $h = 0$ the equation $(*)$ is just $x'(t) = 0$.

Now $F(s)$ is the characteristic function for $x^{(3)}(t) + x'(t) - k\sin(h)x'(t-h) + kx(t) - k\cos(h)x(t-h) = 0$. Converting this to matrix form, we have $z'(t) = A_0 z(t) + A_1 z(t-h)$, where

$$A_0 = \begin{pmatrix} 0 & 1 & 0 \\ 0 & 0 & 1 \\ -k & -1 & 0 \end{pmatrix}, \quad A_1 = \begin{pmatrix} 0 & 0 & 0 \\ 0 & 0 & 0 \\ k\cos(h) & k\sin(h) & 0 \end{pmatrix}.$$

Writing the components of $A_0 z$, $A_1 z$ as inner products and using the Cauchy-Schwarz inequality, one will find that $||A_0|| \le \sqrt{k^2+3}$, $||A_1|| \le |k|$. Thus we can set $d^+ = 2(k^2+3)^{1/2}$, and now with $F_d(s) = F(s+d) = |sI - (A_0 - dI) - e^{-hd}A_1 e^{-hs}|$, we determine the eigenvalues $s_d = \delta_d + i\omega_d$ of the matrix J_d from Section 2. For fixed $d < d^+$ we then examine the minimum of the values of $|F_d(i\omega_d)|$, call it $g(d)$.

Consider this example with $k = 1$, $h = 0.1$. We will display the graph of $y = g(d)$ with these parameters presently. In this graph we must be careful not to prematurely dismiss the zero value of $y = g(\,d)$ at $d = 0$. In fact, with $d = 0$, MATLAB computation will give us $s = \pm i$, $s = \pm 1.0049i$ for the pure imaginary eigenvalues of J_d. Noting the proximity of the two positive imaginary axis eigenvalues, we might think of first determining, algebraically, whether they are distinct, and then evaluating $F(1.0049i)$ if so. However, since $F(i) = 0$, numerical evaluation could give a very small value for $F(1.0049i)$. Instead choosing to return directly to $f(s)$, and evaluating in the vicinity of the two eigenvalues, we find that $|f(s)| \ge 0.97$ for $s = i\omega$, $0.98 \le \omega \le 1.02$. Now we know that the second observed zero of $y = g(d)$, i.e. $d \approx -0.005$, is the stability exponent.

Shown in Figure 8 is the matrix formulation for stability analysis of

$$x'(t) = k\int_{-h}^{0} \sin(u)x(t+u)du$$

with $k = 1$, $h = 0.1$, and $z'(t) = A_0 z(t) + A_1 z(t-h)$.

6 Conclusion

In this chapter we have presented a means of using the possible oscillatory eigenvalues of a linear delay differential equation as a guide in making a computational determination of the system stability exponent and other eigenvalue abscissas. The technique made use of system coefficients to form a linear operator having spectrum

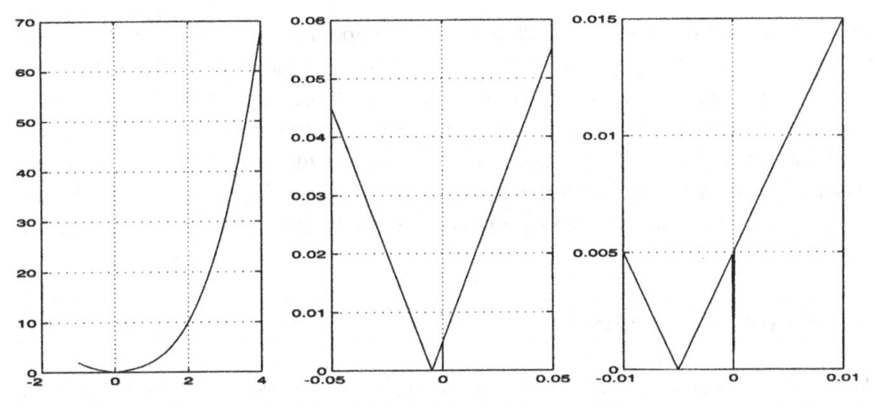

Fig. 8. $y = g(d)$. From the left: $-1 \leq d \leq 4$, $-0.05 \leq d \leq 0.05$, $-0.01 \leq d \leq 0.01$

containing all possible imaginary axis eigenvalues of the delay equation. The information on the delay was fitted by varying a real parameter representing a translation in the characteristic function.

After providing an illustrative example, we examined the technique in the special case of second order scalar equations having a single delay, then giving an example in which we confirm a claim of Abdallah et al by giving a computational determination of the stability exponent for an oscillator with positive delay feedback. Following this, we used the method to examine the behavior of the stability exponent as the delay parameter varies. By direct computation of the stability exponent for various values of the delay parameter, we were able to arrive at an understanding of the behavior of the exponent which could not be provided by conventional wisdom. Finally, using the characteristic function as an intermediary, we gave an example of how a matrix delay system with discrete delay can be used to find the system exponent for a scalar distributed delay equation.

The theorems are given for matrix delay equations having a single delay. There is nothing to prevent the method from carrying over to equations having several commensurate delays. The associated numerical, algebraic, and practical intricacies remain for future investigations. Likewise, there are many questions which remain about the use of the method for distributed delay equations. Among the topics of interest are the kinds of distributed delay equation the method is convenient for, the particular questions involving delays represented by finite Fourier series, and any understanding of distributed delay models that can be deduced from accurate computation of eigenvalue abscissas.

There is one topic worthy of special mention for the sake of practicality. In the presentation given here the system characteristic function is evaluated at the imaginary axis projections of all eigenvalues of the finite dimensional linear operator we associated with the delay equation. In practice it may often be just one, two or four of these eigenvalues which are close to the imaginary axis, and with symmetry considered, we often need to evaluate the characteristic function at just one eigenvalue projection. It may be desirable, from a computational perspective, to classify the eigenvalues of the associated linear operator as sufficiently or not sufficiently close to the imaginary axis to warrant evaluation. A preliminary draft of this chapter was written with this understood, and some corresponding MATLAB programs were written. For the sake of clarity in presenting the idea, eigenvalue classification was not introduced here. In computations for high order delay equations, this kind of classification will make implementation considerably more convenient and feasible.

7 Acknowledgements

This research was supported by NSF (USA) Grant Numbers INT-9818312 and INT-0129062.

References

1. C. Abdallah, P. Dorato, J. Benitez-Read and R. Byrne: "Delayed positive feedback can stabilize oscillatory systems," *Proc. IEEE American Control Conference*, San Francisco, CA, pp. 3106–3107, 1993.
2. J. Chen, G. Gu and C.N. Nett: "A new method for computing delay margins for stability of linear delay systems," *Systems and Control Letters*, vol. 26, pp. 107–117, 1995.
3. R. Datko: "Lyapunov functionals for certain linear delay differential equations in a Hilbert space," *Journal of Mathematical Analysis and Applications*, vol. 76, pp. 37–57, 1980.
4. K. Engelborghs and D. Roose: "On stability of LMS—methods and characteristic roots of delay differential equations," *SIAM J. Num. Analysis*, vol. 40, pp. 629–650, 2002.
5. J. Louisell: "Numerics of the stability exponent and eigenvalue abscissas of a matrix delay system," in *Stability and Control of Time-Delay Systems*, Lecture Notes in Control and Information Sciences No. 228, Eds. L. Dugard and E. I. Verriest, Springer-Verlag, pp. 140–157, 1997.
6. J. Louisell: "Accurate determination of the stability exponent and eigenvalue abscissas in a linear delay-differential system," *Proceedings of the 4th European Control Conference*, ECC 909, Brussels, 1997.
7. J. Louisell : "A matrix method for determining the imaginary axis eigenvalues of a delay system," *IEEE Transactions on Automatic Control*, vol. 46, pp. 2008–2012, 2001.
8. J. E. Marshall, H. Gorecki, K. Walton, and A. Korytowski: *Time-Delay Systems: Stability and Performance Criteria with Applications*, Ellis Horwood, New York, 1992.
9. S.-I. Niculescu: "Stability and hyperbolicity of linear systems with delayed state: a matrix-pencil approach," *IMA Journal of Mathematical Control and Information*, vol. 15, pp. 331–347, 1998.
10. H. Ozbay : *Introduction to Feedback Control Theory* , CRC Press, Boca Raton, FL, 1999.
11. I. M. Repin: "Quadratic Liapunov functionals for systems with delays," *Prikl. Math. Mech.*, vol. 29, pp. 564–566, 1965.
12. C. Ulus: "Numerical computation of inner-outer factors for a class of retarded delay systems,"*Int. J. Systems Sci.*, vol. 28, pp. 897–904, 1997.

The Effect of Approximating Distributed Delay Control Laws on Stability

Wim Michiels[1], Sabine Mondié[2], Dirk Roose[1], and Michel Dambrine[3]

[1] K.U. Leuven, Department of Computer Science, Celestijnenlaan 200A, B-3001 Heverlee, Belgium {Wim.Michiels,Dirk.Roose}@cs.kuleuven.ac.be
[2] CINVESTAV-IPN, Departemento de Control Automático, Av. IPN 2508, Col. Zacatenco, 07360 México D.F., México smondie@ctrl.cinvestav.mx
[3] LAIL, UPRESA CNRS 8021, Ecole Centrale de Lille, B.P. 48 Villeneuve d'Ascq-cedex, France Michel.Dambrine@ec-lille.fr

Summary. An overview of stability results on the implementation of distributed delay control laws, arising in the context of finite spectrum assignment, is given. First the case where distributed delays are approximated with a finite sum of point-wise delays is considered. The instability mechanism is briefly discussed and conditions for a safe implementation are presented. Secondly modifications of the control law to remove the limitations, imposed by these conditions, are outlined. Throughout the chapter eigenvalue plots are used to provide an intuitive explanation for the phenomena and results.

1 Introduction

Consider the linear finite-dimensional system with input delay

$$\dot{x}(t) = Ax(t) + Bu(t-\tau),\ x \in \mathbb{R}^d,\ u \in \mathbb{R}, \tag{1}$$

where we assume that the matrix A is not Hurwitz and the pair (A, B) is stabilizable. An approach for the stabilization and control of (1), called *finite spectrum assignment* [21,37], can be interpreted as follows: a prediction of the state variable over one delay interval is generated first and then a feedback of the predicted state is applied, thereby compensating the effect of the time-delay. This results in a closed-loop system with a finite number of eigenvalues, which can be freely assigned. Mathematically, with the feedback law

$$\begin{aligned} u(t) &= K^T x_p(t, t+\tau) \\ &= K^T \left(e^{A\tau}x(t) + \textstyle\int_0^\tau e^{A\theta}Bu(t-\theta)d\theta\right), \end{aligned} \tag{2}$$

where $x_p(t_1, t_2)$ is the prediction of $x(t)$ at $t = t_2$, based on values of x and u for $t \leq t_1$, the characteristic equation of the closed-loop system is given by

$$\det\left(\lambda I - A - BK^T\right) = 0. \tag{3}$$

This elimination of the delay is employed in the so-called process model control techniques [38], as for example, the celebrated Smith Predictor [34]. When applied to (1) and (2) it can also be interpreted as the effect of a model transformation [16]. For generalizations of the finite spectrum assignment approach to a broader class of time-delay systems than (1), we refer to [1,21,39].

A difficulty in applying the control law (2) consists of the practical implementation of the integral term. Obtaining this term as the solution of a differential equation must be discarded because it involves an unstable pole-zero cancellation when A is unstable, see [21]. It is then suggested to realize the control law by means of a numerical computation of the integral term *on-line*, which involves some approximation.

In this chapter we mainly consider the approximation of the distributed delay with a *sum of point-wise delays* by applying a numerical quadrature rule. To state this more precisely, define $\mathcal{C}([0,\ \tau],\mathbb{C}^d)$ as the space of continuous functions from $[0,\ \tau] \subset \mathbb{R}$ to $\mathbb{C}^d$, equipped with the supremum norm. A quadrature rule on $[0,\ \tau]$ is a sequence of maps $\{\mathcal{I}_n\}_{n\geq 1}$ from $\mathcal{C}([0,\ \tau],\mathbb{C}^d) \to \mathbb{C}$, defined as

$$\mathcal{I}_n(f) = \sum_{j=1}^{n} h_{j,n} f(\theta_{j,n}),\ \ h_{j,n} > 0,\ \theta_{j,n} \in [0,\ \tau], \tag{4}$$

where we assume that the following convergence property is satisfied:

$$\forall f \in \mathcal{C}([0,\ \tau],\mathbb{C}^d),\ \forall \varepsilon > 0,\ \exists \overline{n} \in \mathbb{N} \text{ s.t. } |I_n(f) - \textstyle\int_0^\tau f(\theta)d\theta| < \varepsilon,\ \forall n \geq \overline{n}. \tag{5}$$

When the quadrature formulae (4) are used to approximate the integral term in (2), we end up with a sequence of control laws

$$u(t) = K^T \left(e^{A\tau} x(t) + \textstyle\sum_{j=1}^{n} h_{j,n} e^{A\theta_{j,n}} Bu(t - \theta_{j,n}) \right). \tag{6}$$

The effect of this semi-discretization of the control law on the closed-loop stability will be analyzed in detail.

The structure of the text reflects our main goal, i.e. giving an overview of existing stability results on the implementation of distributed delay control laws. After some preliminaries we comment on a possible instability mechanism when using the control law (6), reported in [6,27,35,36]. Then we discuss conditions for a safe implementation [23,28] and outline modifications of the control law to remove the resulting restrictions [8,23,29,32,33,36]. For the derivation of the stability results and mathematical details we refer to the bibliography. Plots of eigenvalues for a numerical example are extensively used throughout the chapter in order to make the main ideas apparent.

2 Preliminaries

The initial data for both the system (1)-(2) and the system (1) and (6) are $x(0) \in \mathbb{R}^d$, $u_0 \in \mathcal{C}([-\tau,\ 0],\mathbb{R})$. For $t \in [0\ \tau]$, the closed-loop system becomes

$$\dot{x}(t) = Ax(t) + B\,u_0(t-\tau).$$

For $t \geq \tau$, we have

$$\begin{aligned} Bu(t-\tau) &= BK^T\left(e^{A\tau}x(t-\tau) + \int_0^\tau e^{A\theta}Bu(t-\theta-\tau)d\theta\right) \\ &= BK^T\left(e^{A\tau}x(t-\tau) + \int_0^\tau e^{A\theta}(\dot{x}(t-\theta) - Ax(t-\theta))d\theta\right), \end{aligned}$$

where the right-hand derivative of x should be taken at time zero, and we can write the system (1) and (2) in the form

$$\frac{d}{dt}\mathcal{L}(x_t) = Ax(t) + BK^Te^{A\tau}x(t-\tau) - BK^TA\int_0^\tau e^{A\theta}x(t-\theta)d\theta,$$

where the map $\mathcal{L}: \,\mathcal{C}([-\tau\; 0],\mathbb{R}^d) \to \mathbb{R}^d$ is defined as

$$\mathcal{L}(\phi) = \phi(0) - BK^T\int_0^\tau e^{A\theta}\phi(-\theta)d\theta.$$

Because (1)-(2) is a Volterra equation of the second kind, the growth of its solutions is determined by the roots of its characteristic equation (see [21])

$$\det\left\{\lambda\left(I - BK^T\int_0^\tau e^{(A-\lambda I)\theta}d\theta\right) - A - BK^Te^{(A-\lambda I)\tau} + BK^TA\int_0^\tau e^{(A-\lambda I)\theta}d\theta\right\} = 0, \quad (7)$$

which can be simplified to (3). This makes the finite spectrum assignment property apparent.

Analogously, with the approximated control law (6) and for $t \geq \tau$ the closed-loop system can be written as

$$\frac{d}{dt}\mathcal{N}_n(x_t) = Ax(t) + BK^Te^{A\tau}x(t-\tau) - BK^TA\sum_{j=1}^n h_{j,n}e^{A\theta_{j,n}}x(t-\theta_{j,n}), \quad (8)$$

where the map $\mathcal{N}_n: \,\mathcal{C}([-\tau_n,\,0],\mathbb{R}^d) \to \mathbb{R}^d$ is defined as

$$\mathcal{N}_n(\phi) = \phi(0) - BK^T\sum_{j=1}^n h_{j,n}e^{A\theta_{j,n}}\phi(-\theta_{j,n}).$$

with $\tau_n = \max_j \theta_{j,n}$.

Equation (8) is a neutral functional differential equation(NFDE). Under mild assumptions on the integration rule (4), the map $\mathcal{N}_n$ is atomic at zero[4], guaranteeing existence and uniqueness of solutions for initial conditions $\phi \in \mathcal{C}([-\tau_n,\,0],\mathbb{R}^d)$. Let $T^n(t)$ be the solution operator, mapping initial data onto the state at time t, i.e.

[4] This property makes the system (1) and (6) causal, and allows to write the control input u at the present time as a function of the present state and *past* inputs.

$$(T^n(t)(\phi))(\theta) = x_t(\theta;\phi) = x(t+\theta;\phi),\ \theta \in [-\tau_n,\ 0],$$

which is a strongly continuous semi-group. The associated difference equation of (8) is given by $\mathcal{N}_n(x_t) = 0$, i.e.

$$x(t) = BK^T \sum_{j=1}^{n} h_{j,n} e^{A\theta_{j,n}} x(t-\theta_{j,n}). \tag{9}$$

For any initial condition $\phi \in \mathcal{C}_D([-\tau_n,\ 0], \mathbb{R}^d)$, where

$$\mathcal{C}_D([-\tau_n,\ 0], \mathbb{R}^d) = \left\{\phi \in \mathcal{C}([-\tau_n,\ 0], \mathbb{R}^d) :\ \mathcal{N}_n(\phi) = 0\right\},$$

a solution of (9) is uniquely defined. Let $T_D^n(t)$ be the corresponding solution operator.

The asymptotic behavior of the solutions and, thus, stability of the neutral equation (8) is determined by the spectral radius $r(T^n(t))$, satisfying

$$r(T^n(1)) = e^{\alpha},\ \ \alpha = \sup\left\{\Re(\lambda) :\ \det\left(\Delta^n(\lambda)\right) = 0\right\},$$

where the characteristic matrix Δ^n is given by

$$\Delta^n(\lambda) = \left(\lambda \Delta_D^n(\lambda) - A - BK^T e^{(A-\lambda I)\tau} + BK^T A \sum_{j=1}^{n} h_{j,n} e^{(A-\lambda I)\theta_{j,n}}\right)$$

and

$$\Delta_D^n(\lambda) = \left(I - BK^T \sum_{j=1}^{n} h_{j,n} e^{(A-\lambda I)\theta_{j,n}}\right).$$

In a similar way, stability of the difference equation (9) is determined by the spectral radius

$$r(T_D^n(1)) = e^{\beta},\ \ \beta = \sup\left\{\Re(\lambda) :\ \det\left(\Delta_D^n(\lambda)\right) = 0\right\}.$$

An important property in the stability analysis of a NFDE is the relation

$$r_e(T^n(1)) = r(T_D^n(1)), \tag{10}$$

where $r_e(.)$ denotes the radius of the essential spectrum, see e.g. [11].

In the rest of the chapter, we will call the roots of the characteristic equation

$$\det\left(\Delta^n(\lambda)\right) = 0$$

the *eigenvalues* of the neutral equation (8) (in fact they are the eigenvalues of the infinitesimal generator of $T^n(t)$, which determines the evolution when (8) is written as an abstract ordinary differential equation over $\mathcal{C}([-\tau_n,\ 0], \mathbb{R}^d)$). Similarly, we will call the roots of

$$\det\left(\Delta_D^n(\lambda)\right) = 0$$

the eigenvalues of the difference equation (9).

3 Instability Mechanism

A starting point of the research on the implementation of distributed delay control laws was the paper [35], which illustrated that the closed-loop system (1) and (6) may be unstable for *arbitrarily large* values of n, even when the ideal closed-loop system (1) and (2) is exponentially stable, the latter expressed by the Hurwitz stability of matrix $A + BK^T$. This paradox can intuitively be explained with the occurrence of unstable eigenvalues with a large modulus for the approximated closed-loop system. When the approximation becomes better, some eigenvalues tend to the eigenvalues of the limit case, while the others move off to infinity. When some eigenvalues do so without leaving the right half plane, instability persists. This is now illustrated with an example.

Example 1. Consider the scalar system

$$\dot{x}(t) = x(t) + u(t-1), \tag{11}$$

and the control law

$$u(t) = -2\, x_p(t, t+1) = -2 \left(e\, x(t) + \int_0^1 e^{\theta} u(t-\theta) d\theta \right), \tag{12}$$

which assigns one closed-loop eigenvalue $\lambda = -1$. When the integral term in (12) is discretized using the forward rectangular rule, i.e. using (4) with

$$\theta_{j,n} = \frac{j-1}{n}, \; h_{j,n} = \frac{1}{n}, \; j = 1 \ldots n,$$

the control law becomes

$$u(t) = -2 \left(ex(t) + \frac{1}{n} \sum_{j=1}^{n} e^{\frac{j-1}{n}} u\left(t - \frac{j-1}{n}\right) \right). \tag{13}$$

In Fig. 1 (above) the eigenvalues of the closed-loop system(11) and (13) are shown for $n = 40$ and $n = 60$. As $n \to \infty$, one eigenvalue converges to the assigned eigenvalue $\lambda = -1$, while all the eigenvalues, introduced by the approximation, move off to infinity. However, stability is not obtained. The sequences of eigenvalues, whose imaginary parts tend to infinity, yet whose real parts have a finite limit, are explained by the neutral type of the closed-loop system. In Fig. 1 (below) we show the eigenvalues of the associated difference equation

$$u(t) = -2 \left(\frac{1}{n} \sum_{j=1}^{n} e^{\frac{j-1}{n}} u\left(t - \frac{j-1}{n}\right) \right). \tag{14}$$

As expected from theoretical considerations (related to property (10)), the closed-loop eigenvalues with a large modulus, but small real part, are well approximated by eigenvalues of the difference equation [8].

The instability mechanism is due to the *neutral type* of the approximated closed-loop system (1) and (6), in contrast with the retarded type of the ideal closed-loop system (1)-(2). Hence, for any n the approximation involves a non-compact perturbation of the solution semi-group, associated with (1) and (2), which introduces an essential spectrum, see [23]. Having the radius of this essential spectrum larger than one results in the sensitivity of stability. Alternatively, a frequency domain interpretation, including links with (lack of) w-stability [9], is presented in [27] : the integral in (2) has a smoothing effect on its input, unlike any finite sum approximation. This is reflected in the property that the sequence

$$\left\{ \sup_{\omega \geq 0} \left| \int_0^T e^{(A-j\omega I)\theta} d\theta - \sum_{j=1}^{n} h_{j,n} e^{(A-j\omega I)\theta_{j,n}} \right| \right\}_{n \geq 1}$$

does not converge to zero, as $n \to \infty$. The high-frequency error is indeed responsible for the unstable closed-loop eigenvalues with large imaginary parts.

4 Stability Conditions

Because the Hurwitz stability of $A + BK^T$ does not imply the stability of (1) and (6) for large values of n, additional conditions are needed to guarantee that also the eigenvalues, due to the approximation, are in the open left half plane. A comparison of the spectral plots in Fig. 1 suggests that hyperbolic stability properties of the closed-loop system (1) and (6) are tightly related to stability properties of the difference equation (9). This is indeed the case, leading to very simple and easy-to-check stability criteria, which we now review.

4.1 Necessary Condition

The well known result, stating that a necessary condition for the stability of the neutral equation (8), or equivalently (1) and (6), is given by the stability of the difference equation

$$\mathcal{N}_n(x_t) = x(t) - \sum_{j=1}^{n} BK^T e^{A\theta_{j,n}} x(t - \theta_{j,n}) = 0, \tag{15}$$

see e.g. [8], and the relation of (15) with

$$\mathcal{L}(x_t) = x(t) - \int_0^T BK^T e^{A\theta} x(t - \theta) d\theta = 0. \tag{16}$$

lead to the following necessary condition for a safe implementation, which slightly generalizes [28, Theorem 1]:

Theorem 1. *Consider the system (1)-(2) and a quadrature rule satisfying (5). Assume that the closed-loop system (1) and (6) is asymptotically stable for large values of n. Then the characteristic equation of (16),*

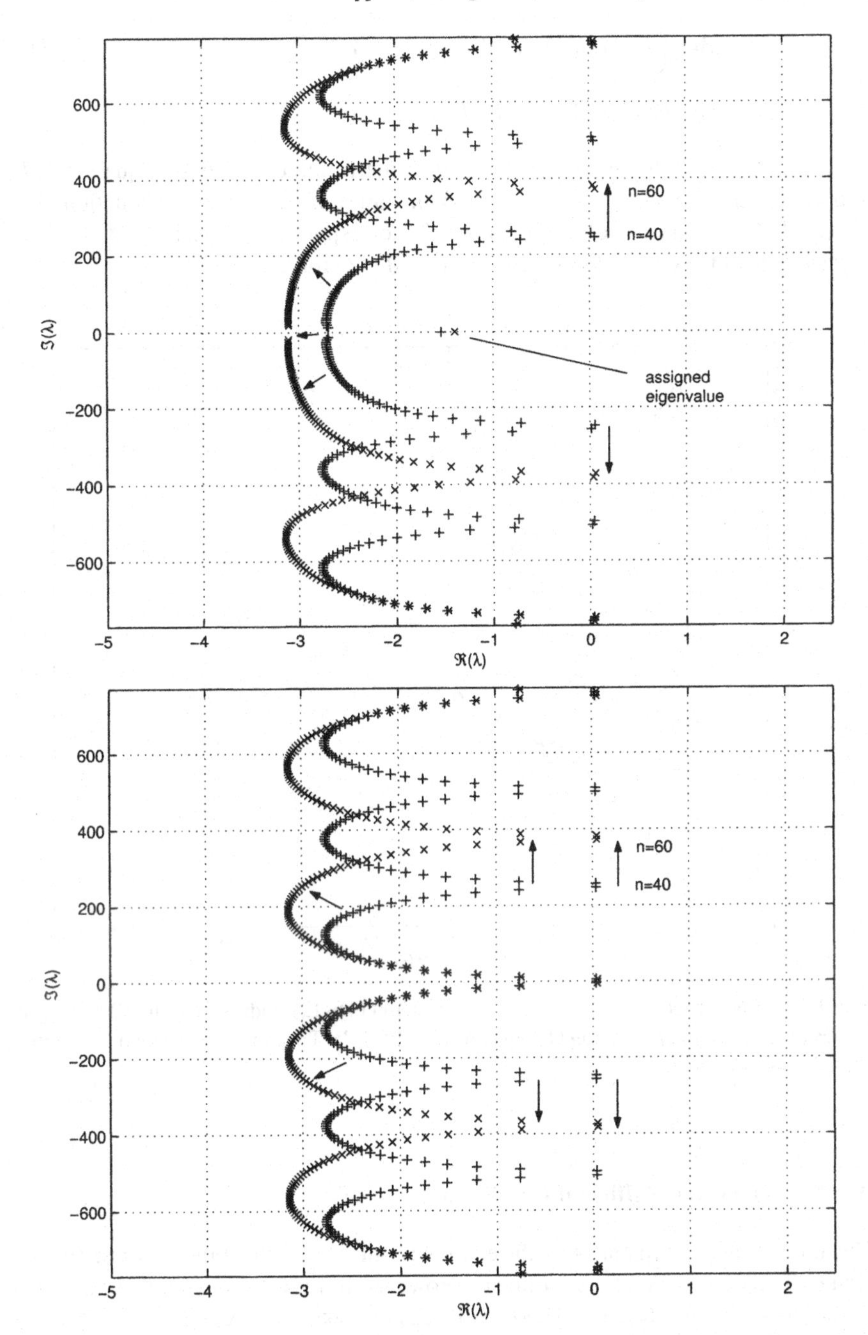

Fig. 1. (above) Closed-loop eigenvalues of the system (11) and (13) for $n = 40$ ('+') and $n = 60$ ('x'). (below) Eigenvalues of the difference equation (14).

$$\det\left\{I - BK^T(\lambda I - A)^{-1}\left(I - e^{-(\lambda I - A)\tau}\right)\right\} = 0, \tag{17}$$

has all its roots in the closed left half plane.

As an illustration we compare in Fig. 2 the rightmost roots of Equation (17), applied to the example (11)-(12), with eigenvalues of the difference equation (14). Notice that the real parts of the roots of (17) correspond to the position of chains of eigenvalues of the closed-loop system (11) and (13) for large n.

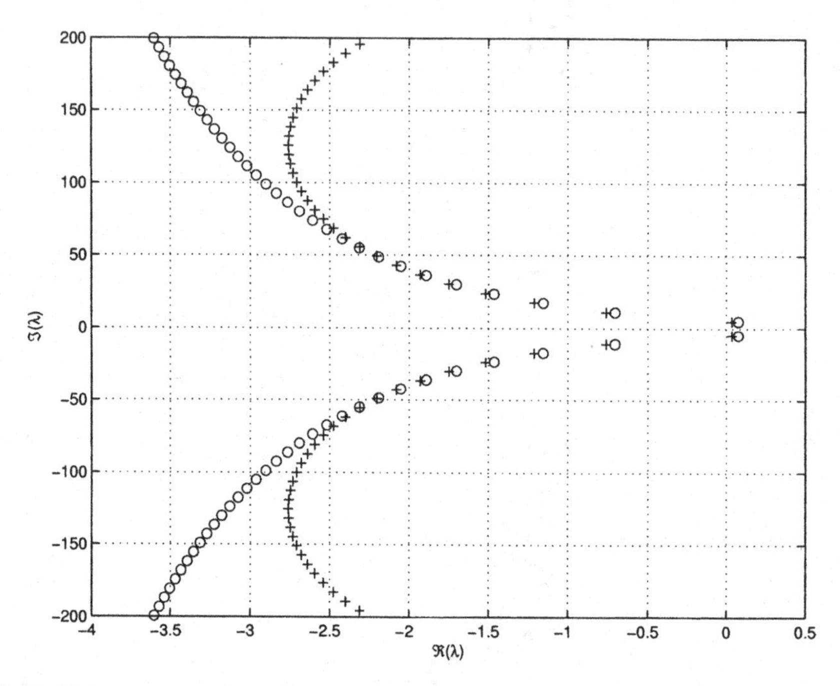

Fig. 2. Rightmost roots of (17), applied to the system (11)-(12) (indicated with 'o') and eigenvalues of the difference equation (14), where $n = 40$ ('+'). The latter are shown on a different scale in Fig. 1 (below).

4.2 Necessary and Sufficient Condition

The numerical experiments with the example could suggest that the necessary condition of Theorem 1 is close to sufficient. However, in [23] it is proven that this is generally *not* the case, because Theorem 1 does not take into account the fact that the radius of the essential spectrum of the solution semigroup, associated with the neutral equation (8), i.e. $r_e(T^n(1))$, is *not* continuous in the delays $\theta_{j,n}$. See [2, 11, 12, 24] and the references therein for more information on delay sensitivity of neutral equations and related questions. A consequence for the system (1) and (6) is that closed-loop stability may depend on the type of integration rule used and be sensitive to infinitesimal perturbations of the abscissa $\theta_{j,n}$, as we now illustrate.

Example 2. We revisit Example 55 and discretize the control law (12) as

$$u(t) = -2\left(e\,x(t) + \sum_{j=1}^{n} h_{j,n} e^{\theta_{j,n}} u\,(t - \theta_{j,n})\right), \tag{18}$$

where

$$\theta_{j,n} = \begin{cases} \frac{j-1}{n}, & j \text{ even} \\ \frac{j-4/5}{n}, & j \text{ odd} \end{cases}, \; h_{j,n} = \tfrac{1}{n},\; j = 1 \ldots n. \tag{19}$$

Notice that the modified integration rule with parameters (19) also satisfies the convergence property (5). In Fig. 3 we show the eigenvalues of the closed-loop system (11) and (18), as well as the eigenvalues of the associated difference equation

$$u(t) = -2\left(\sum_{j=1}^{n} h_{j,n} e^{\theta_{j,n}} u(t - \theta_{j,n})\right), \quad h_{j,n}, \theta_{j,n} \text{ given by (19).} \tag{20}$$

and roots of (17). Although the rightmost roots of (17) are also well approximated by eigenvalues of (20), they do no longer determine the stability of (20) and of the closed-loop system (11) and (18) for large n.

By making infinitesimal perturbations of the abscissa $\theta_{j,n}$ in (19), the rightmost eigenvalues of the difference equation can have their real part arbitrarily close to the value α_n, indicated on the figure, but no larger than $\alpha_n + \varepsilon$, for any $\varepsilon > 0$. In fact, such a value α_n determines stability of the difference equation when subject to small variations in the delays, called *strong stability* in [12]. In [23, Section 4] it is described how α_n can be computed analytically. Furthermore, it is shown that $\lim_{n\to\infty} \alpha_n$ exists and is *independent* of the type of quadrature rule.

Remark 1. The sensitivity of stability w.r.t. infinitesimal perturbations, as well as the high-frequency instability mechanism, described in Sect. 3, are phenomena, which are related to the sensitivity of stability of some boundary controlled hyperbolic PDEs, feedback controlled descriptor systems and neutral type systems against small delays in the control loop, as reported in e.g. [3–5, 13, 17–20]. Sensitivity stability w.r.t. infinitesimal modelling errors also occurs in the Smith predictor control scheme [34], see [26, 31] and the references therein.

Because arbitrarily small perturbations of the abscissa $\theta_{j,n}$ may destroy stability of the closed-loop system (1) and (6), yet are inevitable in any practical application, they should be taken into account in a definition of a safe implementation. Therefore, we say:

Definition 1. *Consider the system (1) and (2), where $A + BK^T$ is Hurwitz, and the quadrature rule (4). Then the implementation (6) of the control law (2) is safe if the following two conditions are satisfied:*

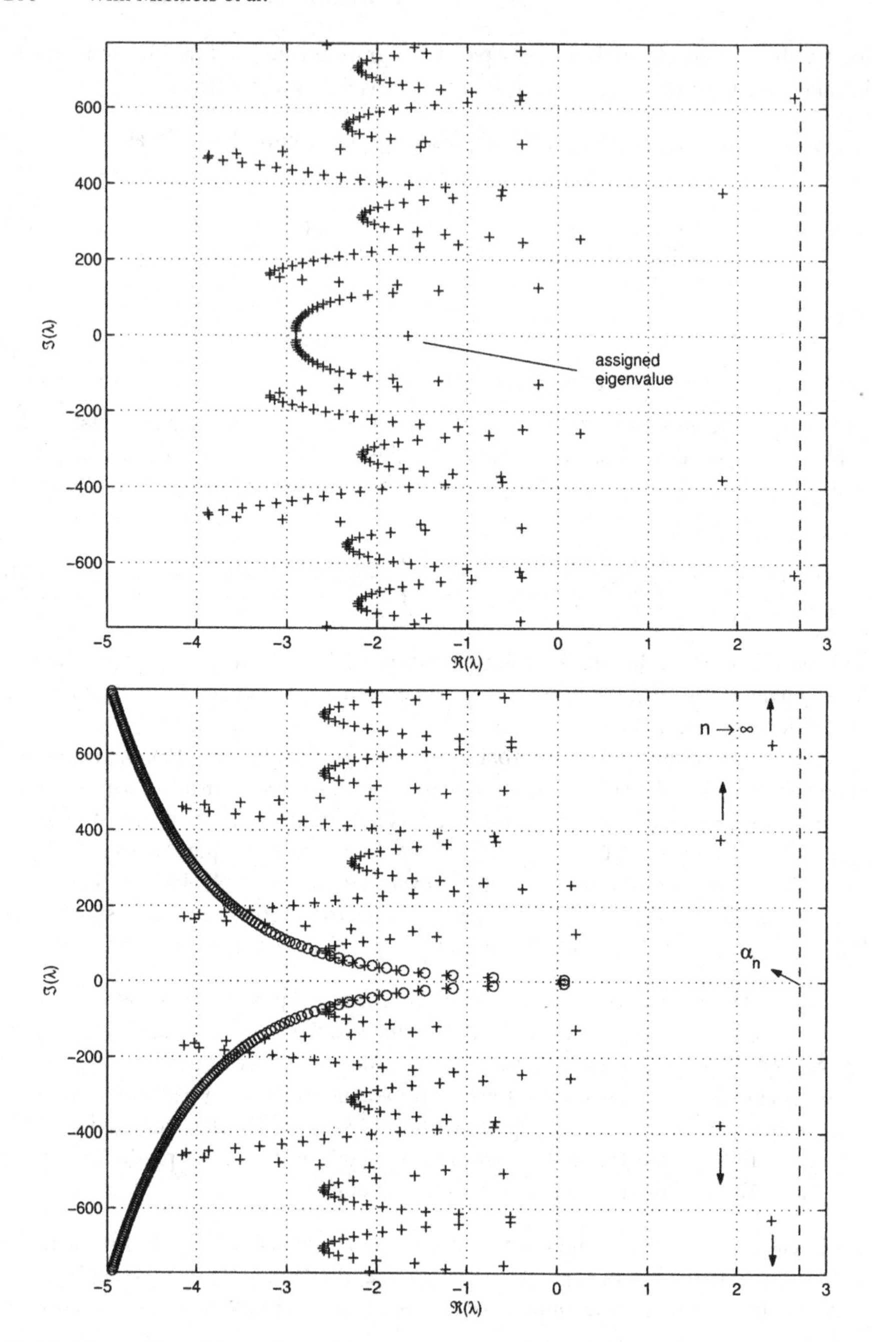

Fig. 3. (above) Closed-loop eigenvalues of the system (11) and (18)-(19) for $n = 40$. (below) Eigenvalues of the associated difference equation (20) ('+') and roots of (17) ('o'). Unlike the situation displayed in Fig. 1, stability of the difference equation for large n is no longer determined by the rightmost roots of (17).

i) There exists a number $\overline{n} \in \mathbb{N}$ such that the closed-loop system (1) and (6) is asymptotically stable for all $n \geq \overline{n}$.
ii) For each $n \geq \overline{n}$, there exist constants $\Delta\theta_{j,n} > 0$ such that the control law

$$u(t) = K^T \left(e^{A\tau} x(t) + \sum_{j=1}^{n} h_{j,n} e^{A(\theta_{j,n}+\delta\theta_{j,n})} Bu(t - (\theta_{j,n} + \delta\theta_{j,n})) \right)$$

achieves asymptotic stability for all $|\delta\theta_{j,n}| < \Delta\theta_{j,n}$.

In the sense of this definition, an almost necessary and sufficient stability condition is given in [23, Theorem 1], which does *not* depend on the type of integration rule. Essentially it corresponds to a strong stability requirement for the difference equation (9) for large n:

Theorem 2. *Consider the system (1) and (2), and assume that $A + BK^T$ is Hurwitz. Let*

$$S = \int_0^T |K^T e^{A\theta} B| d\theta. \tag{21}$$

If $S < 1$, then the control law (2) can be safely implemented as (6), in the sense of Definition 1.
If $S > 1$, then the control law (2) cannot be safely implemented.

In the multiple input case, a *sufficient* condition for a safe implementation is given by

$$\int_0^T ||K^T e^{A\theta} B|| d\theta < 1,$$

see [23, Section 6].

5 Removing restrictions

The main advantage of the control law (2) lies in the fact that all the closed-loop eigenvalues can be freely assigned. A disadvantage is the difficulty of computing the control law on-line, which involves the evaluation of the integral. In particular, for the implementation with a sum of point-wise delays the stability condition of Theorem 2, i.e. $S < 1$, puts severe restrictions on stabilizability and performance, which are shown in [23] to be comparable to the case of a static, non-predictive, state feedback controller, $u(t) = K^T x(t)$. In this section we briefly comment on possible modifications of the control law (6) to remove these restrictions.

5.1 Adding a Low-Pass Filter

The instability mechanism, as explained in Sect. 3, is a *high-frequency* mechanism, related to the occurrence of unstable eigenvalues with *arbitrarily large* imaginary parts. A closer look at the problem reveals that the latter are caused by the throughput at infinity of past inputs in (6) and, therefore, can be avoided by including a *low-pass filter* in the control dynamics. This is now illustrated.

Example 3. We reconsider the system (11) and (13) and modify the control law, by adding a first order low-pass filter, to

$$\begin{cases} \dot{z}(t) = -fz(t) - 2f\left\{ex(t) + \frac{1}{n}\sum_{j=1}^{n} e^{\frac{j-1}{n}} u\left(t - \frac{j-1}{n}\right)\right\}. \\ u = z(t) \end{cases} \tag{22}$$

Due to the filter, the closed-loop system is of *retarded* type. Its eigenvalues are shown in Fig. 3 for different values of n. For sufficiently large values the closed-loop system is asymptotically stable.

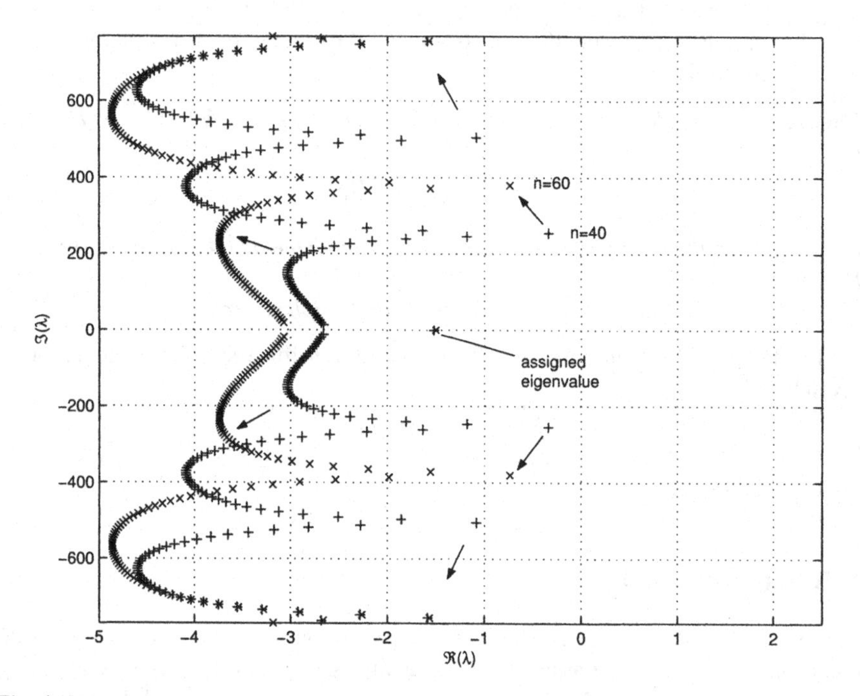

Fig. 4. Closed-loop eigenvalues of (11) and (22) for $f = 100$ and $n = 40$ ('+'), respectively $n = 60$ ('x'). Due to the low-pass filter the closed-loop system is of retarded type and unlike the situation shown in Fig. 1 (above), sequences of eigenvalues, whose imaginary parts grow unbounded, yet whose real parts have a finite limit, do not occur. The low-pass filter puts an upper bound on the imaginary parts of the unstable eigenvalues, which is independent of n. Therefore, the real parts of the introduced eigenvalues move to the left half plane as $n \to \infty$ (actually their real parts move off to minus infinity) and stability is obtained.

This idea is generalized in [29]. Since *any strictly proper linear system* represented by (A_f, B_f, C_f) has a *low-pass* filtering property, the dynamic control law

$$\begin{cases} \dot{z}(t) = A_f z(t) + B_f \underbrace{\left(e^{Ah}x(t) + \int_0^h e^{A\theta} Bu(t-\theta)d\theta\right)}_{x_p(t,t+h)} \\ u(t) = C_f z(t). \end{cases} \tag{23}$$

is suggested. It is shown that the closed-loop eigenvalues are equal to the eigenvalues of the finite-dimensional system

$$\begin{cases} \dot{x}(t) = Ax(t) + BC_f z(t) \\ \dot{z}(t) = A_f z(t) + B_f x(t) \end{cases}, \tag{24}$$

hence, also assignable using standard design methods for ordinary differential equations. Furthermore, a discretization using (4) preserves stability for large values of n.

5.2 Piece-Wise Constant Input

In [36], [32, Section 3] and [33, Section 3] the input $u(t)$ is kept *piecewise constant* in time-intervals of length Δ, inspired by the implementation with a digital controller. Then, at the sampling times, the system (1) is completely equivalent with a discrete system. When $\Delta = \tau/p,\ p \in \mathbb{N}$, this discrete system takes the form

$$\overline{x}(k+1) = A_d \overline{x}(k) + B_d \overline{u}(k-p), \tag{25}$$

where $\overline{x}(k) = x(k\Delta)$, $\overline{u}(k) = u(k\Delta)$ and

$$A_d = e^{A\Delta}, \quad B_d = \left(\int_0^\Delta e^{As}\, ds\right) B.$$

As in the continuous time case, the delay can be compensated using a prediction: for the control law

$$\begin{aligned} \overline{u}(k) &= K_d^T \overline{x}_p(k, k+p) \\ &= K_d^T \left(A_d^p \overline{x}(k) + \textstyle\sum_{n=1}^p A_d^{n-1} B_d \overline{u}(k-n)\right), \end{aligned} \tag{26}$$

the characteristic equation of the closed-loop system is given by

$$\det\left(zI - (A_d + B_d K_d^T)\right) = 0.$$

Because the system (25) and (26) is fully discrete, the maximal possible frequency is given by $1/(2\Delta)$ (the celebrated Nyquist-Shannon criterion) and, therefore, sensitivity of stability w.r.t. arbitrarily small perturbations of the parameters of (26) is not possible.

Note that the control-law (26) can be considered as a full discretization of the continuous control law (2), using a particular quadrature rule. However, in [36] it is also illustrated that one cannot take any quadrature rule, satisfying (5), and obtain stability for sufficiently small values of Δ after a full discretization. For instance, applying Simpson's rule may lead to instability. The instability mechanism is related to the one in the continuous time case, where approximations lead to eigenvalues with arbitrarily large frequencies. In the discrete time-case, unstable modes may occur with the maximal frequency[5] $1/(2\Delta)$, which tends to infinity as $\Delta \to 0$.

Acknowledgements

This work presents research results of the Belgian program on Interuniversity Poles of Attraction, initiated by the Belgian State, Prime Minister's Office for Science, Technology and Culture (IAP P5) and is supported by Conacyt, Mexico, project 41276-Y. Wim Michiels is a postdoctoral fellow of the Fund for Scientific Research -Flanders (Belgium).

References

1. Artstein, Z., "Linear systems with delayed control: a reduction," *IEEE Transactions on Automatic Control*, vol. 27: 869–879, 1982.
2. Avellar C.E. and Hale J.K: "On the zeros of exponential polynomials," *Mathematical analysis and applications*, vol. 73:434–452, 1980.
3. Datko, R., "Two examples of ill-posedness with respect to time delays revisited," *IEEE Transactions on Automatic Control*, vol. 42: 434–452, 1997.
4. Datko, R., "Not all feedback stabilized hyperbolic systems are robust with respect to small time delays in their feedbacks," *SIAM Journal on Control and Optimization*, vol. 26: 697–713, 1988.
5. Datko R., Lagnese J., and Polis M.: "An example on the effect of time delays in boundary feedback stabilization of wave equations," *SIAM Journal on Control and Optimization*, vol. 24:152–156, 1986.
6. Engelborghs K., Dambrine M. and Roose D.: "Limitations of a class of stabilization methods for delay equation, *IEEE Transactions on Automatic Control*, vol. 46(2), 336-339, 2001.
7. Engelborghs, K., Luzyanina T. and Samaey, G., "DDE-BIFTOOL v. 2.00: a Matlab package for bifurcation analysis of delay differential equations," *Technical Report TW-330*, Department of Computer Science, K.U.Leuven, Leuven, Belgium, 2001 (Available from `www.cs.kuleuven.ac.be/~koen/delay/ddebiftool.shtml`).
8. Fattouh, A., Sename, O. and Dion, J.M., "Pulse controller design for linear time-delay systems," *Proceedings of the IFAC Workshop on System, Structure and Control*, Prague, the Czech Republic, 2001.
9. Georgiou, T.T. and Smith, M.C., "Graphs, Causality and Stabilizability: Linear, Shift-Invariant Systems on $\mathcal{L}_2([0,\infty))$ *Math. Control Signals Systems*, vol. 6:195–223, 1993.
10. Hale J.K. and Verduyn Lunel, S.M.: *Introduction to functional differential equations*, vol. 99 of *Applied Mathematical Sciences*. Springer-Verlag, 1993.

[5] Such a mode corresponds to a negative real eigenvalue of the fully discretized system.

11. Hale, J.K., "Effects of delays on dynamics," in *Topological methods in differential equations and inclusions*, (A. Granas, M. Frigon, G. Sabidussi, Eds.), Kluwer Academic Publishers, 191–238, 1995.
12. Hale, J.K. and Verduyn Lunel, S.M, "Strong stabilization of neutral functional differential equations," *IMA Journal of Mathematical Control and Information,* vol. 19: 5–23, 2002.
13. Hannsgen, K.B., Renardy, Y. and Wheeler, R.L., "Effectiveness and robustness with respect to time delays of boundary feedback stabilization in one-dimensional viscoelasticity," *SIAM Journal on Control and Optimization*, vol. 26: 1200-1234, 1988.
14. Hardy G.H. and Wright E.M.: *An introduction to the theory of numbers*. Oxford University Press, 1968.
15. Kolmanovskii, V.B. and Nosov, V.R., *Stability of functional differential equations*, vol. 180, Mathematics in Science and Engineering, Academic Press, 1986.
16. Kwon, W.H. and Pearson, A.E., "Feedback stabilization of linear systems with delayed control," *IEEE Transactions on Automatic Control*, vol. 25: 266–269, 1980.
17. Logemann H.: "Destabilizing effects of small time-delays on feedback-controlled descriptor systems," *Linear Algebra and its Applications*, vol. 272:131–153, 1998.
18. Logemann, H. and Rebarber, R., "The effect of small time-delays on the closed-loop stability of boundary control systems," *Math. Control Signals Systems*, vol. 9: 123-151, 1996.
19. Logemann H., Rebarber R. and Weiss G.: "Conditions for robustness and nonrobustness of the stability of feedback control systems with respect to small delays in the feedback loop," *SIAM Journal on Control and Optimization*, vol. 34(2):572–600, 1996.
20. Logemann, H. and Townley, S., "The effect of small delays in the feedback loop on the stability of neutral systems," *Systems & Control Letters*, vol. 27: 267–274, 1996.
21. Manitius A. Z. and Olbrot A.W.: "Finite spectrum assignment problem for systems with Delays," *IEEE Trans. Autom. Contr.*, vol. AC-24, No. 4, 541-553, 1979.
22. Michiels W., Mondié S. and Roose, D.: "Robust stabilization of time-delay systems with distributed delay control laws: necessary and sufficient conditions for a safe implementation," *Automatica*, 2002, submitted.
23. Michiels, W., Mondié, S. and Roose, D.: "Robust stabilization of time-delay systems with distributed delay control laws: necessary and sufficient conditions for a safe implementation," *Technical Report TW-363* Department of Computer Science, K.U.Leuven, Belgium, 2003.
24. Michiels W., Engelborghs K., Roose D. and Dochain D.: "Sensitivity to infinitesimal delays in neutral equations," *SIAM Journal on Control and Optimization*, 40(4):1134–1158, 2002.
25. Michiels W., Engelborghs K., Vansevenant P., and Roose D.: "The continuous pole placement method for delay equations," *Automatica*, 38(5):747–761, 2002.
26. Michiels W. and Niculescu S.-I.: "On the delay sensitivity of Smith predictors," *International Journal of System Sciences*, invited paper in special issue on control theory of time-delay systems, 2003 (to appear).
27. Mirkin, L. and Zhong, Q.-C., "Are distributed delay control laws intrinsically unapproximable," in *Proceedings of the 4th IFAC Workshop on Time-Delay Systems*, INRIA Rocquencourt, France, 2003.
28. Mondié S., Dambrine M. and Santos, O.: "Approximations of control laws with distributed delays: a necessary condition for stability," *Kybernetica*, vol. 38: 541–551, 2002.
29. Mondié, S. and Michiels, W., "Finite spectrum assignment of unstable time-delay systems with a safe implementation," *IEEE Transactions on Automatic Control*, 2003 (accepted).
30. Niculescu S.-I.: *Delay effects on stability: A robust control approach*, Springer: Heidelberg, 2001.

31. Palmor, Z.J., "Time-delay compensation- Smith predictor and its modifications," in *The Control Handbook*, (CRC and IEEE Press, New York), (chapter 10), 224–237, 1996.
32. Rasvan, V. and Popescu, D, "Control of systems with input delay: An elementary approach," (this volume, part III).
33. Rasvan, V. and Popescu, D., "Feedback stabilization of systems with delays in control," *Control Engineering and Applied Informatics* vol. 3: 62–66, 2001.
34. Smith, O.J., "Closer control of loops with dead time," *Chemical Engineering Progress*, vol. 53: 217–219, 1957.
35. Van Assche V., Dambrine M., Lafay J.-F. and Richard J.-P.: "Some problems arising in the implementation of distributed-delay control law," *Proceedings of the 39th IEEE Conference on Decision and Control*, Phoenix, AZ, December 1999.
36. Van Assche V., Dambrine M., Lafay J.-F. and Richard J.-P.: "Implementation of a distributed control law for a class of systems with delay," *Proceedings of the 3rd Workshop on Time Delay Systems*, 266–271, Santa Fe, NM, December 2001.
37. Wang, Q.G., Lee, T.H. and Tan, K.K, *Finite Spectrum Assignment for Time-Delay Systems*, Lecture Notes in Control and Information Sciences, vol. 239, Springer-Verlag, 1999.
38. Watanabe, K. and Ito, M, "A process model control for linear systems with delay," *IEEE Transactions on Automatic Control*, vol. 26: 1261–1268, 1981
39. Wanatabe, K., "Finite spectrum assignment and observer for multivariable systems with commensurate delays," *IEEE Transactions on Automatic Control*, vol. 31: 543–550, 1986.

Part V

Partial Differential Equations, Nonlinear and Neutral Systems

Synchronization Through Boundary Interaction

Jack K. Hale

School of Mathematics, Georgia Institute of Technology, Atlanta, GA 30332, USA

Summary. The dynamics of a physical system can change when exposed to an environment in which it interacts with other systems. In many situations, such interaction can lead to synchronization in the sense that the dynamics of all systems are essentially the same. Some results and references can be found in Hale (1997) for ode and certain types of pde. Other interesting classes of problems occur when the equations arise from the interaction of systems whose dynamics are defined by a pde on a given domain and the interaction of the systems is through the boundary. We give an illustration of how this can occur for lossless transmission lines which interact through resistive coupling at the end of the lines. The problem will be solved using the equivalent formulation in terms of a set of partial neutral functional differential equations.

1 A Model Problem with Transmission Lines

Consider the lossless transmission line

$$L_s \frac{\partial i}{\partial t} = -\frac{\partial v}{\partial x} \tag{1a}$$

$$C_s \frac{\partial v}{\partial t} = -\frac{\partial i}{\partial x} \tag{1b}$$

with $x \in (0,1)$ with the boundary conditions

$$0 = E - v(0,t) - R_0 i(0,t) \tag{2a}$$

$$-C\frac{d}{dt}v(1,t) = -i(1,t) + g(v(1,t)). \tag{2b}$$

The constant L_s represents the series inductance and C_s the capacitance per unit length of the line. The form of the boundary condition denotes a specific type of circuit across the line consisting of battery source E, a resistor R_0 and a nonlinear resistor circuit g and a capacitance C. We assume that g is at least C^1.

Wu and Xia (1996) have considered N such transmission lines with currents i_k and voltages v_k on a circle with identical resistive coupling R to their nearest neighbors at the boundary point $x = 1$ to obtain the system

$$L_s \frac{\partial i_k}{\partial t} = -\frac{\partial v_k}{\partial x} \tag{3a}$$

$$C_s \frac{\partial v_k}{\partial t} = -\frac{\partial i_k}{\partial x} \tag{3b}$$

$$\tag{3c}$$

with $x \in (0,1)$ with the boundary conditions

$$0 = E - v_k(0,t) - R_0 i_k(0,t) \tag{4a}$$

$$-C\frac{d}{dt}v_k(1,t) = -i_k(1,t) + g(v_k(1,t)) - \frac{1}{R}[v_{k+1} - 2v_k + v_{k-1}](1,t), \tag{4b}$$

where $k = 1, 2, \ldots, N$.

Following Hale (1994), if we now suppose that $\frac{1}{R} = \frac{KC}{h^2}$, where h is the spacing between the lines and $K > 0$ is a constant independent of N, then, as $h \to 0$, we obtain the partial differential equation

$$L_s \frac{\partial i(x,y,t)}{\partial t} = -\frac{\partial v(x,y,t)}{\partial x} \tag{5a}$$

$$C_s \frac{\partial v(x,y,t)}{\partial t} = -\frac{\partial i(x,y,t)}{\partial x}, \quad x \in (0,1), \tag{5b}$$

with the boundary conditions

$$0 = E - v(0,y,t) - R_0 i(0,y,t), \tag{6a}$$

$$-C\frac{\partial v(1,y,t)}{\partial t} = -i(1,y,t) - KC\frac{\partial^2 v(1,y,t)}{\partial y^2}, \quad y \in S^1, \tag{6b}$$

where S^1 is the planar unit circle with center zero.

The variables $(v(x,y,t), i(x,y,t))$ satisfy the telegraph equation in $x \in (0,1)$ and a parabolic equation in $y \in S^1$ at the boundary point $x = 1$. We choose the space of initial data from the space $X \equiv [L^2(0,1) \times H^1(S^1)]^2$. We will show that the asymptotic properties of the solutions of this equation do not depend upon $y \in S^1$ if the coupling constant K is sufficiently large; that is, we have synchronization through the boundary conditions.

To be more precise, we suppose that the equation defines a semigroup $\tilde{T}_{K^{-1}}(t)$, $t \geq 0$ on X for which there is the *compact global attractor* $\tilde{A}_{K^{-1}}$; that is, $\tilde{A}_{K^{-1}}$ is a compact set which is invariant under $\tilde{T}_{K^{-1}}(t)$ for $t \geq 0$ ($\tilde{T}_{K^{-1}}(t)\tilde{A}_{K^{-1}} = \tilde{A}_{K^{-1}}$ for $t \geq 0$) and, for any bounded set $B \subset X$,

$$\lim_{t\to\infty} \operatorname{dist}_X(\tilde{T}_{K^{-1}(t)}B, \tilde{A}_{K^{-1}}) = 0.$$

Definition 1. *We say that system (5), (6) is synchronized if each element of the compact global attractor $\tilde{A}_{K^{-1}}$ is independent of $y \in S^1$.*

The main result is

Theorem 1. *There is a constant $K_1 > 0$ such that, for each $K \geq K_1$, system (5), (6) is synchronized.*

Remark 1. For the discrete system (3), (4), the space X should be $[L^2(0,1)]^2 \times \mathbb{R}^N$. Synchronization can be defined for (3), (4) by saying that the elements of the attractor must be of the form $(v(x,t), \ldots, v(x,t)) \in \mathbb{R}^N$; that is, lies on the diagonal in $\mathbb{R}^N$. The same conclusion as in Theorem 1 holds for (3), (4) if K is sufficiently large.

It would be interesting to prove Theorem 1 by working directly on the partial differential equations. However, it is not known at this time if this is possible. The proof below goes through an auxiliary neutral functional differential equation.

It has been known for a long time (see Hale and Verdyun-Lunel (1993) or Wu (1996)for references) that, if $V(t) = v(1,t)$, then (1), (2) is equivalent to a neutral functional differential equation (NFDE)

$$\frac{d}{dt} D(q)V_t = f(V_t), \tag{7}$$

on the space $C([-r,0],\mathbb{R})$, where

$$D(q)\varphi = \varphi(0) - q\varphi(-r), \tag{8a}$$

$$r = 2\sqrt{L_s C_s}, \quad q = \frac{\sqrt{L_s/C_s} - R_0}{\sqrt{L_s/C_s} + R_0} \tag{8b}$$

and

$$f(V_t) = \frac{\alpha}{C} - \frac{2}{rC} V(t) - \frac{2q}{rC} V(t-r) - D(q)g(V_t) \tag{9}$$

and $\alpha = 2E/((L_s/C_s)^{1/2} + R_0)$.

The reduction of (1), (2) to (7), (8), (9) makes use of the boundary conditions and the represention of the solutions of (1) as

$$v(x,t) = \varphi(x - st) + \psi(x + st) \tag{10a}$$

$$i(x,t) = \frac{1}{(L_s/C_s)^{1/2}} [\varphi(x - st) - \psi(x + st)]. \tag{10b}$$

The same procedure can be applied to (5), (6) to obtain the partial neutral functional differential equation (PNFDE)

$$\frac{\partial}{\partial t} D(q)u_t = K \frac{\partial^2}{\partial x^2} D(q)u_t + f(u_t). \tag{11}$$

with periodic boundary conditions.

Let $Y = C([-r,0], H^1(S^1))$. If $\varphi \in Y$, then there is a unique solution $u(t,\varphi)(x)$ of (8) with initial value φ on $[-r,0]$; that is, $u(\theta,\varphi)(x) = \varphi(\theta)(x)$, $\theta \in [-r,0]$, $x \in S^1$, or, equivalently, $u_0(\cdot,\varphi) = \varphi$. This solution is defined on a maximal interval $[-r, \alpha_\varphi)$ and $u_t(\cdot,\varphi) \in Y$ for $t \in [0, \alpha_\varphi)$ (see Hale (1994), Wu (1996)). Furthermore, $u(t,\varphi)$ is as smooth in φ as the function f.

We suppose that all solutions of (11) are defined for all $t \geq 0$ and let $T_{K^{-1}}(t)$, $t \geq 0$, be the semigroup defined by $T_{K^{-1}}(t)\varphi = u_t(\cdot, \varphi)$. As remarked above, if f is a C^k-function (or analytic), then $T_{K^{-1}}(t)\varphi$ is C^k (or analytic) in φ.

With the operator $D(q)$ defined as in (8), define the space

$$C_D([-r,0],\mathbb{R}) = \{\varphi \in C([-r,0],\mathbb{R}) : D(q)\varphi = 0\} \tag{12}$$

and consider the homogeneous difference equation

$$D(q)y_t = 0. \tag{13}$$

If $|q| < 1$, then $D(q)$ is exponentially stable; that is, there are positive constants α, β such that, for any $\varphi \in C_D([-r,0],\mathbb{R})$, the solution $y(t,\varphi)$, $y_0(\cdot,\varphi) = \varphi$, of the difference equation (13) satisfies

$$|y_t(\cdot,\varphi)| \leq \beta e^{-\alpha t}|\varphi|, \quad t \geq 0. \tag{14}$$

For the special D that we are considering, $e^{-\alpha} = |q|$.

We recall that a *compact global attractor* $\mathcal{A}$ for the semigroup $T_{K^{-1}}(t)$ on Y is a compact set which is *invariant* ($T_{K^{-1}}(t)\mathcal{A} = \mathcal{A}$ for all $t \geq 0$) and $\mathcal{A}$ *attracts* bounded sets of Y; that is,

$$\lim_{t\to\infty} \text{dist}_Y(T_{K^{-1}}(t)B, \mathcal{A}) = 0$$

for each bounded set $B \subset Y$.

Suppose that there is a compact global attractor $\mathcal{A}_{K^{-1}} \subset Y$ for $T_{K^{-1}}(t)$, $t \geq 0$, for each $K \in (0,\infty)$. This will be the case if the original equation (5), (6) has the compact global attractor.

If S^1 is represented by the interval $[0,\pi]$ with 0 and π identified, then the eigenvalues of the Laplacian $K\partial^2/\partial x^2$ with periodic boundary conditions are $-Kn^2$ for $n = 0, 1, 2, \ldots$. Therefore, if K is very large, it is to be expected that the attractor $\mathcal{A}_{K^{-1}}$ of (11) should be very close to the attractor $\mathcal{A}_0$ of the equation (7) (which we assume to exist).

We can prove the following theorem which is essentially contained in Hale (1997).

Theorem 2. *Suppose that there is a positive constant* K_1 *such that the attractors* $\{\mathcal{A}_{K^{-1}}, K \geq K_1\} \cup \mathcal{A}_0$ *of (11) are uniformly bounded. Then there is a* $K_2 \geq K_1$ *such that, for* $K \geq K_2$,

$$\mathcal{A}_{K^{-1}} = \mathcal{A}_0,$$

the compact global attractor of (7); that is, the system is synchronized.

We outline the proof. Let $Y = Y_0 \oplus Y_1$, where Y_0 consists of functions which are independent of the spatial variable and Y_1 are all functions in Y which are orthogonal to the constant functions. For any $\varphi \in Y$, we have

$$\varphi = \varphi_0 + \varphi_1$$
$$\varphi_0 = \frac{1}{\pi}\int_0^{\pi} \varphi(x)dx,$$
$$\int_0^{\pi} \varphi_1(x)dx = 0$$

If we let a solution of (11) be written as

$$u_t = v_t + w_t, \quad v_t \in Y_0, w_t \in Y_1$$

then v_t, w_t satisfy the equations

$$\frac{\partial}{\partial t}D(q)v_t = f(v_t) + \pi^{-1}\int_0^{\pi} [f(v_t + w_t(\cdot, x)) - f(v_t)]dx. \tag{15a}$$

$$\begin{aligned}\frac{\partial}{\partial t}D(q)w_t &= K\frac{\partial^2}{\partial x^2}D(q)w_t + f(v_t + w_t) - \pi^{-1}\int_0^{\pi} [f(v_t + w_t(\cdot, x))dx. \\ &\equiv K\frac{\partial^2}{\partial x^2}D(q)w_t + F(v_t, w_t)\end{aligned} \tag{15b}$$

We first observe that solutions on the attractors must satisfy some special properties. If we let $B = d^2/dx^2$ with periodic boundary conditions, then each solution on any of the attractors $\{\mathcal{A}_{K^{-1}}, K \geq K_1\}$ must satisfy

$$Dw_t = \int_{-\infty}^{t} e^{KB(t-s)}F(v_s, w_s)ds \tag{16}$$

for all $t \in \mathbb{R}$.

Since $F(\varphi_0, 0) = 0$ for each $\varphi_0 \in Y_0$, there is a constant k_0 such that

$$|F(\varphi_0, \varphi_1)|_X \leq k_0|\varphi_1| \;\; \forall \varphi_0 + \varphi_1 \in \{\mathcal{A}_{K^{-1}}, K \geq K_1\} \cup \mathcal{A}_0. \tag{17}$$

From (17), (16), the fact that

$$|e^{KBt}| \leq e^{-Kt}, \quad t \geq 0, \tag{18}$$

D is stable and w_t is bounded for $t \in \mathbb{R}$, it can be shown that $w_t = 0$ for all $t \in \mathbb{R}$ provided that K is sufficiently large. This proves the Theorem.

2 Proof of Theorem 1

Let us now show how the conclusion in Theorem 2 implies the conclusion in Theorem 1. Let $(i(x, y, t), v(x, y, t))$ be a solution which belongs to the attractor $\tilde{A}_{K^{-1}}$.

Theorem 2 implies that $v(1,y,t)$ is independent of $y \in S^1$. The boundary condition (6) implies that $i(1,y,t)$ is independent of $y \in S^1$. From (10), for all $\tau \in \mathbb{R}$,

$$v(x,y,\frac{x}{s}+\tau)+(L_s/C_s)^{1/2}i(x,y,\frac{x}{s}+\tau)=2\varphi(-s\tau,y).$$

Evaluation at $x=1$ implies that $\varphi(\theta,y)$ is independent of y for all $y \in S^1$. Also, from (10),

$$v(x,y,-\frac{x}{s}+\tau)=\varphi(-2x-s\tau,y)+\psi(s\tau,y).$$

Evaluation at $x=1$ implies that $\psi(\theta,y)$ is independent of $y \in S^1$. Again, using (10), we conclude that $(i(x,y,t),v(x,y,t))$ is independent of $y \in S^1$. This proves Theorem 1.

3 Almost Synchronization

If we allow the individual systems to be governed by the same PDE (1) and keep the same boundary condition at $x=0$, but have different nonlinear resistor circuit in the boundary condition at $x=1$, then we obtain the spatially dependent equation

$$\frac{\partial}{\partial t}D(q)u_t=K\frac{\partial^2}{\partial x^2}D(q)u_t+f(x,u_t) \tag{19}$$

with periodic boundary conditions.

In this case, one would expect that the attractors $\mathcal{A}_{K^{-1}}$ would be close to the attractor $\mathcal{A}_0$ of the 'averaged' neutral FDE

$$\frac{\partial}{\partial t}D(q)y_t=\overline{f}(y_t), \quad y_t \in C([-r,0],\mathbb{R}). \tag{20}$$

where

$$\overline{f}(\varphi)=\frac{1}{\pi}\int_0^{\pi} f(x,\varphi)dx, \tag{21}$$

This is actually the case as stated in the following

Theorem 3. *Suppose that there is a positive constant K_1 such that the attractors $\{\mathcal{A}_{K^{-1}}, K \geq K_1\} \cup \mathcal{A}_0$ of (19) are uniformly bounded. Then there is a $K_2 \geq K_1$ such that,for $K \geq K_2$, $\mathcal{A}_{K^{-1}}$ is a smooth graph over the constant functions in Y which approaches $\mathcal{A}_0$ as $K \to \infty$, where $\mathcal{A}_0$ is the compact global attractor for (20); that is, the system is almost synchronized.*

The proof follows in spirit the proof of Theorem 2. We make the same decomposition of solutions $u_t=v_t+w_t$ to obtain the equations

$$\frac{\partial}{\partial t}D(q)v_t = \tilde{f}(v_t)$$
$$+\pi^{-1}\int_0^\pi [f(x, v_t + w_t(\cdot, x)) - f(x, v_t)]dx. \tag{22a}$$
$$\frac{\partial}{\partial t}D(q)w_t = K\frac{\partial^2}{\partial x^2}D(q)w_t$$
$$+f(x, v_t + w_t) - \pi^{-1}\int_0^\pi [f(x, v_t + w_t(\cdot, x))dx.$$
$$\equiv K\frac{\partial^2}{\partial x^2}D(q)w_t + F(x, v_t, w_t) \tag{22b}$$

Under the hypotheses of the theorem, one can show the existence of k_0 and k_1 such that

$$|F(x, \varphi_0, \varphi_1)|_Y \leq k_0|\varphi_1| + k_1 \quad \forall\varphi_0 + \varphi_1 \in \{\mathcal{A}_{K^{-1}}, K \geq K_1\} \cup \mathcal{A}_0, x \in S^1 \tag{23}$$

Using (23), (16) and (18), we have

$$|Dw_t| \leq \frac{k_0}{K} \sup_{s\in(-\infty,t]} |w_s| + \frac{k_1}{K}. \tag{24}$$

Since D is exponential stable and w_t must be bounded on $\mathbb{R}$, we conclude that

$$|w_t| \leq \frac{k_1}{\alpha K}, \quad t \in \mathbb{R} \tag{25}$$

for some $\alpha > 0$. This relation shows that elements on the attractor must approach spatially independent functions as $K \to \infty$, which is almost synchronization.

To prove the more precise statement in the theorem, we make use of the theory of invariant manifolds which we do not discuss.

4 Further Remarks

Other interesting properties of the equation (19) (as well as more general ones) is that the solution operator $T_{K^{-1}}(t)$ can be written as

$$T_{K^{-1}}(t) = S(t) + U(t) \tag{26}$$

where $S(t)$ is a linear semigroup on Y and there are positive constants k_2, α_2 such that

$$||S(t)|| \leq k_2 e^{-\alpha_2 t}, \quad t \geq 0 \tag{27}$$

and the operator $U(t)$ is completely continuous for $t > 0$ (see Hale (1994)).

As a consequence, we obtain the existence of the compact global attractor if $\{T_{K^{-1}}(t)B, t \geq 0\}$ is bounded for each bounded set $B \subset X$ and $T_{K^{-1}}(t)$ is point dissipative; that is, there is a bounded set $U \subset X$ such that, for any $\varphi \in X$, there is a $t_0 > 0$ such that $T_{K^{-1}}(t)\varphi \in U$ for $t \geq t_0$ (see Hale (1988)).

Following Hale and Scheurle (1985), one also can show that each solution $u(t), t \in \mathbb{R}$, which belongs to a compact invariant set (in partiuclar, the compact global attractor) is as smooth in t as the function f. Therefore, all periodic orbits must be smooth manifolds and one can define characteristic multipliers of periodic orbits, hyperbolicity, etc. The effects of perturbations in f on periodic orbits has not been discussed completely, but it should be possible to do so using the techniques in Hale and Weederman (2002).

References

1. Hale, J.K. (1994) Coupled oscillators on a circle. *Resenhas IME-USP 1*, 441-457.
2. Hale, J.K. (1997) Diffusive coupling, dissipation, and synchronization. J. Dyn. Diff. Equ. 9, 1-52.
3. Hale, J.K. (1988) *Asymptotic Behavior of Dissipative Systems*. Math. Surveys and Monographs, Vol. 25, American Math. Soc.
4. Hale, J.K. and J. Scheurle (1985) Smoothness of bounded solutions of nonlinear evolution equations. *J. Differential Eqns. 56*, 142-163.
5. Hale, J.K. and S.M. Verduyn Lunel (1993) *Introduction to Functional Differential Equations*. Springer-Verlag.
6. Hale, J.K. and M. Weedermann (2002) On perturbations of delay differential equations with periodic orbits. *J. Differential Eqns.*
7. Wu, J. (1996) *Theory and Applications of Partial Neutral Functional Differential Equations*. Springer-Verlag.
8. Wu, J. and H. Xia (1996) Self-sustained oscillations in a ring array of transmission lines. *J. Differential Eqns. 124*, 247-248.

Output Regulation of Nonlinear Neutral Systems

Emilia Fridman[1]

Department of Electrical Engineering, Tel-Aviv University
Ramat-Aviv, Tel-Aviv 69978, Israel
emilia@eng.tau.ac.il

Summary. Output regulation of neutral type nonlinear systems is considered. Regulator equations are derived, which generalize Francis-Byrnes-Isidori equations to the case of neutral systems. It is shown that, under standard assumptions, the regulator problem is solvable if and only if these equations are solvable. In the linear case, the solution of these equations is reduced to linear matrix equations.

1 Introduction

One of the most important problems in control theory is that of controlling the output of the system so as to achieve asymptotic tracking of prescribed trajectories. This problem of output regulation has been studied by many authors (see e.g. a survey paper by Byrnes and Isidori [2] and the references therein). In the linear case, Francis [4] showed that the solvability of a multivariable regulator problem corresponds to the solvability of a system of two linear matrix equations. In the nonlinear case, Isidori and Byrnes [11] proved that the solvability of the output regulation problem is equivalent to the solvability of a set of partial differential and algebraic equations. This set of partial differential and algebraic equations is now known as the *regulator equations* or *Francis-Isidori-Byrnes equations*.

For linear *infinite-dimensional* control systems a solution of the regulator problem was introduced by Schumacher [13] and Byrnes *et al.* [3], where a *Hilbert* space was used as a state space. The case of the bounded input and output operators was considered. In the case of systems with time-delay it means that there are *no discrete delays* in the *control input, controller output* and *measured output*. The solution was given in terms of the operator regulator equations.

The solution of the output regulation problem for retarded type systems was obtained recently in [6], where a *Banach* space was used as a state-space. In the present chapter we generalize the results of [6] to the neutral type case. Our solution is based on the application of the center manifold theory. The existence, smoothness and the attractiveness of the center manifold for neutral type systems were proved by Hale [8] (see also [8], chapter 10.2). A partial differential equation for the function, determining the center manifold for such system was derived in [14], [5], [1]. In the present chapter, we consider output regulation of *nonlinear* systems with *state, controller output* and *measured output delays.* As for the systems of retarded type [6], the problem is solvable iff certain regulator equations are solvable. These equations consist of partial differential equations for a center manifold of the closed-loop neutral system and of an algebraic equation. In the linear case the solution of these equations is reduced to linear matrix equations.

Notations. R^m is the Euclidean space with the norm $|\cdot|$ and $C^m[a,b]$ is the Banach space of continuous functions $\phi : [a,b] \to R^m$ with the supremum norm $||\cdot||$.

A function $f : X \to Y$, where X and Y are Banach spaces, is a C^k function if it has k continuous Frechet derivatives.

Denote by $x_t(\theta) = x(t+\theta)$ $(\theta \in [-h; 0])$.

$L_2([-h,0], R^n)$ is the Hilbert space of square integrable R^n valued functions with the corresponding norm.

$W^{1,2}([-h,0], R^n)$ is the Sobolev space of absolutely continuous R^n valued functions on $[-h,0]$ with square integrable derivatives.

The transpose of a matrix M is written M'.

2 Problem Formulation

We consider a nonlinear system modelled by equations of the form

$$\tfrac{d}{dt}Dx_t = f(x_t, u(t), w(t)), \quad e(t) = g(x_t, w(t)) \tag{1a,b}$$

where $x(\theta) = \phi(\theta), \theta \in [-h, 0]$, with state $x(t) \in R^n$, initial function $\phi \in C^n[-h,0]$, control input $u(t) \in R^m$, exogenous input $w(t) \in R^r$ and tracking error $e(t) \in R^p$. The linear bounded operator $D : C^n[-h,0] \to R^n$ is represented in the form of Stieltjes integral [8]:

$$D\phi = \phi(0) - \int_{-h}^{0} d[\xi(\theta)]\phi(\theta),$$

with $n \times n$-matrix function ξ of bounded variations.

We assume

H0: The following conditions hold:

(i) ξ is nonatomic at zero, i.e. $Var_{[-s,0]}\xi(\cdot) \to 0$ for $t \to 0$;

(ii) D is the stable operator, i.e. the equation $Dx_t = 0$ is asymptotically stable.

The exogenous input is generated by an autonomous dynamical system of the form

$$\dot{w}(t) = s(w(t)) \tag{2}$$

The functions $f : V \to R^n$, $s : W \to R^r$, $g : Y \to R^p$ are smooth (i.e. C^∞) mappings, where $V \subset C^n[-h,0] \times R^m \times R^r$, $W \subset R^r$, $Y \subset C^n[-h,0] \times R^r$

are some neighborhoods of the origin of the corresponding spaces. We assume that $f(0,0,0) = 0$, $s(0) = 0$, $g(0,0) = 0$. Thus, for $u = 0$, the system (1a) has an equilibrium state $(x, w) = (0,0)$ with zero error (1b).

A solution of (1) with initial value $x_0 \in C^n[-h,0]$ is a continuous function taking $[-h, A), A > 0$ into R^n such that $D(x_t)$ is continuously differentiable and satisfies (1) for $t \in (0, A)$. Assumption H0 (i) guarantees the existence and the uniqueness of the solution to initial value problem for (1), where $u(t)$ and $w(t)$ are continuous functions [8]. Assumption H0 (ii) guarantees that the characteristic equation corresponding to the linear system

$$\frac{d}{dt}Dx_t = Lx_t,$$

where $L : C^n[-h,0] \to R^n$ is a linear bounded operator, has a finite number of roots with nonnegative real part.

We consider both, a state-feedback and an error-feedback regulator problems.

Problem 1 (State-Feedback Regulator Problem): Find a state-feedback control law

$$u(t) = \alpha(x_t, w(t)), \tag{3}$$

where $\alpha : Y \to R^m$ is a $C^k (k \geq 2)$ function and $\alpha(0,0) = 0$ such that :

1a) the equilibrium $x(t) \equiv 0$ of

$$\frac{d}{dt}Dx_t = f(x_t, \alpha(x_t, 0), 0),$$

is exponentially stable;

1b) there exists a neighborhood $Y \subset C^n[-h,0] \times W$ of the origin such that, the solution of the closed-loop system

$$\frac{d}{dt}Dx_t = f(x_t, \alpha(x_t, w(t)), w(t)), \quad \dot{w}(t) = s(w(t)) \tag{4}$$

satisfies

$$\lim_{t\to\infty} g(x_t, w(t)) = 0. \tag{5}$$

Problem 2 (Error-Feedback Regulator Problem): Find an error-feedback controller

$$u = \Theta(z_t), \quad \frac{d}{dt}\overline{D}z_t = \eta(z_t, e(t)), \quad z(t) \in R^\nu \tag{6}$$

with C^k functions $\eta : Z_0 \to R^\nu$ and $\Theta : Z_1 \to R^m$, where $Z_0 \subset C^\nu[-h,0] \times R^p$, $Z_1 \subset C^\nu[-h,0]$ are some neighborhoods of the origin, such that:

2a) the equilibrium $(x(t), z(t)) \equiv 0$ of

$$\frac{d}{dt}Dx_t = f(x_t, \Theta(z_t), 0), \quad \frac{d}{dt}\overline{D}z_t = \eta(z_t, g(x_t, 0))$$

is exponentially stable;

2b) there exists a neighborhood $Z \subset C^n[-h,0] \times C^\nu[-h,0] \times W$ of the origin such that, the solution of the closed-loop system

$$\tfrac{d}{dt}Dx_t = f(x_t, \Theta(z_t), w(t)), \quad \tfrac{d}{dt}\overline{D}z_t = \eta(z_t, g(x_t, w(t))), \quad \dot{w}(t) = s(w(t)) \tag{7}$$

satisfies (5).

3 Linearized Problem and Assumptions

Using Taylor expansion in the neighborhood of the origin of the Banach space $C^n[-h,0] \times R^m \times R^r$, we obtain the following approximation of the smooth function f:

$$f(x_0, u, w) = \mathrm{A}x_0 + Bu + Pw + O(x_0, u, w)^2,$$

where the linear bounded operator $[\mathrm{A},\ B,\ P] : C^n[-h,0] \times R^m \times R^r \to R^n$ is a Frechet derivative of f at the origin. The function $O(\cdot)^2$ vanishes at the origin with its first-order Frechet derivative. Similarly, smooth functions g, α, Θ and η can be represented in the form

$$\begin{array}{l} g(x_0, w) = \mathrm{C}x_0 + Qw + O(x_0, w)^2, \\ \alpha(x_0, w) = \mathrm{K}x_0 + Lw(t) + O(x_0, w)^2, \\ \Theta(z_0) = \mathrm{H}z_0 + O(z_0)^2, \quad \eta(z_0, e) = \mathrm{F}z_0 + Ge + O(z_0, e)^2, \end{array}$$

where the functions $O(\cdot)^2$ vanish at the origin with their first-order Frechet derivatives. The linear bounded operators $\mathrm{A} : C^n[-h,0] \to R^n$ and $\mathrm{C} : C^n[-h,0] \to R^p$ by Riesz theorem can be represented in the form of Stieltjes integrals [8]:

$$\mathrm{A}\phi = \int_{-h}^{0} d[\mu(\theta)]\phi(\theta), \quad \mathrm{C}\phi = \int_{-h}^{0} d[\zeta(\theta)]\phi(\theta), \tag{8}$$

with $n \times n$ and $p \times n$-matrix functions μ and ζ of bounded variations. A similar representation can be written for the linear bounded operators $\mathrm{K} : C^n[-h,0] \to R^m$, $\mathrm{H} : C^\nu[-h,0] \to R^m$ and $\mathrm{F} : C^\nu[-h,0] \to R^\nu$.

The linearized system is given by

$$\tfrac{d}{dt}Dx_t = \mathrm{A}x_t + Bu(t) + Pw(t), \quad \dot{w}(t) = Sw(t), \quad e(t) = \mathrm{C}x_t + Qw(t). \tag{9a-c}$$

The linearized state-feedback and error-feedback controllers have the form

$$u(t) = \mathrm{K}x_t + Lw(t) \tag{10}$$

and

$$u(t) = \mathrm{H}z_t, \quad \frac{d}{dt}\overline{D}z_t = \mathrm{F}z_t + Ge(t). \tag{11}$$

respectively.

Similarly to the case without delay [11] we assume the following:

H1. The exosystem (2) is neutrally stable (i.e. Lyapunov stable in forward and backward time, and thus S has all its eigenvalues on the imaginary axis).

H2. The triple $\{D, \mathrm{A}, B\}$ is stabilizable, i.e. there exists a linear bounded operator $\mathrm{K} : C^n[-h, 0] \to R^m$ such that the system

$$\frac{d}{dt} D x_t = (\mathrm{A} + B\mathrm{K}) x_t \tag{12}$$

is asymptotically stable.

H3. The pair

$$\begin{bmatrix} \mathrm{A} & P \\ 0 & S \end{bmatrix}, \quad [\mathrm{C} \quad Q]$$

is detectable in the following sense: there exists a $(n + r) \times p$-matrix G such that the system

$$\frac{d}{dt} \begin{bmatrix} D\tilde{z}_{1t} \\ \tilde{z}_2(t) \end{bmatrix} = \left\{ \begin{bmatrix} \mathrm{A} & P \\ 0 & S \end{bmatrix} + G[\mathrm{C} \quad Q] \right\} \begin{bmatrix} \tilde{z}_{1t} \\ \tilde{z}_2(t) \end{bmatrix}, \tag{13}$$

where $\overline{z}_1(t) \in R^n$, $\overline{z}_2(t) \in R^r$, is asymptotically stable.

We note that H2 is equivalent to the following condition [10]:

H2'. $rank \left[\lambda[I - \int_{-h}^{0} d[\xi(\theta)] e^{\lambda\theta}] - \int_{-h}^{0} d[\mu(\theta)] e^{\lambda\theta},\ B \right] = n$ for all $\lambda \in C$ with $Re\lambda \geq 0$.

Similar condition equivalent to H3 can be written for the case of $Cx_t = C_0 x(t)$, where C_0 is a constant matrix. Some sufficient conditions for H2 and for finding a stabilizing controller $u(t) = K_0 x(t)$ or $u(t) = K_1 x(t - h)$ may be found e.g. in [7] (see also references therein) in terms of linear matrix inequalities. Similar sufficient conditions may be derived for H3.

4 Solution of the Regulator Problems

4.1 Center manifold of the closed-loop system

The solution of the output regulation problem is based on the center manifold theory [8], [8].

Lemma 1. *Let H0 hold. Assume that all eigenvalues of S are on the imaginary axis and that for some $\alpha(x_t, w)$ condition 1a) holds. Then the closed-loop system (4) has a local center manifold $x_t(\theta) = \pi(w(t))(\theta)$, $\theta \in [-h, 0]$, where $\pi : W_0 \to C^n[-h, 0]$ ($0 \in W_0 \subset W \subset R^r$) is a C^k mapping with $\pi(0)(\theta) \equiv 0$. The center manifold is locally attractive, i.e. satisfies*

$$||x_t - \pi(w(t))|| \leq M e^{-at} ||x_0 - \pi(w(0))||, \quad M > 0,\ a > 0 \tag{14}$$

for all $x_0, w(0)$ sufficiently close to 0 and all $t \geq 0$.

Proof: The closed-loop system (4) has the form

$$\begin{aligned} \dot{w}(t) &= Sw(t) + O(w(t))^2, \\ \tfrac{d}{dt}Dx_t &= (\mathrm{A} + B\mathrm{K})x_t + (P + BL)w(t) + O(x_t, w(t))^2. \end{aligned} \tag{15a,b}$$

By assumption, the zeros of the characteristic equation corresponding to (12) are in C^-, and the eigenvalues of the matrix S are on the imaginary axis.

It is well-known (see e.g. [8]) that according to this dichotomy, the space $R^r \times C^n[-h, 0]$ of the initial values of the linear system

$$\dot{w}(t) = Sw(t), \quad \frac{d}{dt}Dx_t = (\mathrm{A} + B\mathrm{K})x_t + (P + BL)w(t), \tag{16}$$

can be decomposed as a direct sum $R^r \times C^n[-h, 0] = \mathcal{P} \oplus \mathcal{Q}$, where $\mathcal{P}$ and $\mathcal{Q}$ are invariant sub-spaces of the solutions of (16), in the sense that for all initial conditions from $\mathcal{P}$ ($\mathcal{Q}$), solutions of (16) satisfy $\{w(t), x_t\} \in \mathcal{P}$ ($\{w(t), x_t\} \in \mathcal{Q}$) for all $t \geq 0$. Moreover, $\mathcal{P}$ is an r-dimensional and corresponds to solutions of (16) of the form $p(t)e^{\lambda t}$, where $p(t)$ is a polynomial in t and λ is an eigenvalue of S. The space $\mathcal{Q}$ corresponds to exponentially decaying solutions of (16). By Theorem 2.1 of [8] (p. 314) the system (15) has a local smooth center manifold $x_0 = \pi(w)$. The flow on this manifold is governed by (15a). By Theorem 2.2 of [8] (p.216) this manifold is locally attractive. □

The function π which determines a center manifold of (4) can be considered as a function of one variable $\pi : W_0 \to C^n[-h, 0]$ in the Banach space or a function of two variables $\pi : W_0 \times [-h, 0] \to R^n$ in the Euclidean space. Further we find relation between the smoothness properties in both considerations by introducing two classes of functions:

Class $\mathcal{M}_1$ of C^1 functions $\pi : W_0 \to C^n[-h, 0] (W_0 \subset R^r)$, satisfying the following conditions:
(i) For each $w \in W_0$ there exists a continuous in $\theta \in [-h, 0]$ partial derivative $\frac{\partial \pi(w)(\theta)}{\partial \theta} \triangleq \gamma(w)(\theta)$;
(ii) The function $\gamma : W_0 \to C^n[-h, 0]$ is continuous.

Class $\mathcal{M}_2$ of functions $\psi : W_0 \to C^n[-h, 0]$ such that the functions $\overline{\psi}(w, \theta) \triangleq \psi(w)(\theta)$, $\overline{\psi} : W_0 \times [-h, 0] \to R^n$ are continuously differentiable.

Proposition 1. *[6] $\mathcal{M}_1 = \mathcal{M}_2$.*

Lemma 2. *Assume H0. A C^1 mapping $\pi : W_0 \to C^n[-h, 0]$, $\pi(0) = 0$ defines a center manifold $x_t(\theta) = \pi(w(t))(\theta)$, $\theta \in [-h, 0]$ of (4) if and only if $\pi \in \mathcal{M}_1$ and $\forall w \in W_0$, $\forall \theta \in [-h, 0]$ it satisfies the following system of partial differential equations*

$$\begin{aligned} \frac{\partial \pi(w)(\theta)}{\partial w} s(w) &= \frac{\partial \pi(w)(\theta)}{\partial \theta} \\ \frac{\partial [D\pi(w)]}{\partial w} s(w) &= f(\pi(w), \alpha(\pi(w), w), w). \end{aligned} \tag{17a,b}$$

Proof. Note that for a C^1 mapping $\pi : W_0 \to C^n[-h,0]$ and for $w(t)$, satisfying (2), we find that for each $\theta \in [-h,0]$

$$\frac{d}{dt}[\pi(w(t))(\theta)] = \frac{\partial \pi(w(t))(\theta)}{\partial w} s(w(t)). \tag{18}$$

Necessity: Let a C^1 mapping $\pi : W_0 \to C^n[-h,0]$ determine a center manifold of (15). Then there exists $\delta > 0$ such that $x_t(\theta) = \pi(w(t))(\theta)$ satisfies (4) for $t \in [-\delta, \delta]$ and, hence

$$\begin{aligned} &\tfrac{\partial x_t(\theta)}{\partial t} = \tfrac{\partial x_t(\theta)}{\partial \theta}, \quad x_0 = \phi, \quad \theta \in [-h,0], \quad t \in [-\delta,\delta], \\ &\tfrac{\partial D x_t}{\partial t} = f(x_t, \alpha(x_t, w(t)), w(t)), \quad \dot{w}(t) = s(w(t)). \end{aligned} \tag{19}$$

Substituting $x_t = \pi(w(t))$, $w(0) = w$, $t \in [-\delta, \delta]$ into (19) and setting further $t = 0$, we obtain that for all $w \in W_0$, $\pi(w)(\theta)$ is differentiable in $\theta \in [-h,0]$ and π satisfies (17). The function $\frac{\partial \pi}{\partial \theta} : W_0 \to C^n[-h,0]$ is continuous since the left hand side of (17a) has the same property.

Sufficiency: let a C^1 mapping $\pi : W_0 \to C^n[-h,0]$ satisfy (17). Substitute $w = w(t)$ into (17), where $w(t)$ is a solution of (2), then $x_t = \pi(w(t))$ satisfies (19) (and thus (4)) and therefore π determines the invariant manifold of (4). □

Remark 1. Approximate solution to (17) can be found in a form of series expansions in the powers of w (similarly to [8], [14], [1]).

4.2 State-feedback regulator problem

Applying Lemmas 1 and 2, we obtain regulator equations by using arguments of [11].

Lemma 3. *Under H0 and H1 assume that for some $\alpha(x_t, w)$ condition 1a) holds. Then, condition 1b) is also fulfilled iff there exists a $C^k (k \geq 2)$ mapping $\pi : W_0 \to C^n[-h,0]$, $\pi(0) = 0$ satisfying (17) and the algebraic equation*

$$g(\pi(w), w) = 0. \tag{20}$$

Proof is similar to [6].

Theorem 1. *Under H0, H1 and H2, the state-feedback regulator problem is solvable if and only if there exist $C^k (k \geq 2)$ mappings $x_0(\theta) = \pi(w)(\theta)$, with $\pi \in \mathcal{M}_1$, $\pi(0)(\theta) = 0$, and $u = c(w)$, with $c(0) = 0$, both defined in a neighborhood $W \subset R^r$ of the origin, satisfying the conditions $\forall w \in W_0$, $\forall \theta \in [-h,0]$*

$$\begin{aligned} &\tfrac{\partial \pi(w)(\theta)}{\partial w} s(w) = \tfrac{\partial \pi(w)(\theta)}{\partial \theta}, \\ &\tfrac{\partial [D\pi(w)]}{\partial w} s(w) = f(\pi(w), c(w), w), \\ &g(\pi(w), w) = 0. \end{aligned} \tag{21a-c}$$

Suppose that π and c satisfy (21), then the state-feedback

$$u = \alpha(x_t, w(t)) = c(w(t)) + K[x_t - \pi(w(t))], \tag{22}$$

where K is a stabilizing gain which is defined in H2, solves the state-feedback regulator problem.

Proof. The necessity follows immediately from Lemma 3. For the sufficiency consider the state-feedback (22). This choice satisfies 1a), since

$$f(x_t, \alpha(x_t, 0), 0) = (\mathrm{A} + B\mathrm{K})x_t + O(x_t)^2.$$

Moreover, by construction

$$\alpha(\pi(w), w) = c(w)$$

and therefore, (21a), (21b) reduce to (17). From (21c) by Lemma 2 it follows that condition 1b) is also fulfilled. □

4.3 Error-feedback regulator problem

Applying Lemmas 1 and 2 to the system (7), we obtain the following:

Lemma 4. *Let H0 hold. Assume that all eigenvalues of S are on the imaginary axis and that for some $\theta(z_t)$ and $\eta(z_t, e)$ condition 2a) holds. Then*
(i) *the closed-loop system (7) has a local center manifold $x_t(\theta) = \pi(w(t))(\theta)$, $z_t(\theta) = \sigma(w(t))(\theta)$, where $\pi : W_0 \to C^n[-h, 0]$, $\sigma : W_0 \to C^\nu[-h, 0]$ $(0 \in W_0 \subset W \subset R^r)$ are C^k mappings with $\pi(0)(\theta) \equiv 0$, $\sigma(0)(\theta) \equiv 0$;*
(ii) *the center manifold is locally attractive, i.e. satisfies*

$$\begin{aligned}&||x_t - \pi(w(t))|| + ||z_t - \sigma(w(t))||\\ \leq &Me^{-at}(||x_0 - \pi(w(0))|| + ||z_0 - \sigma(w(0))||), \qquad M > 0,\ a > 0\end{aligned} \tag{23}$$

for all $x_0, z_0, w(0)$ sufficiently close to 0 and all $t \geq 0$.
(iii) *C^1 mappings $\pi : W_0 \to C^n[-h, 0]$, $\pi(0)(\theta) = 0$, $\sigma : W_0 \to C^\nu[-h, 0]$, $\sigma(0)(\theta) = 0$ define a center manifold $x_t(\theta) = \pi(w(t))(\theta)$, $z_t(\theta) = \sigma(w(t))(\theta)$, $\theta \in [-h, 0]$ of (7) if and only if $\pi : W_0 \times [-h, 0] \to R^n$, $\sigma : W_0 \times [-h, 0] \to R^\nu$ are continuously differentiable functions and $\forall w \in W_0$, $\forall \theta \in [-h, 0]$ they satisfy the following system of partial differential equations*

$$\begin{aligned}&\frac{\partial \pi(w)(\theta)}{\partial w} s(w) = \frac{\partial \pi(w)(\theta)}{\partial \theta}, \qquad \frac{\partial \sigma(w)(\theta)}{\partial w} s(w) = \frac{\partial \sigma(w)(\theta)}{\partial \theta},\\ &\frac{\partial [D\pi(w)]}{\partial w} s(w) = f(\pi(w), \theta(\sigma(w)), w), \qquad \frac{\partial [\bar{D}\sigma(w)]}{\partial w} s(w) = \eta(\sigma(w), 0).\end{aligned} \tag{24a-d}$$

Remark 2. In the case when $z(t) = col\{z_1(t), z_2(t)\}$, where z_2 appears in (7) without delay and thus $col\{z_{1t}(\theta), z_2(t)\} = col\{\sigma_1(w(t))(\theta),\ \sigma_2(w(t))\}$, (24b) holds only for $\sigma = \sigma_1$.

Similarly to Lemma 3, the following lemma can be proved

Lemma 5. *Under H0 and H1, assume that for some $\Theta(z_t)$ and $\eta(z_t, e)$ condition 2a) holds. Then, condition 2b) is also fulfilled iff there exist $C^k (k \geq 2)$ mappings $\pi : W_0 \to C^n[-h, 0]$, $\pi(0) = 0$, $\sigma : W_0 \to C^\nu[-h, 0]$, $\sigma(0) = 0$ satisfying (24) and the algebraic equation (20).*

From the latter lemmas we deduce a necessary and sufficient condition for the solvability of the error-feedback regulator problem

Theorem 2. *Under H0-H3, the error-feedback regulator problem is solvable if and only if there exist $C^k(k \geq 2)$ mappings $x_0(\theta) = \pi(w)(\theta)$, with $\pi \in \mathcal{M}_1$, $\pi(0)(\theta) = 0$, and $u = c(w)$, with $c(0) = 0$, both defined in a neighborhood $W \subset R^r$ of the origin, satisfying the conditions (21) $\forall w \in W$, $\forall \theta \in [-h, 0]$.*

Suppose that π and c satisfy (21), and that a linear bounded operator $H : C^n[-h, 0] \to R^m$ is such that the system

$$\frac{d}{dt}Dx_t = (A + BH)x_t \tag{25}$$

is asymptotically stable. Then the error-feedback (6), where

$$\begin{array}{l} z(t) = col\{z_1(t), z_2(t)\},\ \eta = col\{\eta_1, \eta_2\},\ \overline{D} = diag\{D, I\}, \\ u = \Theta(z_t) = c(z_2(t)) + H[z_{1t} - \pi(z_2(t))], \\ \eta_1(z_{1t}, z_2(t), e(t)) = f(z_{1t}, \Theta(z_t), z_2(t)) - G_1(h(z_{1t}, z_2(t)) - e(t)), \\ \eta_2(z_{1t}, z_2(t), e(t)) = s(z_2(t)) - G_2(h(z_{1t}, z_2(t)) - e(t)), \end{array} \tag{26}$$

and where $G = col\{G_1, G_2\}$ is defined in H3, solves the regulator problem.

Proof. The necessity follows immediately from Lemma 5. For the sufficiency we note, that there exist a linear bounded operator H : $C^\nu[-h, 0] \to R^m$ and a matrix $G = col\{G_1, G_2\}$ such that (25) and (13) are asymptotically stable. A standard calculation shows that for any $m \times r$-matrix K, the characteristic quasipolynomial that corresponds to the system

$$\begin{bmatrix} \frac{d}{dt}Dx_t \\ \frac{d}{dt}\overline{D}z_{1t} \\ \dot{z}_2(t) \end{bmatrix} = \begin{bmatrix} A & BH & BK \\ G_1C & A + BH - G_1C & P + BK - G_1Q \\ G_2C & -G_2C & S - G_2Q \end{bmatrix} \begin{bmatrix} x_t \\ z_{1t} \\ z_2(t) \end{bmatrix} \tag{27}$$

is equal to the product of the characteristic quasipolynomials that correspond to (25) and (13) respectively. Therefore, (27) is asymptotically stable.

Consider the error-feedback controller of (6), (26). The linearized system corresponding to the closed-loop system (7) has exactly the form of (27), where

$$K = \left[\frac{\partial c}{\partial w}\right]_{w=0} - \mathrm{H}\left[\frac{\partial \pi}{\partial w}\right]_{w=0}.$$

Thus requirement 2a) is satisfied. By construction $z_2(t)$ appears in (7) without delay and thus (21a)-(21b) imply (24) with $\sigma(w) = col\{\sigma_1(w), \sigma_2(w)\} = col\{\pi(w), w\}$, where in (24b) $\sigma = \sigma_1$. Thus requirement 2b) follows from Lemma 5. □

5 Linear Case

5.1 Linear regulator equations

Consider the linear regulator problem (9). In the linear case the center manifold has a form $x_t = \Pi(\theta)w(t)$, where Π is an $n \times r$ matrix function continuously differentiable in $\theta \in [-h, 0]$. From Theorems 1 and 2 it follows, that the linear problem (9) is solvable iff there exists Π and an $m \times r$-matrix Γ that satisfy the following system

$$\begin{array}{l} \dot{\Pi}(\theta) = \Pi(\theta)S, \quad \theta \in [-h,0], \\ (D\Pi)S = \int_{-h}^{0} d[\mu(\theta)]\Pi(\theta) + B\Gamma + P, \\ \int_{-h}^{0} d[\zeta(\theta)]\Pi(\theta) + Q = 0. \end{array} \tag{28a-c}$$

Eq. (28a) yields $\Pi(\theta) = \Pi(0)\exp S\theta$. Substituting the latter into (28b) and (28c), we obtain the following linear algebraic system for initial value $\Pi(0)$:

$$\begin{array}{l} [\Pi(0) - \int_{-h}^{0} d[\xi(\theta)]\Pi(0)e^{S\theta}]S = \int_{-h}^{0} d[\mu(\theta)]\Pi(0)e^{S\theta} + B\Gamma + P, \\ \int_{-h}^{0} d[\zeta(\theta)]\Pi(0)e^{S\theta} + Q = 0. \end{array} \tag{29}$$

The latter system is a generalization of Francis equations [4] to the case of neutral systems.

We consider now a particular, but important in applications case of (9) with

$$\begin{array}{l} Dx_t = x(t) - \sum_{i=1}^{k} D_i x(t-h_i) - \int_{-h}^{0} D_d(\theta)x(t+\theta)d\theta, \\ \mathrm{A}x_t = \sum_{i=0}^{k} A_i x(t-h_i) + \int_{-h}^{0} A_d(\theta)x(t+\theta)d\theta, \\ \mathrm{C}x_t = \sum_{i=0}^{k} C_i x(t-h_i) + \int_{-h}^{0} C_d(\theta)x(t+\theta)d\theta, \end{array} \tag{30}$$

where $0 = h_0 < h_1 < \ldots < h_k \le h$, D_d, A_d and C_d are piecewise continuous matrix functions and where D_i, A_i and C_i are constant matrices of the appropriate dimensions. In this case (29) has the form:

$$\begin{array}{l} [\Pi(0) - \sum_{i=1}^{k} D_i\Pi(0)e^{-Sh_i} - \int_{-h}^{0} D_d(\theta)\Pi(0)e^{S\theta}d\theta]S = \sum_{i=0}^{k} A_i\Pi(0)e^{-Sh_i} \\ + \int_{-h}^{0} A_d(\theta)\Pi(0)e^{S\theta}d\theta + B\Gamma + P, \\ \sum_{i=0}^{k} C_i\Pi(0)e^{-Sh_i} + \int_{-h}^{0} C_d(\theta)\Pi(0)e^{S\theta}d\theta + Q = 0. \end{array} \tag{31}$$

Theorem 3. *Under H0-H2, the linear state-feedback regulator problem (9) ((9) and (30)) is solvable if and only if there exist $n \times r$ and $m \times r$-matrices $\Pi(0)$ and Γ which solve the linear matrix equations (29) ((31)).*

In the case of error-feedback regulator problem, the similar result holds under H0-H3.

Consider the case of (30) with the general controller output. We assume that the regulator problem for (9) without delay, i.e. for

$$\begin{array}{l} (I - \sum_{i=1}^{k} D_i)\dot{x}(t) = (\sum_{i=0}^{k} A_i)x(t) + Bu(t) + Pw(t), \\ \dot{w}(t) = Sw(t), \\ e(t) = (\sum_{i=0}^{k} C_i)x(t) + Qw(t) \end{array}$$

is solvable for all P and Q. This is equivalent (see e.g. [4]) to the following assumption

A1. $det\mathcal{G}_0(\lambda) \neq 0$ for all eigenvalues λ of S, where

$$\mathcal{G}_0(\lambda) = (\sum_{i=0}^{k} C_i)[\lambda(I - \sum_{i=1}^{k} D_i) - \sum_{i=0}^{k} A_i]^{-1}B.$$

Under A1 the linear regulator equations

$$(I - \sum_{i=1}^{k} D_i)\Pi_0 S = (\sum_{i=0}^{k} A_i)\Pi_0 + B\Gamma + P, \quad (\sum_{i=0}^{k} C_i)\Pi_0 + Q = 0,$$

where Π_0 and Γ are constant matrices, are solvable for all P and Q. Then, by the implicit function theorem for all small enough $h > 0$ (31) is solvable. We have:

Proposition 2. *Under H0-H2 and A1, the output regulation of (9) with (30) via state-feedback of (10) is achievable and the regulator equations (31) are solvable for all small enough h.*

6 Conclusions

The geometric theory of output regulation is generalized to nonlinear neutral type systems. It is shown that the state-feedback and the error-feedback regulator problems are solvable, under the standard assumptions on stabilizability and detectability of the linearized system, if and only if a set of regulator equations is solvable. This set consists of partial differential and algebraic equations. In the linear case these equations are reduced to the linear matrix equations.

The solvability of the nonlinear regulator equations and the approximate solutions to these equations are issues for the future study.

References

1. M. Ait Babram, O. Arino and M. Hbid, Computational scheme of a center manifold for neutral functional differential equations, J. Math. Anal. Appl. 258 (2001) 396-414.
2. C.I.Byrnes and A. Isidori, Output regulation for nonlinear systems: an overview, Int. J. of Rob. and Nonlin. Cont. 10 (2000) 323-337.
3. C.I.Byrnes, I. G. Lauko, D. S. Gilliam and V. I. Shubov, Output Regulation for Linear Distributed parameter systems, IEEE Trans. On Aut. Cont., 45 (2000) 2236-2252.
4. B.A. Francis, The Linear Multivariable Regulator Problem, SIAM J. on Cont. and Optim., 15 (1977) 486-505.
5. E. Fridman, Asymptotics of integral manifolds and decomposition of singularly perturbed systems of neutral type, Differential equations 26 (1990) 457-467.
6. E. Fridman, Output regulation of nonlinear systems with delay, Systems & Control Letters 50 (2003) 81-93.
7. E. Fridman and U. Shaked, A descriptor system approach to H_∞ control of time-delay systems, IEEE Trans. on Automat. Contr., 47 (2002) 253-270.

8. J. Hale, Critical cases for neutral functional differential equations, *J. Differential Eqns.* 10 (1971) 59-82.
9. J. Hale and S. Verduyn Lunel, Introduction to Functional Differential Equations, Springer-Verlag, New York, 1993 .
10. J. Hale and S. Verduyn Lunel, Strong stabilization of neutral functional differential equations, *IMA Journal of Mathematical Control and Information*, 19 (2002) 5-23.
11. A. Isidori and C.I. Byrnes, Output regulation of nonlinear systems, IEEE Trans. On Autom. Control 35 (1990) 131-140.
12. K. Murakami, Bifurcated periodic solutions for delayed Van der pol equation, Neural, Parallel & Scientific Computations 7 (1999) 1-16.
13. J. Schumacher, Finite-dimensional regulators for a class of infinite-dimensional systems, Systems & Control Letters, 3 (1983) 7-12.
14. V. Strygin and E. Fridman, Asymptotics of integral manifolds of singularly perturbed differential equations with retarded argument, Math. Nachr., 117 (1984) 83-109.

Robust Stability Analysis of Various Classes of Delay Systems

Catherine Bonnet[1] and Jonathan R. Partington[2]

[1] INRIA Rocquencourt, Domaine de Voluceau, B.P. 105, 78153 Le Chesnay cedex, France.
Catherine.Bonnet@inria.fr
[2] University of Leeds, School of Mathematics, Leeds LS2 9JT, U.K.
J.R.Partington@leeds.ac.uk

Summary. This chapter is a review of some work of the authors on the robust stabilization of retarded and neutral delay systems, including the case of fractional delay systems. BIBO-stability and nuclearity conditions are derived and the question of parametrization of all BIBO-stabilizing controllers is addressed.

1 Introduction

This chapter presents an overview of some work we have done on the robust stabilization of various linear SISO continuous-time delay systems [3]– [7]. Each of these studies has been motivated by practical problems that we have encountered or found in the literature, however we will not present them here concentrating on the methodology as it can be found in the original papers. In the same spirit, no proofs will be given.

Here the notion of stability is that of input-output stability; that is, an input signal in L^p must produce an output signal in L^q where $1 \leq p, q \leq \infty$. We restrict ourselves to the linear case (that is a convolution operator between input and output signals) although we have also considered the stabilisation of nonlinear delay systems in [8] and [9] using the framework introduced in [15] and [34].

Usually, people consider the case $p = q = 2$, which corresponds to finite energy signals and leads to an analysis of the transfer function using the H_∞ norm. We refer particularly to [13], [11] for a study of robust stabilization of delay systems in this case. Note that the case $p = 2$ and $q = \infty$ (or $p = 1$ and $q = 2$) would induce an analysis of the transfer function using the H_2 norm. We consider here the case $p = q = \infty$, which (equivalently to the case $p = q = 1$) allows one to analyse the transfer function of the system in the Wiener algebra setting.

The chapter is organized as follows. We first describe in Section 2 the setting chosen to analyse robust stabilization. Next, Section 3 is concerned with robust stabilization in this setting of retarded delay systems, including the case of fractional delay systems. Section 36 deals with the case of neutral (standard and fractional) delay systems, the main questions addressed here being the characterization of the BIBO-stability of these systems with easy frequential conditions as well as a parametrization of all stabilizing controllers. Finally, Section 5 deals briefly with the nuclearity of these systems.

Similar questions have been widely studied in the literature. We refer among others to [29], [10], [28], [30] for general considerations, to [35], [13], [14], [15], [33], [24] for questions of stability, and to [31], [17], [18], [19], [27], [25], [23], [26], [22] for other results on delay and fractional systems.

2 Preliminaries and Definitions

For $x \in \mathbb{R}$, $[x]$ denotes the integer part of x and $\{x\}$ the fractional part, so $x = [x] + \{x\}$.

$\mathbb{R}_-$ denotes the negative real axis $\{x \in \mathbb{R} : x \leq 0\}$.

L^∞ denotes the complex-valued measurable functions on the nonnegative real axis such that $\operatorname{ess\,sup}_{t \in \mathbb{R}_+} |f(t)| < \infty$.

$L^1(\mathbb{R}^+)$ or L^1 denotes the complex-valued measurable functions on the nonnegative real axis such that $\int_0^\infty |f(t)|dt < \infty$, and $L^1(\mathbb{R})$ denotes the complex-valued measurable functions on the real axis such that $\int_{-\infty}^\infty |f(t)|dt < \infty$.

H_∞ is the space of bounded analytic functions on the right half-plane $\mathbb{C}_0^+ = \{s \in \mathbb{C} : \operatorname{Re} s > 0\}$.

RH_∞ denotes the space of rational H_∞ functions.

$\mathcal{A}$ denotes the space of distributions of the form $h(t) = h_a(t) + \sum_{i=0}^{\infty} h_i \delta(t - t_i)$ where $t_i \in [0, \infty)$, $0 \leq t_0 < t_1 < \cdots$, $\delta(t - t_i)$ is a delayed Dirac function, $h_i \in \mathbb{C}$, $h_a \in L^1$ and $\sum_{i=0}^{\infty} |h_i| < \infty$.

The norm on $\mathcal{A}$ is defined by $||h||_{\mathcal{A}} = ||h_a||_{L^1} + \sum_{i=0}^{\infty} |h_i|$.

The subspace of $\mathcal{A}$ of distributions in $L^1 + \mathbb{C}\delta$ is particularly interesting as we have that $\overline{RH_\infty}^{\hat{\mathcal{A}}} = \mathcal{L}(L^1 + \mathbb{C}\delta)$ where $\mathcal{L}$ denotes the Laplace transform. In other words, these systems can be approximated by finite-dimensional systems as closely as we like in the given norm.

Furthermore, $\overline{RH_\infty}^{H_\infty} = A(\mathbb{C})$, the space of functions which are analytic and bounded in the right half-plane and continuous on the extended imaginary axis.

The symbol $\hat{\mathcal{A}}$ denotes the space of Laplace transforms of functions in $\mathcal{A}$, which is a linear subspace of $A(\mathbb{C})$.

We write $\hat{\mathcal{A}}_-(0) = \{\hat{f} : \; f \in \mathcal{A}_-(\beta) \text{ for some } \beta < 0\}$.

Also $\hat{\mathcal{A}}_\infty(0) = \{\hat{f} \in \mathcal{A}_-(0) : \inf_{s\in\mathbb{C}_0^+, |s|>\rho} |\hat{f}(s)| > 0 \text{ for some } \rho > 0\}$.

Finally the *Callier–Desoer class* is $\hat{\mathcal{B}}(0) = \{f/g : f \in \hat{\mathcal{A}}_-(0), g \in \hat{\mathcal{A}}_\infty(0)\}$.

We recall that *BIBO-stability* of a system P with convolution kernel h (with vanishing singular part) is defined as $\sup_{x\in L^\infty, x\neq 0} \frac{\|h * x\|_{L^\infty}}{\|x\|_{L^\infty}} < \infty$, which is equivalent to $\|h\|_{\mathcal{A}} = \|P\|_{\hat{\mathcal{A}}} < \infty$. It is well known that this implies that P lies in H_∞.

A transfer function P analytic in $\{\mathrm{Re}\, s > 0\}$ and continuous on $\imath\mathbb{R}$ is said to be *proper* on $\{\mathrm{Re}\, s \geq 0\}$ if, for sufficiently large ρ, $\sup_{\{\mathrm{Re}\, s\geq 0, |s|\geq\rho\}} |P(s)| < \infty$.

Similarly a transfer function P analytic in $\{\mathrm{Re}\, s > 0\}$ and continuous on $\imath\mathbb{R}$ is said to be *strictly proper* on $\{\mathrm{Re}\, s \geq 0\}$ if $\lim_{\rho\longrightarrow\infty} \left(\sup_{\{\mathrm{Re}\, s\geq 0, |s|\geq\rho\}} |P(s)| \right) = 0$.

A transfer function P analytic in $\{\mathrm{Re}\, s > 0\}$ and continuous on $\imath\mathbb{R}$ is said to have a limit at infinity in $\{\mathrm{Re}\, s \geq 0\}$ if there exists a complex constant P_∞ such that $\lim_{\rho\longrightarrow\infty} \left(\sup_{\{\mathrm{Re}\, s\geq 0, |s|\geq\rho\}} |P(s) - P_\infty| \right) = 0$.

Let P be a function that is meromorphic in $\mathbb{C} \setminus \mathbb{R}_-$ and has a branch point at $s = 0$. The point $s = 0$ is defined to be a pole of fractional order $\alpha > 0$ of P if there is a non-zero constant c such that $f(s) = s^{-\alpha}(c + o(1))$ as $s \to 0$ in $\mathbb{C} \setminus \mathbb{R}_-$. It is easy to see that this definition is independent of our choice of a branch of $s^{-\alpha}$ in $\mathbb{C} \setminus \mathbb{R}_-$.

P is said to have a *coprime factorization* (N, D) over $\hat{\mathcal{A}}$ if $P = ND^{-1}$, $D \neq 0$, N, $D \in \hat{\mathcal{A}}$ and there exist *Bézout factors* X, $Y \in \hat{\mathcal{A}}$ such that $-NX + DY = 1$.

Below is a useful necessary and sufficient condition to characterize the coprimeness of a given pair. This condition is also valid for scalar functions in H_∞.

Theorem 1. *Let N, D be in $\hat{\mathcal{A}}$. Then (N, D) is a coprime factorization over $\hat{\mathcal{A}}$ if and only if* $\inf_{\{\mathrm{Re}\, s>0\}} (|N(s)| + |D(s)|) > 0$.

If P admits a coprime factorization over $\hat{\mathcal{A}}$ (respectively H_∞) then the set of all stabilizing controllers is given by the *Youla parametrization:*

$$\mathrm{stab}(P) = \{(X + DQ)(Y + NQ)^{-1} : Q \in \hat{\mathcal{A}} \ (\text{resp. } H_\infty)\}.$$

Let us note that these controllers are not necessarily proper.

The robustness is considered here relatively to coprime factor perturbations (robustness in the BIBO gap topology) and in the sub-case $L^1 + \mathbb{C}\delta$, the optimal robustness margin is given by

$$b^{opt}(G) = \frac{1}{\inf_{Q\in H_\infty} \left\| \begin{pmatrix} X \\ Y \end{pmatrix} + \begin{pmatrix} D \\ N \end{pmatrix} Q \right\|_\infty}$$

and the optimal controller by $C^{opt} = \frac{X + DQ^{opt}}{Y + NQ^{opt}}$. This setting offers nice convergence properties when considering approximation of systems (convergence of the controllers in the BIBO gap topology). Let us also mention that if P is strictly proper, a small change in the delay (in state, input or output) is also a small variation in the BIBO gap topology.

It is well-known that P is stabilizable over H_∞ if and only P admits a coprime factorization over H_∞. However, for $\hat{A}$ we only know that if P admits a coprime factorization over $\hat{A}$ then it is BIBO-stabilizable. A parametrization of all stabilizing controllers of systems which do not necessarily admit a coprime factorization has recently been given in [33].

A system is said to be *nuclear* if its sequence (σ_n) of Hankel singular values satisfies $\sum \sigma_n < \infty$ (see, for example [16]).

We now give three theorems which represent the main tools to establish the results that follow. The first is due to Wiener and the second to Peller, with alternative versions due to Coifman and Rochberg, and Bonsall and Walsh. The third is due to Hardy and Littlewood.

Theorem 2. *(See [21, Theorem 4.18.6].) Let f be in $\hat{A}$. Then f has an inverse in $\hat{A}$ if and only if* $\inf_{\{\mathrm{Re}\, s \geq 0\}} |f(s)| > 0$.

Theorem 3. *(See [29].) Let P be an H_∞ transfer function. Then P is nuclear if and only if*

$$\int\int_{\mathbb{C}_+} |P''(s)|\, dA(s) < \infty,$$

where the integral is with respect to standard plane measure.

Theorem 4. *(See [20].) Let $r \in L^1_{loc}$ have a Laplace transform $\hat{r}(s)$ that is defined (as an absolutely convergent integral) in the open half-plane $\{\mathrm{Re}\, s > 0\}$. Moreover, suppose that $\hat{r}$ is bounded and has a bounded continuous extension to the closed right half-plane $\{\mathrm{Re}\, s \geq 0\}$, and that the boundary function $\tilde{r}(\omega) = \lim_{\sigma \longrightarrow 0} \hat{r}(\sigma + i\omega)$ is locally absolutely continuous and satisfies $\tilde{r}' \in L^1(\mathbb{R})$. Then, $r \in L^1(\mathbb{R}^+)$, and $\|r\|_{L^1(\mathbb{R}^+)} \leq \frac{1}{2}\|\tilde{r}'\|_{L^1(\mathbb{R})}$.*

3 Robust Stabilization of Retarded Delay Systems

3.1 The standard case

As is well-known (see for example [1] or [30]), the poles of delay systems—other than systems of type $e^{-sT}R(s)$ with R rational—lie in infinite chains, which can be of three types: retarded, neutral or advanced. The most easily analysed systems are those with pole chains of only retarded type, and we begin with these.

Thus we consider the class of delay systems with scalar transfer function given by

$$P(s) = \frac{\sum_{i=0}^{n_2} q_i(s)e^{-\beta_i s}}{\sum_{i=0}^{n_1} p_i(s)e^{-\gamma_i s}} = \frac{h_2(s)}{h_1(s)} \tag{1}$$

where $0 = \gamma_0 < \gamma_1 \cdots < \gamma_{n_1}$, $0 \leq \beta_0 < \beta_1 \cdots < \beta_{n_2}$, the p_i and q_i being polynomials satisfying the following condition :

Condition 1 (retarded delay systems): deg $p_0 >$ deg p_i for $i = 1, \cdots, n_1$ and deg $p_0 >$ deg q_i for $i = 0, \cdots, n_2$.

It is well-known [1] that these systems have a finite number of poles in any right half-plane, indeed are in the Callier–Desoer class $\hat{\mathcal{B}}_-(0)$ and possess a coprime factorisation. For these systems, the stability condition is the same than for finite dimensional systems:

Proposition 1. *Let P be defined as in (1) and satisfying Condition 1. P is H_∞ or BIBO-stable if and only if P has no poles in $\{\mathrm{Re}\, s \geq 0\}$.*

Coprime factorisations of retarded delay systems have been known for a long time. We next give an extension to the case where h_1 and h_2 have common unstable zeros and provide an algorithm for the calculation of the Bézout factors in this case.

Proposition 2. *([4]) Let $P(s) = \frac{h_2(s)}{h_1(s)}$ where $h_1(s)$ and $h_2(s)$ defined as above. There exists a rational function $r(s)$ such that $\left(\frac{h_2(s)}{r(s)}, \frac{h_1(s)}{r(s)}\right)$ is a coprime factorization of P over $\hat{A}$. If h_1 and h_2 have no more than δ_0 common unstable zeroes then r can be taken to be a polynomial.*

A suitable choice of Bézout factors is given in the next result.

Theorem 5. *([4]) Let m be the number of unstable zeroes σ of $h_1(s)$ (which are not zeroes of h_2) counted with their multiplicity.*
Let $X(s) = \frac{-\mu(s)}{u(s)}$ and $Y(s) = \frac{r(s) - \frac{\mu(s)h_2(s)}{u(s)}}{h_1(s)}$,
where μ is a polynomial of degree $m-1$ chosen such that

$$\left(r(s) - \frac{\mu(s)h_2(s)}{u(s)}\right)^{(k)} = 0$$

at $s = \sigma$ for $k = 0, \ldots, m_i - 1$ if σ is a zero of multiplicity $m_i \geq 1$, u is a polynomial chosen such that its inverse is in $\hat{A}$ and X is proper ($\deg u \geq \deg \mu$).
Then X and Y are Bézout factors corresponding to the coprime factorizations $\left(\frac{h_2(s)}{r(s)}, \frac{h_1(s)}{r(s)}\right)$ of P over $\hat{A}$.

Example 1. Let $P(s) = \dfrac{e^{-sT}}{s-\sigma}$ with $\sigma \in \mathbb{R}$ and $\gamma = \sqrt{1+\sigma^2}$. Then the pair $(N(s), D(s)) = \left(\dfrac{e^{-sT}}{s+\gamma}, \dfrac{s-\sigma}{s+\gamma}\right)$ is a coprime factorization of P over $\hat{A}$. Corresponding Bézout factors are given by

$$X(s) = -e^{T\sigma}(\sigma+\gamma), \qquad Y(s) = 1 + (\sigma+\gamma)\frac{1-e^{-T(s-\sigma)}}{s-\sigma}.$$

Using this, a family of controllers is directly implementable in Scilab/Scicos when $\sigma = 0$ and implementable after some minor manipulations in Scicos when $\sigma \neq 0$.

The coprime factors in this example have been chosen to be *normalized*, that is, they satisfy $|N(s)|^2 + |D(s)^2| = 1$ on $i\mathbb{R}$. This notion is important in robust control theory [10]. However, for general delay systems the calculation of normalized coprime factors is much more difficult to perform explicitly, as explained in [32].

3.2 The fractional case

We consider the class of fractional systems with scalar transfer function given by

$$P(s) = \frac{\sum_{i=0}^{n_2} q_i(s)e^{-\beta_i s}}{\sum_{i=0}^{n_1} p_i(s)e^{-\gamma_i s}} = \frac{h_2(s)}{h_1(s)} \tag{2}$$

where $0 = \gamma_0 < \gamma_1 \cdots < \gamma_{n_1}$, $0 \leq \beta_0 < \beta_1 \cdots < \beta_{n_2}$, the p_i being polynomials of the form $\sum_{k=0}^{l_i} a_k s^{\alpha_k}$ with $\alpha_k \in \mathbb{R}_+$ and the q_i being polynomials of the form $\sum_{k=0}^{m_i} b_k s^{\delta_k}$ with $\delta_k \in \mathbb{R}_+$.

We will assume throughout that h_2 and h_1 have no common zeroes in $\{\mathrm{Re}\, s \geq 0\} \setminus \{0\}$.

Note that, for $s \neq 0$ and $\delta \in \mathbb{R}$, we define s^δ to be $\exp(\delta(\log|s| + i \arg s))$, and a continuous choice of $\arg s$ in a domain leads to an analytic branch of s^δ. In this work we shall normally make the choice $-\pi < \arg s < \pi$, for $s \in \mathbb{C} \setminus \mathbb{R}_-$.

We will need later to characterize the behaviour of h_1 and h_2 at zero and infinity, so let us remark that there exist constants α, $\beta \geq 0$ and c_1, $c_2 \neq 0$, such that for $s \in \{\mathrm{Re}\, s \geq 0\}$

$$h_1(s) = s^\alpha(c_1 + o(1)) \qquad \text{near } s = 0, \tag{3}$$

$$h_2(s) = s^\beta(c_2 + o(1)) \qquad \text{near } s = 0. \tag{4}$$

Let us write also

$$\gamma = \deg p_0 > 0, \qquad \text{and} \quad \delta = \max_{i=0,\dots n_2} \deg q_i \geq 0.$$

and let us denote $\mu = \min(\alpha, \beta)$.

As for the classical delay systems, we shall consider the systems which satisfy Condition 1. We can see from the next two results that they behave exactly the same as the standard delay systems.

Theorem 6. *([6]) Let P be defined as in (2) and satisfying Condition 1. Then P is BIBO stable if and only if it has no poles in $\{\mathrm{Re}\, s \geq 0\}$ (in particular, no poles of fractional order at $s = 0$).*

Proposition 3. *([6]) The system P defined as in (2) has a finite number of poles in any cut right half-plane, that is, the set*

$$\{s_0 \in \mathbb{C} \setminus \mathbb{R}_- : \quad \mathrm{Re}\, s_0 \geq a \quad \textit{and } s_0 \textit{ is a pole of } P\}$$

is finite for all $a \in \mathbb{R}$

As for the standard retarded delay case, coprime and Bézout factors can be found in terms of the behaviour of h_1 at infinity. Moreover, it is also necessary to take into account here the behaviour of h_1 and h_2 at zero.

Proposition 4. *([5]) A coprime factorization (N, D) in $\hat{A}$ of P defined as (2) and satisfying Condition 1 is given by*

$$N(s) = \frac{(s+1)^{[\mu]}(s^{\{\mu\}}+1)h_2(s)}{s^{\mu}(s+1)^{[\gamma]}(s^{\{\gamma\}}+1)},$$

$$D(s) = \frac{(s+1)^{[\mu]}(s^{\{\mu\}}+1)h_1(s)}{s^{\mu}(s+1)^{[\gamma]}(s^{\{\gamma\}}+1)}.$$

The next result gives a formula for some corresponding Bézout factors, but it is considerably more complicated.

Theorem 7. *([5]) Let $\sigma_1, \cdots, \sigma_m$ be the m nonzero unstable zeroes of h_1 and let*

$$T_1(s) = s^{\mu}(s+1)^{[\gamma]}(s^{\{\gamma\}}+1),$$
$$T_2(s) = (s+1)^{[\mu]}(s^{\{\mu\}}+1)h_2(s),$$
$$T_3(s) = (s+1)^{[\mu]}(s^{\{\mu\}}+1)h_1(s).$$

Now, let us define

$$Y(s) = \frac{T_1(s) + T_2(s)X(s)}{T_3(s)} \quad \textit{and}$$

$$X(s) = \frac{f_0 + f_{\lambda_1}s^{\lambda_1} + \cdots f_{\lambda_n}s^{\lambda_n}}{(s+1)^M} + \frac{f_{M-m+1}s^{M-m+1} + \cdots f_M s^M}{(s+1)^M}$$

where $\lambda_n \in \mathbb{R}$ *and* $M \in \mathbb{N}$ *is chosen such that* $M > \lambda_n + m$, *the coefficients* $f_0, f_{\lambda_1}, \cdots, f_{\lambda_n}$ *are chosen in order to satisfy that* $T_1(s)+T_2(s)X(s)$ *is of fractional order* α *at zero, and the coefficients* $f_{M-m+1}, \cdots, f_M$ *are chosen so that* $T_1(\sigma_i) + T_2(\sigma_i)X(\sigma_i) = 0$ *for* $1 \leq i \leq m$.

Then (X,Y) are Bézout factors associated to the coprime factors N and D of P.

4 Robust Stabilization of Delay Systems of Neutral Type

The case of neutral systems is much more difficult. Such systems will be defined by (1) or (2) and will satisfy Condition 2 below.

Condition 2: deg $p_0 \geq$ deg p_i for $i = 1, \cdots, n_1$ (with equality for at least one polynomial p_i) and deg $p_0 >$ deg q_i for $i = 0, \cdots, n_2$.

4.1 The standard case

Delay systems that are of purely neutral type have their poles in a band $\{-a \leq \text{Re}\, s \leq a\}$ about the imaginary axis, although there is in general no asymptotic formula for them. Moreover, the location of the poles does not determine the stability of a neutral delay system: for example the transfer function $P_0(s) = 1/(s+1+se^{-s})$ has no poles in the closed right half-plane, but does not lie in H_∞, and thus is not BIBO-stable either. In fact every band $\{-\delta < \text{Re}\, s < 0\}$ contains infinitely many poles of P; they approach the imaginary axis as their modulus tends to infinity.

On the other hand $P_1(s) = P_0(s)/(s+1)$ is in H_∞, and the system $P_5(s) = P_0(s)/(s+1)^5$ is even BIBO-stable. For more on this and similar examples we refer to the analysis in [7]. It should be pointed out that there remain several unsolved questions in this area, since the known tests are not subtle enough to decide the question of BIBO stability for every neutral delay system.

Thus the following well-known proposition gives a sufficient but not necessary condition for stability.

Proposition 5. *Let P be defined as in (1) and satisfying Condition 2. If there exists* $a < 0$ *such that P has no poles in* $\mathbb{C} \cap \{\text{Re}\, s > a\}$ *then P is BIBO-stable.*

4.2 The fractional case

We consider here systems (2) satisfying Condition 2. The following result extends Proposition 5.

Proposition 6. *[6] Let P be defined as in (2) and satisfying Condition 2. If there exists* $a < 0$ *such that P has no poles in* $(\mathbb{C} \setminus \mathbb{R}_-) \cap \{\text{Re}\, s > a\} \cup \{0\}$ *then P is BIBO-stable.*

5 Nuclearity of Delay Systems

Nuclearity is an important notion when considering model reduction, for example the standard techniques of Hankel-norm approximation and truncated balanced realizations are guaranteed to converge in the nuclear case and, in general, not otherwise (see [16]). Note that it is often simpler to perform control design on a reduced-order model (in particular, on a finite-dimensional system), rather than directly on an infinite-dimensional system. We refer to [30] for more on this topic in the context of delay systems.

Nuclearity is characterized by Theorem 3. Moreover, the nuclearity of a system with transfer function P implies that

$$\int_{-\infty}^{\infty} |P'(iy)|\, dy < \infty,$$

which in turn is a sufficient condition for BIBO stability.

The following elementary lemma is a useful tool for analysing the nuclearity of systems.

Lemma 1. *[6] Suppose that $f \in H_\infty(\mathbb{C}_+)$ and $|f(s)| = O(|s|^{-\alpha})$ as $|s| \to \infty$ in $\mathbb{C}_+$. Then*

$$\text{(i)} \qquad \int\!\!\int_{\mathbb{C}_+} |f(s)|\, dA(s) < \infty \qquad \text{if } \alpha > 2;$$

$$\text{(ii)} \qquad \int\!\!\int_{\mathbb{C}_+} |f(s)|e^{-s}\, dA(s) < \infty \qquad \text{if } \alpha > 1;$$

On combining this with Theorem 3 one arrives at a test for nuclearity of (fractional) delay systems, which can be expressed in the following form.

Proposition 7. *[6] Let P be defined as in (2) and satisfying Condition 1. Suppose that P is BIBO-stable. Then P is nuclear if the following condition is satisfied:*

- *Any delayed term of the form $r(s)e^{-\lambda s}$ occurring in one of h_2'', $h_2'h_1'$, h_2h_1'' or $h_2h_1'h_1'$ is such that the degree of the fractional polynomial part $r(s)$ is less than $\gamma - 1$, $2\gamma - 1$, $2\gamma - 1$ or $3\gamma - 1$, respectively.*

References

1. R. Bellman and K. L. Cooke. *Differential-difference equations.* Academic Press, New York, 1963.
2. D. Brethe and J.-J. Loiseau. Stabilization of linear time-delay systems. *JESA–RAIRO–APII*, 6, 1025–1047, 1997.
3. C. Bonnet and J. R. Partington. Robust stabilization in the BIBO gap topology. *Int. J. Robust Nonlin. Control*, 7, 429–447, 1997.
4. C. Bonnet and J. R. Partington. Bézout factors and L^1-optimal controllers for delay systems using a two-parameter compensator scheme. *IEEE Trans. Automat. Control*, 44, 1512–1521, 1999.

5. C. Bonnet and J. R. Partington. Stabilization of fractional exponential systems including delays. *Kybernetica*, 37 (3), 345–353, 2001.
6. C. Bonnet and J. R. Partington. Analysis of fractional delay systems of retarded and neutral type. *Automatica*, 38 (7), 1133–1138, 2002.
7. C. Bonnet and J. R. Partington. H_∞ and BIBO stabilization of delay systems of neutral type. *Systems Control Lett.*, to appear.
8. C. Bonnet, J. R. Partington and M. Sorine. Robust control and tracking of a delay system with discontinuous nonlinearity in the feedback. *Int. J. Control* 72 (15), 1354–1364, 1999.
9. C. Bonnet, J. R. Partington and M. Sorine. Robust stabilization of a delay system with saturating actuator or sensor. *Int. J. Robust Nonlin. Control* 10 (7), 579–590, 2000.
10. R. F. Curtain and H. Zwart. *An introduction to infinite-dimensional linear systems theory.* Texts in Applied Mathematics, 21. Springer-Verlag, New York, 1995.
11. H. Dym and T. T. Georgiou and M. C. Smith. Explicit formulas for optimally robust controllers for delay systems. *IEEE Trans. Automat. Control*, 40, 656–669, 1995.
12. C. Foias, H. Özbay and A. Tannenbaum. *Robust control in infinite dimensional systems - Frequency domain methods.* Springer Lecture Notes in Control and Inform. Sci., no. 209, 1996.
13. T. T. Georgiou and M. C. Smith. Robust stabilization in the gap metric: controller design for distributed plants. *IEEE Trans. Automat. Control*, 37, 1133–1143, 1992.
14. T. T. Georgiou and M. C. Smith. Metric uncertainty and nonlinear feedback stabilization. In *Feedback control, nonlinear systems, and complexity.* Springer Lecture Notes in Control and Inform. Sci., no. 202, 88–98, 1995.
15. T. T. Georgiou and M. C. Smith. Robustness analysis of nonlinear feedback systems: an input-output approach. *IEEE Trans. Automat. Control*, 42 (9), 1200–1221, 1997.
16. K. Glover, R. F. Curtain, and J. R. Partington. Realization and approximation of linear infinite dimensional systems with error bounds. *SIAM J. Control Optimiz.*, 26, 863–898, 1988.
17. K. Glover, J. Lam and J. R. Partington Rational approximation of a class of infinite-dimensional systems I: Singular values of Hankel operators *Math. Control Sig. Sys.*, 3, 325–344, 1990.
18. K. Glover, J. Lam and J. R. Partington Rational approximation of a class of infinite-dimensional systems II: Optimal convergence rates of L_∞ approximants *Math. Control Sig. Sys.*, 4, 233–246, 1991.
19. H. Glüsing-Luerssen. A behavioral approach to delay-differential systems. *SIAM J. Control Optimiz.*, 35 (2), 480–499, 1997.
20. G. Gripenberg, S.-O. Londen and O. Staffans. *Volterra integral and functional equations.* Cambridge University Press, Cambridge, 1990.
21. E. Hille and R. S. Phillips. *Functional analysis and semi-groups.* Amer. Math. Soc, 1957.
22. R. Hotzel. Some stability conditions for fractional delay systems. *J. Mathematical Systems Estimation Control*, 8 (4), 1–19, 1998.
23. J.-J. Loiseau and H. Mounier. Stabilisation de l'équation de la chaleur commandée en flux. In *Systèmes Différentiels Fractionnaires, Modèles, Méthodes et Applications*, ESAIM proceedings, 5, 131–144, 1998.
24. P. M. Mäkilä and J. R. Partington. Robust stabilization-BIBO stability, distance notions and robustness optimization. *Automatica*, 23, 681–693, 1993.
25. P. M. Mäkilä and J. R. Partington. Shift operator induced approximations of delay systems. *SIAM J. Control Optimiz.*, 37 (6), 1897–1912, 1999.
26. D. Matignon. Stability properties for generalized fractional differential systems. In: *Systèmes Différentiels Fractionnaires, Modèles, Méthodes et Applications.* Vol. 5. ESAIM proceedings, 145–158, 1998.

27. D. C. McFarlane and K. Glover. *Robust controller design using normalized coprime factor descriptions*. Springer-Verlag, 1989.
28. H. Özbay. *Introduction to feedback control theory*. CRC Press, 1999.
29. J. R. Partington. *An introduction to Hankel operators*. Cambridge University Press, 1988.
30. J. R. Partington. *Linear operators and linear systems*. Cambridge University Press, to appear.
31. J. R. Partington and K. Glover. Robust stabilization of delay systems by approximation of coprime factors. *Systems Control Lett.*, 14 (4), 325–331, 1990.
32. J. R. Partington and G. K. Sankaran. Algebraic construction of normalized coprime factors for delay systems. *Math. Control Sig. Sys.* 15 (1) 1–12, 2002.
33. A. Quadrat. On a generalization of the Youla–Kucera parametrization. Part I: the fractional ideal approach to SISO systems. *Systems Control Lett.*, 50, 135–148, 2003.
34. M. S. Verma. Coprime fractional representations and stability of non-linear feedback systems. *Int. J. Control*, 48, 897–918, 1988.
35. M. Vidyasagar. *Control System Synthesis*. MIT Press, 1985.

On Strong Stability and Stabilizability of Linear Systems of Neutral Type

Rabah Rabah[1], Grigory M. Sklyar[2], and Alexandr V. Rezounenko[3]

[1] IRCCyN UMR 6597, 1 rue de la Noë, BP 92101, F-44321 Nantes Cedex 3, France. rabah@emn.fr
[2] Institute of Mathematics, University of Szczecin, 70–451 Szczecin, Wielkopolska 15, Poland. sklar@sus.univ.szczecin.pl
[3] Department of Mechanics and Mathematics, Kharkov University, 4 Svobody sqr., Kharkov, 61077, Ukraine. rezounenko@univer.kharkov.ua

Summary. For linear stationary systems, the infinite dimensional framework allows one to distinguish different notions of stability: weak, strong or exponential. The purpose of this chapter is to investigate the problem of strong stability, i.e. asymptotic non-exponential stability for linear systems of neutral type in order to use this characterization in the study of the stabilizability problem for this type of systems. An important tool in this investigation is the Riesz basis property of generalized eigenspaces of the neutral system, because that the generalized eigenvectors do not form, in general, a Riesz basis. This allows one to describe more precisely asymptotic non-exponential stability of neutral systems. For a particular case, conditions of strong stabilizability of neutral type systems are given with a feedback law without derivative of the delayed state.

1 Introduction and Preliminary Results

In this note we present a new approach for studying the strong stability and stabilizability properties of the functional differential equation of neutral type

$$\dot{x}(t) = A_{-1}\dot{x}(t-1) + \int_{-1}^{0} A_2(\theta)\dot{x}(t+\theta)d\theta + \int_{-1}^{0} A_3(\theta)x(t+\theta)d\theta \qquad (1)$$

where A_{-1} is constant $n \times n$-matrix, $\det A_{-1} \neq 0$, A_2, A_3 are $n \times n$-matrices whose elements belong to $L_2(-1,0)$.

This equation occurs, for example, when a system of neutral type is stabilized. Even if the initial system contains pointwise delays only, then the set of natural feedback laws contains distributed delays (see e.g., [15,17]), so the corresponding closed-loop system takes the form (1).

The problem of exponential stability of systems like (1) is well studied [8, 10]. Our purpose is to analyze more subtle properties of stability (and stabilizability), namely strong non-exponential asymptotic stability (see e.g. [4]). One needs to consider an operator model generated by the system (1) in some infinite dimensional space. It is well-known that for neutral type systems the choice of the phase-space is crucial (in contrast to the case of retarded functional differential equations where solutions are more smooth than the initial data).

In [6, 8], the framework is based on the description of neutral type systems in the space of continuous functions $C([-1,0];\mathbf{C}^n)$. The essential result in this framework is that the exponential stability is characterized by the condition that the spectrum of the system belongs to the open left-half plane. The problem of asymptotic (non-exponential) stability is much more complicated.

Following [25] we treat our system as a system in the Hilbert space $M_2 = \mathbf{C}^n \times L_2(-1,0;\mathbf{C}^n)$. This fact is important for us since it allows to use deep ideas and technique of the operator theory in Hilbert space [1, 7] and results [19] on the existence of Riesz basises (see Section 2 for more details) in the analysis the following operator model (see [25]) of the system (1)

$$\frac{\mathrm{d}}{\mathrm{d}t}\begin{pmatrix} y(t) \\ z_t(\cdot) \end{pmatrix} = \mathcal{A}\begin{pmatrix} y(t) \\ z_t(\cdot) \end{pmatrix} = \begin{pmatrix} \int_{-1}^{0} A_2(\theta)\dot{z}_t(\theta)\mathrm{d}\theta + \int_{-1}^{0} A_3(\theta)z_t(\theta)d\theta \\ \mathrm{d}z_t(\theta)/\mathrm{d}\theta \end{pmatrix}, \tag{2}$$

where the domain of $\mathcal{A}$ is given by

$$\mathcal{D}(\mathcal{A}) = \{(y, z(\cdot)) : z \in H^1(-1,0;C^n), y = z(0) - A_{-1}z(-1)\} \subset M_2.$$

Theorem 1. *The operator $\mathcal{A}$ defined in (2) is the infinitesimal generator of a C_0-semigroup denoted $T(t) \equiv e^{\mathcal{A}t}, t \geq 0$ on the Hilbert space $M_2 = \mathbf{C}^n \times L_2(-1,0;\mathbf{C}^n)$.*

If additionally $\det A_{-1} \neq 0$, *then the operator $\mathcal{A}$ is the generator of a group $e^{\mathcal{A}t}, t \in \mathbf{R}$ on M_2.*

Let us denote by $\mu_1, ..., \mu_\ell$, $\mu_i \neq \mu_j$ if $i \neq j$, the eigenvalues of A_{-1} and the dimensions of their rootspaces by $p_1, ..., p_\ell$, $\sum_{k=1}^{\ell} p_k = n$. Consider the points $\lambda_m^{(k)} \equiv \ln|\mu_m| + i(\arg\mu_m + 2\pi k), m = 1,..,\ell; k \in Z$ and the circles $L_m^{(k)}$ of fixed radius $r \leq r_0 \equiv \frac{1}{3}\min\{|\lambda_m^{(k)} - \lambda_i^{(j)}|, (m,k) \neq (i,j)\}$ centered at $\lambda_m^{(k)}$.

Theorem 2. *The spectrum of $\mathcal{A}$ consists of the eigenvalues only which are the roots of the equation* $\det \Delta(\lambda) = 0$, *where*

$$\Delta_{\mathcal{A}}(\lambda) = \Delta(\lambda) \equiv -\lambda I + \lambda e^{-\lambda}A_{-1} + \lambda\int_{-1}^{0} e^{\lambda s}A_2(s)ds + \int_{-1}^{0} e^{\lambda s}A_3(s)ds. \tag{3}$$

The corresponding eigenvectors of $\mathcal{A}$ are $\varphi = \begin{pmatrix} C - e^{-\lambda}A_{-1}C \\ e^{\lambda\theta}C \end{pmatrix}$, with $C \in Ker\Delta(\lambda)$.

There exists N_1 such that for any k, such that $|k| \geq N_1$, the total multiplicity of the roots of the equation $\det \Delta(\lambda) = 0$, *contained in the circle $L_m^{(k)}$, equals p_m.*

To describe the location of the spectrum of $\mathcal{A}$ we use Rouche theorem. More precisely, for sufficiently large k and any m we show that $|f_1(\lambda)| > |f_2(\lambda)|$ for any $\lambda \in L_m^{(k)}$ and $f_1(\lambda) \equiv \det(A_{-1} - e^{-\lambda}I)$, $f_2(\lambda) \equiv \det(A_{-1} - e^{-\lambda}I) - \det\left((A_{-1} - e^{-\lambda}I) + e^{\lambda}\int_{-1}^{0} e^{\lambda s}A_2(s)ds + e^{\lambda}\lambda^{-1}\int_{-1}^{0} e^{\lambda s}A_3(s)ds\right)$.

Thus, $f_1 - f_2$ has the same number of roots inside $L_m^{(k)}$ as function f_1.

We start our analysis of the stability properties of the system (2), using the classical technique (see e.g. [5]), and prove the following

Theorem 3. *Assume that the spectrum satisfies $\sigma(\mathcal{A}) \subset \{\lambda : \Re\lambda < 0\}$ and*

$$||A_{-1}|| + ||A_2||_{L_2(-1,0)} < 1. \tag{4}$$

Then the system (2) is exponentially stable.

The proof of Theorem 3 is based on the fact that for a C_0-semigroup $T(t)$ on *Hilbert* space one has $s_0(A) = \omega_0(T)$, where A is the generator of T, $s_0(A)$ is the *abscissa of uniform boundedness of the resolvent* and

$$\omega_0(T) \equiv \inf\{\omega \in R : \exists M > 0 : ||T(t)|| \leq Me^{\omega t}, \forall t \geq 0\}$$

See [24] for definitions and details. This fact is crucial for the following theorem which we use to prove Theorem 3.

Theorem 4. *[5, p.222, th.5.1.5] Let A be the infinitesimal generator of the C_0-semigroup $T(t)$ on Hilbert space Z. Then $T(t)$ is exponentially stable if and only if $(sI - A)^{-1} \in H_\infty(\mathcal{L}(Z))$. Here $H_\infty(\mathcal{L}(Z))$ is Hardy space of bounded holomorphic functions on $\{z \in C : \Re z > 0\}$ with values in $\mathcal{L}(Z)$.*

It is easy to see that Theorem 3 can not be applied to the case when there exists $\mu \in \sigma(A_{-1})$ such that $|\mu| = 1$ (see (4)). Moreover, Theorem 2 shows that under the assumptions of Theorem 3 one has $\sigma(A_{-1}) \subset \{\mu : |\mu| \leq \delta < 1\}$ or equivalently $\sigma(\mathcal{A}) \subset \{\lambda : \Re\lambda \leq -\varepsilon < 0\}$. For the system (1) this is the case of exponential stability.

Further on we are mainly interested in studying the case when the system is strongly asymptotically (non-exponentially) stable. Let us give an illustration of such a situation. We consider the scalar system (1)

$$\dot{x}(t) = a\dot{x}(t-1) + \int_{-1}^{0} \varphi_2(\theta)\dot{x}(t+\theta)d\theta + \int_{-1}^{0} \varphi_3(\theta)x(t+\theta)d\theta \tag{5}$$

where a is a constant such that $|a| = 1$, and φ_2, φ_3 are any functions belonging to $L_2(-1,0)$.

Proposition 1. *The system (5) is strongly asymptotically (non-exponentially) stable if and only if $\sigma(\mathcal{A}) \subset \{\lambda : \Re\lambda < 0\}$ (see (2) for the definition of $\mathcal{A}$ with φ_2, φ_3 instead of A_2, A_3).*

To prove this and the other statements of this chapter on the strong stability and stabilizability properties of neutral type systems we develop a new approach which we describe in details in the following sections.

First, we involve the classical theorem from the theory of semigroups (see Section 1) which is proposed as the main tool for the analysis of the stability properties. Second, we discuss a new result [19] on the existence of Riesz basises and, particularly, use it (see Section 2) to prove Proposition 1. Section 3 is devoted to the study of the stabilizability properties, when the control action is $Bu(t)$. For neutral type systems, exponential stabilizability by feedback requires, in general, the delayed derivative in the feedback [15]. Our method allows to use only state feedback but we obtain asymptotic (non-exponential) stability of the closed loop system. To this end we recall first some deep results on strong stabilizability of linear systems in Hilbert spaces. In the last section one presents conclusions and perspectives.

2 Strong Stability

Apparently, the first result which may be considered as a basis of the investigation on strong stability is the following one [23, p.102].

Theorem 5. *Let be given a complete nonunitary contraction T in the Hilbert space H such that*

$$\mathrm{mes}(\sigma(T) \cap S_0(1)) = 0,$$

where $S_0(1) = \{\lambda \in C : |\lambda| = 1\}$ and $\mathrm{mes}(.)$ *is a Lebesgue measure in $S_0(1)$. Then for each $x \in H$ we have*

$$\lim_{n\to\infty} T^n x = 0 \quad \textit{and} \quad \lim_{n\to\infty} T^{*n} x = 0.$$

Let us recall that a contraction T is said to be completely nonunitary if there does not exist a subspace $H_1 \subset H$, invariant by T, such that $T|_{H_1}$ is an unitary operator.

In the Theorem 5, the notion of stability is not explicitly involved. However, with this important theorem one can obtain the following result on asymptotic stability of a semigroup in Hilbert space.

Definition 1. *A C_0-semigroup $T(t), t \geq 0$ is called to be contractive semigroup if $||T(t)|| \leq 1, t \geq 0$.*

Definition 2. *A semigroup $T(t), t \geq 0$ is said to be unitary if $\forall x \in H$ and $t \geq 0$ we have*

$$||T(t)x|| = ||x|| = ||T(t)^* x||.$$

The semigroup $T(t)$ is said to be completely nonunitary if $\forall x \in H, x \neq 0$, there exists $t \geq 0$ such that

$$||T(t)x|| < ||x|| \quad \textit{or} \quad ||T(t)^* x|| < ||x||.$$

Let us recall that a semigroup is contractive if and only if the infinitesimal generator A of the semigroup is maximal dissipative: $\Re\langle Ax, x\rangle \leq 0$ for all $x \in \mathcal{D}(A)$, and unitary if and only if the infinitesimal generator A is skew-adjoint.

Theorem 6. *[12, §A3 Strong Stability of Evolution Equations]. Let A be the infinitesimal generator of a contractive completely nonunitary semigroup $e^{At}, t \geq 0$ in the Hilbert space H and*

$$\text{mes}(\sigma(A) \cap (i\mathbf{R})) = 0, \tag{6}$$

where $i\mathbf{R}$ is the imaginary axis and $\text{mes}(.)$ a Lebesgue measure on this set. Then for all $x \in H$ we have $e^{At}x \to 0, \quad t \to \infty$.

The proof of Theorem 6 is based on Theorem 5 and the introduction of the cogenerator of the semigroup [23], i.e. the operator

$$F = (A + I)(A - I)^{-1}.$$

The condition (6) is essential for the stability of the semigroup. For practical use, an important particular case is the condition: the set $(\sigma(A) \cap (i\mathbf{R}))$ is at most countable.

It turns out that when this condition is satisfied, then the semigroup $\{e^{At}\}, t \geq 0$ is completely nonunitary if and only if the operator A has no pure imaginary eigenvalues. This gives a simple formulation of the Theorem 6.

Moreover, with this assumption the result on strong asymptotic stability may be extended to the case of Banach space. Namely one has the following criteria of strong asymtotic stability.

Theorem 7. *Let e^{At}, $t \geq 0$ be a C_0-semigroup in the Banach space X and A be the infinitesimal generator of the semigroup. Assume that $(\sigma(A) \cap (i\mathbf{R}))$ is at most countable and the operator A^* has no pure imaginary eigenvalues. Then e^{At} is strongly asymptotically stable (i.e. $e^{At}x \to 0$, $t \to +\infty$ as $x \in X$) if and only if one of the following conditions is valid:*

i) There exists a norm $\|\cdot\|_1$, equivalent to the initial one $\|\cdot\|$, such that the semigroup e^{At} is contractive according to this norm: $\|e^{At}x\|_1 \leq \|x\|_1$, $\forall x \in X, t \geq 0$;

ii) The semigroup e^{At} is uniformly bounded: $\exists C > 0$ such that $\|e^{At}\| \leq C, t \geq 0$.

The Theorem 7 was obtained first in [20] for the case of bounded operator A, then generalized in [2, 14] for the general case. The development of this theory concerns a large class of differential equations in Banach space (see [24] and references therein).

3 Riesz Basis Property

We notice that the condition "$\sigma(A) \cap (i\mathbf{R})$ is at most countable" (see Theorem 7) can be easily verified for many concrete systems which arise from applications. For example, in the case when A has compact resolvent, one has that $\sigma(A)$ itself is at most countable and consists of the point spectrum only. The location of eigenvalues of A for some systems can be easily described while for others this question needs a careful investigation, using, for example, Rouche theorem and perturbation analysis (see e.g. Theorem 2). It is well known that the property $\sigma(A) \subset \{\lambda : \Re\lambda < 0\}$ is necessary but not sufficient for the strong asymptotic stability of e^{At}.

Taking this into account, we arrive at the necessity to have an efficient method to check the property i) (or equivalently ii)) of Theorem 7. Working in a Hilbert space , we get a powerful tool to study the property i), namely, the concept of Riesz basis. The simplest case is a Riesz basis of vectors. Let us remind the definition.

Definition 3. *A basis $\{\psi_j\}$ of a Hilbert space H is called a Riesz basis if there are an orthonormal basis $\{\phi_j\}$ of H and a linear bounded invertible operator R, such that $R\psi_j = \phi_j$.*

To the best of our knowledge, the main source of abstract results on Riesz basises is the monograph [7]. The most desired situation for concrete systems is to have a Riesz basis formed by eigenvectors of A or, at least, by generalized eigenvectors [7,12,21]. In more general situations, one studies the existence of basises formed by subspaces. We remind that *a sequence of nonzero subspaces $\{V_k\}_i^\infty$ of the space V is called basis (of subspaces) of the space V, if any vector $x \in V$ can be uniquely presented as $x = \sum_{k=1}^{\infty} x_k$, where $x_k \in V_k$, $k = 1, 2, ..$* We say that the basis $\{V_k\}_i^\infty$ is orthogonal if V_i is orthogonal to V_j when $i \neq j$. As in the case of a basis of vectors we can introduce the following definition.

Definition 4. *[7] A basis $\{V_k\}$ of subspaces is called a basis equivalent to orthogonal (a Riesz basis) if there are an orthogonal basis of subspaces $\{W_k\}$ and a linear bounded invertible operator R, such that $RV_k = W_k$.*

The best "candidates" to form the basis of subspaces are generalized eigenspaces of the generator of a semigroup, but there are simple examples (see Example 1 below) showing that generalized eigenspaces do not form such a basis in the general case.

One of the crucial ideas of our approach is to construct a Riesz basis of finite-dimensional subspaces which are invariant for the generator of the semigroup (see (2)). The existence of such basises essentially simplifies, for example, the verification of the property i) of Theorem 7.

In [19] we obtained the following general result.

Theorem 8. *There exists a sequence of invariant for $\mathcal{A}$ (see (2)) finite-dimensional subspaces which constitute a Riesz basis in M_2.*

More precisely, these subspaces are $\{V_m^{(k)}, |k| \geq N, m = 1, .., \ell\}$ and a $2(N+1)n$-dimensional subspace spanned by all eigen- and rootvectors, corresponding to all eigenvalues of $\mathcal{A}$, which are outside of all circles $L_m^{(k)}$, $|k| \geq N, m = 1, .., \ell$.

Here $V_m^{(k)} \equiv P_m^{(k)} M_2$, where

$$P_m^{(k)} M_2 = \frac{1}{2\pi i} \int_{L_m^{(k)}} R(\mathcal{A}, \lambda) d\lambda$$

are spectral projectors; $L_m^{(k)}$ are circles defined before.

We emphasize that the operator $\mathcal{A}$ may not possess in a Riesz basis of generalized eigenspaces. We illustrate this on the following

Example 1. Consider the particular case of the system (1):

$$\dot{x}(t) = A_{-1}\dot{x}(t-1) + A_0 x(t), \qquad A_{-1} = \begin{pmatrix} 1\,1 \\ 0\,1 \end{pmatrix}, \quad A_0 = \begin{pmatrix} \alpha\; 0 \\ 0\; \beta \end{pmatrix}. \tag{7}$$

One can check that the characteristic equation is $\det \Delta(\lambda) = (\alpha - \lambda + \lambda e^{-\lambda})(\beta - \lambda + \lambda e^{-\lambda}) = 0$ and for $\alpha \neq \beta$ there are two sequences of eigenvectors, such that $||v_n^1 - v_n^2|| \to 0$, as $n \to \infty$. By the definition, such vectors can not form a Riesz basis.

To prove Proposition 1 we notice that for the particular case of the system (5), Theorem 8 gives that $\ell = n = 1$, all the subspaces $V_1^{(k)}, |k| \geq N$, are one-dimensional and together with the $2(N+1)$-dimensional subspace form a Riesz basis of the space M_2. Using this property, we consider the operator R (see Def. 3) which maps the eigen- and possibly finite number of root-vectors of $\mathcal{A}$ to an orthonormal basis of M_2. It is now easy to check that the new norm $||\cdot||_1 \equiv ||R\cdot||$ is equivalent to $||\cdot||$. This, together with the property $\sigma(\mathcal{A}) \subset \{\lambda : \Re\lambda < 0\}$, allow us to apply Theorem 7 to prove Proposition 1.

Let us precise that for the general multivariable system (1) the stability conditions is more complicated. A complete analysis for the general case will be given in one of our forcoming papers.

4 Strong Stabilizability

4.1 The abstract theory

The problem of strong stabilizability of control systems in infinite dimensional spaces has been intensively studied since seventies. The basic abstract formulation of this problem is the following one.

Consider the linear system

$$\dot{x} = Ax + Bu, \quad x \in H, u \in U, \tag{8}$$

where H, U are Hilbert spaces, the operator A is the infinitesimal generator of a C_0-semigroup of contractions $\{e^{At}\}$, $t \geq 0$, i.e. $||e^{At}|| \leq 1$, $t \geq 0$ or, what is the same, such that A is a maximal dissipative operator. The operator B is usually assumed to be a bounded linear operator from U to X. The problem under investigation is: if the feedback control law $u = -B^*x$ is a stabilizing control, i.e. $e^{(A-BB^*)t}x \to 0$, $t \to +\infty$, $\forall x \in H$.

Under some additional assumptions this problem was studied in [22] using the Lyapunov method. In [13] and other works by N. Levan (cf. references in [16]) a framework based on the decomposition of the contractive semigroup and the harmonic analysis of operators was developed. Based on this framework the partial answers to the problem of strong stabilizability were given. Necessary and sufficient conditions of the strong stabilizability under the assumption (6) were given in the PhD thesis of G. M. Sklyar (Kharkov, 1983). In a slightly weaker formulation this result was published in [11]. The complete result was included to the book [12] (§A3. Strong Stabilizability of Evolution Equations). Let us recall it:

We denote for the system (8)

$$L_r = \overline{\sum_{t\geq 0} e^{At}BU} \quad \text{and} \quad L_{*r} = \overline{\sum_{t\geq 0} e^{A^*t}BU}.$$

For the contractive semigroup e^{At}, $t \geq 0$, the canonical decomposition [23] holds:

$$H = V \oplus W,$$

where the restriction $e^{At}|_V$ is an unitary semigroup, while the restriction $e^{At}|_W$ is completely nonunitary.

Theorem 9. *Let for the system (8) the condition*

$$\text{mes}(\sigma(A) \cap (i\mathbf{R})) = 0$$

be valid. Then the system (8) is strongly stabilizable if and only if

$$V \cap L_r^{\perp} \cap L_{*r}^{\perp} = \{0\}. \tag{9}$$

The stabilizing control law is then given by $u = -B^*x$.

Let us note that the condition (9) becomes much simpler when the set $(\sigma(A) \cap (i\mathbf{R}))$ is at most countable. It is equivalent to the following condition: *There does not exist an eigenvector $x \in H$ of the operator A, with pure imaginary eigenvalue such that $B^*x = 0$.*

In this formulation the result was found one more time in [3], where the authors used Theorem 7. An extensive investigation of the strong stabilizability can be found [16] (see also references therein). In [21] the problem of the description of a large class of (strong) stabilizing control laws of the type $u = Px$ is given. The main tools are the Theorem 7 and the technique of the characterization of equivalent norms for which the operator $A + BP$ is dissipative. For the particular case when the operator A is skew-adjoint with separated discrete spectrum this class was identified. The problem of robustness in this class was also investigated. The perspective is to develop this framework to the case of unbounded operator P in the feedback $u = Px$.

The abstract theory of stabilizability given here may be applied to the system of neutral type.

4.2 Systems of neutral type

Let us present some results on the stabilizability for the particular case of the system (2).

For simplicity we consider a control neutral type system with one delay in the state

$$\dot{x}(t) = A_0 x(t) + A_1 x(t-1) + A_{-1}\dot{x}(t-1) + Bu(t), \tag{10}$$

$x \in R^n, u \in R^r, A_j, j = -1, 0, 1$ are $n \times n$-matrices, B is a $n \times r$-matrix.

The stabilizability problem consists in determination of linear feedback control $u = p(x(\cdot))$ such that the closed-loop system

$$\dot{x}(t) = A_0 x(t) + A_1 x(t-1) + A_{-1}\dot{x}(t-1) + Bp(x(\cdot))$$

becomes a stable one. The abstract functional model of the system (10) uses the operator $\mathcal{A} : \mathcal{D}(\mathcal{A}) \to M_2$ defined by (c.f. (2))

$$\mathcal{A}\begin{pmatrix} y \\ z(\cdot) \end{pmatrix} = \begin{pmatrix} A_0 y + (A_1 + A_0 A_{-1}) z(-1) \\ \frac{\partial}{\partial\theta} z(\cdot) \end{pmatrix},$$

where $\mathcal{D}(\mathcal{A})$ is defined as before (see (2)).

With these notations (10) can be rewritten as

$$\frac{d}{dt}\begin{pmatrix} y(t) \\ z_t(\cdot) \end{pmatrix} = \mathcal{A}\begin{pmatrix} y(t) \\ z_t(\cdot) \end{pmatrix} + \mathcal{B}u(t), \tag{11}$$

where $\mathcal{B} = \begin{pmatrix} B \\ 0 \end{pmatrix}$ is a linear operator $\mathcal{B} : \mathbf{C}^n \to M_2$.

The spectrum $\sigma(\mathcal{A})$ is the set

$$\sigma(\mathcal{A}) = \sigma = \{\lambda |\det(\lambda I - A_{-1}\lambda e^{-\lambda} - A_0 - A_1 e^{-\lambda})\} = 0.$$

and consists of eigenvalues only. Denote further by $\sum$ the set of all nonzero eigenvalues of matrix A_{-1}. Then for any $\mu \in \sum$ the set σ includes a family of eigenvalues

$$\sum\nolimits^{\mu} = \{\lambda_k^{\mu} = \log|\mu| + i(\operatorname{Arg}\mu + 2\pi k) + \overline{o}(1), \quad k \in \mathbf{Z}\}, \tag{12}$$

where $\overline{o}$ is meant as $k \to \pm\infty$.

We assume that the following assumptions are satisfied.

$(a1)$ $\sum \subset \{w : |w| \leq 1\}$ and there exists $\mu \in \sum : |\mu| = 1$.

$(a2)$ All the eigenvalues $\mu \in \sum$ such that $|\mu| = 1$ are simple in the sense that there are no Jordan chains corresponding to such eigenvalues.

$(a3)$ Finite-dimensional system

$$\dot{x}(t) = A_0 x(t) + Bu(t), \quad x \in R^n, u \in R^r \tag{13}$$

is controlable, i.e. $\operatorname{rank}(B, A_0 B, ..., A_0^{n-1} B) = n$. In particular, this means that (3) is stabilizable, i.e. there exists a linear feedback control $u = P_0^0 x$ such that $\Re\,\sigma(A_0 + BP_0^0) < 0$.

$(a4)$ $\operatorname{rank}(A_1 + A_0 A_{-1}, B) = \operatorname{rank} B$.

As it is discussed in [18], we are mainly interested in the controls which are bounded with respect to operator $\mathcal{A}$ (for the definition and details see e.g. [9]) i.e. controls of the form

$$u = \mathcal{P}(x(\cdot)) = \int_{-1}^{0} \tilde{P}(\theta)\dot{x}(t+\theta)d\theta + \int_{-1}^{0} \hat{P}(\theta)x(t+\theta)d\theta,$$

where $\tilde{P}(\theta), \hat{P}(\theta), \theta \in [-1,0]$ are square-integrable $(r \times n)$-matrix-functions. This is the natural choice of controls to achieve the non-exponential stabilizability.

We have the following result.

Theorem 10. *Let the system (10) satisfy the assumptions* $(a1) - (a4)$. *Then this system is strongly stabilizable with the aid of feedback controls which are bounded with respect to operator* $\mathcal{A}$ *if and only if for an arbitrarily chosen matrix* P_0 *such that*

$$\sigma(A_0 + BP_0) \cap \log(\sum) \cap (i\mathbf{R}) = \emptyset$$

there do not exist an eigenvector g *of* A_{-1} *corresponding to an eigenvalue* $\mu \in \sum$, $|\mu| = 1$ *and* $k \in \mathbf{Z}$ *such that*

$$B^* R^*_{\lambda_k^\mu}(A_0 + BP_0)g = 0,$$

where λ_k^μ *is given by (12).*

Under this condition the strong stabilization can be achieved by the choice of control:

$$u = Q_1 x(t) + Q_2 x(t-1) + \int_{-1}^{0} Q_3(\theta)x(\theta)d\theta,$$

where Q_1, Q_2 *are constant* $(r \times n)$*-matrices,* Q_3 *is* $(r \times n)$*-matrix which elements belong to* $L_2(-1,0)$.

The proof is based on the analysis of the system (11) and using the Theorem 9.

5 Conclusions and Perspectives

In this note we presented a new approach for investigation the strong stability and stabilizability for systems of neutral type. We describe the main ideas and facts which form the background of our research. These facts are collected from the theory of differential equation with retarded argument (see e.g. [5, 6, 8]), the theory of semigroups as well as general operator theory (see e.g. [7, 9, 23, 24] and recent results from [18, 19]). We give examples emphasizing that our approach is widely applicable, perspective and extends the classical stability theory.

Acknowledgements. This work was realized with the financial support of Region Pays de la Loire, Ecole Centrale de Nantes (France) and Polish KBN Grant 5 PO3A 030 21.

References

1. Akhiezer N. I., Glazman I. M. (1993) Theory of linear operators in Hilbert space. Translated from the Russian and with a preface by Merlynd Nestell. Reprint of the 1961 and 1963 translations. Two volumes. Dover Publications, New York.
   Theory of linear operators in Hilbert space. Vol. I, II. Transl. from the 3rd Russian ed. Monographs and Studies in Mathematics, 9, 10. Edinburgh. Boston - London - Melbourne: Pitman Advanced Publishing Program. XXXII, 552 p.
2. Arendt W., Batty C. J. (1988) Tauberian theorems and stability of one-parameter semigrouos. Trans. Amer. Math. Soc. 306:837–852
3. Batty C. J., Phong V. Q. (1990) Stability of individual elements under one-parameter semigroups. Trans. Amer. Math. Soc. 322:805–818.
4. Brumley W. E. (1970) On the asymptotic behavior of solutions of differential-difference equations of neutral type. J. Differential Equations 7, 175-188.
5. Curtain R. F., Zwart H. (1995) An introduction to infinite-dimensional linear systems theory. Springer-Verlag, New York.
6. Diekmann O, van Gils S, Verduyn Lunel S. M., Walther H-O (1995) Delay equations. Functional, complex, and nonlinear analysis. Applied Mathematical Sciences, 110. Springer-Verlag, New York.
7. Gohberg I. C., Krein M. G. (1969) Introduction to the theory of linear nonselfadjoint operators (English) Translations of Mathematical Monographs. 18. Providence, RI: AMS. XV, 378 p.
8. Hale J., Verduyn Lunel S. M. (1993) Theory of functional differential equations. Springer-Verlag, New York.
9. Kato T (1980) Perturbation theory for linear operators. Springer Verlag.
10. Kolmanovskii V, Myshkis A (1999) Introduction to the theory and applications of functional differential equations., Mathematics and its Applications (Dordrecht). 463. Dordrecht: Kluwer Academic Publishers.
11. Korobov V. I., Sklyar G. M. (1984) Strong stabilizability of contractive systems in Hilbert space. Differentsial'nye Uravn. 20:1862–1869.
12. Krabs W., Sklyar G. M. (2002) On Controllability of Linear Vibrations. Nova Science Publ. Huntigton, N.Y.
13. Levan N., Rigby I. (1979) Strong stabilizability of linear contractive systems in Banach space. SIAM J. Control, 17:23–35.
14. Lyubich Yu. I., Phong V. Q. (1988) Asymptotic stability of linear differential equation in Banach space. Studia Math. 88:37–42.
15. O'Connor D. A, Tarn T. J. (1983) On stabilization by state feedback for neutral differential equations. IEEE Transactions on Automatic Control. Vol. AC-28, n. 5, 615–618.
16. Oostvenn J (1999) Strongly stabilizable infinite dimensional systems. Ph.D. Thesis. University of Groningen.
17. Pandolfi L. (1976) Stabilization of neutral functional differential equations. J. Optimization Theory and Appl. 20, n. 2, 191–204.
18. Rabah R., Sklyar G. M., On a class of strongly stabilizable systems of neutral type, (submitted).
19. Rabah R., Sklyar G. M., and Rezounenko A.V (2003) Generalized Riesz basis property in the analysis of neutral type systems, Comptes Rendus de l'Académie des Sciences, Série Mathématiques. To appear. (See also the extended version, Preprint RI02-10, IRCCyN, Nantes, France).
20. Sklyar G.M., Shirman V.Ya. (1982) On Asymptotic Stability of Linear Differential Equation in Banach Space. Teor, Funk., Funkt. Analiz. Prilozh. **37**: 127–132.

21. Sklyar G, Rezounenko A. (2001) A theorem on the strong asymptotic stability and determination of stabilizing control. C.R.Acad. Sci. Paris, Ser. I. 333:807–812.
22. Slemrod M (1973) A note on complete controllability and stabilizability for linear control systems in Hilbert space. SIAM J. Control 12:500–508.
23. Sz.-Nagy, B., Foias, C. (1970) Harmonic Analysis of Operators on Hilbert Space. Budapest: Akadémiai Kiadó; Amsterdam-London: North-Holland Publishing Company. XIII, 387 p.
24. van Neerven J. (1996) The asymptotic behaviour of semigroups of linear operators, in "Operator Theory: Advances and Applications", Vol. 88. Basel: Birkhauser.
25. Yamamoto Y, Ueshima S (1986) A new model for neutral delay-differential systems. Internat. J. Control 43(2):465–471.

Robust Delay Dependent Stability Analysis of Neutral Systems

Salvador A. Rodriguez[1], Jean-Michel Dion[1], and Luc Dugard[1]

[1]Laboratoire d'Automatique de Grenoble (INPG CNRS UJF) ENSIEG, BP 46, 38402, St. Martin d'Hères, FRANCE. Luc.Dugard@inpg.fr

Summary. This chapter focuses on the delay-dependent robust stability of linear neutral delay systems. The systems under consideration are described by functional differential equations, with norm bounded time varying nonlinear uncertainties in the "state", in the delayed "state" and norm bounded time varing quasilinear uncertainties in the difference operator. Two unknown constant delays, in the delayed "state" and in the difference operator, lead to consider a more general delay-dependent robust stability problem. The analysis is performed via Lyapunov-Krasovskii functional approach. The main difference with respect to [18] is that we obtain sufficient conditions for robust stability given in terms of the existence of positive definite solutions of LMIs. The proposed stability analysis extends some previous results on the subject.

1 Introduction

A great variety of systems can be modeled by time-delay systems [13], i.e. the "future" states depend not only on the "present" states, but also on the "delayed" states. Indeed, the delay naturally occurs in the dynamical behavior of systems in many fields: mechanics, physics, etc. Even if the systems themselves do not have internal delays, closed loop systems may involve delay phenomena, because of actuators, sensors and computation time.

Among systems with delays, the class of neutral systems is characterized by the fact that the delay argument occurs in the "state" and also in the derivative of the difference operator applied to the "state variable" $D(t)x_t$. Some examples of such neutral systems are given in [14], [1].

Several works have been concerned with the stability analysis of neutral systems either in the time domain approach, see for example: [9], [19], or in the frequency domain approach, see for example: [5], [19]. In these studies, the attention was mainly focused in giving conditions for delay independent stability, which are conservative when the delays are unknown. It is then of interest to consider delay-dependent stability analysis, see [5], [11].

In practice, the model parameters are not precisely known, leading to study the robustness of the stability w.r.t. uncertainties [12]. In this case, neutral systems can be represented by uncertain models, see for instance [14] (for example, lossless transmission line models may have uncertain parameters). Sufficient robust stability conditions have been obtained by the authors for neutral systems, but with only one delay parameter and with restrictive hypothesis on the neutral part [17] and in a much more general case in [18] but leading to non linear very complex conditions, difficult to check.

The objective of this chapter is to study the stability analysis of linear neutral systems in a delay-dependent framework incorporating robustness issues. The delays are assumed to be unknown and constant and the uncertainties may be time varying and nonlinear. Here we avoid the differentiability condition on the "state" (in general we suppose it is not differentiable) we transform a system with pointwise delays into a system with distributed delays by applying the Leibnitz's rule. One obtains sufficient delay-dependent stability conditions via the Lyapunov-Krasovskii functional approach. The systems under consideration are described by functional differential equations, with norm bounded time varing nonlinear uncertainties in the "state", in the delayed "state" and norm bounded time varing quasilinear uncertainties in the difference operator. The main contribution here with respect to [18] is that we obtain delay-dependent stability result expressed in terms of LMI.

This chapter is organized as follows: Section 2 gives some preliminaries and states the problem. Model transformation is discussed in section 3. The main stability result is given in section 4. In section 5, two examples are proposed to show the interest of the approach. Some final remarks end the chapter.

1.1 Notations

I_m is the identity matrix of dimension $m \times m$. $x \in \mathbb{R}^n$, $\|\cdot\|$ denotes the Euclidean norm of x. For a real number $r > 0$, $\mathcal{C}_r = \mathcal{C}([-r,0],\mathbb{R}^n)$ is the Banach space of continuous vector functions $\varphi : [-r,0] \to \mathbb{R}^n$ with the supremum norm $\|\varphi\|_{\mathcal{C}_r} = \sup_{-r \leq t \leq 0} \|\varphi(t)\|$. $\mathcal{C}_{r,\nu}$ denotes the open set $\varphi \in \mathcal{C}_r$ with $\|\varphi\|_{\mathcal{C}} < \nu$. $\mathcal{C}_r^1 = \mathcal{C}^1([-r,0],\mathbb{R}^n)$ denotes the Banach space of continuous differentiable functions $\varphi : [-r,0] \to \mathbb{R}^n$ with the norm $\|\varphi\|_{\mathcal{C}_r^1} = \|\varphi\|_{\mathcal{C}_r} + \|\dot{\varphi}\|_{\mathcal{C}_r}$. The function x_t denotes the restriction of x to the interval $[t-r,t]$ so that x_t is an element of $\mathcal{C}_r$ defined by $x_t(\theta) = x(t+\theta)$ for $-r \leq \theta \leq 0$.

1.2 Problem statement

Consider the following system, written in the form proposed by J.K. Hale and M.A. Cruz [8], [9], [14]:

$$\frac{d}{dt}[D(t)x_t] = f(t,x_t), t \geq \sigma, \tag{1}$$

$$D(t)\varphi := [\varphi(0) - g(t,\varphi)], \tag{2}$$

with

$$x_\sigma \equiv \phi, \phi \in C_r, \tag{3}$$

where the state x_t is a functional in C_r and $f, g : [\sigma, \infty) \times C_r \to \mathbb{R}^n$ are continuous.

Since the initial value problem for (1)-(3), in general, does not have a solution [8], system (1)-(3) is defined as a neutral system if $g(t, \varphi)$ is non-atomic at zero. The concept insures that the function $g(t, \varphi)$ does not depend very strongly upon $\varphi(0)$. General existence, uniqueness and continuous dependence theorems have been given in [8] for System (1)-(3) under the hypothesis that $g(t, \varphi)$ is non-atomic at zero.

Definition 1. *[8]Suppose that $\mathcal{H}$ is an open set in $\mathbb{R}^+ \times C([\sigma, \infty), \mathbb{R}^n)$. A neutral system is a system of the form (1)-(3) for which $f, g : \mathcal{H} \to \mathbb{R}^n$ are continuous and g is nonatomic at zero on $\mathcal{H}$.*

In order to define a neutral system, suppose also that f, g are uniformly bounded in t for φ in closed bounded sets of C_r, that g is linear in the second argument and that there are a $n \times n$ matrix $\mu(t, \theta)$, $t \in [\sigma, \infty)$, $\theta \in [-r, 0]$, of bounded variation in θ and a scalar continuous function $l(s)$ nondecreasing for $s \in [0, r]$, $l(0) = 0$ such that

$$\begin{array}{l} g(t, \varphi) = \int_{-r}^{0} d_\theta[\mu(t, \theta)]\varphi(\theta), \\ \| \int_{-s}^{0} d_\theta[\mu(t, \theta)]\varphi(\theta)\| \le l(s) \sup_{-s \le \theta \le 0} \|\varphi(\theta)\|, \end{array} \tag{4}$$

for all t, φ in C.

We denote the solutions of the system (1)-(3) by $x(\sigma, \phi)$ where $x_\sigma(\sigma, \phi) \equiv \phi$. The value of $x(\sigma, \phi)$ at t is denoted by $x(t) = x(t; \sigma, \phi)$.

Let us recall some definitions of stability.

Definition 2. *[9] The zero solution of the system (1)-(3) is said to be stable if, for each $\varepsilon > 0$ and $\sigma \ge 0$, there exists $\delta = \delta(\sigma, \varepsilon) > 0$ such that if $\|\phi\|_{C_r} < \delta$, then $\|x(t; \sigma, \phi)\| < \varepsilon$ for all $t \ge \sigma$. The zero solution of (1)-(3) is asymptotically stable if it is stable and the value δ can be chosen independently of σ such that, in addition, $\|x(t; \sigma, \phi)\| \to 0$ as $t \to +\infty$ holds.*

Definition 3. *[6] Consider $H \in C([\sigma, \infty), \mathbb{R}^n)$ and the equation*

$$D(t)x_t = D(t)\phi + H(t) - H(\sigma), x_\sigma = \phi, t \ge \sigma. \tag{5}$$

Suppose that $\mathcal{H}$ is a subset of $C([\sigma, \infty), \mathbb{R}^n)$. The operator $D(t)$ is said to be uniformly stable with respect to $\mathcal{H}$ if there are constants K, Λ such that for any $\phi \in C_r$, $\sigma \in \mathbb{R}^+$, and $H \in \mathcal{H}$, the solution $x(t; \sigma, \phi, H)$ of (5) satisfies

$$\|x_t(\sigma, \phi, H)\|_{C_r} \le K\|\phi\|_{C_r} + \Lambda \sup_{\sigma \le \tau \le t} | H(\tau) - H(\sigma) |, t \ge \sigma. \tag{6}$$

Let $V : \mathbb{R}^+ \times C_r \to \mathbb{R}^+$ be a continuous functional; the upper right-hand derivative of V along the solution of the System (1)-(3) is defined by

$$\dot{V}(t, \phi) = \lim_{h \to 0+} \sup \frac{1}{h}[V(t + h, x_{t+h}(t, \phi)) - V(t, \phi)] \tag{7}$$

We will use the following Lyapunov-Krasovskii functional approach:

Theorem 1. *[6] Consider the neutral system (1)-(3). Assume that $D(t)$ is uniformly stable with respect to C_r and that there exist non decreasing continuous functions $v_i : \mathbb{R}^+ \to \mathbb{R}^+$, $i = 1, 2, 3$ such that $v_i(0) = 0$ and $v_i(s) > 0$, for all $s > 0$ and $i = 1, 2, 3$. Then, the zero solution of (1)-(3) is asymptotically stable if there exists a continuous functional $V : \mathbb{R}^+ \times C_r \to \mathbb{R}^+$ such that:*

i) $v_1(\|D(t)\varphi\|) \le V(t, \varphi) \le v_2(\|\varphi\|_{C_r})$ *and*
ii) $\dot{V}(t, x_t) \le -v_3(\|D(t)x_t\|)$, *for all* $t \ge \sigma$.

The present chapter considers the following fairly general class of uncertain neutral systems (1)-(3), $f(t, x_t) = Ax(t) + Bx(t - r_2) + \Delta_A(t, x_t) + \Delta_B(t, x_t)$, $g(t, x_t) = Cx(t - r_1) + \Delta_C(t, x_t)$ (assumed non-atomic at zero), linear neutral differential equations that include continuous quasilinear and nonlinear uncertainties. The delays r_1 and r_2 are assumed to be non negative constants and unknown,

$$\begin{gathered} \tfrac{d}{dt}[D(t)x_t] = Ax(t) + Bx(t - r_2) + F(t, x_t), t \ge \sigma \\ D(t)\varphi := [\varphi(0) - C\varphi(-r_1) - \Delta_C(t, \varphi)], \\ F(t, \varphi) := \Delta_A(t, \varphi) + \Delta_B(t, \varphi), \end{gathered} \tag{8}$$

with

$$x_\sigma \equiv \phi, \{\phi, \varphi\} \in C_r, \tag{9}$$

where $x_t = \{x(t + \theta) : \theta \in [-r, 0], r := \max\{r_1, r_2\}\}$. The known matrices A, B, and C are constant and the unknown nonlinear mappings $\Delta_A, \Delta_B : \mathbb{R} \times C_{r,\nu} \to \mathbb{R}^n$, and the linear mapping in the second argument $\Delta_C : \mathbb{R} \times C_{r,\nu} \to \mathbb{R}^n$ take closed bounded sets into bounded sets. They are described by

$$\begin{gathered} \Delta_A(t, \varphi(0)) := E_A \delta_A(t, \varphi(0)), \\ \delta_A^T(t, \varphi(0))\delta_A(t, \varphi(0)) \le \varphi^T(0) W_A^T W_A \varphi(0), \\ \Delta_B(t, \varphi(-r_2)) := E_B \delta_B(t, \varphi(-r_2)), \\ \delta_B^T(t, \varphi(-r_2))\delta_B(t, \varphi(-r_2)) \le \varphi^T(-r_2) W_B^T W_B \varphi(-r_2), \\ \Delta_C(t, \varphi(-r_1)) := \delta_C(t)\varphi(-r_1), \\ W_C + \delta_C(t) \ge 0, W_C - \delta_C(t) \ge 0 \\ \forall (t, \varphi) \in \mathbb{R}^+ \times C_{r,\nu}, \end{gathered} \tag{10}$$

where the matrices E_A, E_B are known, the matrices W_A, W_B and W_C are real given weighting matrices, and the unknown mappings δ_A and δ_B satisfy

$$\delta_A(t, 0) \equiv 0, \delta_B(t, 0) \equiv 0, \tag{11}$$

so that $x = 0$ is a solution of the neutral differential equation (8)-(9). Notice that the uncertainties on Δ_C are unstructured, this will allow us to get stability, results using LMI tools.

Remark 1. Note that $r_1 > 0$ implies $C\varphi(-r_1) + \Delta_C(t, \varphi)$ depends only upon values of $\varphi(0)$, conditions (10) allow to satisfy (4), then $Cx(t - r_1) + \Delta_C(t, \varphi)$ is nonatomic at zero in $\mathcal{H}$ [8].

If it is required to have smooth solutions of (8)-(9), then assume that the initial function ϕ is in C_r^1 and satisfies the sewing condition

$$\dot{\phi}(0^-) = A\phi(0) + B\phi(-r_2) + C\dot{\phi}(-r_1) + \Delta_A(\sigma,\phi) + \Delta_B(\sigma,\phi) + \dot{\Delta}_C(\sigma,\phi) \quad (12)$$

This condition implies the continuity of $\dot{x}(t)$ for $t \geq \sigma$ and then $x(t)$ is differentiable on $(\sigma - r, \infty)$.

Two important stability problems are associated to the neutral system (8)-(11) (see also [17]):

Robust stability problem: Find conditions, if they exist, to ensure the asymptotic stability (*independent of the delays*) of the neutral system (8)-(11) for all unknown functions Δ_A, Δ_B and Δ_C

In [10] a similar simpler problem was presented with no uncertainty on C, $(\Delta C(\cdot,\cdot) \equiv 0)$.

Delay-dependent stability problem: Find bounds r_1^*, r_2^*, if they exist, on the delays r_1, r_2, such that the asymptotic stability of (8)-(11) when Δ_A, Δ_B and Δ_C are identically zero, is preserved for $r_1 \leq r_1^*$ and for $r_2 \leq r_2^*$.

In [11] and [17] the delay-dependent stability problem was performed only in the delay r_2, not in r_1.

Now consider the scalar neutral system [18]

$$\frac{d}{dt}[x(t) - r_1 c x(t-r_1)] = -ax(t) - bx(t-r_2), t \geq \sigma, x_\sigma \equiv \phi, \phi \in C_r. \quad (13)$$

where a, b, $c \neq 0$, and r_1, r_2 are constant. Of course, since $x_t \in C_r$, this system can be written in the form (8)-(11) by the Riesz representation theorem but in this case the difference operator depends on r_1. In the particular case $r = r_1 = r_2$, the stability of the difference operator $Dx_t := x(t) - rcx(t-r)$ is directly linked to some upper bound on the delay $r^* = |r| < 1/|c|$. In this case the stability will depend on r.

Remark 2. In example (13) the stability of the difference operator does depend on r because the coefficient of the "delay differentiated state" contains the delay r_1. When it is not the case, it is well known that the delay-stability depends only in the delay of the "delay state", r_2 [11].

In this chapter, we consider the following mixed problem:

Delay-dependent robust stability problem: Find bounds r_1^*, r_2^* if they exist, on r_1, r_2, and conditions to ensure the asymptotic stability of the neutral system (8)-(11), for $r_1 \leq r_1^*$ and for $r_2 \leq r_2^*$ for any Δ_A, Δ_B, Δ_C.

Different methods have been considered to study the stability of the solutions of such systems [14] and among them, the direct Lyapunov method (Razumikhin or

Krasovskii approaches) [9], [14]. It reduces the stability problem to the construction of appropriate functionals V, defined along the systems solutions. We use this methodology to study robust stability and delay-dependent stability problems simultaneously. The delay-dependent robust stability conditions are given in terms of the existence of positive definite solutions for some matrix inequality.

2 Model Transformation

Some authors introduce a transformation to prove stability, for example transform a time delay system $\dot{x}(t) = f(t, x_t)$ into a neutral time delay system written in the Hale's form by integration over one delay interval [15]:

$$\dot{x}(t) = f(t, x_t) \rightarrow \frac{d}{dt}\left[x(t) - \int_{-r}^{0} f(t+\theta, x_{t+\theta})\, d\theta\right] = f(t-r, x_{t-r}).$$

Or sometimes, when possible, transform a neutral system [8]:

$$\dot{x}(t) = G'_t(t, x_t) + G'_{x_t}(t, x_t)\, \dot{x}_t + F(t, x_t),$$

into a neutral system written in the Hale's form:

$$\frac{d}{dt}\left[x(t) - G(t, x_t) - \int_{-r}^{0} F(t+\theta, x_{t+\theta})\, d\theta\right] = F(t-r, x_{t-r}).$$

Or finally the Leibnitz'rule in the state x_t [11]

$$x(t) - x(t-r) = \int_{-r}^{0} \dot{x}(t+\theta)\, d\theta.$$

However, in all these transformations, the "state" is supposed to be differentiable. In this chapter, we avoid this restriction.

Let us consider now the neutral system (8)-(11) :

$$\frac{d}{dt}D(t)x_t = Ax(t) + Bx(t-r_2) + F(t, x_t). \tag{14}$$

From the Leibnitz's rule in the difference operator, $D(t)x_t - D(t-r_2)x_{t-r_2} = \int_{-r_2}^{0} d_\theta[D(t+\theta)x_{t+\theta}]$, holds for $t \geq r + r_2$

$$\begin{gathered} x(t-r_2) = x(t) - Cx(t-r_1) - \Delta_C(t, x_t) \\ + Cx(t-r_1-r_2) + \Delta_C(t-r_2, x_{t-r_2}) - \int_{-r_2}^{0} \tfrac{d}{dt}[D(t+\theta)x_{t+\theta}]d\theta \end{gathered} \tag{15}$$

since $D(t)x_t$ is continuously differentiable.

Remark 3. Here we state the problem with respect to the space $\mathcal{C}([-r, 0], \mathbb{R}^n)$, but the previous transformation holds in other particular spaces, for example in the Sobolev space $\mathcal{W}^{1,p}([-r, 0], \mathbb{R}^n)$, proposed in [10]. Or even in more general cases, for example using the product space $\mathbb{R}^n \times \mathcal{L}^p([-r, 0], \mathbb{R}^n)$, proposed by [4], where Dx_t is absolutely continuous on $[0, \infty]$ (with the integral in (15) in the sense of the Lebesgue Integral [7]).

Then equation (8) can be rewritten in the new variable ξ as (see [12], [17])

$$\begin{aligned} \tfrac{d}{dt}[D(t)\xi_t] = (A+B)\xi(t) - BC\xi(t-r_1) - B\Delta_C(t,\xi_t(-r_1)) \\ +BC\xi(t-r_1-r_2) + B\Delta_C(t-r_2,\xi_{t-r_2}(-r_1)) \\ -B\int_{-r_2}^{0}[A\xi(t+\theta) + B\xi(t+\theta-r_2) + F(t+\theta,\xi_{t+\theta})]d\theta + F(t,\xi_t). \end{aligned} \tag{16}$$

It is not very difficult to check that every solution of neutral system (8) is also solution of the equation (16), then the stability of (16) implies the stability of (8) [12].

If a function satisfies (8) on some interval and has a sufficiently smooth derivative (assume that the initial function ϕ is in C_r^1 and the sewing condition (12) is fulfilled), then carrying out the differentiation in (8) leads to

$$\frac{d}{dt}[D(t)\varphi] = \dot{\varphi}(0) - C\dot{\varphi}(-r_1) - \Delta'_{C,t}(t,\varphi) - \Delta'_{C,\varphi}(t,\varphi)\psi \tag{17}$$

where $\Delta'_{C,t}(t,\varphi)$ is the partial derivative with respect to "t" and $\Delta'_{C,\varphi}(t,\varphi)$ is the derivative with respect to "φ".

In the next section, the stability analysis of the transformed model (16) is performed in terms of the variable x.

3 Delay-Dependent Robust Stability

In this section, as in [18], one uses the Lyapunov-Krasovskii functional approach, with the help of theorem 1 to prove the main result of this chapter, given in the next theorem:

Theorem 2. *Consider the Neutral System (8)-(11). If the following conditions are satisfied:*

i) $A_1 := A + B$ *is a Hurwitz stable matrix;*
ii) The difference operator $D(t)\varphi := [\varphi(0) - C\varphi(-r_1) - \Delta_C(t,\varphi)]$ *is linear in* φ, *continuous and uniformly stable with respect to* C_r *and* $\Delta_C(t,\varphi)$ *is nonatomic at zero;*
iii) there exist a real positive number r_2^* *and positive definite matrices* P, $S_i > 0$, $i = \overline{1,7}$ *such that the following inequality holds:*

$$\Gamma := \begin{pmatrix} Q(r_2^*) & \Omega_{12} & 0 & \Omega_{14} & \overline{S}(r_2^*) \\ \Omega_{12}^T & \Omega_{22} & 0 & 0 & 0 \\ 0 & 0 & \Omega_{33} & 0 & 0 \\ \Omega_{14}^T & 0 & 0 & \Omega_{44} & 0 \\ \overline{S}^T(r_2^*) & 0 & 0 & 0 & R \end{pmatrix} < 0, \tag{18}$$

where

$$Q(r_2^*) := \Omega_{11} + \overline{S}(r_2^*) R^{-1} \overline{S}^{\mathsf{T}}(r_2^*), \tag{19}$$

$$\Omega_{11} := PA_1 + A_1^{\mathsf{T}} P + 2S - \overline{S}(r_2^*) R^{-1} \overline{S}^{\mathsf{T}}(r_2^*) \tag{20}$$

$$S := W_A^T W_A + \sum_{i=1}^{2} S_i + r_2^* \sum_{i=3}^{5} S_i, \tag{21}$$

$$\overline{S}\,(r_2^*) := \left(\overline{S}_1\ \overline{S}_2\,(r_2^*)\right), \tag{22}$$

$$\overline{S}_1 := \left(PA_1,\ \sqrt{2}PB,\ PE_A,\ PE_B\right), \tag{23}$$

$$\overline{S}_2\,(r_2^*) := \sqrt{r_2^*}\left(PBE_A,\ PBE_B,\ PBA,\ PB^2\right), \tag{24}$$

$$R^{-1} := \begin{pmatrix} -I_{6n} & 0 & 0 \\ 0 & -S_5^{-1} & 0 \\ 0 & 0 & -S_6^{-1} \end{pmatrix}, \tag{25}$$

$$\Omega_{12} := \left(PA + W_A^\top W_A + \sum_{i=1}^{2} S_i + r_2^* \sum_{i=3}^{5} S_i\right) C, \tag{26}$$

$$\Omega_{22} := 2W_C^\top W_C - S_1 + S_7 + C^\top SC + 3W_C^\top SW_C \tag{27}$$

$$\Omega_{33} := W_B^\top W_B - S_2 + r_2^* S_6, \tag{28}$$

$$\Omega_{44} := W_C^\top W_C - S_7, \tag{29}$$

$$\Omega_{14} := PBC, \tag{30}$$

Then the Neutral System (8)-(11) is robustly delay-dependent asymptotically stable for any $r_2 \leq r_2^*$.

Remark 4. In [18] only in the case where $\Delta_C \equiv 0$ the main result can be checked in terms of a LMI, however in this work condition 3: (18)-(21) can be checked always by using LMI Tools [2]. Notice that condition 1 is necessary and directly follows from the satisfaction of condition 3. Condition 2 consists in checking the uniformly stability of $D(t)$ with respect to $\mathcal{C}([\sigma,\infty),\mathbb{R}^n)$ see Definition 3 [6].

Remark 5. $C + W_C < I$ is a sufficient condition to have an operator $D(t)$ uniformly stable with respect to $\mathcal{C}([\sigma,\infty),\mathbb{R}^n)$.

Proof: Consider the Neutral System (8)-(11) and the Lyapunov-Krasovkii functional

$$V(t,\varphi) := V_1(t,\varphi) + V_2(\varphi) + V_3(\varphi), \varphi \in \mathcal{C}_r, \tag{31}$$

where

$$V_1(t,\varphi) := D^T(t)\varphi P D(t)\varphi, \tag{32}$$

$$V_2(\varphi) := \textstyle\sum_{i=1}^{2} \int_{-r_i}^{0} \varphi^T(\theta) S_i \varphi(\theta) d\theta + \int_{-r_2}^{0} [\int_{\theta}^{0} \varphi^T(\vartheta) S_3 \varphi(\vartheta) d\vartheta] d\theta, \tag{33}$$

$$\begin{aligned} V_3(\varphi) := &\textstyle\int_{-r_2}^{0} [\int_{\theta-r_2}^{0} \varphi^T(\vartheta) S_4 \varphi(\vartheta) d\vartheta] d\theta + \int_{-r_1-r_2}^{-r_1} \varphi^T(\theta) S_7 \varphi(\theta) d\theta, \\ &+ \textstyle\sum_{i=5}^{6} \int_{-r_2}^{0} [\int_{\theta}^{0} \psi_i^T(\vartheta) S_i \psi_i(\vartheta) d\vartheta] d\theta \end{aligned} \tag{34}$$

and the functionals, ψ_i, are $\psi_5 := \varphi$, $\psi_6(\vartheta) := \varphi(\vartheta - r_2)$. For the functional V, we can construct $\dot{V}$ along the trajectories of (16) in terms of x, if φ is replaced by x_t in the right hand side of $V(\varphi)$, pass to x, differentiate in t (so that $\dot{D}(t)x_t$ appears only for the current (not delayed) value of t), substitute $\dot{D}(t)x_t$ from (16), pass from x to x_t, and replace x_t by φ. With this procedure, the expressions for V and $\dot{V}$ are often straightforwardly written in terms of x, keeping in mind the transition from x to φ, as proposed by [14]. Then we have:

$$\dot{V}_1(t,x_t) = D^T(t)x_tP\frac{dD(t)x_t}{dt} + \frac{dD^T(t)x_t}{dt}PD(t)x_t, \tag{35}$$

$$\begin{aligned}\dot{V}_2(x_t) = \textstyle\sum_{i=1}^{2}[x^T(t)S_ix(t) - x^T(t-r_i)S_ix(t-r_i)] \\ + \textstyle\int_{-r_2}^{0}[x_t^T(0)S_3x_t(0) - x_{t+\theta}^T(0)S_3x_{t+\theta}(0)]d\theta,\end{aligned} \tag{36}$$

$$\begin{aligned}\dot{V}_3(x_t) = \textstyle\int_{-r_2}^{0}[-x_{t+\theta}^T(-r_2)S_4x_{t+\theta}(-r_2) + x_t^T(0)S_4x_t(0)]d\theta \\ +x^T(t-r_1)S_7x(t-r_1) - x^T(t-r_1-r_2)S_7x(t-r_1-r_2) \\ + \textstyle\int_{-r_2}^{0}[x_t^T(0)S_5x_t(0) - x_{t+\theta}^T(0)S_5x_{t+\theta}(0)+ \\ +x_t^T(-r_2)S_6x_t(-r_2) - x_{t+\theta}^T(-r_2)S_6x_{t+\theta}(-r_2)]d\theta\end{aligned} \tag{37}$$

Substituting (16) in (35) and setting $A_1 := A + B$ leads to:

$$\begin{aligned}\dot{V}_1(t,x_t) = D^T(t)x_tPA_1x(t) + x^T(t)A_1^TPD(t)x_t \\ +2D^T(t)x_tP[E_A\delta_A(t,x_t) + E_B\delta_B(t,x_t)] \\ -D^T(t)x_tPBC[x(t-r_1) - x(t-r_1-r_2)] \\ -[x^T(t-r_1) - x^T(t-r_1-r_2)]C^TB^TPD(t)x_t \\ -2D^T(t)x_tPBE_C[\delta_C(t)x(t-r_1) - \delta_C(t-r_2)x(t-r_1-r_2)] \\ -2D^T(t)x_tPB\textstyle\int_{-r_2}^{0}[Ax_{t+\theta}(0) + Bx_{t+\theta}(-r_2) \\ +E_A\delta_A(t+\theta,x_{t+\theta}(0)) + E_B\delta_B(t+\theta,x_{t+\theta}(-r_2))]d\theta.\end{aligned} \tag{38}$$

Using the following well known inequality:

$$-2a^Tb \leq \inf_{X>0}\{a^TXa + b^TX^{-1}b, \forall a,b \in \mathbb{R}^n\}, \tag{39}$$

assumption (10) and positive square matrices R_1 and R_2, we have directly the following inequalities

$$2D^T(t)x_tPE_B\delta_B(t,x_t) \leq D^T(t)x_tPE_BE_B^TPD(t)x_t + x_t^T(-r_2)W_B^TW_Bx_t(-r_2) \tag{40}$$

$$\begin{aligned}-2D^T(t)x_tPB\textstyle\int_{-r_2}^{0}Bx_{t+\theta}(-r_2)d\theta \leq \\ \textstyle\int_{-r_2}^{0}[D^T(t)x_tPB^2R_2^{-1}(B^2)^TPD(t)x_t \\ +x_{t+\theta}^T(-r_2)R_2x_{t+\theta}(-r_2)]d\theta,\end{aligned} \tag{41}$$

and similar inequalities for δ_A and R_1,

$$\begin{aligned}-2D^T(t)x_tPB\textstyle\int_{-r_2}^{0}E_A\delta_A(t+\theta,x_{t+\theta}(0))d\theta \leq \\ \textstyle\int_{-r_2}^{0}[D^T(t)x_tPBE_AE_A^TB^TPD(t)x_t \\ +x_{t+\theta}^T(0)W_A^TW_Ax_{t+\theta}(0)]d\theta,\end{aligned} \tag{42}$$

similar inequalities for δ_B and R_2.

The inequalities (40)-(42) allow to get a bound for $\dot{V}_1$. Now we choose $R_1 := S_5$, $R_2 := S_6$,

$$\begin{cases} S_3 = W_A^T W_A, \text{if } W_A^T W_A > 0, \\ \quad S_3 > W_A^T W_A, \text{otherwise}, \end{cases} \tag{43}$$

$$\begin{cases} S_4 = W_B^T W_B, \text{if } W_B^T W_B > 0, \\ \quad S_4 > W_B^T W_B, \text{, otherwise} \end{cases} \tag{44}$$

and using equalities (36) and (37), we derive the following bound for $\dot{V}$:

$$\begin{gathered} \dot{V}(t,x_t) \leq D^T(t)x_t PA_1 x(t) + x^T(t-r_1)S_7 x(t-r_1) \\ +x^T(t)A_1^T PD(t)x_t - x^T(t-r_1-r_2)S_7 x(t-r_1-r_2) \\ -D^T(t)x_t PBC[x(t-r_1) - x(t-r_1-r_2)] \\ -[x^T(t-r_1) - x^T(t-r_1-r_2)]C^T B^T PD(t)x_t \\ +D^T(t)x_t PE_A E_A^T PD(t)x_t + x^T(t)W_A^T W_A x(t) \\ +D^T(t)x_t PE_B E_B^T PD(t)x_t + x^T(t-r_2)W_B^T W_B x(t-r_2) \\ +2D^T(t)x_t PBE_C E_C^T B^T PD(t)x_t + x^T(t-r_1)W_C^T W_C x(t-r_1) \\ +x^T(t-r_1-r_2)W_C^T W_C x(t-r_1-r_2) \\ +\int_{-r_2}^0 D^T(t)x_t PBAS_5^{-1}A^T B^T PD(t)x_t d\theta \\ +\int_{-r_2}^0 D^T(t)x_t PB^2 S_6^{-1}(B^2)^T PD(t)x_t d\theta \\ +\int_{-r_2}^0 D^T(t)x_t PBE_A E_A^T B^T PD(t)x_t d\theta \\ +\int_{-r_2}^0 D^T(t)x_t PBE_B E_B^T B^T PD(t)x_t d\theta \\ +\sum_{i=1}^2 [x^T(t)S_i x(t) - x^T(t-r_i)S_i x(t-r_i)] \\ +r_2 \sum_{i=3}^5 x^T(t)S_i x(t) + r_2 x^T(t-r_2)S_6 x(t-r_2). \end{gathered} \tag{45}$$

Using the following identities with $S = W_A^T W_A + \sum_{i=1}^2 S_i + r_2 \sum_{i=3}^5 S_i$, we can rewrite each expression containing the vector function $x(t)$ in (45) as an expression containing $\mathcal{D}x_t$ and $x(t-r_1)$:

$$\begin{gathered} x^T(t)Sx(t) = D^T(t)x_t SD(t)x_t + D^T(t)x_t SCx(t-r_1)+ \\ +x^T(t-r_1)C^T SD(t)x_t + x^T(t-r_1)C^T SCx(t-r_1) \\ +2D^T(t)x_t S\Delta_C(t,x_t) + 2x^T(t-r_1)C^T S\Delta_C(t,x_t) \\ +\Delta_C^T(t,x_t)S\Delta_C(t,x_t), \end{gathered} \tag{46}$$

$$\begin{gathered} D^T(t)x_t PA_1 x(t) + x^T(t)A_1^T PD(t)x_t = \\ D^T(t)x_t(PA_1 + A_1^T P)D(t)x_t + D^T(t)x_t PA_1 Cx(t-r_1) \\ +x^T(t-r_1)C^T A_1^T PD(t)x_t + 2D^T(t)x_t PA_1\Delta_C(t,x_t), \end{gathered} \tag{47}$$

and we can overbound the terms

$$\begin{gathered} 2D^T(t)x_t S\Delta_C(t,x_t) \leq D^T(t)x_t SD(t)x_t \\ +x^T(t-r_1)W_C^T SW_C x(t-r_1), \end{gathered} \tag{48}$$

$$\Delta_C^T(t,x_t)S\Delta_C(t,x_t) \leq x^T(t-r_1)W_C^T SW_C x(t-r_1), \tag{49}$$

similarly for $2x^T(t-r_1)C^T S\Delta_C(t,x_t)$, $2D^T(t)x_t PA_1\Delta_C(t,x_t)$. Note that (49) is obtained directly from the fact that $(W_C+\delta_C(t))^T S(W_C-\delta_C(t))\geq 0$ by assumption (10).

With all these inequalities and identities, if there exists a real positive number $r_2^*\geq r_2$ such that the Matrix Inequality (18) holds, then (18) is equivalent to:

$$\dot{V}(t,x_t)\leq \omega^T\Omega\omega<0, \tag{50}$$

where the vector ω is

$$\omega:=\begin{pmatrix} D(t)x_t \\ x(t-r_1) \\ x(t-r_2) \\ x(t-r_1-r_2)\end{pmatrix} \tag{51}$$

and the matrix $\Omega:=(\Omega_{i,j}), i=1,...,4, j=1,...,4$ is defined in Theorem 2. This fact follows via an appropriate Schur transformation [2].

The relation (50) means $\Omega<0$, $\Omega_{1,1}<0$ and that there exists some $\gamma>0$ such that $\dot{V}(t,x_t)\leq-\gamma\|D(t)x_t\|$ for all $t\geq\sigma$. D is stable by Assumption (2), see [6], then, the robust asymptotic stability of (8)-(11) is ensured by Theorem 1 for all delay $r_2\leq r_2^*$, [6].

Remark 6. When conditions of Theorem 2 are satisfied, the robustly delay-dependent asymptotically stability of (8)-(11) is ensured for all $r_2\leq r_2^*$ and for any r_1 positive. However Theorem 2 can be applied to analyze stability of (13) where the neutral coefficient cr_1 depends on r_1. In this case, the stability will depend on r_1; so, r_1 should be such that condition 2 of Theorem 2 is satisfied.

Remark 7. When the uncertainties $\Delta A(\cdot,\cdot)$, $\Delta B(\cdot,\cdot)$, ΔC are zero, we can choose the Lyapunov-Krasovskii functional (31) $V(\varphi):=V_1(\varphi)+V_2(\varphi)+V_3(\varphi)$ with V_1 defined in (32) and

$$V_2(\varphi):=\int_{-r}^{0}\varphi^T(\theta)S_1\varphi(\theta)d\theta+\int_{-r_2}^{0}[\int_{\theta}^{0}\varphi^T(\vartheta)S_3\varphi(\vartheta)d\vartheta]d\theta, \tag{52}$$

$$V_3(\varphi):=\int_{-r_2}^{0}[\int_{\theta-r_2}^{0}\varphi^T(\vartheta)S_4\varphi(\vartheta)d\vartheta]d\theta+\int_{-r_1-r_2}^{-r_1}\varphi^T(\theta)S_7\varphi(\theta)d\theta, \tag{53}$$

leading to the delay dependant result without uncertainties of [11].

Remark 8. If we are interested only in the delay independent robust stability of Neutral System (8)-(11), the model transformation can be avoided; choosing $V(\varphi):=V_1(\varphi)+V_2(\varphi)$ with V_1 defined in (32) and

$$V_2(\varphi):=\sum_{i=1}^{2}\int_{-r_i}^{0}\varphi^T(\theta)S_i\varphi(\theta)d\theta, \tag{54}$$

leads to similar results proposed in [10].

Remark 9. If the difference operator is defined by $D(t)=D$, $\Delta_C(t,\varphi)=\Delta C\varphi(t-r_1)$, ΔC a constant matrix and if we suppose φ is right-hand differentiable ($\varphi\in C_r^1$), then we recover the results of [17].

4 Examples

We will illustrate robust stability results on two simple neutral systems with no uncertainty in the C matrix.

4.1 Consider the following linear neutral system:

$$\frac{d}{dt}\left[x(t) - \begin{pmatrix} 0.1 & 0 \\ 0 & 0.1 \end{pmatrix} x(t-r_1)\right] = -\begin{pmatrix} 2 & 0 \\ 0 & 0.9 \end{pmatrix} x(t) - \begin{pmatrix} 1 & 0 \\ 1 & 1 \end{pmatrix} x(t-r_2) \quad (55)$$

$$y(t) = \begin{pmatrix} 0 & 1 \end{pmatrix} x(t-r_2) \quad (56)$$

Note first that, when $C \equiv 0$, the time delay system is not asymptotically stable independently of the size of the delay r_2 if $A+B$ is stable and $A-B$ is not (condition obtained for the system free of delay). Equivalently, in the neutral system case, for $r_1 \equiv 0$, if the matrix $(I-C)^{-1}(A+B)$ is Hurwitz stable and $(I-C)^{-1}(A-B)$ is unstable, then the system cannot be stable independently of the delay r_2.

In this example, let us consider $r_1 = r_2$. For uncertainties characterized by $W_A = W_B = 0.15 I_2$, the sufficient condition provided by Theorem 2 leads to $r_2^* = 0.5s$, a feasible solution for the LMI (18) of Theorem 2. In Figure 1 a simulation is presented for system (55)-(56) with initial condition $\phi_1(\theta) = -0.75\theta + 1$, $\phi_2(\theta) = -16.625\theta + 1$. Under this initial condition, the sewing condition (12) holds if $W_A = W_B = 0$, and x is differentiable for $t > 0$; however, in the uncertain case there is a small jump at $t = 0$.

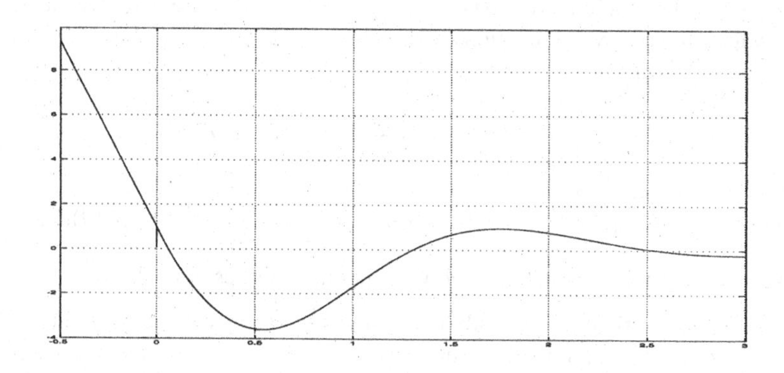

Fig. 1. $-4 \leq y(t) \leq 10$, $-0.5 \leq t \leq 3$

When the case without uncertainties is considered ($W_A = W_B = 0$) the solution given by the theorem is r_1^*, any finite positive bound, and $r_2^* = 0.894s$.

4.2 The second example is a realistic neutral system problem motivated by the small PEEC (Partial Element Equivalent Circuit) model studied in [1]: $\frac{d}{dt}[x(t) -$

$Cx(t-r)] = Ax(t) + Bx(t-r)$. It is characterized by the following A, B, C matrices:

$$\frac{A}{100} = \begin{pmatrix} -7 & 1 & 2 \\ 3 & -9 & 0 \\ 1 & 2 & -6 \end{pmatrix}, \frac{B}{100} = \begin{pmatrix} 1 & 0 & -3 \\ -0.5 & -0.5 & -1 \\ -0.5 & -1.5 & 0 \end{pmatrix}, C = \frac{1}{72}\begin{pmatrix} -1 & 5 & 2 \\ 4 & 0 & 3 \\ -2 & 4 & 1 \end{pmatrix} \tag{57}$$

$$y(t) = \begin{pmatrix} 0 & 0 & 1 \end{pmatrix} x(t - r_2). \tag{58}$$

In [11], the system stability is computed for $r^* = r_1^* = r_2^* = 0.43s$. In [1], the system stability is computed for $r = 1$ using a contractive continuous RK-method. However in our case, we guarantee the system stability for all the delays less than or equal to r^*. Applying Theorem 2, using LMI-Toolbox, we find that the inequality is feasible for $r_2^* = 100s$. In fact, the inequality is satisfied for larger values of r_2^*. The following simulation is shown in Fig. 2 for $r_2^* = 10s$. The initial condition is given between -10s and 0. Notice that there are discontinuities in the derivative of $f(t)$ for $t = 0, 10, ..., k \times 10$.

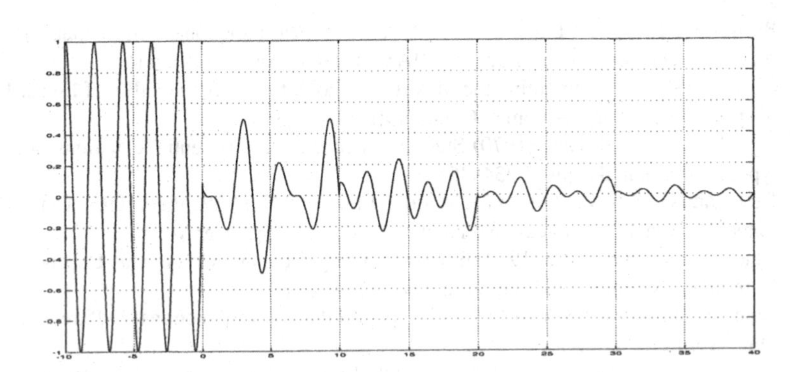

Fig. 2. $-1 \leq y(t) \leq 1, -10 \leq t \leq 40$

If we only use V_1 given in equation (32) for the system without uncertainties, it leads to the delay independent condition [19]

$$\begin{pmatrix} A^T P + PA + S & P(AC + B) + SC \\ C^T S + (C^T A^T + B^T) & C^T SC - S \end{pmatrix} < 0, \tag{59}$$

$$C^T SC - S < 0, \tag{60}$$

where matrices $P > 0$ and $S > 0$ are symmetric positive definite. The system (57) does satisfy this last LMI, it is then asymptotically stable, independent of the delay $r = r_1 = r_2$.

5 Concluding Remarks

In this chapter, we considered the delay dependent robust stability of a fairly general class of neutral systems. A key point of this work is the fact that we give sufficient conditions checkable in the LMI framework. We proposed a Lyapunov-Krassovskii functional which includes, as particular cases, functionals developed in previous works. Along the same lines, more general classes of neutral systems including quasilinearities in the operator D of (1) satisfying the atomicity condition should be considered.

References

1. Bellen A., N. Guglielmi and A.E. Ruehli (1999) Methods for Linear Systems of Circuits Delays Differential Equations of Neutral Type. IEEE Trans. Circuits and Sys. 46, 1:212–216.
2. Boyd S., L.El Ghaoui, E. Feron and V. Balakrishnan (1994) Linear Matrix Inequalities in System and Control Theory. Society for Industrial and Applied Mathematics.
3. Brayton R. (1976) Nonlinear Oscillations in a Distributed Network, Quart. Appl. Math., 24: 289-301.
4. Burns J.A. Herdman T.L and Stech H.W (1983) Linear Functional Differential Equations As Semigroups On Products Spaces. SIAM. J. Math. Anal., 14:98–116.
5. Chen J. (1995) On Computing the Maximal Delay Intervals for Stability of Linear Delay Systems. IEEE Trans. Automat. Contr., 40:1087–1093.
6. Cruz M.A. and J.K. Hale (1970) Stability of functional differential equations of neutral type. J. Differential Eqns. 7:334–355.
7. Diekmann O., S.A. van Gils, S.M. Verduyn Lunel and H.-O. Walther (1995). Delay Equations Functional-, Complex-, and Nonlinear Analysis. Springer Verlag.
8. Hale J.K. and M.A. Cruz (1970) Existence, Uniqueness and Continuous Dependence for Hereditary Systems. Ann. Mat. Pura Appl. 85, 4:63–82.
9. Hale J.K. and S.M. Verduyn Lunel (1993) Introduction to Functional Differential Equations. Springer-Verlag.
10. Henry D. (1974) Linear Autonomous Neutral Functional Differential Equations. J. Diff. Equ., 15:106–128.
11. Ivanescu D., S. Niculescu, L. Dugard, J.M. Dion and E.I. Verriest (2003) On Delay Dependent Stability for Linear Neutral Systems. Automatica. 39:255–261.
12. Kharitonov V.L. (1998) Robust Stability Analysis of Time-Delay Systems: a survey. Proc. IFAC Syst. Struct. & Contr., Nantes, France:1–12.
13. Kolmanovskii V.B. (1996) The stability of Hereditary Systems of neutral type. J. Appl. Maths. Mechs. 60, 2:205–216.
14. Kolmanovskii V.B. and A.D. Myshkis (1999) Introduction to the Theory and Applications of Functional Differential Equations. Kluwer Academic Publishers.
15. Kolmanovskii V.B. and J.P. Richard (1998) Stability of systems with pure, discrete multi-delays. IFAC Conference System Structure and Control Nantes, France:13–18.
16. Niculescu S.I. (2001) On Robust Stability of Neutral Systems, Special issue On Time-Delay Systems. Kybernetica 37:253–263.
17. Rodriguez S.A., J.M. Dion, L. Dugard and D. Ivănescu (2001) On delay-dependent robust stability of neutral systems. 3rd IFAC Workshop on Time Delay Systems, Santa Fe USA:101–106.

18. Rodriguez S.A., J.M. Dion and L. Dugard. (2002) Robust Stability Analysis of Neutral Systems Under Model Transformation. 41st IEEE Conference on Decision and Control, Las Vegas, Nevada, USA:1850–1855.
19. Verriest E.I. and S.I. Niculescu (1997) Delay-Independent Stability of LNS: A Riccati Equation Approach, in Stability and Control of Time-Delay Systems. (L. Dugard and E.I. Verriest, Eds.),Springer-Verlag L. 228:92–100.

Part VI

Applications

On Delay-Based Linear Models and Robust Control of Cavity Flows

Xin Yuan[1], Mehmet Önder Efe[2], and Hitay Özbay[3]

[1] Collaborative Center of Control Science, Department of Electrical Engineering, The Ohio State University, Columbus, OH 43210, U.S.A., yuanx@ee.eng.ohio-state.edu
[2] Collaborative Center of Control Science, Department of Electrical Engineering, The Ohio State University, Columbus, OH 43210, U.S.A. onderefe@ieee.org
[3] Department of Electrical and Electronics Engineering, Bilkent University, Bilkent, Ankara, TR-06800, Turkey; on leave from The Ohio State University, ozbay@ee.eng.ohio-state.edu

Summary. Design and implementation of flow control problems pose challenging difficulties as the flow dynamics are governed by coupled nonlinear equations. Recent research outcomes stipulate that the problem can be studied either from a reduced order modeling point of view or from a transfer function point of view. The latter identifies the physics of the problem on the basis of separate components such as scattering, acoustics, shear layer etc. This chapter uses the transfer function representation and demonstrates a good match between the real-time observations and a well-tuned transfer function can be obtained. Utilizing the devised model, an H_∞ controller based on Toker-Özbay formula is presented. The simulation results illustrate that the effect of the noise can be eliminated significantly by appropriately exciting the flow dynamics.

1 Introduction

Aerodynamic flow control is a core issue aiming to reduce skin friction thereby increasing the maneuverability of aerial vehicles and reducing the fuel expenditure. The research towards this goal is in its infancy, however, some major problems have been identified. These particularly include the development of a dynamic model for a given flow geometry, and describing the best control scheme in some sense of optimality. In this chapter we discuss cavity flow problem.

One of the two branches of research towards the model development for cavity flow stipulate the use of proper orthogonal decomposition techniques to remedy the problem of infinite dimensionality, [1], [2]. These procedures yield a set of ordinary differential equations, which are autonomous. The underlying idea is to extract the most dominant features (modes) containing the essential part of the flow energy. Although the modeling issue has well been addressed, the control design is still dependent upon models that explicitly include the control input. The other viewpoint exploits the strength of representing the physical properties by dynamical models in transfer function forms, [3], [4], [5]. Cited studies demonstrate that the shear layer, scattering, cavity acoustics and receptivity can be represented dynamically as transfer functions. Due to the reflections from the upstream wall of the cavity, after some propagation delay time, the reflections interact with the oncoming flow and a delay-based coupled dynamics arise. It must be noted that the devised form of the transfer function matches the frequency content of the data obtained from Navier-Stokes equations.

In this chapter, we optimize the parameters of the model developed in [3], [4], [5] to match the magnitude and frequencies of the resonant peaks and use this model to synthesize an H_∞ controller.

It is a well known fact that H_∞ controller design scheme is particularly well suited if the model involves uncertainties. This study demonstrates that an optimal controller can be determined by utilizing the framework presented in [6].

The chapter is organized as follows: Section 2 introduces the transfer function based model of the cavity flows with its sub-components. The third section is devoted to the parameter tuning issues. The frequency response match is presented in that section. Following this, the methodology to design an H_∞ controller is discussed together with the simulation results. Concluding remarks are made in the last section.

2 Delay-Based Models of Cavity Flow

The process shown in Figure 1 is for cavity flows, which constitute the simplest geometry for studying aerodynamic flow control problems. Basically, this representation captures major dynamic phenomena inside the flow field. For the shear layer, we have,

$$G(s) = G_0(s)e^{-s\tau_s}, \tag{1}$$

where $\tau_s = L/(\kappa U)$ (with L being the length of the cavity, U being the freestream velocity and κ is a known constant). In (1), $G_0(s) = \frac{\omega_0^2}{s^2+2\zeta\omega_0 s+\omega_0^2}$ with ω_0 and ζ are the natural frequency and damping ratio, respectively.

The acoustics term is given as

$$A(s) = \frac{e^{-s\tau_a}}{1 - re^{-2s\tau_a}}, \tag{2}$$

where

$$r(s) = \frac{r}{1 + s/\omega_r}, \tag{3}$$

is the attenuation factor of the reflection process. Denoting the speed of sound by a, the time delay representing the acoustic lag between the trailing edge and the leading edge is given as $\tau_a = L/a$. Here, the Mach number can be calculated as $M = U/a$. Furthermore, the numerical values of R (receptivity), S_c (scattering), V (actuator) and S_m (sensor) are assumed to be available in [4] and denote them by K_R, K_S, K_v and K_m respectively. In the view of these, the cavity transfer function P can be formed as given below:

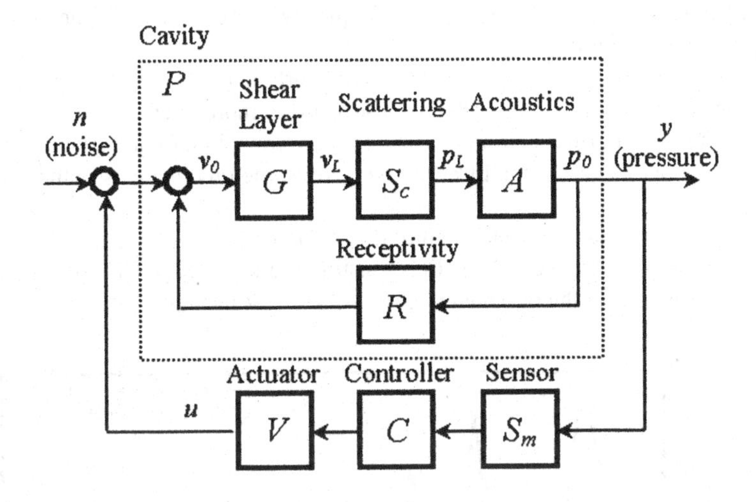

Fig. 1. Block representation of the cavity flow and the control loop [3], [4]

$$P(s) = \frac{AS_cG}{1 - RAS_cG}. \tag{4}$$

Although the system is a linear one, at some Mach numbers unstable limit cycling appears. The focus of this chapter is to study how a robust controller can be devised under dynamic uncertainties.

The goal of the controller is to reduce the peak value of $|P(j\omega)S(j\omega)|$, where $S(j\omega)$ is the sensitivity function, so that the effect of the noise at the output is reduced. An example of achieving such a goal can be found in [4], in which the controller is composed of a filter followed by a gain and a time delay. In the next section, we outline the effect of each parameter on the frequency response characteristics.

3 Parameter Tuning for the Flow Dynamics

Our studies have demonstrated that the flow dynamics developed by Rowley *et al*, [4], [5] exhibit certain degrees of flexibility to match the frequency response obtained from the real-time data with that of the input-output model. In order to analyze this, we have performed several tests to see which parameter is responsible for introducing what sort of modification into the frequency content. Following is a list summarizing our conclusions in this respect:

As ω_0 increases, the dominant peak moves towards higher frequencies. The frequency domain picture is stretched to the right.

As ζ increases, the values of the peaks get lowered, and the frequency content becomes more flattened.

An increase in K_S lifts up the entire frequency domain picture while magnifying the peak values slightly.

If r is increased, more peaks appear particularly in the higher frequencies.

As τ_s increases, the frequency response acquires more fluctuations (peaks) in the low frequencies. Further increments lead to more wavy low frequency behavior.

Change in τ_a causes small translations with some tiny changes in the peak magnitudes.

As K_R increases, the peak magnitudes get larger.

Apparently, the above information constitutes a knowledge base for us, and lets us know how to tune the parameters given some real-time data.

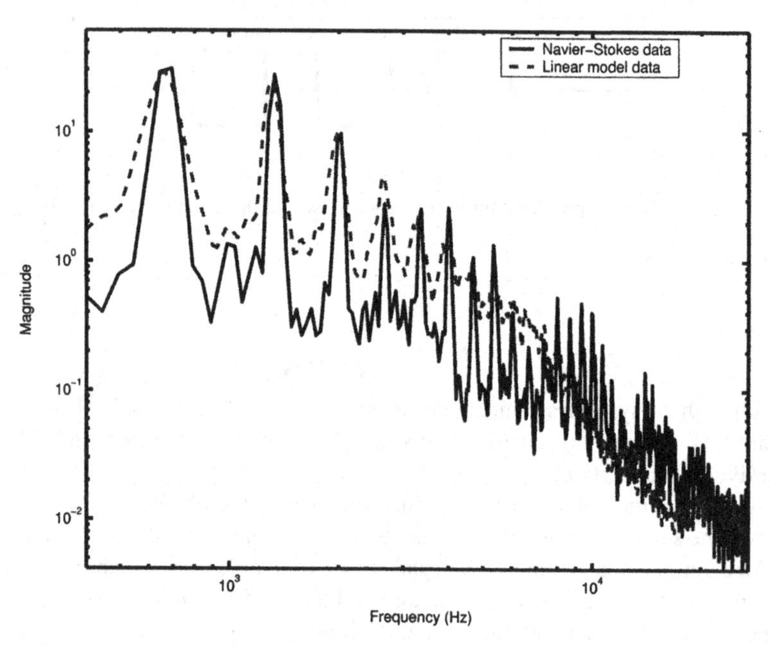

Fig. 2. Power spectral density comparison of pressure data obtained from full order Navier-Stokes simulation and data from the model in (4)

In this chapter, we use trial and error method to match the frequency content and the result of the model match is illustrated by Figure 2, from which we can see that the power spectral density of the simulation output of the linear model matches that of the simulation outputs based on Navier-Stokes equations very well. The data for the linear system are given in Table 1, on which the robust controller design is based.

Table 1. Parameters of the Linear Model

Parameter	Value
ω_0	200 rad/sec
ζ	0.95
τ_s	0.0195 sec
τ_a	0.001 sec
r	0.1
ω_r	100 rad/sec
K_R	0.408
K_S	1.65
K_v	1
K_m	1

4 Robust Controller Design

Inserting the transfer functions of the shear layer and the acoustics into the plant transfer function, we get

$$\begin{aligned} P(s) &= \frac{\frac{\omega_0^2}{s^2+2\zeta\omega_0 s+\omega_0^2} e^{-s\tau_s} K_s \frac{e^{-s\tau_a}}{1-re^{-2s\tau_a}}}{1 - K_R \frac{\omega_0^2}{s^2+2\zeta\omega_0 s+\omega_0^2} e^{-s\tau_s} K_s \frac{e^{-s\tau_a}}{1-re^{-2s\tau_a}}}, \\ &= \frac{K_S \omega_0^2 e^{-s(\tau_s+\tau_a)}}{(s^2+2\zeta\omega_0 s+\omega_0^2)(1-re^{-2s\tau_a}) - K_R K_S \omega_0^2 e^{-s(\tau_s+\tau_a)}}, \\ &= \frac{K_S \omega_0^2 e^{-s(\tau_s+\tau_a)}}{(s^2+2\zeta\omega_0 s+\omega_0^2) - r(s^2+2\zeta\omega_0 s+\omega_0^2)e^{-2s\tau_a} - K_R K_S \omega_0^2 e^{-s(\tau_s+\tau_a)}}. \end{aligned} \tag{5}$$

Let us factorize $P(s)$ into the form $P(s) = N_{o1}(s)N_{o2}(s)M_n(s)$, where

$$N_{o2}(s) = K_S G_0(s) = \frac{K_S}{1 + 2\zeta s/\omega_0 + s^2/\omega_0^2}, \tag{6}$$

$$M_n(s) = e^{-h_1 s}, \quad \text{where } h_1 = \tau_s + \tau_a, \tag{7}$$

$$N_{o1}(s) = (1 - K_R N_{o2}(s) M_n(s) - r(s) M_2(s))^{-1}, \tag{8}$$

where $M_2(s) = e^{-2\tau_a s}$. Plant is stable for the numerical values determined in this case. For different numerical values it is possible to have an unstable plant, then finitely many unstable modes may appear from the roots of $1/N_{o1}(s) = 0$. This situation can be handled in our approach as well. At this stage, we propose to use Toker-Özbay formula (see [6]) to design the controller. The optimal robust performance is defined by

$$\gamma_{opt} := \inf_{C \in \Omega} \left\| \begin{pmatrix} W_1 S \\ W_2 T \end{pmatrix} \right\|_\infty, \tag{9}$$

where Ω is the set of all compensators stabilizing P. It is known that $S = (1 + PC)^{-1}$ and $T = 1 - S$ are the sensitivity and complementary sensitivity functions and W_1 and W_2 are the performance and stability weighting functions. Since the goal of the controller is to reduce the peak value of $|P(j\omega)S(j\omega)|$, we choose the performance weighting function $W_1(s)$ as given in (11), such that the oscillation magnitude at the dominating modes is suppressed. To take care of the uncertainties in the high frequency, we set the complementary sensitivity weighting function $W_2(s)$ as given by (13). In Figure 3, Bode plots of these weighting functions and the plant are depicted.

$$W_{1o}(s) = k_1 \frac{(1 + s/\omega_{1n})}{(1 + s/\omega_{1d})}, \tag{10}$$

$$W_1(s) = W_{1o}(s)(1 - r(s)M_2(s))N_{o1}(s), \tag{11}$$

$$W_{2o}(s) = \epsilon_2 s(1 + s/\omega_{1n}), \tag{12}$$

$$W_2(s) = W_{2o}(s)(1 - r(s)M_2(s)). \tag{13}$$

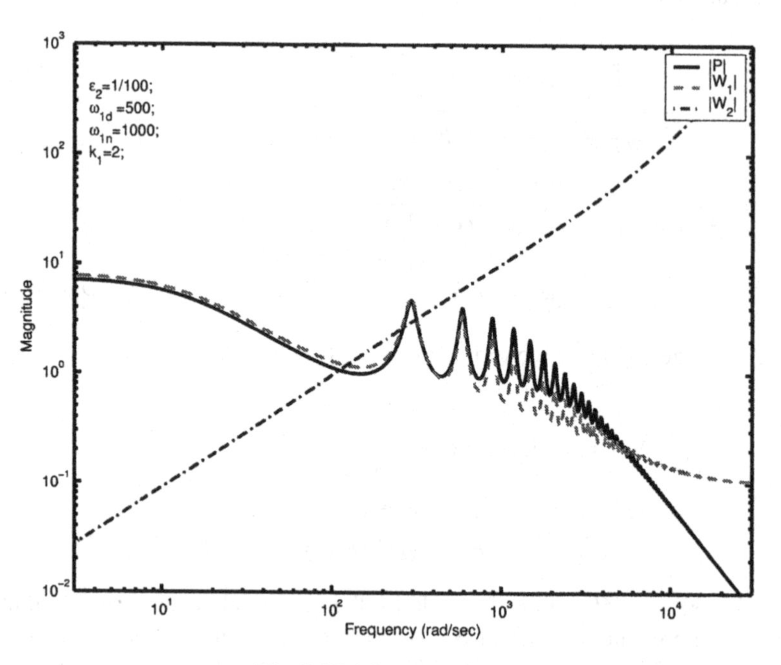

Fig. 3. Weighting functions

Define the following functions

$$P_2 = N_{o2}M_n, \tag{14}$$
$$S_2 = (1 + P_2C_2)^{-1}, \tag{15}$$
$$T_2 = 1 - S_2. \tag{16}$$

By inverting the outer part of the plant, we see that the H_∞ controller has to be in the form $C(s) = C_2(s)(1 - r(s)M_2(s)) + K_R$, where C_2 is designed for

$$\gamma = \inf_{C_2 \text{ stabilizes} P_2} \left\| \begin{pmatrix} W_{1o}S_2 \\ W_{2o}T_2 \end{pmatrix} \right\|_\infty \tag{17}$$

By definition, it can be seen that the performance specifications of both systems are equivalent, i.e.,

$$|W_1(j\omega)S(j\omega)| = |W_{1o}(j\omega)S_2(j\omega)|, \tag{18}$$

For the robustness of the system, it can be shown that if

$$\left|\frac{\Delta_{P_2}}{P_2}\right| < \left|\gamma^{-1}W_{2o}\right|, \tag{19}$$
$$\left|\Delta_{K_R}\right| < \left|\gamma^{-1}\frac{W_{1o}N_{o1}(1 - rM_2)^2}{N_{o2}}\right|. \tag{20}$$

then the system associated with plant P and controller C is robustly stable, where Δ_{P_2} denote the uncertainty of the plant P_2 and Δ_{K_R} be the uncertainty of the variable K_R. The problem is significantly simplified such that C_2 can be computed explicitly by hand calculations. The optimal controller C_2 for P_2 is in the form

$$C_2(s) = \left(\frac{\gamma}{\gamma_{min}} - \frac{\gamma_{min}}{\gamma}\right) \frac{N_{o2}^{-1}(s)}{(1 + as + bs^2)} \left(\frac{1}{1 + H(s)}\right), \tag{21}$$

To compute the optimal performance level γ, define

$$\gamma_{min} := k_1 \frac{\omega_{1d}}{\omega_{1n}}, \tag{22}$$
$$\gamma_{max} := k_1, \tag{23}$$

and

$$x = \sqrt{\frac{k_1^2 - \gamma^2}{\gamma^2 - \gamma_{min}^2}}, \tag{24}$$
$$b = \frac{\epsilon_2\sqrt{1 - (\gamma_{min}/\gamma)^2}}{k_1\omega_{1d}}, \tag{25}$$
$$a = \sqrt{2b + \epsilon_2^2(k_1^{-2} - \gamma^{-2})}. \tag{26}$$

Taking the largest value of γ satisfying the below equality (27) in the allowable range: $\gamma_{max} > \gamma > \gamma_{min}$ will give us the optimal γ needed in the optimal controller formula (21). For the current problem, we obtain $\gamma = 1.9484$.

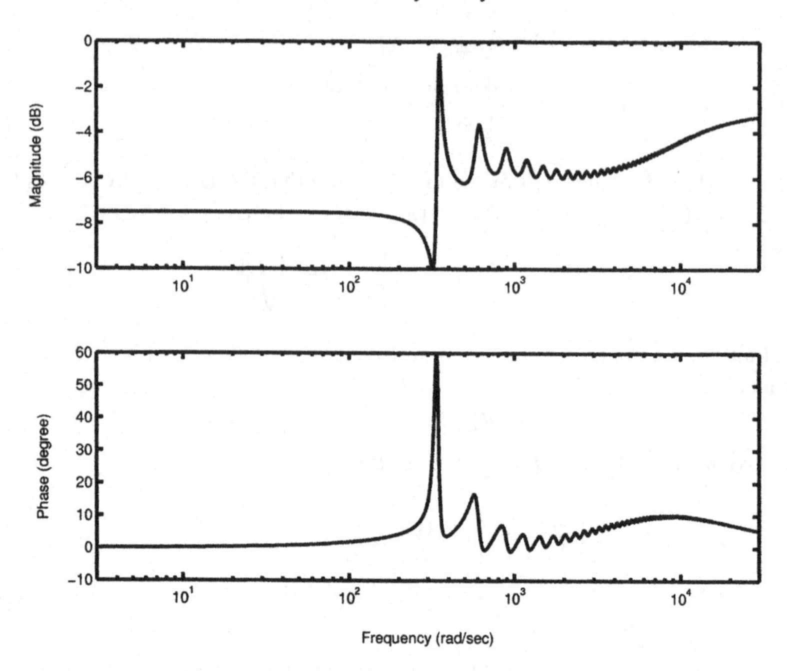

Fig. 4. Bode plot of the optimal controller

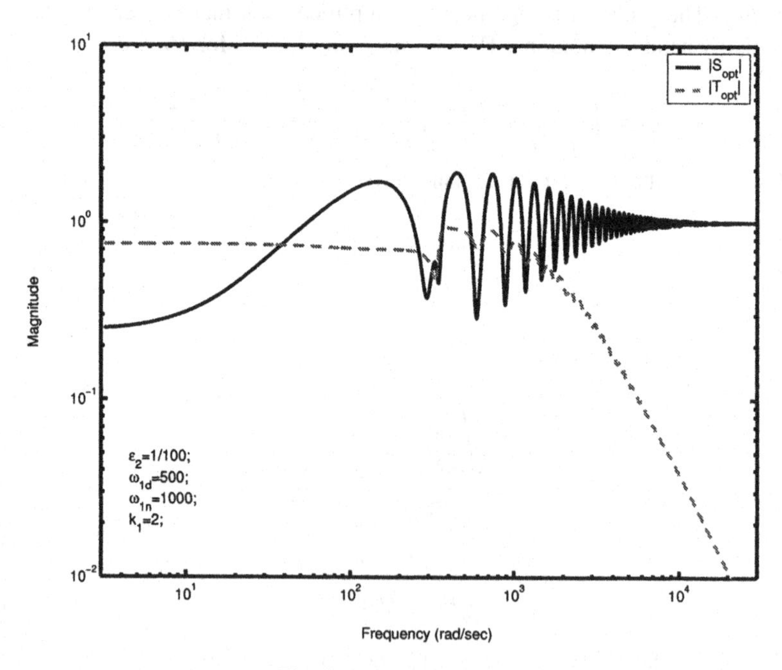

Fig. 5. Optimal S and T

$$\pi = h_1\omega_{1d}x + tan^{-1}x + tan^{-1}\frac{\omega_{1d}x}{\omega_{1n}} + tan^{-1}\frac{a\omega_{1d}x}{1 - b\omega_{1d}^2x^2}. \quad (27)$$

To implement the controller in real-time, $H(s)$ is expanded as a finite impulse response (FIR) filter and an exponential decay term, i.e. $H(s) = H_{FIR}(s) + H_{IIR}(s)$, which are described in (28) and (29). The magnitude of $H(s)$ and its constituents are shown in Figure 7.

$$H_{FIR}(s) = \frac{(\omega_{1d} + \omega_{1n})s + \omega_{1n}\omega_{1d} - \omega_x^2 + \frac{\gamma}{\gamma_{min}}\omega_{1d}^2(1 + x^2)(d_1 s + d_0)e^{-h_1 s}}{s^2 + \omega_x^2}, \quad (28)$$

$$H_{IIR}(s) = \frac{\gamma}{\gamma_{min}}\left(\frac{\omega_{1d}^2(1 + x^2)(c_1 s + c_0) - 1}{1 + as + bs^2}\right)e^{-h_1 s}, \quad (29)$$

in which c_0, c_1, d_0, d_1 and ω_x are defined as below:

$$c_0 = \frac{(b(1 - b\omega_x^2) - a^2)}{a}d_1, \quad (30)$$

$$c_1 = -bd_1, \quad (31)$$

$$d_0 = d_1\frac{(b\omega_x^2 - 1)}{a}, \quad (32)$$

$$d_1 = \frac{-a}{\omega_x^2a^2 + (1 - b\omega_x^2)^2}, \quad (33)$$

$$\omega_x = \omega_{1d}x. \quad (34)$$

It has been demonstrated that the impulse response of $H_{FIR}(s)$ is restricted to the time interval $[0, h_1]$, which is shown in Figure 6. Hence, $H_{FIR}(s)$ can be realized as a FIR filter of duration h_1. The discrete-time realization of $H_{FIR}(s)$ requires only h_1/T_s states, where T_s is the sampling period. Theoretically, the infinite dimensional controller can be implemented through a finite impulse response (FIR) filter approximation.

In Figure 4, we demonstrate the Bode plot of the controller, which has been discussed above. Figure 8 illustrates the Bode plot of the closed loop control system compared to the open loop plant. As the figure suggests, the controller modifies the frequency content of the open loop system significantly at the dominating modes. The resonant peaks in the frequency response of the open loop system are suppressed and the improvement is obvious. Figure 9 shows the time domain simulation result of the the closed loop system.

5 Conclusions

In this study, we focus on the transfer function based models of cavity flows. Due to the delays inherited from the physics of the problem, the problem is an infinite

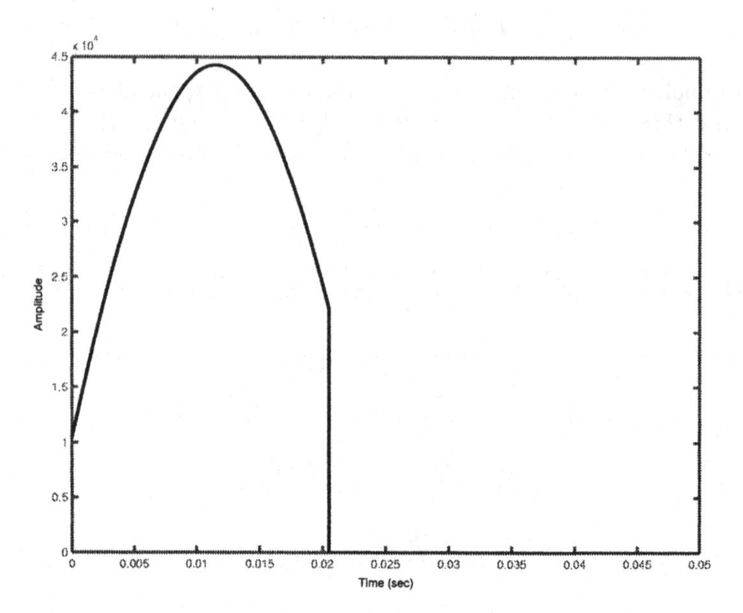

Fig. 6. Impulse response of $H_{FIR}(s)$

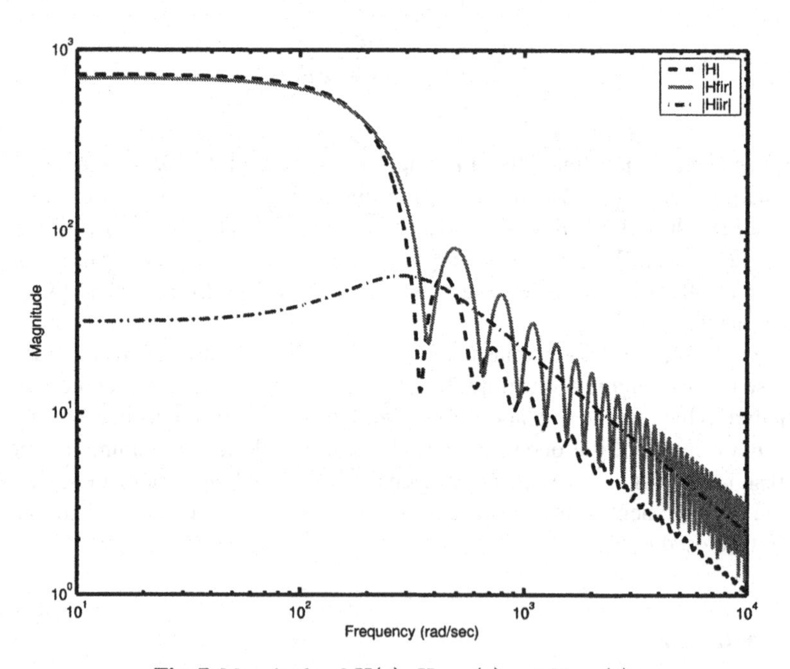

Fig. 7. Magnitude of $H(s)$, $H_{FIR}(s)$ and $H_{IIR}(s)$

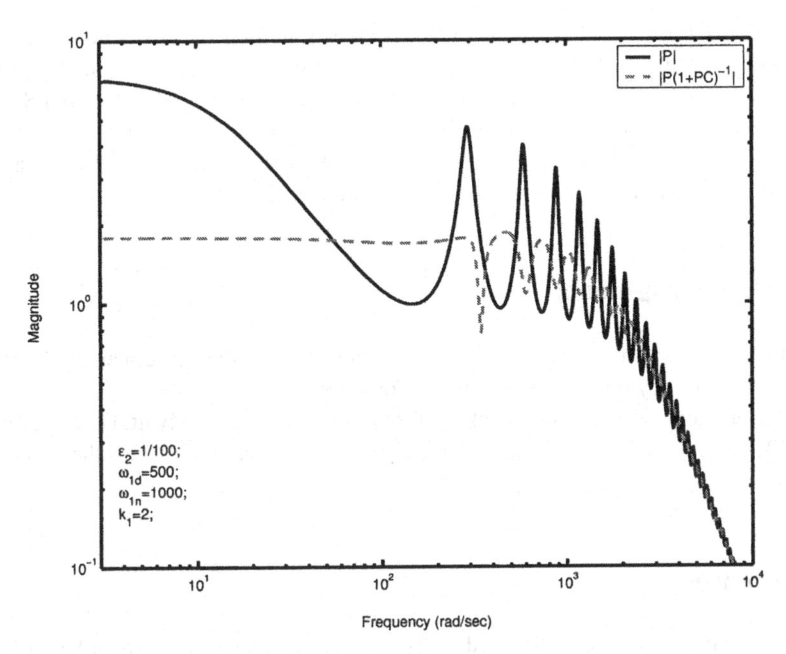

Fig. 8. Bode plot of the closed loop system compared wiht that of the open loop system

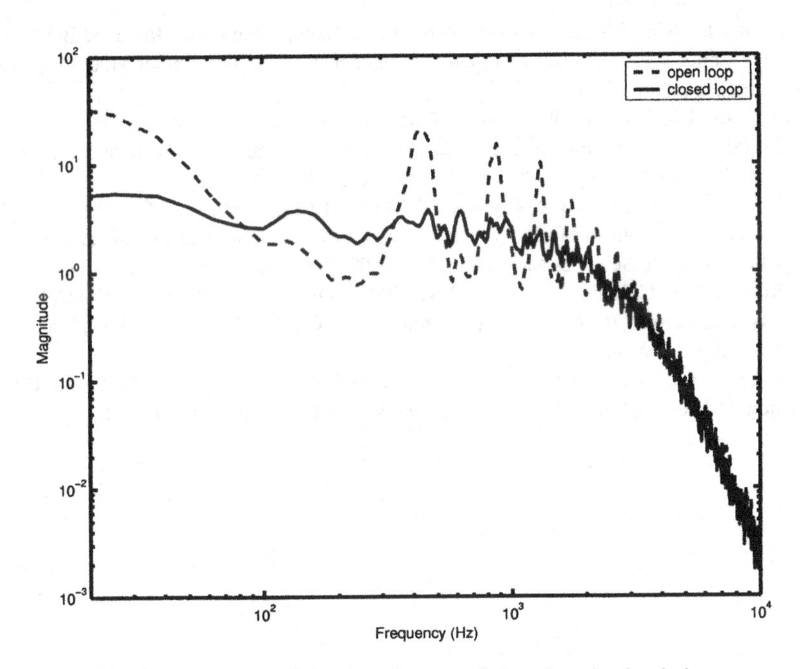

Fig. 9. Power spectral density of output of time domain simulation

dimensional one. We demonstrate that a previously studied form of delay-based flow model can be tuned so as to capture the resonant peaks appearing in the frequency response. We present how a H_∞ based controller can be devised for such a Single-Input-Single-Output system. The observed results demonstrate that the controller performs well under the presence of uncertainties. The undesired resonant peaks of the open loop system have been suppressed fairly well.

6 Acknowledgments

This work was supported in part by AFRL/AFOSR under the agreement no. F33615-01-2-3154, and by the National Science Foundation.

The authors would like to thank Prof. M. Samimy, Dr. J.H. Myatt, Dr. J. DeBonis, Dr. M. Debiasi, and E. Caraballo for fruitful discussions in devising the presented work.

References

1. Ravindran SS (2000). A Reduced Order Approach for Optimal Control of Fluids Using Proper Orthogonal Decomposition. International Journal for Numerical Methods in Fluids, 34:425-488.
2. Atwell JA, King BB (2001). Proper Orthogonal Decomposition for Reduced Basis Feedback Controllers for Parabolic Equations. Mathematical and Computer Modelling of Dynamical Systems, 33:1-19.
3. Williams DR, Rowley CW, Colonius T, Murray RM, MacMartin DG, Fabris D, Albertson J (2002). Model Based Control of Cavity Oscillations Part I: Experiments. 40th Aerospace Sciences Meeting (AIAA 2002-0971), Reno, NV.
4. Rowley CW, Williams DR, Colonius T, Murray RM, MacMartin DG, Fabris D (2002). Model Based Control of Cavity Oscillations Part II: System Identification and Analysis. 40th Aerospace Sciences Meeting (AIAA 2002-0972), Reno, NV.
5. Rowley CW, Colonius T, Murray RM (2001). Dynamical Models for Control of Cavity Oscillations. 7th AIAA/CEAS Aeroacoustics Conf. (AIAA 2001-2126), May 28-30, Maastricht, The Netherlands.
6. Toker O, Özbay H (1995). H_∞ Optimal and Suboptimal Controllers for Infnite Dimensional SISO Plants. IEEE Transactions on Automatic Control, 40:751-755.

Active-adaptive Control of Acoustic Resonances in Flows

Anuradha M. Annaswamy

Department of Mechanical Engineering, Massachusetts Institute of Technology, Cambridge, MA 02139, USA aanna@mit.edu

Summary. Several fluid flow problems related to propulsion and power generation exhibit strong acoustic resonances. Produced due to interactions of the acoustics with other underlying unsteady mechanisms such as unsteady heat-release or shear flow instability, these resonances manifest as large and sustained pressure oscillations. In addition to the obvious undesirable effect of high ambient noise and acoustic fatigue, these oscillations are coupled with other damaging effects such as excessive vibrations, high burn rates, lift-loss, and ground erosion. Compromises made in order to reduce these oscillations lead to departures from the desired operating conditions and can in turn result in suboptimal performance with reduced heat-output, increased emissions, or decreased efficiency. Over the past few years, active control technology has been increasingly sought after to realize the desired performance metrics in these problems without encountering resonant behavior. In order to provide guaranteed and uniform performance over a large range of operating conditions in the presence of various system uncertainties, it has been demonstrated in these problems that a model-based approach to designing the control strategy is feasible and scalable, and leads to a reliable and improved pressure reduction at the desired operating conditions. In this chapter, two examples of such fluid flow problems, combustion-instability and impingement-tones in supersonic flows, and their active control will be discussed. Models of the resonant mechanisms using both physically-based and system-identification principles are presented. In active-adaptive control of combustion systems, Posi-cast control methods and their closed-loop performance in practical combustors are discussed. In active-adaptive control of supersonic impingement tones, a POD-based active control strategy and the corresponding experimental results from a Short Takeoff Vertical Landing (STOVL) supersonic jet facility at Mach 1.5 are presented.

1 Introduction

Problems in power generation and propulsion concern the organized motion of a fluid past various boundaries under varied conditions. Because of the boundary conditions and operating conditions, often the fluid flow experiences acoustic resonances, which correspond to large and undesirable pressure oscillations. Occuring due to the combined presence of several mechanisms, some of which are coupled in feedback, these oscillations have been observed to be effectively curtailed through the use of active control. In particular, careful articulation of the input conditions in the flow such as the pulsing of a fuel flow or an air flow or adding external flow at receptive locations, have been found to make a drastic change in the system behavior with very little external input.

An important feature of the flow resonances is the presence of time-delays in the system. Given that the instability is imbedded in a flow problem which is governed by flow particles being convected from one location to another, it is not surprising that a good part of the flow dynamics, including instability, is affected by convective delays. Two examples of such fluid flow problems include combustion-instability [1] and impingement-tones in supersonic flows [2], in both of which time-delay plays a dominant role, in the modeling of the physical process and in the active control design of such systems. Over the past five years, we have developed linear and non-linear time-delay models of the resonant mechanisms based on both physically-based and system-identification methods. In active-adaptive control of combustion systems, we have developed a new controller referred to as an Adaptive Posi-cast controller and have successfully implemented it in a number of rigs that exhibited fairly large time-delays. In active-adaptive control of impingment tones, we have implemented an active optimization strategy that makes use of POD modes from on-line measurements and MEMS devices. In the following sections, highlights of these results are presented.

2 Acoustic Resonances in Combustion Systems

Continuous combustion systems, common in power generation and propulsion applications, are susceptible to the phenomenon known as thermoacoustic instability. This instability is due to a self-sustained coupling between the acoustic field of the combustion chamber, and the heat release rate. Pressure oscillations inside the combustor cause fluctuations in the heat-release rate, which in turn produces an energy input into the acoustics, generating a feedback-loop. Under certain conditions the pressure and heat-release fluctuations are in phase, causing this feedback to be destabilizing. The resulting instability is undesirable because the large amplitude pressure and heat release rate oscillations lead to high levels of acoustic noise and vibration, as well as structural damage.

Several mechanisms contribute to combustion dynamics including acoustics, heat-release dynamics, hydrodynamics, and mixing (see Figure 1). Of these, the first two mechanisms are better understood and have been modeled through both physically based [3, 4] and system-identification based approaches [5]. In both of these cases, the model is invariably of the form

$$W(s) = \frac{y}{v} = W_0(s)e^{-\tau s} \tag{1}$$

where y is the unsteady pressure, v is the voltage that drives the fuel-injector valve, $W_0(s)$ is an nth order transfer function whose zeros are in the left-half of the complex plane, with a relative degree one or two, and with the high frequency gain known. The time-delay τ corresponds to the transport lag, which is the distance between the injection of the fuel, and the burning zone. Typically, $W_0(s)$ has a few complex poles, at least two of which are in the right-half of the complex plane, or are very lightly damped. The problem is one of deriving a controller that determine $v(t)$ which guarantees that the closed-loop remains stable even with partial knowledge of the plant-model parameters.

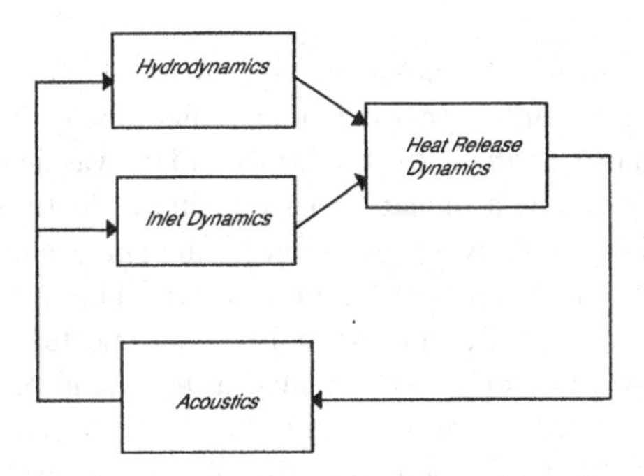

Fig. 1. Schematic of acoustics, hydrodynamics and heat-release dynamics interactions in combustion systems.

2.1 The adaptive Posi-cast controller

The controller chosen to stabilize the above class of systems is motivated by time-delay controllers in [6,7] and has been discussed in detail in [3,8]. The general form of this controller is given by

$$\begin{aligned}\dot{\omega}_1 &= \Lambda_0\omega_1 + \ell u(t-\tau)\\ \dot{\omega}_2 &= \Lambda_0\omega_2 + \ell y(t)\\ u &= \theta_1^T(t)\omega_1 + \theta_2^T(t)\omega_2 + r(t) + \overline{\lambda}^T(t)\overline{u}(t)\\ \dot{\tilde{\theta}}(t) &= -(y(t)-y_m(t))\omega(t-\tau).\end{aligned} \tag{2}$$

where $\theta_1, \omega_1 : \mathbb{R}^+ \to \mathbb{R}^n$, $\theta_2, \omega_2 : \mathbb{R}^+ \to \mathbb{R}^n$, Λ is an asymptotically stable matrix and $\det(sI - \Lambda) = \lambda(s)$, is a Hurwitz polynomial. $\overline{u}_i$ is the ith sample of $u(t)$ in the interval $[t-\tau, t)$, $i = 1, \ldots, p$, and p is suitably chosen as dictated by the required accuracy and the affordable complexity. It has been established in [8] that starting from initial conditions whose magnitude depends on τ, the closed-loop signals will remain bounded. In [3], a lower-order controller whose order depends on the relative degree of the plant rather than its order has been shown to be sufficient for stabilization as well, for a small τ.

2.2 Experimental results

The controller described above has been implemented in several rigs, all of which exhibit acoustic resonances. We briefly describe two such rigs below, the corresponding control input, and the nature of the uncontrolled combustion dynamics. We then show the impact of the controller in closed-loop.

A dump combustor

Experiments were conducted using an axi-symmetric dump combustor at the University of Maryland. The inlet consisted of a circular tube with a diameter

4.1 cm and length 2.24 m. The upstream boundary was defined by a choked orifice and the downstream by a sudden-expansion dump. At 1m upstream of the inlet dump plane, a choked nozzle was used to inject ethylene in a transverse manner into the air flow. The inlet flow simulated the prevaporized premixed reactants entering a ramjet combustor. A set of secondary pulsed fuel injectors, which were pointed at 45° into the flow direction, were mounted at the dump plane. They supplied a small amount of liquid ethanol directly into the combustor continuously or at prescribed frequencies. The total amount of ethanol injection was controlled by pulse duration and injection frequencies. In a typical case, about 15% of total combustion enthalpy was provided through the controlled fuel injection; the remaining 85% was supplied through the steady premixed inlet flow. The associated time delay in the secondary fuel injection to pressure modulation was 2 ms. The combustor was naturally unstable at an equivalence ratio of 0.6-0.7 and air flow rate of 23 g/s, which corresponded to a Reynolds number of 30,000. For these conditions, a fundamental mode was observed at 39-40 Hz with pressure amplitude of up to 160 dB and 10 kPa peak to peak values.

With the secondary fuel injection as the input, the output pressure response to a white-noise input was determined, and using the input-output pairs, a system-identification model of the combustor was determined, and was of the form

$$W(s) = \frac{0.460s^6 + 182s^5 + 6.12\cdot 10^5 s^4 + 8.63\cdot 10^7 s^3 + 1.44\cdot 10^{11} s^2 +}{s^7 + 8.93s^6 + 1.62\cdot 10^6 s^5 + 7.06\cdot 10^6 s^4 + 4.44\cdot 10^{11} s^3 +}$$
$$\frac{5.83\cdot 10^{12} s + 1.52\cdot 10^{15}}{1.39\cdot 10^{12} s^2 + 2.26\cdot 10^{16} s + 3.16\cdot 10^{16}} e^{-0.002s} \quad (3)$$

We then implemented a Posi-cast controller of the form in Eq. (2), with a sampling frequency of 500 Hz. This led to a satisfactory suppression of the pressure response which is shown in time-domain and frequency-domain in Figures 2 and 3 respectively.

A Swirl-stabilized Combustor

The experimental facility discussed in this section is a generic combustor designed to model the fuel injection/premix ducts of a Rolls-Royce RB211-DLE industrial

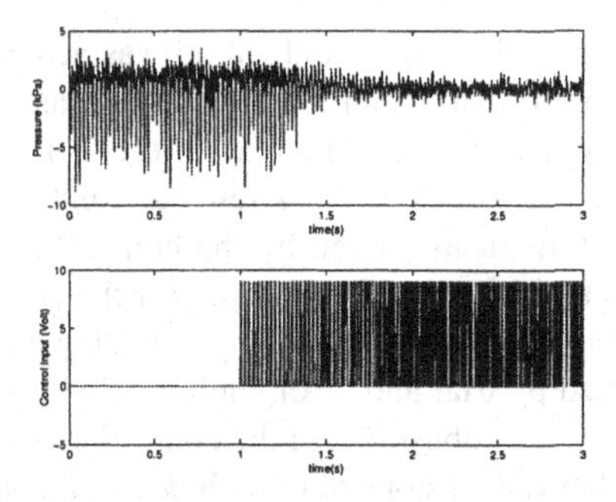

Fig. 2. Time plots of pressure and controller output with the adaptive Posi-cast controller. Controller was turned on at t=1*sec*.

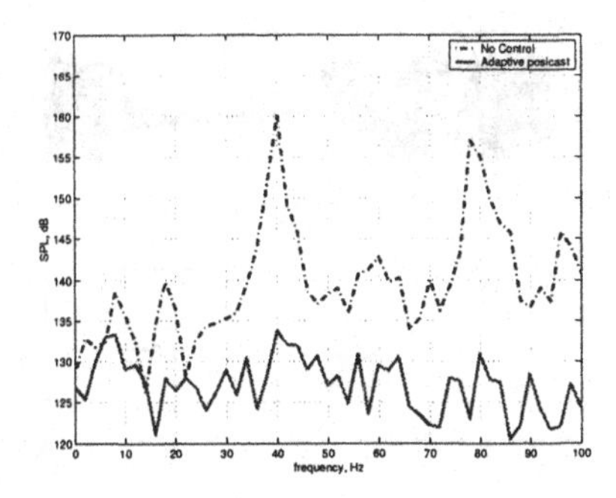

Fig. 3. Pressure spectrum in the dump combustor without control and with adaptive Posi-cast controller.

gas turbine. The swirler unit is a scale model, however, where the geometry of the plenum and combustor have been reduced to simple cylindrical pipes. Fuel is injected upstream into the annular channels through eight cylindrical bars each fitted with two exit holes of 1.0 mm diameter. Actuation for control was achieved using a Direct Drive Valve (DDV) from Moog to modulate the fuel flow rate into the swirler and hence produce variations in equivalence ratio in the premix ducts. In the current set of feedback control tests the experimental data has been obtained in the ranges m_a=0.03-0.05 kg/s and m_f=1.6-2.5 g/s, resulting in ϕ=0.5-0.75. The rig exhibits a 207 Hz plenum mode instability with up to 165 dB and 20 kPa peak to peak without control. This leads to unsteady combustion with a significant time delay (9.6 ms) that needs to be accounted for in the control strategy.

To accurately reproduce pressure oscillations with affordable computation time, a sampling frequency was chosen to be 2 kHz, which resulted in p=19 in the Posi-cast controller in Eq. (2), leading to twenty-one adjustable parameters in the controller. Figure 4 shows the closed-loop control result with the Posi-cast controller. The controller was turned on at 0 s. and stabilization was achieved in about seven seconds. All of the 21 control parameters started at zero and converged to constant values. Of these, the parameter k_1, is shown in Figure 4. The zero initial conditions simply imply that the controller does not know the combustion dynamics initially and these values are automatically tuned by the controller to give optimal performance. Figure 5 shows typical pressure spectrum for the control on/off cases where there is a reduction of approximately 15 dB by the delay controller, whose control parameters were obtained by trial and error, and 30 dB for the Posi-cast controller at the 207 Hz instability. The robustness of the controller was also tested and results showed that the controller retains control for a 20% change in frequency and a 23% change in air mass flow rate [9].

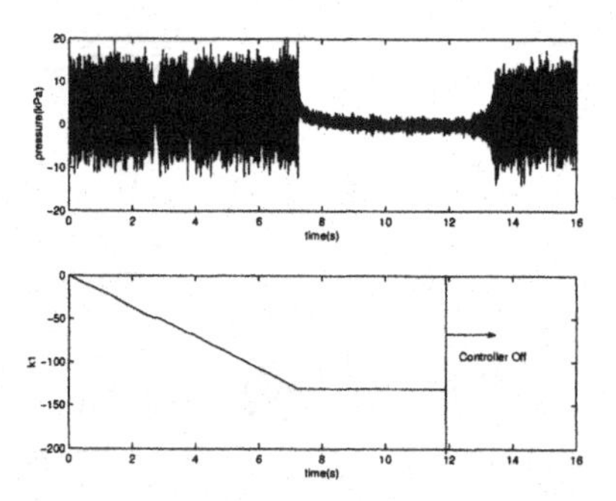

Fig. 4. Time series of the pressure fluctuations in the combustor with the Posi-cast controller

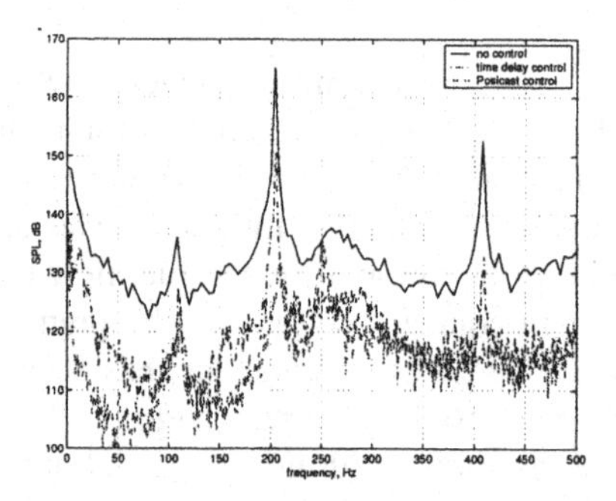

Fig. 5. SPL spectra showing the reduction in noise when the Posi-cast and the delay controllers are turned on.

3 Acoustic Resonances in Impinging Jets

Several acoustic resonances have their origin in the instability of certain fluid motions. One of these motions is in the context of impinging high-speed jets. Experienced by STOVL aircraft while hovering in close proximity to the ground, impingement tones, which are discrete, high-amplitude acoustic tones, are produced due to interactions between high speed jets emanating from the STOVL aircraft nozzle and the ground [10]. The high-amplitude impingement tones are undesirable not only due to the associated high ambient noise, but also because of the accompanied highly unsteady pressure loads on the ground plane and on nearby surfaces. While the high noise levels can lead to structural fatigue of the aircraft surfaces in the vicinity of the nozzles, the high dynamic loads on the impingement surface can lead to an increased erosion of the landing surface as well as a dramatic lift-loss during hover. In an effort to reduce or eliminate these tones, several passive and active control methods have been attempted over the years to interrupt the feedback loop that is the primary cause of the impingement tones. Of these, the technique proposed by Alvi *et al.* [11] appears most promising from the point of view of efficiency, flexibility, and robustness. This method introduces microjets along the periphery of the nozzle exit which interrupt the shear-layer at its most receptive location thereby efficiently impacting the impingement tones. Due to their small size, these microjets can be optimally distributed along the circumference and can also be introduced on-demand.

Alvi *et al.* [11] showed that an open-loop control strategy that employs the microjets is effective in suppressing the impingement tones. It was also observed that the amount of suppression is dependent to a large extent on the operating conditions [11]. For example, it was observed in experimental studies that the amount of reduction that was achieved varied with the height of the lift-nozzle from the ground-plate as well as with the flow conditions. Since in practice, the operating conditions are expected to change drastically, a more attractive control strategy is one that employs feedback and has the ability to control the impingement tones over a large range of desired operating conditions.

3.1 Closed-loop control of impinging jets

In [12], it was shown that a possible reduced-order model of the impinging jets is of the form

$$\dot{T}(t) = \breve{A}(u)T(t) \tag{4}$$

where T corresponds to the states of the system, u is the control input, and A has lightly damped eigenvalues. Since the details of the A matrix were unknown, a systematic control method that guarantees stabilization was not implemented. However, it was observed that a closed-loop control strategy of the form

$$u = k\phi(\theta) \tag{5}$$

where the control input u is the microjet pressure distribution along the nozzle, ϕ is the most dominant Proper-Orthogonal-Decomposition(POD) mode of pressure measurements p and k is a calibration gain was quite successful in suppressing the pressure oscillations. The complete closed-loop procedure consisted of collecting pressure measurements $p(t)$, expanding them using POD modes, determining the dominant mode ϕ, and matching u to this dominant mode as in Eq. (5). That is, the control input adapts to the on-line eigenmode ϕ, which may vary with flow-conditions. This active-adaptive control strategy, which is denoted as "mode-matched" control, was used to determine the control input in the experimental investigations using the STOVL facility at Florida State University.

3.2 Experimental results

The mode-matched control strategy described above was implemented at the STOVL supersonic jet facility of the Fluid Mechanics Research Laboratory, FSU (see [13] for details). Four banks of microjets were distributed around the nozzle exit, while pressure fluctuations were sensed using six $Kulite^{TM}$ tranducers placed symmetrically around the nozzle periphery plate, at $r/d = 1.3$, from the nozzle centerline where d is the nozzle throat diameter, and r is the radial distance of the transducer from the center. The jets were fabricated using 400 μm diameter stainless tubes and are oriented at approximately 20° with respect to the main jet axis. The supply for the microjets was provided from compressed nitrogen cylinders through a main and four secondary plenum chambers. In this manner, the supply pressures to each bank of microjets could be independently controlled. The control experiment was performed for a range of heights of the nozzle above ground.

4 Summary

Problems of acoustic resonances in flows and their active control were discussed in this chapter. These problems include continuous combustion processes, which exhibit thermoacoustic instability as a result of feedback interactions between acoustics and heat-release, and supersonic impinging flows, which exhibit acoustic resonances due to feedback coupling between shear-layer dynamics and acoustics. In both cases, introducing external inputs such as pulsed fuel and additional air-flow via microjets produce a drastic change in the underlying dynamics, and cause the resonances to be damped out. In the combustion dynamics, it is observed that time-delays play a dominant role and pose a significant challenge to the control design. It is shown that by using an adaptive Posi-cast controller, even large delays can be accommodated and satisfactory pressure suppression can be obtained.

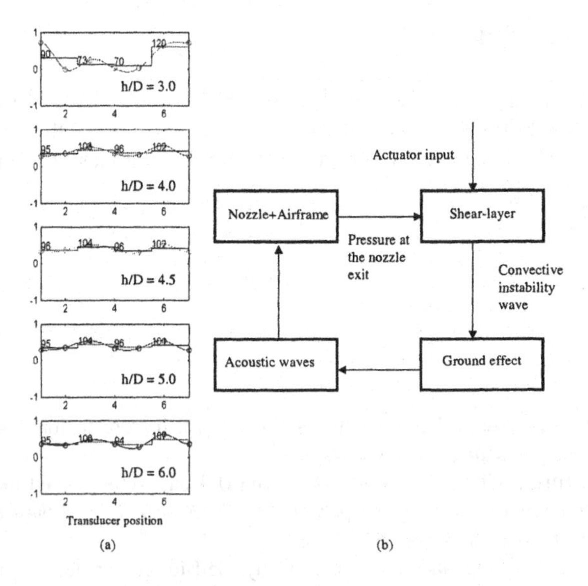

Fig. 6. (a) The first mode shape and suggested microjet pressure distribution for each height. h is the height of the lift-plate from ground and D is the diameter of the lift-plate. (b) Block diagram of the closed-loop control program of impingement tones.

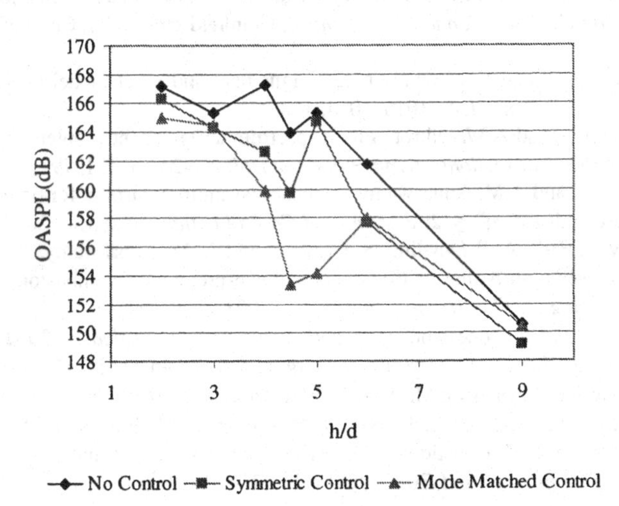

Fig. 7. Overall sound pressure levels (OASPL) without control, with open-loop control and with closed-loop control strategies at NPR=3.7.

5 Acknowledgements

This work was supported in part by the Propulsion Program of the Office of Naval Research, grant No. N00014-99-0448 and in part by the Air Force Office of Scientific Research, through the Unsteady Aerodynamics and Hypersonics program, with Dr. Schmisseur as the Program Manager.

References

1. A. M. Annaswamy and A. F. Ghoniem. Active control in combustion systems. *IEEE Control Systems Magazine*, 22(6):37–54, 2002.
2. A.M. Annaswamy, J. Choi, D. Sahoo, F. Alvi, and H. Lou. Active closed-loop control of supersonic impinging jet flows using pod models. In *Proceedings of the Conference on Decision and Control*, Las Vegas, NV, 2002.
3. S. Evesque, A.P. Dowling, and A.M. Annaswamy. Self-tuning regulators for combustion oscillations. *Proc. Royal Society, Series A*, 459:1709–1749, 2003.
4. P. Mehta, A. Banaszuk, M. Soteriou, and I. Mezic. On Reduced Order Modeling and Control of a Ducted Bluff Body Flame. In *Proceedings of the CDC*, Maui, HI, USA, December 2003.
5. S. Park, B. Pang, K. Yu, A. M. Annaswamy, and A. F. Ghoniem. Performance of an Adaptive posi-cast Controller in a Liquid Fueled Dump Combustor. In *International Colloquium on Combustion and Noise Control*, Cranfield University, Cranfield, UK, August 2003.
6. K. Ichikawa. Frequency-domain pole assignement and exact model-matching for delay systems. *Int. J. Control*, 41:1015–1024, 1985.
7. A.Z. Manitius and A.W. Olbrot. Finite spectrum assignement problem for systems with delays. *IEEE Transactions on Automatic Control*, AC-24 no. 4, 1979.
8. S. Niculescu and A.M. Annaswamy. An adaptive smith-controller for time-delay systems with relative degree $n^* \leq 2$. *Systems and Control Letters*, to appear, 2003.
9. A. Riley, S. Park, A. P. Dowling, S. Evesque, and A. M. Annaswamy. Adaptive Closed-loop Control on an Atmospheric Gaseoous Lean-premixed Ccombustor. *ASME Journal of Engineering for Gas Turbines and Power*, 2003 (to appear).
10. A. Powell. On edge tones and associated phenomena. *Acoustica*, 3:233–243, 1953.
11. F. S. Alvi, R. Elavarasan, C. Shih, G. Garg, and A. Krothapalli. "Control of Supersonic Impinging Jet Flows using Microjets". In *AIAA 2000-2236*, 2000.
12. A.M. Annaswamy, J. Choi, D. Sahoo, O. Egungwu, H. Lou, and F.S. Alvi. Active-adaptive control of acoustic resonances in supersonic impinging jets. In *AIAA Fluid Dynamics Conference and Exhibit*, Orlando, FL, June 2003.

13. H. Lou, F.S. Alvi, C. Shih, J. Choi, and A. Annaswamy. Active control of supersonic impinging jets: Flowfield properties and closed-loop strategies. Technical Report 2002-2728, AIAA Paper, 2002.
14. C. Shih, F. S. Alvi, H. Lou, G. Garg, and A. Krothapalli. "Adaptive Flow Control of Supersonic Impinging Jets". In *AIAA 2001-3027*, 2001.

Robust Prediction-Dased Control for Unstable Delay Systems

Rogelio Lozano[1], Pedro Garcia Gil[2], Pedro Castillo[1], and Alejandro Dzul[3]

[1] Heudiasyc-UTC, UMR CNRS 6599, B.P. 20529, Compiègne, France
rlozano@hds.utc.fr castillo@hds.utc.fr

[2] Dept. of Systems Engineering and Control, Universidad Politecnica de Valencia, P. O. Bax 22012, E-46071 Valencia, Spain
pggil@isa.upv.es

[3] Divisi6n de Estudios de Posgrado e Investigaci6n, Instituto Tecnol6gico da la Laguna 27000 Torre6n, Coahuila, México
dzul@hds.utc.fr

Summary. We present a discrete-time prediction based state-feedback controller. It is shown that this controller stabilizes possibly unstable continuous-time delay systems. The stability is shown to be robust with respect to uncertainties in the knowledge on the plant parameters, the system delay and the sampling period. The proposed prediction based controller has been tested in a real-time application to control the yaw angular displacement of a 4-rotor mini-helicopter.

1 Introduction

The area of control of delay systems has attracted the attention of many researchers in the past few years [10], [11], [18]. This is motivated by the fact that delays are responsible for unstabilities in closed-loop control systems. Delays appear due to transport phenomenons, computation of the control input, time-consuming information processing in measurement devices, etc.. A number of approaches for the control of systems with delay are available as the Smith predictor [8], [9], [15], [13] and its many improved schemes generically named Process-Model Control schemes [16], finite spectrum assignment techniques [6], reduction achieved through transformations and algebraic approaches as those in [12], [4], [3].

A close analysis of these methods show that they all use, in an explicit or implicit manner, prediction of the state in order to achieve the control of the system. A common drawback, linked to the internal unstability of the prediction, is that they fail to stabilize unstable systems.

The well known pole-placement controller proposed by [6] requires the computation of an integral used to predict the state. In the ideal case this control scheme leads to a finite pole-placement. However, arbitrary small errors in the computation of the integral term produce unstability, as shown in [7]. This can also be explained by the fact that in the ideal case the closed-loop behavior is governed by a finite polynomial while in presence of small errors, the closed-loop behavior is given by a quasi-polynomial having an infinite number of roots. Since in practice we normally use a computer to implement the control law [1], it is justified to study whether unstabilities can also appear in discrete-time pole-placement control algorithms. Note that small variations in the sampling period may be such that the closed-loop behavior will be described by a quasi-polynomial in the complex variable z. The zero location of quasi-polynomials are known to be very sensitive to small changes in the polynomial parameters and can easily move from the stable region to the unstable region. Therefore it is important to prove robustness also with respect to small variations of the sampling period. To our knowledge, such type of robustness has not been studied in the literature for discrete-time systems.

In this chapter we present the stability analysis of an hybrid control scheme, *i.e.* when the system representation is given in continuous-time while the controller is expressed in discrete-time. The controller is basically a discrete-time state-feedback control in which the actual state is replaced by the prediction of the state. We present a stability proof based on Lyapunov analysis of the hybrid closed-loop system. Convergence of the state to the origin is insured regardless of whether the original system is stable or not. The stability is established in spite of uncertainties in the knowledge of the plant parameters and the delay. Robusteness is also proved with respect to small variations of the time between sampling instants.

The proposed prediction-based controller has been tested in a real-time application to control the yaw angular displacement of a 4-rotor mini-helicopter [5]. The experimental validation of the proposed algorithm, has been developed on a novel real-time system, MaRTE OS, which allows the implementation of minimum real-time systems according to standard POSIX.13 del IEEE [14].

The chapter is organized as follows: the discrete-time representation of the system including uncertainties is presented in section 2. Section 3 is devoted to present the state predictor. The prediction-based state-feedback controller is given in Section 4. Section 5 presents the stability analysis of the hybrid closed-loop system. In Section 6 we show the experimental results and finally, the conclusions are given in Section 7.

2 Problem formulation

Let us consider the following continuous-time state space representation of a system with input delay

$$\dot{x}(t) = A_c x(t) + B_c u(t-h) \tag{1}$$

where the nominal plant parameter matrices are $A_c \in \Re^{n\times n}$, $B_c \in \Re^{n\times m}$ and h is the plant delay. Usually, in the discrete-time framework, the sampling time instant t_k is defined as $t_k = kT$ where T is the sampling period and k is an integer. However, since we wish to prove robustness of the control scheme with respect to the time elapsed between sampling time instants, we will not define t_k as a multiple of T. We will rather define t_k as the *k-th* sampling instant and such that

$$t_{k+1} - t_k = T + \varepsilon \tag{2}$$

where T is the ideal sampling period and ε is a small variation of the time between sampling instants. Furthermore, we will assume that T and h satisfy

$$h = dT + \epsilon \tag{3}$$

where d is an integer and ϵ is a small uncertainty in the knowledge of the delay h. Both variations ε and ϵ can be positive or negative and even time-varying. However ε and ϵ have to be bounded in such a way that

$$|\epsilon| \leq \overline{\epsilon} \ll T.$$

and

$$|\varepsilon| \leq \overline{\varepsilon} \ll T.$$

We will use the notation $x_k = x(t_k)$. From (1) we obtain the following time response equation

$$x_{k+1} = A_1 x_k + \int_{t_k}^{t_{k+1}} e^{A_c(t_{k+1}-\tau)} B_c u(\tau - dT - \epsilon) d\tau \tag{4}$$

where

$$A_1 = e^{A_c(t_{k+1}-t_k)} = e^{A_c(T+\varepsilon)} \tag{5}$$

We will define A as

$$A = e^{A_c T} \tag{6}$$

and Δ_4 such that

$$A_1 = A + \Delta_4 \tag{7}$$

Since we are interested in implementing the control law in a computer, we will assume that the input u is constant between sampling instants, *i.e.* $u(t) = u_k \; \forall t \in [t_k, t_{k+1})$.

We will next obtain a recursive equation for x_k in which the influence of the uncertainties in the plant parameters A_c and B_c ,the delay h and the ideal sampling period T will appear clearly. We will first study separately the cases when $\epsilon > 0$ and $\epsilon < 0$ and then obtain a general state space recursive expression for x_k.

2.1 Case : $\epsilon > 0$

We can rewrite (4) as

$$
\begin{aligned}
x_{k+1} = A_1 x_k &+ \int_{t_k}^{t_k+\epsilon} e^{A_c(t_{k+1}-\tau)} B_c d\tau \; u(t_k - dT - T) \\
&+ \int_{t_k+\epsilon}^{t_{k+1}} e^{A_c(t_{k+1}-\tau)} B_c d\tau \; u(t_k - dT)
\end{aligned} \tag{8}
$$

or

$$
x_{k+1} = A_1 x_k + B_1 u_{k-d} + \Delta_1 u_{k-d-1} \tag{9}
$$

where

$$
\begin{aligned}
B_1 &= \int_{t_k+\epsilon}^{t_{k+1}} e^{A_c(t_{k+1}-\tau)} B_c d\tau \\
\Delta_1 &= \int_{t_k}^{t_k+\epsilon} e^{A_c(t_{k+1}-\tau)} B_c d\tau
\end{aligned}
$$

For future use we will decompose B_1 as

$$
B_1 = B + \Delta_0 \tag{10}
$$

where

$$
\begin{aligned}
B &= \int_{t_{k+1}-T}^{t_{k+1}} e^{A_c(t_{k+1}-\tau)} B_c d\tau \\
&= \int_{-T}^{0} e^{-A_c s} B_c ds
\end{aligned} \tag{11}
$$

and

$$
\Delta_0 = \int_{t_k+\epsilon}^{t_k+\varepsilon} e^{A_c(t_{k+1}-\tau)} B_c d\tau \tag{12}
$$

2.2 Case : $\epsilon < 0$

We can rewrite (4) as

$$
\begin{aligned}
x_{k+1} = A_1 x_k &+ \int_{t_k}^{t_{k+1}-|\epsilon|} e^{A_c(t_{k+1}-\tau)} B_c d\tau \; u_{k-d} \\
&+ \int_{t_{k+1}-|\epsilon|}^{t_{k+1}} e^{A_c(t_{k+1}-\tau)} B_c d\tau \; u_{k-d+1}
\end{aligned} \tag{13}
$$

or

$$
x_{k+1} = A_1 x_k + B_2 u_{k-d} + \Delta_3 u_{k-d+1} \tag{14}
$$

where

$$B_2 = \int_{t_k}^{t_{k+1}-|\epsilon|} e^{A_c(t_{k+1}-\tau)} B_c d\tau$$
$$\Delta_3 = \int_{t_{k+1}-|\epsilon|}^{t_{k+1}} e^{A_c(t_{k+1}-\tau)} B_c d\tau$$

Let us rewrite B_2 as

$$B_2 = B + \Delta_2' \tag{15}$$

where B is given in (11) and

$$\Delta_2' = -\int_{t_{k+1}-|\epsilon|}^{t_{k+1}} e^{A_c(t_{k+1}-\tau)} B_c d\tau + \int_{t_k}^{t_k+\varepsilon} e^{A_c(t_{k+1}-\tau)} B_c d\tau \tag{16}$$

In general we have

$$\begin{aligned} x_{k+1} &= Ax_k + Bu_{k-d} + \Delta_1 u_{k-d-1} + \Delta_2 u_{k-d} \\ &\quad + \Delta_3 u_{k-d+1} + \Delta_4 x_k \\ &= Ax_k + Bu_{k-d} + \Delta f_k \end{aligned} \tag{17}$$

where

$$\Delta_2 = \begin{cases} \Delta_2' & \epsilon < 0 \\ \Delta_0 & \epsilon > 0 \end{cases}$$

and $\Delta \in \Re^{n \times s}$ and $f_k \in \Re^s$ with $s = 3m + n$ are defined as

$$\begin{aligned} \Delta &= [\Delta_1, \Delta_2, \Delta_3, \Delta_4] \\ f_k &= \left[u_{k-d-1}^T, u_{k-d}^T, u_{k-d+1}^T, x_k^T\right]^T \end{aligned} \tag{18}$$

(17) can be viewed as a general state-space representation for discrete-time systems in which Δ takes into account uncertainties in matrices A_c and B_c, in the delay h and in the ideal sampling period T. We assume that the nominal plant parameters A_c and B_c and the ideal sampling period T are such that (A, B) is a controllable pair. Note that ε and ϵ in (2) and (3) are in general time-varying. Rigorously speaking we should use the notation ϵ_k and ε_k. The same is true for Δ in (17). However, to prove robustness of the control scheme we will mainly use the property that $\Delta \to 0$ as the uncertainties in A_c, B_c, h(*i.e* ϵ) and T(*i.e.* ε) go to zero. Therefore, to simplify the notation in the rest of the chapter we will use Δ without the subscript k.

3 d-step ahead prediction scheme

From (17) the prediction of x_{k+2} is given by

$$\begin{aligned} x_{k+2} &= A(Ax_k + Bu_{k-d} + \Delta f_k) + Bu_{k-d+1} + \Delta f_{k+1} \\ &= A^2 x_k + ABu_{k-d} + Bu_{k-d+1} + A\Delta f_k + \Delta f_{k+1} \end{aligned} \tag{19}$$

Similarly we have

$$\begin{aligned} x_{k+3} &= A(A^2 x_k + ABu_{k-d} + Bu_{k-d+1} + A\Delta f_k + \Delta f_{k+1}) \\ &\quad + Bu_{k-d+2} + \Delta f_{k+2} \\ &= A^3 x_k + A^2 Bu_{k-d} + ABu_{k-d+1} + Bu_{k-d+2} \\ &\quad + A^2 \Delta f_k + A\Delta f_{k+1} + \Delta f_{k+2} \end{aligned} \tag{20}$$

Extending this prediction *d* steps ahead we have

$$\begin{aligned} x_{k+d} &= A^d x_k + A^{d-1} Bu_{k-d} + ... + ABu_{k-2} + Bu_{k-1} \\ &\quad + A^{d-1}\Delta f_k + A^{d-2}\Delta f_{k+1} + ... + \Delta f_{k+d-1} \end{aligned} \tag{21}$$

or

$$\begin{aligned} x_{k+d} &= A^d x_k + A^{d-1} Bu_{k-d} + ... + ABu_{k-2} \\ &\quad + Bu_{k-1} + \overline{\Delta}\, \overline{f}_{k+d-1} \end{aligned} \tag{22}$$

where $\overline{\Delta}$ and $\overline{f}_{k+d-1}$ are given by

$$\overline{\Delta} = \left[A^{d-1}\Delta,\ A^{d-2}\Delta, ..., \Delta\right] \tag{23}$$

and

$$\overline{f}_{k+d-1} = \left[f_k^T,\ f_{k+1}^T, ...,\ f_{k+d-1}^T\right]^T \tag{24}$$

Define x_{k+d}^p as the prediction of the state x_{k+d} at time t_k

$$x_{k+d}^p = A^d x_k + A^{d-1} Bu_{k-d} + ... + Bu_{k-1} \tag{25}$$

Note that x_{k+d}^p can be computed with information available at time t_k.

4 Prediction-based state feedback control

Let us define the following prediction-based control input

$$u_k = K^T x^p_{k+d} \tag{26}$$

or using (25)

$$u_k = K^T (A^d x_k + A^{d-1} B u_{k-d} + ... + B u_{k-1}) \tag{27}$$

From the above and (22) it follows that

$$u_k = K^T (x_{k+d} - \overline{\Delta}\, \overline{f}_{k+d-1}) \tag{28}$$

Introducing the above equation into (17) we obtain

$$x_{k+1} = (A + BK^T) x_k - BK^T \overline{\Delta}\, \overline{f}_{k-1} + \Delta f_k \tag{29}$$

As will be shown next, for small parameter and delay uncertainties, the stability of the above system will be insured if $A + BK^T$ is stable and if we can show that $\overline{f}_{k-1}$ and f_k are linear combinations of the elements of the closed-loop system state

$$z_k = \left[x_k^T, ..., x_{k-d}^T, u_{k-d-1}^T, ..., u_{k-2d-1}^T\right]^T \tag{30}$$

where $z_k \in \Re^l$ with $l = (d+1)(n+m)$. Recall from (17) and (18) that

$$\Delta f_k = \Delta_1 u_{k-d-1} + \Delta_2 u_{k-d} + \Delta_3 u_{k-d+1} + \Delta_4 x_k \tag{31}$$

In the above equation $u_{k-d-1}$and x_k are clearly elements of z_k in (30). Using Equation (27), u_{k-d} above can be expressed in terms of $x_{k-d}, u_{k-2d}, ...,$ and u_{k-d-1} which are elements of z_k. Similarly, u_{k-d+1}can be expressed in terms of $x_{k-d+1}, u_{k-2d+1}, ...$ and u_{k-d}. As before, u_{k-d} can be expressed in terms of elements of z_k. Therefore f_k in (29) can be expressed as a function of the elements of z_k. Note also that we can prove similarly that f_{k-1} is a function of z_k.

From (23) and (24) we have

$$\overline{\Delta}\, \overline{f}_{k-1} = A^{d-1} \Delta f_{k-d} + A^{d-2} \Delta f_{k-d+1} + ... + \Delta f_{k-1} \tag{32}$$

In view of (18), $f_{k-d}, f_{k-d+1}, .., $ and f_{k-2} in the above equation, are functions of z_k in (30). As explained before f_{k-1} is also a function of z_k and we conclude that $\overline{f}_{k-1}$ in (32) is a function of z_k. Therefore the term $-BK^T \overline{\Delta}\, \overline{f}_{k-1} + \Delta f_k$ in (29) can be expressed as

$$-BK^T \overline{\Delta}\, \overline{f}_{k-1} + \Delta f_k = \Delta' z_k \tag{33}$$

where Δ' is a matrix whose elements vanish as Δ goes to zero. From (28) we get

$$u_{k-d} = K^T (x_k - \overline{\Delta}\, \overline{f}_{k-1}) \tag{34}$$

or

$$u_{k-d} = K^T x_k + \Delta'' z_k \tag{35}$$

where Δ'' is a matrix whose elements vanish as Δ goes to zero. From (29), (33) and (35), the closed-loop system can be written as

$$\begin{pmatrix} x_{k+1} \\ x_k \\ \vdots \\ x_{k-d+1} \\ u_{k-d} \\ u_{k-d-1} \\ \vdots \\ u_{k-2d} \end{pmatrix} = \begin{bmatrix} (A+BK^T) & 0 & 0 & \cdots & \cdots & \cdots & \cdots & 0 \\ 1 & 0 & 0 & \cdots & \cdots & \cdots & \cdots & 0 \\ \vdots & \ddots & \ddots & \cdots & \cdots & \cdots & \cdots & \vdots \\ 0 & 0 & 1 & \ddots & \cdots & \cdots & \cdots & 0 \\ K^T & 0 & \ddots & \ddots & \ddots & \cdots & \cdots & 0 \\ \vdots & \cdots & \cdots & \ddots & 1 & \ddots & \cdots & \vdots \\ \vdots & \cdots & \cdots & \cdots & \ddots & \ddots & \ddots & \vdots \\ 0 & 0 & \cdots & \cdots & \cdots & 0 & 1 & 0 \end{bmatrix} \begin{pmatrix} x_k \\ x_{k-1} \\ \vdots \\ x_{k-d} \\ u_{k-d-1} \\ u_{k-d-2} \\ \vdots \\ u_{k-2d-1} \end{pmatrix} + \begin{bmatrix} \Delta' \\ 0 \\ \vdots \\ 0 \\ \Delta'' \\ 0 \\ \vdots \\ 0 \end{bmatrix} z_k \tag{36}$$

With obvious notation we rewrite the above system as

$$z_{k+1} = \overline{A} z_k + \overline{B} z_k \tag{37}$$

where $\overline{B} \to 0$ as $\Delta \to 0$ and $\overline{A} \in \Re^{l \times l}$, $\overline{B} \in \Re^{l \times l}$ with $l = (d+1)(n+m)$. Note that from (3) it follows that $d \to \infty$ as $T \to 0$. This means that as $T \to 0$ the closed-loop system in (36) becomes infinite dimensional. In the following section we present a stability analysis of the closed-loop system (36) when $T \neq 0$ *i.e.* when the dimension of z_k in (36) is finite.

5 Stability of the closed-loop system

We will now prove the stability of the closed-loop system in (36) or (37) and robustness with respect to small uncertainties in A_c, B_c, h and T in the system (1). It can be seen from (36) and (37) that the eigenvalues of $\overline{A}$ are given by the set of the n eigenvalues of $(A + BK^T)$ and $(l - n)$ eigenvalues at the origin. If K is chosen such that $(A + BK^T)$ is a Schur matrix, then $\overline{A}$ is also a Schur matrix *i.e.* $\overline{A}$ has all its eigenvalues strictly inside then unit circle. It then follows that for every $Q > 0$, $\exists\, P > 0$ such that the following Lyapunov equation holds

$$\overline{A}^T P \overline{A} - P = -Q \tag{38}$$

Let us define the candidate Lyapunov function V_k

$$V_k = z_k^T P z_k \tag{39}$$

From (37), (38) and (39) we have

$$\begin{aligned} V_{k+1} &= z_{k+1}^T P z_{k+1} \\ &= (\overline{A} z_k + \overline{B} z_k)^T P (\overline{A} z_k + \overline{B} z_k) \\ &= z_k^T \overline{A}^T P \overline{A} z_k + 2 z_k^T \overline{B}^T P \overline{A} z_k + z_k^T \overline{B}^T P \overline{B} z_k \\ &= V_k - z_k^T Q z_k + z_k^T (2\overline{B}^T P \overline{A} + \overline{B}^T P \overline{B}) z_k \end{aligned} \tag{40}$$

If the uncertainties are small enough such that

$$-Q + \left\| 2\overline{B}^T P \overline{A} + \overline{B}^T P \overline{B} \right\| < -\delta Q \tag{41}$$

then

$$V_{k+1} - V_k < -\delta z_k^T Q z_k \tag{42}$$

It then follows that $z_k \to 0$ exponentially as $k \to \infty$. Given that x and u converge to zero at the sampling instants (see (30)), it follows that $u(t)$ converges to zero $\forall t$ as $t \to \infty$. From (1) it follows that $x(t)$ converges to zero $\forall t$ as $t \to \infty$.

6 Practical application

In this section we show that the proposed controller has a satisfactory behavior when applied to control the yaw displacement of a mini helicopter. We use a mini-helicopter having 4 rotors as shown in figure (1).

In this type of helicopters the front and the rear motors rotate clockwise while the other two rotate counter-clockwise which reduces the gyroscopic phenomena. The 4-rotor helicopter does not have a swatch plate. In fact it does not need any servomechanism. The main thrust is the sum of the thrusts of each motor. Pitch movement is obtained by increasing (reducing) the speed of the rear motor while reducing (increasing) the speed of the front motor.

The roll movement is obtained similarly using the lateral motors. The yaw movement is obtained by increasing (decreasing) the speed of the front and rear motors while decreasing (increasing) the speed of the lateral motors. This is done while keeping the total thrust constant. Delays are introduced to the system due to the position/orientation measuring system and also due to the computation of the control input.

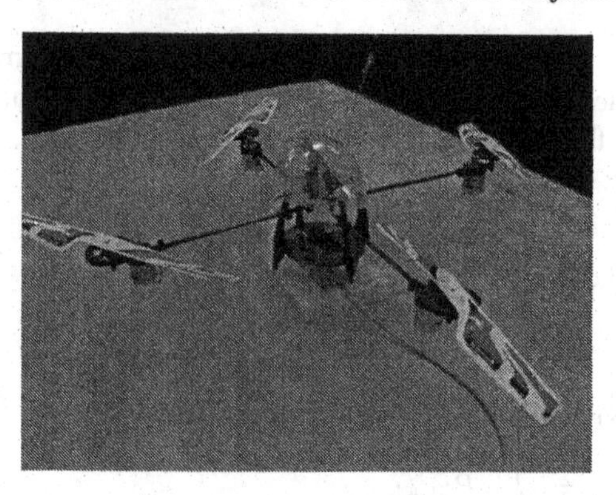

Fig. 1. The 4 rotors helicopter.

We aim at using visual servoing control for the mini-helicopter in future work. We know that image processing will introduce a considerable delay and one of our objectives in this chapter is to show that our prediction-based control algorithm can be used to avoid unstabilities in the position and orientation control of a flying vehicle.

The radio used for radio control is a Futaba Skysport 4. The radio and the PC (INTEL Pentium 3) are connected using data acquisition cards (ADVANTECH PCL-818HG and PCL-726). In order to simplify the experiments, the control inputs can be independently commuted between the automatic and the manual control modes.

The connection in the radio is directly made to the joystick potentiometers for the thrust, yaw, pitch and roll controls. The helicopter evolves freely in a 3D space without any flying stand.

We use the 3D tracker system (POLHEMUS) [17] for measuring the position (x,y,z) and orientation (ϕ, θ, ψ) of the helicopter. The Polhemus is connected via RS232 to the PC.

6.1 Real-Time implementation

We present in this section the characteristics and implementation of real-time control system environment that we have used. We use an embedded system based on the MaRTE OS environment.

MaRTE OS [2] is a real-time kernel for embedded applications that follows the Minimal Real-Time POSIX.13 subset [14], providing both the C and Ada language POSIX interfaces. It allows cross-development of Ada and C real-time applications. Mixed Ada-C applications can also be developed, with a globally consistent scheduling of Ada tasks and C threads.

MaRTE OS works in a cross development environment. The host computer is a Linux PC with the *gnat* and *gcc* compilers. The target platform is any bare machine based on any 386 PC or higher, with a floppy disk (or equivalent) for booting the application, but not requiring a hard disk.

Figure (2) shows the interaction between the system and the external devices.

To design the real time control five main tasks have been defined:

- Control_Task: this periodic task gets information of the helicopter position and calculate the actions to be sent to the motors.
  This task has a period of 80 ms. The control actions are sent to a shared protected object which stores the system information. The actions are not sent directly to the motors.
- Send_Actions: this is a periodic task which is in charge of extracting the information from the control status and send the motor actions using the digital/analog converter. This task can introduce forced delays in the actions to be sent to the motors in order to test different control algorithms. The forced delays are introduced by getting actions calculated in previous periods when the delay is greater than the control period. If the delay is less than the period then an internal delay is executed.
- Monitor: This is a periodic task for control status monitoring. The task gets information from the shared object control status and send it to a RS232 line to be used by the host to visualize the control variables.
- User_Commands_Task: this task reads user commands from the keyboard and execute them. User commands can change the monitoring period, change control parameters or start and stop the control.
- Control_Status: this is a shared protected object where the tasks get or put information about the process.

Several drivers have been implemented to handle the RS-232 serial line, keyboard, and the digital/analog converters.

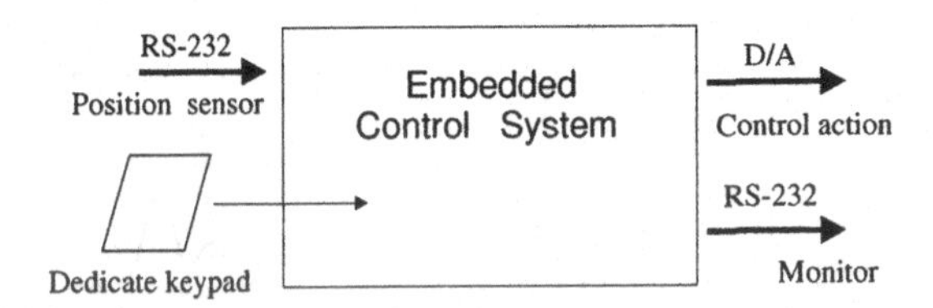

Fig. 2. Interaction between the system and the external devices.

6.2 Experimental results

The transfer function from the yaw-control input to the yaw-displacement has been identified by introducing a pulse input while the mini-helicopter was hovering.

The obtained pulse response is shown in figure 3. We know that the mini-helicopter has an in-built gyro that introduces an angular velocity feedback. The transfer function without the gyro is basically a double integrator. However, the transfer function of the system including the angular velocity feedback, has a pole at the origin and a negative real pole.

We assumed that the system was represented by a second order system with two parameters. Trying different values for the parameters we observed that the following model has a behavior that is close to the behavior of the real system:

$$G(s) = \frac{200}{s(s+4)} \tag{43}$$

A simple controller as

$$u_k = 0.08(y^* - y_k) \tag{44}$$

where y^* is a reference signal, can be used to stabilize the model (43).

However, when there is a delay of 3 sampling periods (0.24 seconds) in the measurement of the yaw angular position, the controller becomes

$$u_k = 0.08(y^* - y_{k-3}) \tag{45}$$

and the closed loop system behavior is unstable as can be seen in figure 4.

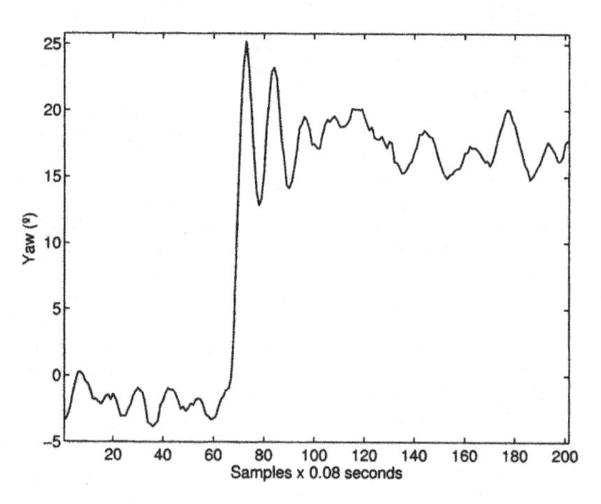

Fig. 3. Pulse response of the system without measurement delay.

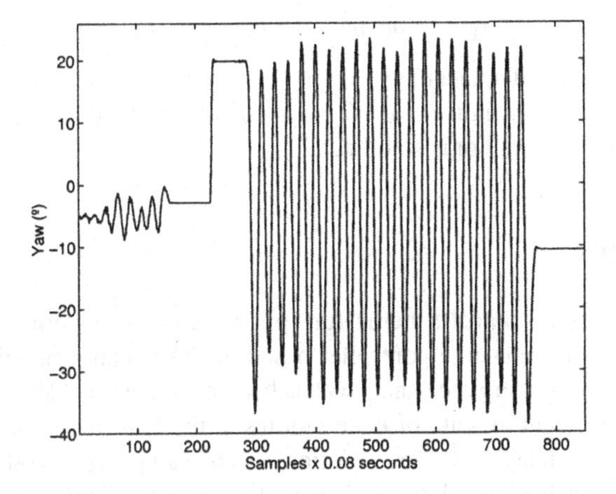

Fig. 4. Output of the delayed system when using the controller (45) without prediction.

Predictor-Based system stabilization

The discrete-time state-space representation for the model in (43) for $T = 0.08$ seconds, is given by:

$$\begin{bmatrix} x_{k+1}^1 \\ x_{k+1}^2 \end{bmatrix} = \begin{bmatrix} 0.7261 & 0 \\ 0.2739 & 1 \end{bmatrix} \begin{bmatrix} x_k^1 \\ x_k^2 \end{bmatrix} + \begin{bmatrix} 0.5477 \\ 0.0923 \end{bmatrix} u_k \tag{46}$$

$$y_k = \begin{bmatrix} 0 & 6.25 \end{bmatrix} \begin{bmatrix} x_k^1 \\ x_k^2 \end{bmatrix} \tag{47}$$

Since the state x_k is not measurable, we use the following observer

$$\begin{bmatrix} \hat{x}_{k+1}^1 \\ \hat{x}_{k+1}^2 \end{bmatrix} = \begin{bmatrix} 0.7261 & -2.7940 \\ 0.2739 & -1.0261 \end{bmatrix} \begin{bmatrix} \hat{x}_k^1 \\ \hat{x}_k^2 \end{bmatrix} \tag{48}$$

$$+ \begin{bmatrix} 0.4470 \\ 0.3242 \end{bmatrix} y_k + \begin{bmatrix} 0.5477 \\ 0.0923 \end{bmatrix} u_{k-3} \tag{49}$$

The prediction of the state 3-steps ahead can be done using the following equation (see also (25))

$$x_p(k) = \begin{bmatrix} 0.3829 & 0 & 0.2888 & 0.3977 & 0.5477 \\ 0.6171 & 1 & 0.3512 & 0.2423 & 0.0923 \end{bmatrix} \begin{bmatrix} \hat{x}_k^1 \\ \hat{x}_k^2 \\ u_{k-3} \\ u_{k-2} \\ u_{k-1} \end{bmatrix} \tag{50}$$

The control law in (44) (see also (26)) becomes:

$$u_k = 0.08(y^* - \begin{bmatrix} 0 \ 6.25 \end{bmatrix} x_p(k)) \tag{51}$$

The yaw angular displacement of the mini-helicopter when using the above control law is shown in figure (5). We have chosen y^* as a square wave function. As it can be seen, the system is stabilized.

7 Conclusions

We have presented a control scheme for continuous-time systems with delay. We have proposed a discrete-time controller based on state feedback using the prediction of the state. A convergence analysis has been presented that shows that the state converges to the origin in spite of uncertainties in the knowledge of the plant parameters, the system delay and even variations of the sampling period. The proposed control scheme has been implemented to control the yaw displacement of a real 4-rotor mini-helicopter. Real-time experiments have shown a satisfactory performance of the proposed control scheme.

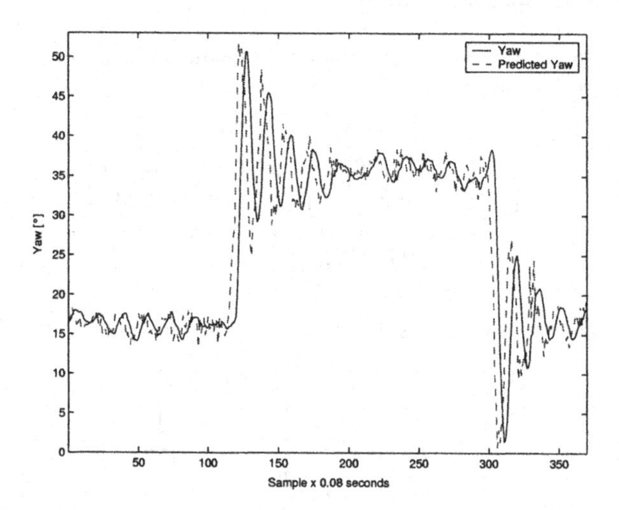

Fig. 5. Closed-loop behavior using the prediction-based controller.

References

1. Astrom K.J. and B. Wittenmark (1997). *Computer-Controlled Systems - Theory and Design*, Pretince-Hall, third edition.
2. Aldea M. and González M. "MaRTE OS: An Ada Kernel for Real-Time Embedded Applications". Proceedings of the International Conference on Realiable Software Technologies, Ada-Europe-2001, Leuven, Belgium, Lecture Notes in Computer Science, LNCS 2043, May 2001.

3. Brethé D. and Loiseau J.J. (1998). An effective algorithm for finite spectrum assignment of single-input systems with delays, Mathematics in computers and simulation, 45, 339-348.
4. Kamen E.W, Khargonekar P.P. and A. Tannenbaum (1986). Proper stable bezout factorizations and feedback control of linear time delay systems, *Int. J. Contr.*, Vol. 43, No. 3, 837-857.
5. R. Lozano, B. Brogliato, O. Egeland, B. Maschke. Passivity-based control system analysis and design. *Springer-Verlag,* Communications and Control Engineering Series, 2000. ISBN 1-85233-285-9.
6. Manitius, A. Z. and A.W Olbrot (1979). Finite Spectrum Assignment problem for Systems with Delays, *IEEE Trans. Autom. Contr.*, Vol. AC-24, No. 4, 541-553.
7. Mondié S., Dambrine M., Santos O. (2001a). Approximation of control laws with distributed delays: a necessary condition for stability, IFAC Conference on Systems, Structure and Control, Prague, Czek Republic.
8. Mondié S., Lozano, R. and Collado J. (2001b). Resetting Process-Model Control for unstable systems with delay, 40th IEEE Conference on Decision and Control, Orlando, Florida.
9. Mondié S., García P., Lozano R. (2002). Resetting Smith Predictor for the Control of Unstable Systems with Delay, IFAC 15th Triennial World Congress, Barcelona, Spain, 2002.
10. Niculescu S. (2001). Delay effects on stability: a robust control approach. Springer-Verlag, Heidelberg, Germany.
11. S. Niculescu, R. Lozano. On the passivity of linear delay systems. *IEEE Transactions on Automatic Control*, pp460-464, April 2001
12. Morse A.S., Ring Models for Delay-Differential Systems, Automatica, Vol. 12, 529-531, 1976.
13. Palmor Z.J. (1996). Time delay Compensation- Smith predictor and its modifications, in The Control Handbook, (W.S. Levine, Eds), CRSC Press, 224-237.
14. POSIX.13 (1998) IEEE Std. 1003.13-1998. Information Technology -Standardized Application Enviroment Profile- POSIX Realtime Application Support (AEP). The Institute of Electrical and Electronics Engineers.
15. Smith O.J.M. (1959). Closer Control of loops with dead time, *Chem. Eng. Prog.*,53, 217-219.
16. Watanabe K., Ito, M. (1981). A process model control for linear systems with delay, IEEE Trans. Autom. Contr., Vol. AC-26, No. 6, 1261-1268.
17. Fastrack 3Space Polhemus, *User's Manual*, Colchester, Vermont, USA.
18. C. Foias, H. Özbay, A. Tannenbaum. Robust Control of Infinite Dimensional Systems: Frecuency Domain Methods, Lecture Notes in Control and Information Sciences, No. 209, *Springer-Verlag,* London, 1996, ISBN 3-540-19994-2.

Robust Stability of Teleoperation Schemes Subject to Constant and Time-Varying Communication Delays

Damia Taoutaou[1], Silviu-Iulian Niculescu[1], and Keqin Gu[2]

[1] HEUDIASYC (UMR CNRS 6599), Université de Technologie de Compiègne, Centre de Recherche de Royallieu, BP 20529, 60205, Compiègne, cedex, France.
E-mail: silviu,taoutaou@hds.utc.fr

[2] Department of Mechanical and Industrial Engineering, Southern Illinois University at Edwarsville, Edwardsville, IL 62026, USA.
E-mail: kgu@siue.edu

Summary. This chapter addresses the robust stability of some bilateral teleoperation control scheme subject to various constant and/or time-varying delays in the communication channel. The stability conditions are derived using *frequency-domain* techniques. More specifically, in the case of constant delays, the stability regions of the systems' parameters are completely characterized. Next, the analysis is extended to the case of time-varying uncertain delay, and we derive sufficient (closed-loop) stability conditions.

1 Introduction

A basic *teleoperation system* consists of a *slave device* and a *master device*. The master is *directly manipulated* by a *human operator*, and the slave is designed to track the master closely. The main purpose of such a master-slave configuration is to manipulate the environment (or space) generally inaccessible to human operators, such as hazardous environment. Such systems are often known as a *bilateral teleoperator* systems.

Time delay plays an important role in the teleoperation systems. Due to the physical distance between the master and slave, as well as the signal processing, the communications involve significant delays. Another source of delay is the reaction of the human operators. In this chapter, we will discuss the effect of the communication delays (constant or time-varying) on the closed-loop stability of such systems.

In this context, we are interested in characterizing the way that delays change performances in communication channels connecting the master and slave sites (*bilateral teleoperation*). It is well known that the *passivity* of the channel (see, e.g., [1, 6, 17, 18, 24]) may be used to guarantee desirable characteristics for the closed-loop schemes (see also [10]). The techniques proposed to perform such an analysis use the scattering transformation [1] or the wave variable transformation [17, 18], if the delays are assumed constant. The case of time-varying or distributed delays was considered in [11, 19] using the wave transformation approach and in [14] but under some assumptions on the delay variation.

Consider the following equations widely used to describe the dynamics of teleoperators [1, 11].

$$\begin{cases} M_m \ddot{x}_m(t) + B_m \dot{x}_m(t) = F_h(t) - F_m(t) \\ M_s \ddot{x}_s(t) + B_s \dot{x}_s(t) = F_s(t) - (1+\alpha_f) Z_e \dot{x}_s(t), \end{cases} \tag{1}$$

where $\dot{x}, M, B$ are the velocities, inertias, and damping coefficients, respectively. The subscripts m and s denote the corresponding quantity is of the master and the slave, respectively. The input F_h denotes the operator force or torque, and Z_e is the environmental impedance. The quantity F_s is the force or torque applied to the slave transmitted from the master, and F_m is the force on the master fed back from the slave.

For an explicit stability analysis, see [6] for various frequency-domain techniques (see also [12]), and [2] for a Lyapunov functional approach. For delay-independent stability, the approach proposed in this chapter is simpler than the one proposed in [6], and the derived conditions are *necessary and sufficient*, and in an *analytical* form.

For delay-independent stability, the main idea is to use a frequency-domain method based on the Tsypkin's criterion [5, 12]. For frequency-sweeping tests applied to various control systems, see, for instance, [3]. Various discussions and comments related to such techniques can be found in [12]. Such an approach was used in [20] for the closed-loop stability analysis of a simple teleoperation control scheme, where delay-independent/delay-dependent stability conditions were derived under the assumption of symmetric delays in the channels ($\tau_1 = \tau_2 = \tau$).

As in [11], consider the control law described by the following equations

$$F_s(t) = K_s \int_0^t (\dot{x}_{sd}(t) - \dot{x}_s(t)) d\theta + B_{s2}(\dot{x}_{sd}(t) - \dot{x}_s(t)), \tag{2}$$

$$F_m(t) = K_m \int_0^t (\dot{x}_m(t) - \dot{x}_{md}(t)) d\theta + B_{m2}(\dot{x}_m(t) - \dot{x}_{md}(t)). \tag{3}$$

Due to communication delays, the most recently available information is used instead, that is, we choose,

$$\dot{x}_{sd}(t) = \dot{x}_m(t - \tau_1), \tag{4}$$

$$\dot{x}_{md}(t) = \dot{x}_s(t - \tau_2), \tag{5}$$

where τ_1 and τ_2 are the delays in the forward and feedback communication channels, respectively.

As mentioned above, we are interested in first finding *analytical conditions* on the system's parameters such that the closed-loop system is asymptotically stable for arbitrary communication delays. For those parameters which do not satisfy such delay-independent stability conditions, we will find the corresponding *delay intervals* such that the closed-loop system is stable. Furthermore, we are also interested in finding conditions for which there is only one delay interval, and computing the corresponding *optimal bounds*. A similar problem, but only with constant and symmetric time-delays (τ_1, τ_2), was considered in [20].

In the case of time-varying delay uncertainty, the idea is to construct an appropriate fictitious transfer function such that the stability of the original closed-loop scheme is reduced to some H_∞-norm property of the corresponding transfer. To the best of the authors' knowledge, such an approach was not considered in the bilateral teleoperation case.

The chapter is organized as follows: Section 3 is devoted to the stability analysis of the closed-loop system using frequency-domain techniques. Constraints on the controller's gain K_s and 'damping' B_{s1} will be given such that the closed-loop scheme is asymptotically stable *independent* of the *communication delays*. Next, the *delay-dependent* stability of the closed-loop system will be considered. Section 4 discusses the case of time-varying delays. Some concluding remarks end the chapter. The notations are standard.

2 Stability Analysis for Constant Delays

2.1 Problem setup

Carrying out the Laplace transform (under zero initial conditions) of the closed-loop system, using the velocities $v_m(t) = \dot{x}_m(t)$ and $v_s(t) = \dot{x}_s(t)$ as the system variables, we obtain

$$\begin{cases} M_m s V_m(s) + B_m V_m(s) = F_h(s) - F_m(s), \\ M_s s V_s(s) + B_s V_s(s) = F_s(s) - (1+\alpha_f) Z_e V_s(s), \end{cases} \tag{6}$$

and

$$F_s(s) = \frac{K_s + B_{s2}s}{s} e^{-\tau_1 s} V_m(s) - \frac{K_s + B_{s2}s}{s} V_s(s). \tag{7}$$

$$F_m(s) = \frac{K_m + B_{m2}s}{s} V_m(s) - \frac{K_m + B_{m2}s}{s} e^{-\tau_2 s} V_s(s). \tag{8}$$

Using the control laws (7) and (8) in the second equation of (6), with the notation $\overline{B}_s = B_s + (1+\alpha_f) Z_e$, it follows that:

$$V_s(s) = \frac{K_s + B_{s2}s}{M_s s^2 + (\overline{B}_s + B_{s2})s + K_s} e^{-\tau_1 s} V_m(s), \tag{9}$$

Let $\tau = \tau_1 + \tau_2$, and use the following notations:

$$\Gamma_1(s) = B_{s2}s + K_s : \text{slave torque}, \tag{10}$$

$$\Gamma_2(s) = M_s s + \overline{B}_s : \text{slave}, \tag{11}$$

$$\Gamma_3(s) = M_m s + B_m : \text{master}, \tag{12}$$

$$\Gamma_4(s) = B_{m2}s + K_m : \text{master torque}, \tag{13}$$

we obtain from the first equation of (6) and (8)

$$V_m(s)\Gamma_3(s) = F_h(s) + \frac{\Gamma_4(s)}{s}\left(\mathrm{e}^{-\tau_2 s}V_s(s) - V_m(s)\right).$$

Using (9) in the above, we obtain

$$V_m(s)\cdot\left(\frac{s\Gamma_3(s)+\Gamma_4(s)}{s} - \mathrm{e}^{-\tau s}\frac{\Gamma_4(s)\Gamma_1(s)}{s(\Gamma_1(s)+s\Gamma_2(s))}\right) = F_h(s) \tag{14}$$

Therefore, the transfer function from F_h to V_m is given by:

$$H_1(s) = \frac{1}{\frac{(s\Gamma_3(s)+\Gamma_4(s))}{s}\left(1 - \frac{\mathrm{e}^{-\tau s}\Gamma_4(s)\Gamma_1(s)}{\Gamma_1(s)+s\Gamma_2(s)}\frac{1}{(s\Gamma_3(s)+\Gamma_4(s))}\right)}. \tag{15}$$

Furthermore, based on the form of $V_s(s)$, the transfer function from F_h to V_s is given by:

$$H_2(s) = H_1(s)\cdot\frac{K_s + B_{s1}s}{M_s s^2 + (\overline{B}_s + B_{s1})s + K_s}\mathrm{e}^{-\tau_1 s}. \tag{16}$$

Since $M_s, B_s, B_{s1}, K_s, \alpha_f, Z_e$ are positive real numbers, $H_1(s)$ and $H_2(s)$ share the right half plane poles. Therefore, to study the stability of the closed-loop system, it is sufficient to study the stability of the transfer function $H_1(s)$. Or, equivalent, one needs only to study the distribution of zeros of the expression:

$$1 - \mathrm{e}^{-\tau s}\frac{\Gamma_4(s)\Gamma_1(s)}{\Gamma_1(s)+s\Gamma_2(s)}\frac{1}{(s\Gamma_3(s)+\Gamma_4(s))}. \tag{17}$$

We will first study asymptotic stability of the closed-loop system when it is *free from delays*. In this case, the zeros of the characteristic function (17) becomes those of the third-order polynomial:

$$P(s) = s\Gamma_2(s)\Gamma_3(s) + \Gamma_3(s)\Gamma_1(s) + \Gamma_4(s)\Gamma_2(s). \tag{18}$$

Using the Routh-Hurwitz stability criterion (see, for example, [8]), it follows that the system free from delays is asymptotically stable if and only if the following inequality holds:

$$\left(\frac{K_m}{M_m} + \frac{K_s}{M_s} + \frac{B_m B_{s2}}{M_m M_s} + \frac{\overline{B}_s(B_{m2}+B_m)}{M_s M_m}\right)\cdot\left(\frac{B_{s2}+\overline{B}_s}{M_s} + \frac{B_{m2}+B_m}{M_m}\right) > \frac{K_s B_m + K_m\overline{B}_s}{M_s M_m}. \tag{19}$$

It is not difficult to show that (19) is always valid for all positive parameters. Therefore, as expected, if the system is free from delay, the controller (2)-(5) guarantees the *asymptotic stability* of the closed-loop system.

2.2 Delay-independent stability

The next step is to find the conditions under which the stability in the closed-loop systems is guaranteed for *arbitrary communication delays* τ_1 and τ_2. First, under a certain parameter constraint, we will find necessary and sufficient conditions for stability. Next, we will provide a *simple* sufficient condition easy to use in practice.

Theorem 1. *Assume the feedback gains K_m and K_s, B_{m2} and B_{s2} are positive constants. Then the closed-loop system is asymptotically stable for all communication delays τ_1, τ_2 if and only if, $\forall \omega > 0$:*

$$| (j\omega)\Gamma_3(j\omega) + \Gamma_4(j\omega) |> \left| \frac{\Gamma_4(j\omega)\Gamma_1(j\omega)}{\Gamma_1(j\omega) + j\omega\Gamma_2(j\omega)} \right|. \tag{20}$$

Proof. In view of the form of $H_1(s)$, since $s\Gamma_3(s) + \Gamma_4(s)$ is Hurwitz stable, it follows that the stability of the closed-loop system (1)-(5) is equivalent to the stability of the unit feedback closed-loop system with the open-loop transfer function

$$H_o(s) = \frac{\Gamma_4(s)\Gamma_1(s)}{(s\Gamma_3(s) + \Gamma_4(s))(\Gamma_1(s) + s\Gamma_2(s))} e^{-s\tau}. \tag{21}$$

Since $(s\Gamma_3(s) + \Gamma_4(s))(\Gamma_1(s) + s\Gamma_2(s))$ is Hurwitz stable, and $H_o(s)$ is strictly proper for $\tau = 0$, then we may apply the Tsypkin's criterion, and the condition (20) follows directly.

Note that for $\omega = 0$,

$$|j\omega\Gamma_3(j\omega) + \Gamma_4(j\omega)| = \left| \frac{\Gamma_4(j\omega)\Gamma_1(j\omega)}{\Gamma_1(j\omega) + j\omega\Gamma_2(j\omega)} \right| = K_m$$

Furthermore, if (20) is verified for $\omega > 0$, then the same inequality holds for $\omega < 0$. The condition (20) in Theorem 1 is a simple *frequency-sweeping test* that can be easily performed if the parameters of the system and the controller are given. To obtain a even simpler criterion than (20), introduce the notation

$$\gamma(K_m, B_{m2}, K_s, B_{s2}) = \sup_{\omega>0} \left| \frac{\Gamma_4(j\omega)\Gamma_1(j\omega)}{\Gamma_1(j\omega) + j\omega\Gamma_2(j\omega)} \right|, \tag{22}$$

which depends continuously on the controller's parameters K_m, B_{m2}, K_s, B_{s2} (they are all real and positive). Then, we have the following natural corollary:

Corollary 1. *The closed-loop system is asymptotically stable for arbitrary communication delays $\tau_1, \tau_2 \geq 0$ if the controller gains K_s, K_m and the "damping coefficients" B_{s2}, B_{m2} are chosen to satisfy*

$$K_m < \frac{(B_m + B_{m2})^2}{2M_m} \tag{23}$$

$$\gamma(K_s, B_{s2}, K_m, B_{m2}) \leq K_m \tag{24}$$

Proof. The result is a straighforward from Theorem 1: The condition (23) ensures that $\mid j\omega\Gamma_3(j\omega) + \Gamma_4(j\omega) \mid$ is a strictly increasing function of ω, which implies $K_m <\mid j\omega\Gamma_3(j\omega) + \Gamma_4(j\omega) \mid$ for all $\omega > 0$. Therefore the condition (20) is implied by (24).

As given in the next Proposition, the condition (24) can be written out explicitly.

Proposition 1. *The closed-loop system is asymptotically stable for all communication delays* $\tau_1, \tau_2 \geq 0$, *if the controller's parameters satisfy:*

$$K_s \leq \frac{M_s K_m^2}{B_{m2}^2}\left(\sqrt{1 + \frac{B_{s2}^2 B_{m2}^2}{M_s^2 K_m^2}\left(\left(1 + \frac{\overline{B}}{B_{s2}}\right)^2 - 1\right)} - 1\right) \tag{25}$$

$$K_m \geq \frac{B_{s2} B_m}{M_s} \tag{26}$$

$$K_m < \frac{(B_m + B_{m2})^2}{2M_m} \tag{27}$$

Proof. We will show that (25) and (26) is necessary and sufficient condition for (24), which will be sufficient to complete the proof. Define

$$f : [0, \infty) \mapsto (0, \infty)$$

$$f(\omega^2) = \frac{\mid \Gamma_4(j\omega) \mid^2 \cdot \mid \Gamma_1(j\omega) \mid^2}{\mid \Gamma_1(j\omega) + j\omega\Gamma_2(j\omega) \mid^2}. \tag{28}$$

Then, $f(\omega^2)$ is in the form of

$$f(\omega^2) = K_m^2 \frac{a\omega^4 + b\omega^2 + 1}{d\omega^4 + e\omega^2 + 1}$$

where the denominator

$$d\omega^4 + e\omega^2 + 1 > 0 \text{ for all } \omega^2 > 0 \tag{29}$$

Therefore, the equation (23), or equivalently, $f(\omega^2) \leq K_m^2$, is equivalent to

$$a\omega^4 + b\omega^2 + 1 \leq d\omega^4 + e\omega^2 + 1 \text{ for all } \omega^2 > 0$$

in view of (29). But the above is satisfied if and only if

$$a \leq d \tag{30}$$

$$b \leq e \tag{31}$$

With the specific parameters substituted, (30) reduces to (26). The condition (31) is a quadratic inequality of K_s, which is satisfied if and only if (25) is satisfied in view of the fact that K_s is positive.

Remark 1 (Tuning parameters). Proposition 1 above gives a very simple way of constructing the controller (1) such that the closed-loop system is guaranteed to be asymptotically stable for all communication delays $\tau_1,\ \tau_2 \geq 0$.

2.3 Delay-dependent stability

If (20) is not satisfied for all $\omega > 0$, the conditions for Theorem 1 do not hold, and there must exist delays such that the system is unstable. Since the system without delays is asymptotically stable, there always exists one or more intervals of delay such that the system is asymptotically stable. We are interested in finding the maximum $\tau^* > 0$ such that the system is asymptotically stable for all $\tau \in [0, \tau^*)$. This can be carried out by solving the equation

$$| (j\omega\Gamma_3(j\omega) + \Gamma_4(j\omega)) |^2 = \left| \frac{\Gamma_4(j\omega)\Gamma_1(j\omega)}{\Gamma_1(j\omega) + j\omega\Gamma_2(j\omega)} \right|^2 \tag{32}$$

This equation can be reduced to a third order polynomial equation of the variable ω^2, and formulas are available to express the solutions explicitly (see, for example, [22]). Clearly, since (20) is not satisfied for all $\omega \geq 0$, and it is clearly satisfied for sufficiently large ω, the equation (32) has at least one real positive solution. Let all the real positive solutions be denoted as ω_i, $i = 1, 2, ..., m$. Clearly, $1 \leq m \leq 3$. Then, we can conclude:

Theorem 2 (Switch characterizations). *If (20) is not satisfied for all $\omega > 0$, let*

$$\tau^* = \min_{\ell \in \mathbf{Z}} \min_{1 \leq i \leq m} \frac{1}{\omega_i} \left[Log \left(\frac{\Gamma_4(j\omega)\Gamma_1(j\omega)}{(j\omega\Gamma_3(j\omega) + \Gamma_4(j\omega))(j\omega\Gamma_2(j\omega) + \Gamma_1(j\omega))} \right) + 2\pi\ell \right] > 0, \tag{33}$$

where "Log" denotes the principal value of the logarithm.
Then, the closed-loop system is asymptotically stable for all $\tau \in [0, \tau^*)$.

Proof. As discussed above, the equation (32) has one to three real positive solutions. If and only if ω is a real positive solution, there exists a τ satisfying the characteristic equation

$$(s\Gamma_3(s) + \Gamma_4(s)) - \mathbf{e}^{-\tau s} \frac{\Gamma_4(s)\Gamma_1(s)}{\Gamma_1(s) + s\Gamma_2(s)} = 0$$

for $s = j\omega$, some simple but tedious computations lead to the smallest $\tau > 0$ in (33). Specific discussions on deciding the stable delay intervals are very similar to [16].

3 Time-Varying Uncertain Delays

Introduce the vector of state variables $x = [x_1, ..., x_4]^T$, where

$$x_1(t) = \int_0^t v_m(\theta)d\theta, \quad x_2(t) = v_m(t) \tag{35}$$

$$x_3(t) = \int_0^t v_s(\theta)d\theta, \quad x_4(t) = v_s(t) \tag{36}$$

Then, the closed-loop system described by (1) to (5) can be written as

$$\dot{x}(t) = Ax(t) + B_1x(t-\tau_1) + B_2x(t-\tau_2) + B_3F_h(t) \tag{37}$$

where:

$$A = \begin{bmatrix} 0 & 1 & 0 & 0 \\ -\frac{K_m}{M_m} & -\frac{B_m+B_{m2}}{M_m} & 0 & 0 \\ 0 & 0 & 0 & 1 \\ 0 & 0 & -\frac{K_s}{M_s} & -\frac{(Bs+B_{s2}+(1+\alpha_f)Z_e)}{M_s} \end{bmatrix}$$

$$B_1 = \begin{bmatrix} 0 & 0 & 0 & 0 \\ 0 & 0 & 0 & 0 \\ 0 & 0 & 0 & 0 \\ \frac{K_s}{M_s} & \frac{B_{s2}}{M_s} & 0 & 0 \end{bmatrix}, \; B_2 = \begin{bmatrix} 0 & 0 & 0 & 0 \\ 0 & 0 & \frac{K_m}{M_m} & \frac{B_{m2}}{M_m} \\ 0 & 0 & 0 & 0 \\ 0 & 0 & 0 & 0 \end{bmatrix} \tag{39}$$

and $B_3 = \begin{bmatrix} 0 \; 1 \; 0 \; 0 \end{bmatrix}^T$.

In the sequel, we will consider the case that the time-delays τ_1 and τ_2 are subject to time-varying uncertainties. Let $\delta_1(t)$ and $\delta_2(t)$ be continuous time-varying bounded functions with bounded derivatives,

$$0 \leq \delta_i(t) \leq \epsilon_i, \;\; \dot{\delta}_i(t) \leq \rho_i, 0 \leq \rho_i < 1 \;\; i = 1, 2. \tag{40}$$

With the delay uncertainty, we write the system as follows:

$$\dot{x}(t) = Ax(t) + B_1x(t-\tau_1-\delta_1(t)) + B_2x(t-\tau_2-\delta_2(t)) \tag{41}$$

We have also omitted the human input term F_h since it does not affect the stability analysis in the state-space form. Although not considered here, it is also possible to allow δ_i to assume both positive and negative values with potential further reduction of conservatism, see [7]. Equation (41) can be written as:

$$\begin{aligned} \dot{x}(t) = {} & Ax + B_1x(t-\tau_1) + B_2x(t-\tau_2) \\ & - B_1\int_{-\delta_1(t)}^{0} \frac{\partial}{\partial\theta}x(t-\tau_1+\theta)d\theta \\ & - B_2\int_{-\delta_2(t)}^{0} \frac{\partial}{\partial\theta}x(t-\tau_2+\theta)d\theta \end{aligned} \tag{42}$$

Use (41) for the terms $\frac{\partial}{\partial\theta}x(t-\tau_1+\theta)$ and $\frac{\partial}{\partial\theta}x(t-\tau_2+\theta)$ in the above equation, (known as the model transformation) and let

$$u_1(t) = A\int_{-\delta_1(t)}^{0} x(t-\tau_1+\theta)d\theta \tag{43}$$

$$u_3(t) = A\int_{-\delta_2(t)}^{0} x(t-\tau_2+\theta)d\theta \tag{44}$$

$$u_2(t) = B_2\int_{-\delta_1(t)}^{0} x(t-\tau_1+\theta-\tau_2-\delta_2(t-\tau_1+\theta))d\theta \tag{45}$$

$$u_4(t) = B_1 \int_{-\delta_2(t)}^{0} x(t - \tau_2 + \theta - \tau_1 - \delta_1(t - \tau_2 + \theta))d\theta \tag{46}$$

Since $B_1B_1 = B_2B_2 = 0$, we can write (41) as :

$$\begin{aligned}\dot{x}(t) = Ax + B_1x(t - \tau_1) + B_2x(t - \tau_2) - B_1u_1(t) \\ -B_1u_2(t) - B_2u_3(t) - B_2u_4(t)\end{aligned} \tag{47}$$

Assuming zero initial conditions, we will estimate the gains from x to u_i, $i = 1, 2, 3, 4$. It is useful to define $\nu_i(\eta) = \eta - \delta_i(\eta)$, $i = 1, 2$. Then,

$$\eta - \varepsilon_i \leq \nu_i(\eta) \leq \eta$$

Also, since $d\nu_i/d\eta = 1 - \delta_i'(\eta) \geq 1 - \rho_i > 0$, ν_i is a strictly increasing function, the inverse function $\eta = \eta(\nu_i)$ is well defined, and

$$\frac{\partial \eta}{\partial \nu_i} = \frac{1}{1 - \delta_i'(\eta)} \leq \frac{1}{1 - \rho_i}$$

Furthermore, due to the range of δ_i, we can easily verify that

$$\nu_i \leq \eta(\nu_i) \leq \nu_i + \epsilon_i$$

Using Jensen's Inequality [23] [7], we can show that:

$$\begin{aligned}&\int_0^t u_4^T(\xi)u_4(\xi)d\xi \\ &\leq \int_0^t \delta_2(\xi)[\int_{-\delta_2(\xi)}^{0} (x^T(\nu_1(\xi - \tau_2 + \theta) - \tau_1)B_1^T \cdot \\ &\qquad B_1x^T(\nu_1(\xi - \tau_2 + \theta) - \tau_1))d\theta]d\xi\end{aligned} \tag{51}$$

Change integration variable from θ to μ, with $\mu = v_1(\xi - \tau_2 + \theta) - \tau_1$. Then, we have

$$\begin{aligned}&\int_{-\delta_2(\xi)}^{0} x^T(\nu_1(\xi - \tau_2 + \theta) - \tau_1)B_1^T \cdot \\ &\qquad B_1x(\nu_1(\xi - \tau_2 + \theta) - \tau_1)d\theta \\ &\leq \int_{\xi - \tau_2 - \varepsilon_2 - \varepsilon_1 - \tau_1}^{\xi - \tau_2 - \tau_1} \frac{1}{1 - \rho_1} x^T(\mu)B_1^TB_1x(\mu)d\mu\end{aligned} \tag{52}$$

Therefore,

$$\int_0^t u_4^T(\xi)u_4(\xi)d\xi \leq \frac{(\varepsilon_1 + \varepsilon_2)\varepsilon_2}{1 - \rho_1} ||B_1||^2 \int_0^t x^T(\mu)x(\mu)d\mu$$

Similarly, we can show

$$\int_0^t u_2^T(\xi)u_2(\xi)d\xi \leq \frac{(\varepsilon_1+\varepsilon_2)\varepsilon_1}{1-\rho_2}||B_2||^2 \int_0^t x^T(\mu)x(\mu)d\mu$$

With a simpler procedure, we can also show

$$\int_0^t u_1^T(\xi)u_1(\xi)d\xi \leq \varepsilon_1^2||A||^2 \int_0^t x^T(\mu)x(\mu)d\mu$$
$$\int_0^t u_3^T(\xi)u_3(\xi)d\xi \leq \varepsilon_2^2||A||^2 \int_0^t x^T(\mu)x(\mu)d\mu$$

With the above discussion, we can write the system described by (47) and (43)-(46) as

$$\begin{aligned} \dot{x}(t) &= Ax(t) - B_1x(t-\tau_1) - B_2x(t-\tau_2) + \hat{B}u \\ y_i(t) &= c_ix(t), \qquad i = 1,2,3,4 \end{aligned} \tag{54}$$

where

$$\begin{aligned} u(t) &= [u_1^T(t)\ u_2^T(t)\ u_3^T(t)\ u_4^T(t)]^T \\ \hat{B} &= [B_1\ B_1\ B_2\ B_2] \end{aligned}$$

and

$$c_1 = \varepsilon_1||A||,\quad c_2 = \sqrt{\frac{(\varepsilon_1+\varepsilon_2)\varepsilon_1}{1-\rho_2}}||B_2||, \tag{55}$$

$$c_3 = \varepsilon_3||A||,\quad c_4 = \sqrt{\frac{(\varepsilon_1+\varepsilon_2)\varepsilon_2}{1-\rho_1}}||B_1|| \tag{56}$$

with feedback $u_i(t) = \Delta_i y_i(t), \quad 1 \leq i \leq 4.$

With the definition of u_i and c_i, it can be easily shown that the gains of the dynamic operator Δ_i is bounded by 1.

Theorem 3. *The closed loop system is uniformally asymptotically stable for any time-varying delay uncertainty $\delta_i(t)$, $i = 1,2,3,4$, satisfiying (40), if there exist scalars α_i, $i = 1,2,3,4$ such that*

$$||\Lambda H(j\omega)\Lambda^{-1}||_\infty < \frac{1}{\epsilon_{max}}$$

*where Λ= **diag** $(\alpha_1 I_n, .., \alpha_4 I_n)$, and*

$$H(s) = \begin{bmatrix} c_1 I_n \\ c_2 I_n \\ c_3 I_n \\ c_4 I_n \end{bmatrix} (sI - A + B_1 e^{-\tau_1 s} + B_2 e^{-\tau_2 s})^{-1}\hat{B} \tag{58}$$

Proof. Use the small gain theorem, as discussed in Chapter 8 of [7].

4 Concluding Remarks

In this chapter, we have been interested in the closed-loop stability of some simple bilateral teleoperation scheme in the hypothesis of the existence of some communication delays. A frequency-domain approach was used to perform the stability analysis in terms of delays. The main advantage of the derived method lies in its simplicity.

5 Acknowledgements

This work is partially supported by ACI: Application de l'Automatique en algorithmique des télécommunications (France) and National Science Foundation (US) Grant INT-9818312

References

1. Anderson, J. A. and Spong, M. W.: Bilateral control of teleoperators with time delay. *IEEE Trans. Automat. Contr.* **AC-34** (1989) 494-501.
2. Anderson, J. A. and Spong, M. W.: Asymptotic Stability for force reflecting teleoperator with time delay. *In. Journal on Robot. Research* (1989) 135-149.
3. Chen, J. and Latchman, H. A.: Frequency sweeping tests for stability independent of delay. *IEEE Trans. Automat. Contr.*, **40** (1995) 1640-1645.
4. Cooke, K. L. and van den Driessche, P.: On zeroes of some transcendental equations. in *Funkcialaj Ekvacioj* **29** (1986) 77-90.
5. El'sgol'ts, L. E. and Norkin, S. B.: *Introduction to the theory and applications of differential equations with deviating arguments* (Mathematics in Science and Eng., **105**, Academic Press, New York, 1973).
6. Eusebi, A. and Melchiori, C.: Force reflecting telemanipulators with time-delay: Stability analysis and control design. *IEEE Trans. Robotics & Automation* **14** (1998) 635-640.
7. Gu, K., Kharitonov, V. L. and Chen, J.: *Stability of Time-Delay Systems*, Berkhauser, Boston, 2003.
8. Jury, E. I.: *Inners and stability of dynamical systems* (2nd Edition, Robert E. Krieger Publ.: Malabar, FL, 1982).
9. Kolmanovskii, V. B. and Myshkis, A. D.: *Applied Theory of functional differential equations* (Kluwer, Dordrecht, The Netherlands, 1992).
10. Lozano, R., Brogliato, B., Egeland, O. and Maschke, B.: *Dissipative systems analysis and control. Theory and applications* (CES, Springer: London, 2000).
11. Lozano, R., Shopra, N. and Spong, M. W.: Passivation of force reflecting bilateral teleoperation with time varying delay. in *Proc. 8th Mechatronics Forum Intl Conf.*, Enschede, The Netherlands (June 2002).
12. Niculescu, S.-I.: *Delay effects on stability. A robust control approach* (Springer-Verlag: Heidelberg, vol. 269, 2001).
13. Niculescu, S. -I. and Abdallah, C. T.: Delay effects on static output feedback stabilization. in *Proc. 39th IEEE Conf. Dec. Contr.*, Sydney, Australia (December 2000).
14. Niculescu, S.-I., Abdallah, C.T. and Hokayem, P.: Some remarks on the wave transformation approach for telemanipulators with time-varying distributed delay. in *Proc. 4th Asian Control. Conf.*, Singapore, September 2002.

15. Niculescu, S. -I., Annaswamy, A. M., Hathout, J. P. and Ghoniem, A. F.: Control of Time-Delay Induced Instabilities in Combustion Systems. in *Proc. 1st IFAC Symp. Syst. Struct. Contr.*, Prague (Czech Republic) (2001).
16. Niculescu, S.-I. and Gu, K.: Robust stability of some oscillatory systems including time-varying delay with applications in congestion control. *Asian Control Conference*, September 2002, Singapore.
17. Niemeyer, G. and Slotine, J.-J. E.: Stable adaptive teleoperation. *IEEE J. Oceanic Eng.* **16** (1991) 152-162.
18. Niemeyer, G. and Slotine, J.-J. E.: Designing force reflecting teleoperators with large time delays to appear as virtual tools. in *Proc. 1997 IEEE ICRA*, Albuquerque, NM (1997) 2212-2218.
19. Niemeyer, G. and Slotine, J. J. E.: Towards force-reflecting Teleoperation Over Internet, in *Proc. 1998 IEEE Int. Conf. Robotics Automation*, 1909-1915, Leuven (Belgium).
20. Niculescu, S.-I., Taoutaou, D., and Lozano, R.: On the closed-loop stability of a teleoperation control scheme subject to communication time-delays. in *Proc. 41st IEEE Conf. Dec. Contr.*, Las Vegas, Nevada, December 2002.
21. Lee, S. and Lee, H.S.: Design of optimal time delayed teleoperator control system. in *Proc. 1994 IEEE Int. Conf. Robotics Automation* 8-13, San Diego, CL.
22. Spiegel M. R. : *Mathematical handbook of formulas and tables* (Shaum's outlines series, MacGraw-Hill: NY, 31 st edition, 1993).
23. A. N. Shiryayev. *Probability*, Springer Verlag, 1996.
24. Ortega, R., Chopra, N. and Spong M.W.: A new passivity formulation for bilateral teleoperation with time delay. in *Proc. CNRS-NSF Wshop Time Delay Syst.*, Paris, France, p. 131-137, January 2003.

Bounded Control of Multiple-Delay Systems with Applications to ATM Networks*

Sophie Tarbouriech[1], Chaouki T. Abdallah[2], and Marco Ariola[3]

[1] LAAS-CNRS, 7 Avenue du Colonel Roche, 31077 Toulouse cedex 4, FRANCE tarbour@laas.fr
[2] Electrical and Computer Engineering, University of New Mexico, Albuquerque, NM 87131, USA chaouki@eece.unm.edu
[3] Dipartimento di Informatica e Sistemistica, Università degli Studi di Napoli Federico II, Napoli, ITALY ariola@unina.it

1 Introduction

The transmission of multimedia traffic on the broadband integrated service digital networks (B-ISDN) has created the need for new transport technologies such as Asynchronous Transfer Mode (ATM) . Briefly, because of the variability of the multimedia traffic, ATM networks seek to guarantee an end-to-end quality of service (QoS) by dividing the varying types of traffic (voice, data, etc.) into short, fixed-size cells (53 bytes each) whose transmission delay may be predicted and controlled. ATM is thus a *Virtual Circuit* (VC) technology which combines advantages of circuit-switching (all intermediate switches are alerted of the transmission requirements, and a connecting circuit is established) and packet-switching (many circuits can share the network resources). In order for the various VC's to share network resources, flow and congestion control algorithms need to be designed and implemented. The congestion control problem is solved by regulating the input traffic rate. In addition, because of its inherent flexibility, ATM traffic may be served under one of the following service classes:

1. The *constant bit rate* (CBR) class, which accommodates traffic that must be received at a guaranteed bit rate, such as telephone conversations, video conferencing, and television.

2. The *variable bit rate* (VBR) which accommodates bursty traffic such as industrial control, multimedia e-mail, and interactive compressed video.

3. The *available bit rate* (ABR) which is a best-effort class for applications such as file transfer or e-mail. Thus, no service guarantees (transfer delay) are required, but the source of data packets controls its data rate, using a feedback signal provided by switches downstream which measure the congestion of the network. Due to the presence of this feedback, many classical and advanced control theory concepts

* This work has been supported by an international agreement CNRS (France)-NSF (USA).

have been suggested to deal with the congestion control problem in the ATM/ABR case [3, 12].

4. The *unspecified bit rate* (UBR) which uses any leftover capacity to accomodate applications such as e-mail.

Note that for the CBR and VBR service categories, a traffic contract is negotiated at the initial stage of the VC setup, and maintained for the duration of the connection. This contract will guarantee the following QoS parameters: 1) Minimum cell rate (MCR), 2) Peak cell rate (PCR), 3) cell delay variation (CDV), 4) maximum cell transfer delay (maxCTD), and 5) cell loss ratio (CLR). This then forces CBR and VBR sources to keep their rate constant regardless of the congestion status of the network. The ABR sources on the other hand, are only required to guarantee an MCR and an PCR, and thus can adjust their rates to accomodate the level available after all CBR and VBR traffic has been accommodated. In order to avoid congestion, the ATM Forum adopted a rate-based ABR control algorithm as opposed to a credit approach whereby the number of incoming cells as opposed to their rate is controlled [6]. This chapter will then concentrate on the ABR service category since ABR sources are the ones to adjust their rates using explicit network feedback. In the original ATM forum specification, an ATM/ABR source is required to send one cell called a resource management (RM) cell for every 32 data cells. Switches along the path from the source to the destination then write into the RM cell their required data rate to avoid congestion. The destination switch then has information about the minimum rate required by all switches along the VC which is then relayed back to the ATM/ABR source as a feedback signal which serves to adjust its own data rate.

The earliest control algorithms for ABR consisted of setting a binary digit in the RM cell by any switch along the VC when its queue level exceeds a certain treshhold [3]. This was then shown to cause oscillations in the closed-loop system. Other controllers were then suggested by various authors [4, 5], to address this problem. Most of these controllers are either complex or did not guarantee the closed-loop stability (in a sense defined later).

In addition, one of the limiting factors of these earlier proposed controllers was that the ABR bandwidth needed to be known in the implementation of the control algorithm. This however poses a problem in multimedia applications where the ABR bandwidth is bursty and is effectively the remaining available bandwidth after the CBR and VBR traffic have been accommodated. In [12] this particular issue was dealt with using a Smith predictor which then considered the available ABR bandwidth as an unknown disturbance . While this controller had many desirable properties, it only guaranteed stability in an appropriately defined sense but had no optimality guarantees. In addition, the delays encountered along with the number of ABR sources were assumed known, although the earlier tech report [6] did not require the delays to be exactly known. In [9], robust controllers were designed when both the number of ABR sources and the delays were uncertain.

In the current chapter, we present a framework which allows us to deal with the ATM/ABR problem with uncertain delays , and number of sources. Moreover, we shall account for the limitations on the rate of traffic and on the speed of change in such rates. Our formulation will allow us to deal with other performance objectives while maintaining a simple controller structure.

Notation. $\Re^+$ is the set of non-negative real numbers. $A_{(i)}$ denotes the ith row of matrix A. $A(i,j)$ denotes the element of the ith row and the jth column of matrix A. I_m denotes the $m-$order identity matrix. 1_m denotes in $\Re^m$ the vector $[1 \dots 1]'$. $\mathcal{C}_\tau = \mathcal{C}([-\tau, 0], \Re^n)$ denotes the Banach space of continuous vector functions mapping the interval $[-\tau, 0]$ into $\Re^n$ with the topology of uniform convergence. $\| \cdot \|$ refers to either the Euclidean vector norm or the induced matrix 2-norm. $\| \phi \|_c = \sup_{-\tau \leq t \leq 0} \| \phi(t) \|$ stands for the norm of a function $\phi \in \mathcal{C}_\tau$. When the delay is finite then "sup" can be replaced by "max". $\mathcal{C}_\tau^v$ is the set defined by $\mathcal{C}_\tau^v = \{\phi \in \mathcal{C}_\tau \, ; \, \| \phi \|_c < v, \; v > 0\}$. Finally, $\mathcal{P}_E(x)$ indicates the entire part of the real number x.

2 The Network Model and the Control Problem

2.1 The network model

There are two philosophically distinct approaches to modeling an ATM network. The first assumes a continuous-time flow of the data and thus results in a delay-differential model of the system [12], while the other one assumes a discrete-time flow and results in a difference equation model [4, 9]. In either model however, the eventual controller needs to be implemented in discrete-time. In this chapter, we choose the delay-differential model and assume for the time being that the controller is also continuous-time with the understanding that a discrete-time controller may be obtained as discussed for example in [6]. As discussed earlier, the considered ABR class is designed as a best-effort class for applications such as file transfer or e-mail. Thus, no service guarantees are required (beyond meeting the MCR and PCR limits), but the source of data packets controls its data rate, using a feedback signal provided by switches downstream which measure the congestion of the network. Due to the presence of this feedback, many classical and advanced control theory concepts have been suggested to deal with the congestion control problem in the ATM/ABR case [3], [12]. In what follows we present the dynamic model of an ATM queue following [12] and [9]. The data cells enter the network from a source node S_i, and are then stored and forwarded along intermediate links to various intermediate nodes. At each node, the process is repeated until a data cell reaches its destination node D_j. Each node stores its data cells to be transmitted in a queue along each one of its outgoing links. The network is thus modeled as a graph consisting of a set of $N = \{1, \cdots, \overline{n}\}$ nodes or switches, connected via a set of $L = \{1, \cdots, l\}$ links. Each node $i \in N$ has a set $I(i) \subset L$ of input links and a set $O(i) \subset L$ of output links. Let t_i (sec) be the transmission time of a cell through a link i and the transmission capacity or bandwidth of the corresponding *link* be $c_i = 1/t_i$ (cells/s). Let $t_{di}(sec)$ be the propagation time delay of link i. Let t_{prj} (sec) be the transmission time of a *node* j denoting the time it takes a cell from the time it arrives at node j to the time it goes into one of outgoing links queues. In the following, t_{prj} is assumed to be small enough so that any congestion is only due to the transmission capacity and not by any processing delays.

At any particular time, let C be the set of active source/destination pairs $(S, D) \in N \times N$. Let n_c be the cardinality of C, and associate with each pair (S, D) a VC and a path $p(S, D)$ specified by the sequence of links that the VC traverse in going from S to its corresponding D.

In order to provide feedback signals to itself, each source node generates a forward RM cell for every 32 data cells. The destination node or intermediate nodes (switches) then returns this RM cell (which then becomes a backward RM cell) to the source. These RM cells contain a field called the explicit rate (ER) feedback field, a congestion indicator (CI) bit, and a no increase (NI) bit. The RM cells then travel the same path as the data cells and flow through a particular switch (node) which then can take one or more of the following actions:

1. Insert feedback control information in the ER field of an RM cell.
2. Provide binary feedback information by marking the CI bit or the NI bit.
3. Set the explicit forward congestion indicator (EFCI) bit in the data cell header, so that the destination can mark the CI bit in the corresponding RM cell.
4. Generate and send its own backward RM cell to the source.

Now, each ABR source has an actual cell rate (ACR) along with its MCR and PCR. The ACR must lie between the lower MCR limit and the upper PCR limit and is adjusted according to the feedback provided though the backwards RM cells. The ER field of a forward RM cell is set by the source at its current ACR, and the source waits until it receives the backward RM cell in order to act according to one of the following scenarios:

1. The CI and NI bits are not set, denoting a no congestion situation. The source node then can increase its ACR by RIF*PCR where RIF is the rate increase factor subject to the new ACR being no greater than the explicit rate specified in the ER field by any of the switches downstream, and of course still less than PCR.
2. The CI bit is set, denoting a congestion situation. The source node will then decrease its ACR by RDF*PCR where RDF is the rate decrease factor subject to the new ACR being no greater than the explicit rate specified in the ER field by any of the switches downstream, and of course still greater than MCR.
3. If the NI bit is set, the source sets its new ACR to be the mimimum of the old ACR and the explicit rate specified in the ER field by any of the switches downstream.

This control approach however leads to oscillatory behavior [9]. In what follows, a deterministic fluid model of the cell flow is assumed, so that the source transmission rate is denoted by the continuous variable $u(t) = ACR$ (cells/sec). Then, each ABR source declares its peak cell rate $c_s = 1/t_s = PCR$ and is assumed to always have a cell to send (i.e. be persistent).

The model we consider is that described in [12] and used in [2]. We then assume that each output link of a given node maintains a First-In-First-Out (FIFO) queue shared by all VCs flowing through the link. Hence we suppose that the flow of packets is conserved and therefore the queue level model for each buffer in the ATM network is given as the following continuous-time differential equation:

$$\dot{x}(t) = -d(t) + \sum_{i=1}^{n} u(t - T_i) \tag{1}$$

with the initial condition:

$$u(t_0 + \psi) = \phi(\psi), \forall \psi \in [-\tau, 0],\ (t_0, \phi) \in \Re_+ \times C_\tau^v,\ \tau = \max_{i=1,\dots,n} T_i \tag{2}$$

where $x(t)$ is the queue level associated with the considered link; n is the number of virtual circuits sharing the queue level associated with the considered link which can be controlled by feedback from the current switch. In other words, there may be many other sources feeding into the current switches but they may be bottlenecked at some other switch (thus cannot increase their own rate due to feedback from the current switch) or are already transmitting at their current PCR (and thus cannot increase their ACR); u is the rate accomodated by the considered link. We assume that all the n virtual circuits which share the link have the same input rate u; T_i is the propagation delay from the ith controlled source to the queue; $d(t) = \mu(t) - r^u(t)$ is the disturbance consisting of the rate of packets leaving the queue $\mu(t)$, minus $r^u(t)$, the rate of all the packets arriving from all uncontrollable sources.

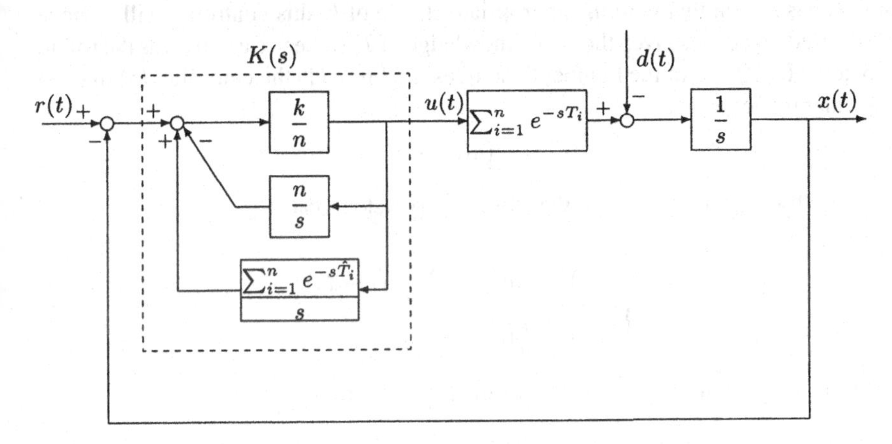

Fig. 1. The controlled system

Furthermore we assume that the following assumptions hold with respect to system (1).

Assumption 3 *The input $u(t)$ (ACR = u) is limited in amplitude as follows:*

$$u(t) \in \Omega_0 = \{u, 0 \leq u(t) \leq u_0, \forall t \geq 0\} \tag{3}$$

with $u_0 > 0$. Note that in this case, we have chosen $MCR = 0$ and $PCR = u_0$ and that the input constraints are supposed to be satisfied by the initial function $\phi(\psi)$,

$\forall \psi \in [-\tau, 0]$. *This assumption basically states that the inflow rate is bounded above and that each source is persistent.*

Assumption 4 *The rate of $u(t)$, that is, its time-derivative $\dot{u}(t)$ is limited in amplitude as follows:*

$$u(t) \in \Omega_1 = \{u, -u_1 \leq \dot{u}(t) \leq u_1\} \tag{4}$$

with $u_1 > 0$. This basically guarantees that no source can change its cell rate instantaneously.

2.2 The control problem

The control objective of this network is to achieve a certain stability property and assure full link utilization as described for example in [12] while simultaneously taking into account the actuator state limitations and external disturbances . Thus, and similarly to [12], let us introduce the fixed-structure controller:

$$u(t) = \frac{k}{n}\left[r_0 - x(t) - \sum_{i=1}^{n} \int_{t-T_i}^{t} u(\tau)d\tau\right] \tag{5}$$

where $r_0 > 0$ represents the queue capacity and k is a positive scalar. The author in [12] has shown that with the appropriate choice of k, this controller will achieve the desired objectives given the exact knowledge of T_i (otherwise known as the round trip delay RTD) and of the number of sources n. ¿From (1) the controller (5) may be equivalently defined as

$$\dot{u}(t) = \frac{k}{n}(d(t) - nu(t)) \tag{6}$$

With this type of controller, the closed-loop system reads:

$$\begin{cases} \dot{x}(t) = -d(t) + \sum_{i=1}^{n} u(t - T_i) \\ \dot{u}(t) = \frac{k}{n}(d(t) - nu(t)) \end{cases} \tag{7}$$

Hence, by defining the new vectors of states z as follows:

$$z(t) = \begin{bmatrix} x(t) \\ u(t) \end{bmatrix} \in \Re^2 \tag{8}$$

the initial closed-loop system (7) reads:

$$\dot{z}(t) = Az(t) + A_d \sum_{i=1}^{n} z(t - T_i) + Bd(t) \tag{9}$$

with

$$A = \begin{bmatrix} 0 & 0 \\ 0 & -k \end{bmatrix} \in \Re^{2\times 2}, \quad B = \begin{bmatrix} -1 \\ \frac{k}{n} \end{bmatrix} \in \Re^{2\times 1}, \quad A_d = \begin{bmatrix} 0 & 1 \\ 0 & 0 \end{bmatrix} \in \Re^{2\times 2} \tag{10}$$

This closed-loop system is defined with the initial condition

$$z(t_0+\psi)=\begin{bmatrix}0\\ \phi(\psi)\end{bmatrix},\forall\psi\in[-\tau,0],\ (t_0,\phi)\in\Re_+\times\mathcal{C}_\tau^v,\quad \tau=\max_{i=1,\ldots,n}T_i \tag{11}$$

Furthermore, due to the form of the closed-loop system (9), we can show that the constraints (3) and (4) can be described in the assumption below.

Assumption 5 *With respect to the closed-loop system (9), the following constraints must be satisfied:*

$$z(t)\in\mathcal{Z}_0=\{z\in\Re^2;0\leq\begin{bmatrix}0\ 1\end{bmatrix}z(t)\leq u_0,\forall t\} \tag{12}$$

$$z(t)\in\mathcal{Z}_1=\{z\in\Re^2;-u_1\leq\begin{bmatrix}0\ 1\end{bmatrix}\dot{z}(t)\leq u_1,\forall t\} \tag{13}$$

The control problem addressed in the chapter can then be re-formulated as follows:

Problem 1. Find a gain k, a set of initial condition $\mathcal{S}_0\subseteq\Re^2$ and a set of admissible disturbances $\mathcal{W}_0\subset\Re$ such that the closed-loop system (9)-(10) exhibits the following properties:

1. Stability . $\forall\phi(\psi)\in\mathcal{S}_0,\forall\psi\in[-\tau,0]$, and $\forall d\in\mathcal{W}_0$ one has:

$$\begin{bmatrix}1\ 0\end{bmatrix}z(t)=x(t)\leq r_0,\forall t\geq 0 \tag{14}$$

Since r_0 corresponds to the queue capacity, this condition allows us to guarantee that no cells are lost, but is not a usual stability requirement. It does however guarantee no oscillation, nor overshoot.

2. Full link utilization . $\forall\phi(\psi)\in\mathcal{S}_0,\forall\psi\in[-\tau,0]$, and $\forall d\in\mathcal{W}_0$ one has:

$$\begin{bmatrix}1\ 0\end{bmatrix}z(t)=x(t)\geq 0,\forall t\geq 0 \tag{15}$$

3. Actuator constraints . The position and rate constraints of the actuators are linearly satisfied.

Remark 1. The satisfaction of condition 2) in problem 1 may be relaxed to $x(t)\geq 0$, $\forall t\geq T_{tr}\geq 0$ [12], [2], where T_{tr} mainly accounts for the transient time of the dynamics. By imposing a linear behavior on the actuator, we avoid the saturation regimes of limited variables.

In the disturbance free case (that is, $d(t)=0$, $\forall t\geq 0$), the resulting nonlinear closed-loop system considering the limitations (12) and (13) possesses a basin of attraction of the equilibrium point $z_e=0$ [13], [15]. Then there exists a subset of this basin of attraction in which the behavior of the closed-loop system remains linear. When $d(t)\neq 0$, it is not possible to strictly define one equilibrium point for the closed-loop system (9) with the time-varying disturbance $d(t)$. At a given time such that $d(t)=d_e$, a corresponding equilibrium point z_e such that $\dot{z}_e=0$

could be computed, implying that associated with any constant disturbance $d_e \in \mathcal{W}_0$, there exists a set of equilibrium points $\mathcal{Z}_e$. Thus, the closed-loop system due to constraints (12) and (13) exhibits local behaviors around these equilibrium points whose study may be very difficult, if not impossible. Recall that we are interested by a linear behavior of the closed-loop system. Thus, an interesting way to overcome these difficulties is to determine a suitable set of admissible initial conditions, $\mathcal{S}_0$ from which the stability of system (9) with respect to the desired equilibrium points is guaranteed.

Thus the set of equilibrium points under consideration can be defined as follows:

$$\mathcal{Z}_e = \left\{ z_e \in \Re^2 ; z_e = \begin{bmatrix} x_e \\ u_e \end{bmatrix}, \dot{z}_e = 0, \forall d_e = \text{ constant } \in \mathcal{W}_0 \right\} \tag{16}$$

Hence, for any admissible constant disturbance d_e, $d_e \in \mathcal{W}_0$, the objective is that the trajectories of system (9) converge towards the equilibrium point $z_e = \begin{bmatrix} x_e \\ u_e \end{bmatrix}$. Hence, for $d(t) = d_e, \forall t \geq 0$, z_e is an equilibrium point for system (9) provided that some conditions are verified (see the next section).

3 Mathematical Preliminaries

3.1 Properties of the model

Since we are interested in the linear behavior of the closed-loop system, that is in avoiding the saturation of z and $\dot{z}$, we state the following lemma [16].

Lemma 1. *The closed-loop system model (9) subject to constraints (12) and (13) is only valid in the region of linearity $\mathcal{Z}_0 \cap \mathcal{Z}_1$. In other words, the closed-loop system model (9) subject to constraints (12) and (13) is only valid, that is, remains linear if and only if the set of initial conditions $\mathcal{S}_0$ is such that*

$$\forall \phi(\psi) \in \mathcal{S}_0, \forall \psi \in [-\tau, 0],\; z(t) \in \mathcal{Z}_0 \cap \mathcal{Z}_1, \forall t$$

When there is a value of $\phi(\psi)$ from which $z(t)$ does not remain in $\mathcal{Z}_0 \cap \mathcal{Z}_1, \forall t$, the closed-loop system resulting from (1), (3), (4), and (5) has to be described by using saturation functions. In this case, the occurrence of saturation on the variables u and $\dot{u}$ has to be investigated and new ways of modeling the resulting closed-loop must be investigated. Such a study will not be considered here, but will be investigated in later research. Note that a solution to such a control problem via statistical learning control was proposed in the case of discrete-time systems with saturation [1].

3.2 Characterization of the equilibrium set

Lemma 2. *Suppose that there exists an equilibrium point $z_e = z(t_e)$ for system (9). Then this equilibrium point satisfies:*

$$\sum_{i}^{n} u(t_e - T_i) = d_e \tag{17}$$

$$u(t_e) = \frac{d_e}{n} \tag{18}$$

$$x_e = r_0 - d_e \left(\frac{1}{k} + \frac{1}{n} \sum_{i=1}^{n} T_i \right) \tag{19}$$

Proof. Relations (17) and (18) are derived by searching $z_e = z(t_e)$ satisfying in (9) $\dot{z}_e = 0$. Relation (19) is derived from (5) by considering that $u = \frac{d_e}{n}$ on the interval $[t_e - T_i, t_e]$. □

Remark 2. Condition (18) is consistent with those in [12] and means that the ABR bandwidth d_e is equally shared by the n VC's. Relation (17) is equivalent to

$$\begin{bmatrix} u(t_e - T_1) \\ \vdots \\ u(t_e - T_n) \end{bmatrix} = \frac{d_e}{n} 1_n + \zeta_e$$

where ζ_e is any vector of $\Re^n$ such as $1_n' \zeta_e = 0$. Hence, a particular solution consists in choosing $\zeta_e = 0$ leading to $u(t_e - T_i) = \frac{d_e}{n}$, $\forall i = 1, ..., n$. Condition (19) is consistent with the value exhibited in [12].

4 Main Results

A natural way for maintaining the system trajectories in a certain set consists of imposing the positive invariance of such a set with respect for the considered system. Hence, part of our results is based on the use of the extended Farkas lemma applied to delay systems: see [8], [14] and references therein.

Let us formulate the following proposition in order to capture the solution to Problem 1. As a first step, we consider that the value of all delays T_i, $i = 1, ..., n$, are exactly known.

Proposition 1. *If the positive values of r_0, u_0, u_1, k, n, T_i and d_0 satisfy:*

$$0 \leq \frac{d_0}{n} \leq u_0 \tag{20}$$

$$ku_0 + \frac{k}{n} d_0 \leq u_1 \tag{21}$$

$$r_0 - \frac{n}{k} u_0 - u_0 \sum_{i=1}^{n} T_i \geq 0 \tag{22}$$

then Problem 1 is solved for the given values of k and any disturbances satisfying

$$d(t) \in \mathcal{D}_0 = \{d; 0 \leq d \leq d_0,\ d_0 > 0\} \tag{23}$$

Proof. To solve Problem 1, we have to satisfy the three requirements of Problem 1, namely, stability , full link utilization , and actuator constraints .

• We must first verify that $x(t) \leq r_0$, $\forall t \geq 0$. ¿From the controller described in (5), one can write the following:

$$r_0 - x(t) = \frac{n}{k}u(t) + \sum_{i=1}^{n} \int_{t-T_i}^{t} u(\tau)d\tau$$

¿From Assumption 3 one gets:

$$0 \leq \sum_{i=1}^{n} \int_{t-T_i}^{t} u(\tau)d\tau \leq u_0 \sum_{i=1}^{n} T_i \tag{24}$$

Therefore, it can be deduced from (3) and (24) that

$$r_0 - x(t) = \frac{n}{k}u(t) + \sum_{i=1}^{n} \int_{t-T_i}^{t} u(\tau)d\tau \geq 0.$$

Furthermore, since from Lemma 2, the trajectories of system (9) may attain its equilibrium point $z(t_e) = z_e$ as defined in (17), (18) and (19) we have to verify that $r_0 - x_e \geq 0$. Thus, from (19) it follows that $0 \leq r_0 - x_e = d_e \left(\frac{1}{k} + \frac{1}{n} \sum_{i=1}^{n} T_i \right) \leq d_0 \left(\frac{1}{k} + \frac{1}{n} \sum_{i=1}^{n} T_i \right)$. Thus, the first requirement of Problem 1 is satisfied for any $u_i(t)$ and $d(t)$ satisfying (3) and (23).

• The second point to verify is the fact that $x(t)$ must be non-negative. Thus, one has to prove that

$$x(t) = r_0 - \frac{n}{k}u(t) - \sum_{i=1}^{n} \int_{t-T_i}^{t} u(\tau)d\tau \geq 0$$

¿From (3) it follows: $x(t) \geq r_0 - \frac{n}{k}u_0 - u_0 \sum_{i=1}^{n} T_i$. Hence, if condition (22) is satisfied one gets $x(t) \geq 0$. This property must also be verified at the equilibrium. Thus, from Lemma 2, if relation (25) is satisfied we have $x_e \geq 0$ for any $u(t)$ and $d(t)$ satisfying (3) and (23).

• The last point consists in verifying the constraints along the trajectories of the linear closed-loop system (9). Recall that, from Lemma 1, system (9) subject to constraints (12) and (13) is only valid in $\mathcal{Z}_0 \cap \mathcal{Z}_1$. Thus, we have to prove that

i) the equilibrium point belongs to this region $\mathcal{Z}_0 \cap \mathcal{Z}_1$ (see Assumption 5).

ii) for any u and d such that $0 \leq u \leq u_0$ and $0 \leq d \leq d_0$ it follows $0 \leq u \leq u_0$ and $-u_1 \leq \dot{u} \leq u_1$ or equivalently $0 \leq u \leq u_0$ and $-u_1 \leq \frac{k}{n}[d(t) - nu(t)] \leq u_1$.

The satisfaction of relations (20) and (22) one gets:

$$r_0 - \frac{d_0}{n}\left(\frac{n}{k} + \sum_{i=1}^{n} T_i\right) \geq 0 \tag{25}$$

With respect to the point 1, we have shown above that $0 \leq x_e \leq r_0$ from the satisfaction of (25). Furthermore, from (20) one can verify that $u(t_e)$ and $u(t_e - T_i)$ satisfy (3).

With respect to the point 2, by using the extended Farkas lemma [8], [14], it follows that if relation (21) is satisfied then there exists a non-negative matrix N such that:

$$\begin{bmatrix} 1 & 0 \\ -k & \frac{k}{n} \\ -1 & 0 \\ k & -\frac{k}{n} \end{bmatrix} = N \begin{bmatrix} 1 & 0 \\ 0 & 1 \\ -1 & 0 \\ 0 & -1 \end{bmatrix} \quad \text{and} \quad N \begin{bmatrix} u_0 \\ d_0 \\ 0 \\ 0 \end{bmatrix} \leq \begin{bmatrix} u_0 \\ u_1 \\ 0 \\ u_1 \end{bmatrix} \tag{26}$$

with

$$N = \begin{bmatrix} 1 & 0 & 0 & 0 \\ k & \frac{k}{n} & 2k & 0 \\ 0 & 0 & 1 & 0 \\ k & \frac{k}{n} & 0 & \frac{2k}{n} \end{bmatrix} \tag{27}$$

□

Remark 3. Relation (21) gives an implicit relation between the bounds u_0 and u_1. Indeed, necessarily, we have to satisfy $\frac{n}{k}u_1 - nu_0 \geq 0$.

In a second stage, we suppose that the delays T_i are uncertain and moreover the value of the number of Virtual Circuits (say n) is unknown. In order to solve our control problem in the case where we have also to provide an estimation of n, we suppose that all the delays T_i satisfy:

$$0 \leq T_i \leq T_{max}, \ \forall i = 1, ..., n \tag{28}$$

Let us now present a solution to our control problem when the values n and T_{max} are not perfectly known.

Proposition 2. *Given $r_0 > 0$, $u_0 > 0$, $u_1 > 0$ and $d_0 > 0$. If there exist positive values X, Y and Z satisfying:*

$$0 \leq d_0 X \leq u_0 \tag{29}$$

$$r_0 - d_0 X - d_0 Z \geq 0 \tag{30}$$

$$u_0 + d_0 X \leq u_1 Y \tag{31}$$

$$r_0 X - u_0 Y - u_0 Z \geq 0 \tag{32}$$

$$\frac{r_0}{u_0(Y+Z)} \geq 1 + \max\left\{\frac{d_0}{u_0}; \frac{d_0}{r_0 - d_0 Z}; \frac{d_0}{u_1 Y - u_0}\right\} \tag{33}$$

then Problem 1 is solved for the values

$$k = \frac{1}{Y} \; and \; n = \mathcal{P}_E\left(\frac{1}{X}\right)$$

and any disturbances $d(t) \in \mathcal{D}_0$ (described in (23)) and for all delays verifying (28) with $T_{max} = Z$.

Proof. Consider relations of Proposition 1 with unknown n, k and T_{max}. In order to have linear conditions in the decision variables we choose $X = \frac{1}{n}$, $Y = \frac{1}{k}$ and $Z = T_{max}$. Hence, relations (20) and (21) directly translate into (29) and (31). Relations (30) and (32) comes from relations (25) and (22) by considering (28). Finally, when all conditions are coherent, one obtains:

$$\frac{u_0(Y+Z)}{r_0} \leq X \leq \min\left\{\frac{u_0}{d_0}; \frac{r_0 - d_0 Z}{d_0}; \frac{u_1 Y - u_0}{d_0}\right\},$$

or equivalently,

$$\frac{r_0}{u_0(Y+Z)} \geq \frac{1}{X} \geq \max\left\{\frac{d_0}{u_0}; \frac{d_0}{r_0 - d_0 Z}; \frac{d_0}{u_1 Y - u_0}\right\}.$$

Thus, in order to pick the entire value of $\frac{1}{X}$ to obtain n, there must exist an entire value in the interval $\left[\max\left\{\frac{d_0}{u_0}; \frac{d_0}{r_0 - d_0 Z}; \frac{d_0}{u_1 Y - u_0}\right\}, \frac{r_0}{u_0(Y+Z)}\right]$. Thus, to ensure this, we have to satisfy condition (33). □

4.1 Numerical examples

Hereafter we provide two numerical examples; one making use of the conditions provided by Proposition 1, the other one making use of the conditions provided by Proposition 2. **Example 1.** Let

$$r_0 = 13\,, \quad u_0 = 1\,, \quad u_1 = 2\,, \quad n = 5\,, \quad T_i = 2 \text{ for } i = 1, \ldots 5.$$

We applied the conditions (20)–(22) of Proposition 1, trying to find the largest value of d_0 such that there exist a feasible value of k. It is easy to show that these conditions can be turned into a Generalized Eigenvalue Problem which can be solved with the aid of the LMI Toolbox [5]. We found that Problem 1 is satisfied with $k = 1.66$ for a maximum value of $d_0 = 1$. In Figure 2 we show the simulation results obtained with this controller using the scheme of Figure 1: all the requirements (stability , full link utilization and actuator constraints) are met. **Example 2.** Let

$$r_0 = 100\,, \quad u_0 = 1\,, \quad u_1 = 2\,, \quad d_0 = 10.$$

We applied the conditions (29)–(33) of Proposition 2, trying to find feasible values for n, k and T_{max}. Again with the aid of the LMI Toolbox, we found that Problem 1 is solved for the values $n = 13$, $k = 0.31$, for all the delays verifying (28) with $T_{max} = 3.86$. In Figure 3 we show the simulation results obtained with this controller using

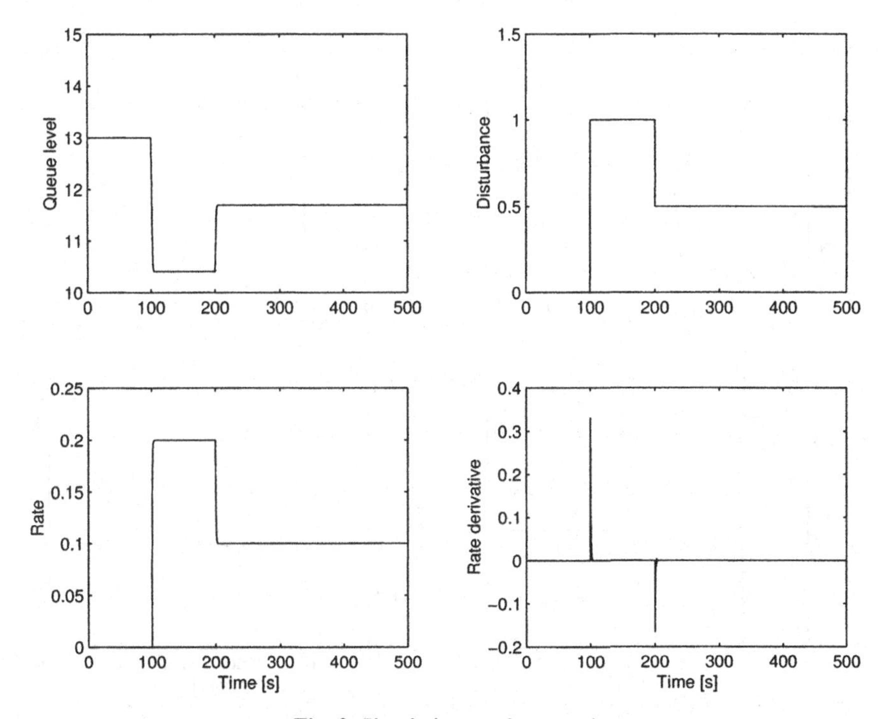

Fig. 2. Simulation results: case 1

the scheme of Figure 1. The controller estimates the maximum time-delay $T_{max} = 3.86$ and the number of the virtual circuits $n = 13$, whereas only 10 virtual circuits were considered and the corresponding time-delays were all chosen less than T_{max}. Also in this case all the requirements (stability , full link utilization and actuator constraints) are met.

5 Conclusions

In this chapter we have provided a new approach to deal with the ATM/ABR control problem keeping in mind requirements of simplicity of the controller structure and allowing for various performance objectives ot be met. Our approach basically leads to polynomial design inequalities to be satisfied. Such inequalities have been studied by the authors and their collaborators in various papers [11]. The statistical learning control approach discussed by the authors in [1, 2] for example, promises to be effective in this setting. While our controller structure is currently derived in continuous time, it is possible to translate such designs into discrete-time as was done for example in [6]. In addition, and while our controller structure is basically the Smith predictor structure of [12], other controller structures are being investigated.

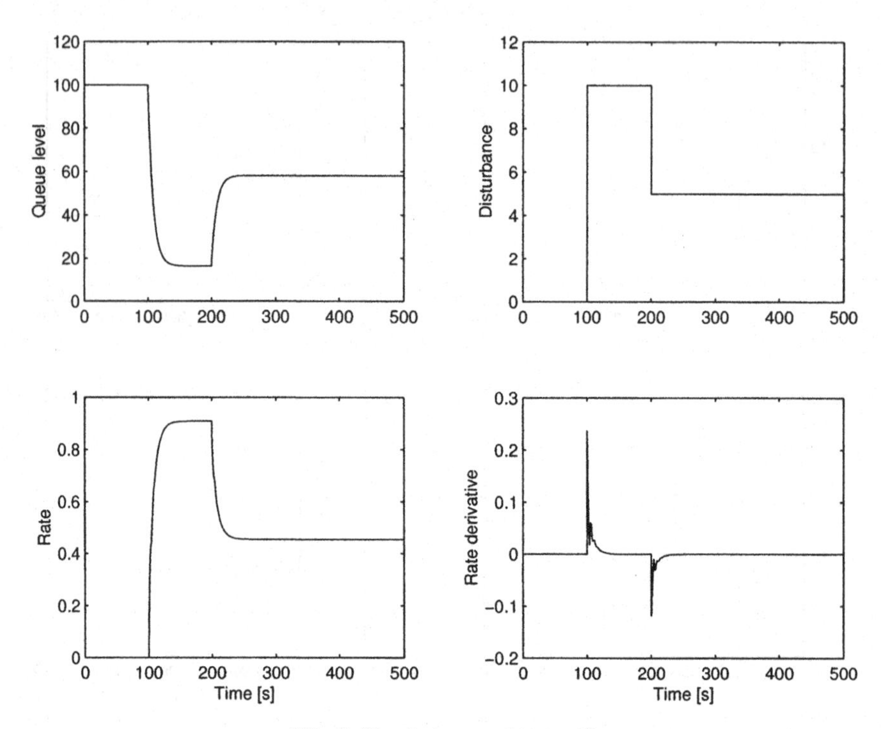

Fig. 3. Simulation results: case 2

References

1. Abdallah CT, Ariola M, Byrne R (2000) Statistical-Learning Control of an ABR Explicit Rate Algorithm for ATM Switches. Proc. of the 39th IEEE Conference on Decision and Control, Sidney, Australia 53–54
2. Abdallah CT, Ariola M, Koltchinskii V (2001) Statistical-Learning Control of Multiple-Delay Systems with applications to ATM Networks. Kybernetica 37(3):355–365
3. ATM Forum Traffic Management Working Group AF-TM-0056.000 (1996) *ATM Forum Traffic Management Specification* Version 4.0
4. Benmohamed L, Wang YT (1998) A Control-Theoretic ABR Explicit Rate Algorithm for ATM Switches with Per-VC Queuing. In: Proceedings Infocom98, San Francisco, CA 183–191
5. Blanchini F, Lo Cigno R, Tempo R (1998) Control of ATM Networks: Fragility and Robustness Issues. Proc. of the American Control Conference, Philadelphia, PA, 2847–2851
6. Cavendish D, Mascolo S, Gerla M (1996) SP-EPRCA: an ATM rate Based Congestion Control Scheme Based on a Smith Predictor. UCLA CS Tech Report 960001, available at: ftp://ftp.cs.ucla.edu/tech-report/
7. Gahinet P, Nemirovski A, Laub AJ, Chilali M (1995) LMI Control Toolbox. The Mathworks Inc, Natick MA
8. Hennet J-C, Tarbouriech S (1997) Stability and stabilization of delay differential systems. Automatica 33(3):347-354

9. Imer OC, Compans S, Başar T, Srikant R (2001) ABR Control in ATM Networks. IEEE Control Systems Magazine 21(1):38–56
10. Imer OC, Compans S, Başar T, Srikant R (2001) Available Bit rate Congestion Control in ABR congestion. IEEE Control Systems Magazine 21(1):38–56
11. Koltchinskii V, Abdallah CT, Ariola M, Dorato P, Panchenko D (2000) Improved Sample Complexity Estimates for Statistical Learning Control of Uncertain Systems. IEEE Trans. Autom. Control 45(12):2383–2388
12. Mascolo S (2000) Smith's Principle for Congestion Control in High-Speed Data Networks. IEEE Trans. Autom. Control 45(2):358-364
13. Saberi A, Lin Z, Teel AR (1996) Control of linear systems with saturating actuators. IEEE Trans. Autom. Control 41(3):368-378
14. Seifert G (1976) Positively invariant closed-loop systems of delay differential equations. J. Differential Equations 22:292–304
15. Tarbouriech S, Garcia G (eds) (1997) Control of uncertain systems with bounded inputs. Lecture Notes in Control and Information Sciences vol.227, Springer-Verlag
16. Tarbouriech T, Gomes da Silva Jr. J-M (2000) Synthesis of controllers for continuous-time delay systems with saturating controls. IEEE Trans. Autom. Control 45(1):105–110

Dynamic Time Delay Models for Load Balancing. Part I: Deterministic Models

J. Douglas Birdwell[1], John Chiasson[1], Zhong Tang[1], Chaouki Abdallah[2], Majeed M. Hayat[2], and Tsewei Wang[3]

[1] ECE Dept, University of Tennessee, Knoxville TN 37996, USA
{birdwell,chiasson,ztang}@utk.edu
[2] ECE Dept, University of New Mexico, Albuquerque NM 87131-1356, USA
{chaouki,hayat}@eece.unm.edu
[3] ChE Dept, University of Tennessee, Knoxville TN 37996, USA
twang@utk.edu

Summary. Parallel computer architectures utilize a set of computational elements (CE) to achieve performance that is not attainable on a single processor, or CE, computer. A common architecture is the cluster of otherwise independent computers communicating through a shared network. To make use of parallel computing resources, problems must be broken down into smaller units that can be solved individually by each CE while exchanging information with CEs solving other problems.

Effective utilization of a parallel computer architecture requires the computational load to be distributed more or less evenly over the available CEs. The qualifier "more or less" is used because the communications required to distribute the load consume both computational resources and network bandwidth. A point of diminishing returns exists.

In this work, a nonlinear deterministic dynamic time-delay systems is developed to model load balancing in a cluster of computer nodes used for parallel computations. This model is then compared with an experimental implementation of the load balancing algorithm on a parallel computer network.

1 Introduction

Parallel computer architectures utilize a set of computational elements (CE) to achieve performance that is not attainable on a single processor, or CE, computer. A common architecture is the cluster of otherwise independent computers communicating through a shared network. To make use of parallel computing resources, problems must be broken down into smaller units that can be solved individually by each CE while exchanging information with CEs solving other problems.

The Federal Bureau of Investigation (FBI) National DNA Index System (NDIS) and Combined DNA Index System (CODIS) software are candidates for parallelization. New methods developed by Wang et al. [7] [1] [8] [39] lead naturally to a parallel decomposition of the DNA database search problem while providing orders of magnitude improvements in performance over the current release of the CODIS software. The projected growth of the NDIS database and in the demand for searches of the database necessitates migration to a parallel computing platform.

Effective utilization of a parallel computer architecture requires the computational load to be distributed more or less evenly over the available CEs. The qualifier "more or less" is used because the communications required to distribute the load consume both computational resources and network bandwidth. A point of diminishing returns exists. The distribution of computational load across available resources is referred to as the *load balancing* problem in the literature. Various taxonomies of load balancing algorithms exist. Direct methods examine the global distribution of computational load and assign portions of the workload to resources before processing begins. Iterative methods examine the progress of the computation and the expected utilization of resources, and adjust the workload assignments periodically as computation progresses. Assignment may be either deterministic, as with the dimension exchange/diffusion [18] and gradient methods, stochastic, or optimization based. A comparison of several deterministic methods is provided by Willeback-LeMain and Reeves [40].

To adequately model load balancing problems, several features of the parallel computation environment should be captured (1) The workload awaiting processing at each CE; (2) the relative performances of the CEs; (3) the computational requirements of each workload component; (4) the delays and bandwidth constraints of CEs and network components involved in the exchange of workloads, and (5) the delays imposed by CEs and the network on the exchange of measurements. A queuing theory [30] approach is well-suited to the modelling requirements and has been used in the literature by Spies [38] and others. However, whereas Spies assumes a homogeneous network of CEs and models the queues in detail, the present work generalizes queue length to an expected waiting time, normalizing to account for differences among CEs, and aggregates the behavior of each queue using a continuous state model. The present work focuses upon the effects of delays in the exchange of information among CEs, and the constraints these effects impose on the design of a load balancing strategy. Preliminary results by the authors appear in [7] with a stability analysis for a proposed *linear* model given in [2]. Here, a nonlinear model is developed to obtain better fidelity and experimental results are presented and compared to that given by the model.

Section 2 presents our approach to modelling the computer network and load balancing algorithms to incorporate the presence of delay in communicating between nodes and transferring tasks. Section 3 presents simulations of the nonlinear model. Section 4 presents experimental data from an actual implementation of a load balancing algorithm which is compared with the simulations. Finally, Section 5 is a summary and conclusion of the present work and a discussion of future work.

2 Models of Load Balancing Algorithms

In this section, a continuous time model in the form of a nonlinear delay-differential system of equations is developed to model load balancing among a network of computers. A modification to the model is presented so that the number of tasks a node distributes to the other nodes is based on their relative load levels.

To introduce the basic approach to load balancing, consider a computing network consisting of n computers (nodes) all of which can communicate with each other. At start up, the computers are assigned an equal number of tasks. However, when a node executes a particular task it can in turn generate more tasks so that very quickly the loads on various nodes become unequal. To balance the loads, each computer in the network sends its queue size $q_j(t)$ to all other computers in the network. A node i receives this information from node j *delayed* by a finite amount of time τ_{ij}, that is, it receives $q_j(t - \tau_{ij})$. Each node i then uses this information to compute its local estimate[4] of the average number of tasks in the queues of the n computers in the network. In this work, the simple estimator $\left(\sum_{j=1}^{n} q_j(t - \tau_{ij})\right)/n$ $(\tau_{ii} = 0)$ is used which is based on the most recent observations is used. Node i then compares its queue size $q_i(t)$ with its estimate of the network average as $q_i(t) - \left(\sum_{j=1}^{n} q_j(t - \tau_{ij})\right)/n$ and, if this is greater than zero, the node sends some of its tasks to the other nodes while if it is less than zero, no tasks are sent (see Figure 1). Further, the tasks sent by node i are received by node j with a delay h_{ij}. The controller (load balancing algorithm) decides how often and fast to do load balancing (transfer tasks among the nodes) and how many tasks are to be sent to each node. As just explained, each node controller (load balancing algorithm) has only *delayed* values of the queue lengths of the other nodes, and each transfer of data from one node to another is

received only after a finite time delay. An important issue considered here is to study the effect of these delays on system performance. Specifically, the continuous time model developed here represents our effort to capture the effect of the delays in load balancing techniques and were developed so that system theoretic methods could be used to analyze them.

2.1 Basic model

The basic mathematical model of a given computing node for load balancing is given by

---

[4] It is an estimate because at any time, each node only has the delayed value of the number of tasks in the other nodes.

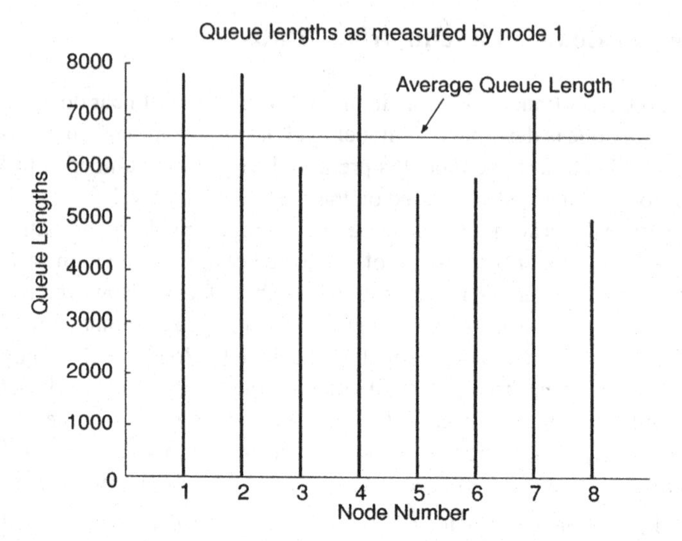

Fig. 1. Graphical description of load balancing. This bar graph shows the load for each computer vs. node of the network. The thin horizontal line is the average load as estimated by node 1. Node 1 will transfer (part of) its load only if it is above its estimate of the average. Also, it only transfers to nodes that it estimates are below the node average.

$$
\begin{aligned}
\frac{dx_i(t)}{dt} &= \lambda_i - \mu_i + u_i(t) - \sum_{j=1}^{n} p_{ij} \frac{t_{p_i}}{t_{p_j}} u_j(t - h_{ij}) \\
y_i(t) &= x_i(t) - \frac{\sum_{j=1}^{n} x_j(t - \tau_{ij})}{n} \\
u_i(t) &= -K_i \mathrm{sat}\,(y_i(t)) \\
p_{ij} &\geqslant 0, p_{jj} = 0, \sum_{i=1}^{n} p_{ij} = 1
\end{aligned}
\tag{1}
$$

where

$$
\begin{aligned}
\mathrm{sat}\,(y) &= y \text{ if } y \geqslant 0 \\
&= 0 \text{ if } y < 0.
\end{aligned}
$$

In this model we have

- n is the number of nodes.
- $x_i(t)$ is the *expected waiting time* experienced by a task inserted into the queue of the ith node. With $q_i(t)$ the number of *tasks* in the ith node and t_{p_i} the average time needed to process a task on the ith node, the expected (average) waiting time is then given by $x_i(t) = q_i(t)t_{p_i}$. Note that $x_j/t_{p_j} = q_j$ is the number of tasks in the node 1 queue. If these tasks were transferred to node i, then the waiting time

transferred is $q_j t_{p_i} = x_j t_{p_i}/t_{p_j}$, so that the fraction t_{p_i}/t_{p_j} converts waiting time on node j to waiting time on node i.

- λ_i is the rate of generation of waiting time on the ith node caused by the addition of tasks (rate of increase in x_i)
- μ_i is the rate of reduction in waiting time caused by the service of tasks at the ith node and is given by $\mu_i \equiv (1 \times t_{p_i})/t_{p_i} = 1$ for all i.
- $u_i(t)$ is the rate of removal (transfer) of the tasks from node i at time t by the load balancing algorithm at node i. Note that $u_i(t) \leq 0$.
- $p_{ij}u_j(t)$ is the rate that node j sends waiting time (tasks) to node i at time t where $p_{ij} \geqslant 0, \sum_{i=1}^{n} p_{ij} = 1$ and $p_{jj} = 0$. That is, the transfer from node j of expected waiting time (tasks) $\int_{t_1}^{t_2} u_j(t)dt$ in the interval of time $[t_1, t_2]$ to the other nodes is carried out with the ith node being sent the fraction $p_{ij}\frac{t_{p_i}}{t_{p_j}}\int_{t_1}^{t_2} u_j(t)dt$ where the fraction t_{p_i}/t_{p_j} converts the task from waiting time on node j to waiting time on node i. As $\sum_{i=1}^{n}\left(p_{ij}\int_{t_1}^{t_2} u_j(t)dt\right) = \int_{t_1}^{t_2} u_j(t)dt$, this results in a removing *all* the waiting time $\int_{t_1}^{t_2} u_j(t)dt$ from node j.
- The quantity $-p_{ij}u_j(t-h_{ij})$ is the rate of increase (rate of transfer) of the expected waiting time (tasks) at time t from node j by (to) node i where h_{ij} $(h_{ii} = 0)$ is the time delay for the task transfer from node j to node i.
- The quantities τ_{ij} $(\tau_{ii} = 0)$ denote the time delay for communicating the expected waiting time x_j from node j to node i.
- The quantity $x_i^{avg} = \left(\sum_{j=1}^{n} x_j(t-\tau_{ij})\right)/n$ is the estimate[5] by the ith node of the average waiting time of the network and is referred to as the *local average* (local estimate of the average).

In this model, all rates are in units of the *rate of change of expected waiting time*, or *time/time* which is dimensionless. As $u_i(t) \leq 0$, node i can only send tasks to other nodes and cannot initiate transfers from another node to itself. A delay is experienced by transmitted tasks before they are received at the other node. The control law $u_i(t) = -K_i\text{sat}(y_i(t))$ states that if the ith node output $x_i(t)$ is above the local average $\left(\sum_{j=1}^{n} x_j(t-\tau_{ij})\right)/n$, then it sends data to the other nodes, while if it is less than the local average nothing is sent. The jth node receives the fraction $\int_{t_1}^{t_2} p_{ji}u_i(t)dt$ of transferred waiting time $\int_{t_1}^{t_2} u_i(t)dt$ delayed by the time h_{ij}.

2.2 Constant p_{ij}

The model (1) is the basic model but one important detail remains unspecified, namely the exact form p_{ji} for each sending node i. One approach is to choose them as constant and equal

$$p_{ji} = \frac{1}{n-1}\delta_{ji} \tag{2}$$

[5] This is an only an estimate due to the delays.

where δ_{ji} is the standard Kronecker delta function. It is clear that $p_{ji} \geqslant 0$, $\sum_{j=1}^{n} p_{ji} = 1$.

Remark If the p_{ij} are specified by (2) and the saturation functions in (1) are removed, the following *linear time invariant* model results

$$\begin{aligned}
\frac{dx_i(t)}{dt} &= \lambda_i - \mu_i + u_i(t) - \sum_{j \neq i} p u_j(t - h_{ij}) \\
y_i(t) &= x_i(t) - \frac{\sum_{j=1}^{n} x_j(t - \tau_{ij})}{n} \qquad (3) \\
u_i(t) &= -K_i y_i(t),\ p = \frac{1}{n-1}.
\end{aligned}$$

When $u_i(t) = -K_i y_i(t) < 0$, this operates as in (1) in that the tasks are immediately removed and sent to the other nodes where each of those nodes experiences a delay (h_{ij}) in getting these tasks. However, a fundamental problem with this linear model is that when $y_i(t) < 0$ the controller (load balancing algorithm) $u_i(t) = -K_i y_i(t) > 0$ so that the node is *instantaneously* taking on waiting time (tasks) from the other nodes before those tasks are removed from the other nodes' queues. That is, it is accepting the waiting times (tasks) $pu_j(t)$ from each of the other nodes. There is a finite time delay associated with this transfer of tasks, and this model ignores this fact. In spite of this fact, it is still of value to consider the system (3) because it can be completely analyzed with regards to stability, and it does capture the oscillatory behavior of the $y_i(t)$. A stability analysis of this linear model is presented in [2].

2.3 Non constant p_{ij}

It could be useful to use the local information of the waiting times $x_i(t), i = 1, .., n$ to set the values of the p_{ij}. Recall that p_{ij} is the fraction of $u_j(t)$ that node j allocates (transfers) to node i at time t, and conservation of the tasks requires $p_{ij} \geqslant 0, \sum_{i=1}^{n} p_{ij} = 1$ and $p_{jj} = 0$. The quantity $x_i(t - \tau_{ji}) - x_j^{avg}$ represents what node j estimates[6] the waiting time in the queue of node i is with respect to the local average of node j. If queue of node i is above the local average, then node j does not send tasks to it. Therefore $\text{sat}\big(x_j^{avg} - x_i(t - \tau_{ji})\big)$ is an appropriate measure by node j as to how much node i is *below* the local average. Node j then repeats this computation for all the other nodes and then portions out its tasks among the other nodes according to the amounts they are below the local average, that is,

$$p_{ij} = \frac{\text{sat}\left(x_j^{avg} - x_i(t - \tau_{ji})\right)}{\sum_{i \ni i \neq j} \text{sat}\left(x_j^{avg} - x_i(t - \tau_{ji})\right)}. \qquad (4)$$

A p_{ij} is defined to be zero if the denominator $\sum_{i \ni i \neq j} \text{sat}\left(x_j^{avg} - x_i(t - \tau_{ji})\right) = 0$.

[6] Again, the term "estimates" is used because node j does not know the current value of $x_i(t)$, but only its earlier value $x_i(t - \tau_{ij})$.

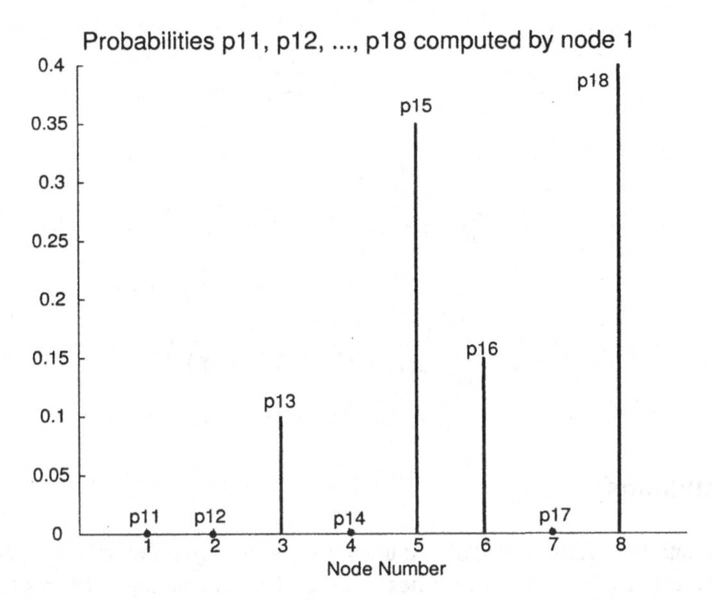

Fig. 2. Illustration of a hypothetical distribution p_{i1} of the load at some time t from node 1's point of view. Node 1 will send data out to node i in proportion p_{i1} it estimates node i is below the average where $\sum_{i=1}^{n} p_{i1} = 1$ and $p_{11} = 0$

Remark If the denominator $\sum\limits_{i \ni i \neq j} \text{sat}\big(x_j^{avg} - x_i(t - \tau_{ji})\big)$ is zero, then $x_j^{avg} - x_i(t - \tau_{ji}) < 0$ for all $i \neq j$. However, by definition of the average, $\sum\limits_{i \ni i \neq j} \big(x_j^{avg} - x_i(t - \tau_{ji})\big) + x_j^{avg} - x_j(t) = \sum\limits_{i} \big(x_j^{avg} - x_i(t - \tau_{ji})\big) = 0$ which implies $x_j^{avg} - x_j(t) = -\sum\limits_{i \ni i \neq j} \big(x_j^{avg} - x_i(t - \tau_{ji})\big) > 0$. That is, if the denominator is zero, the node j is below the local average so that $u_j(t) = -K_j \text{sat}(y_j(t)) = 0$ and is therefore not sending out any tasks.

With the definition of the p_{ij} given by (4), a load balancing algorithm which portions out the tasks in proportion to the amounts they are below the local average, is given by the following nonlinear differential-delay system

$$\frac{dx_i(t)}{dt} = \lambda_i - \mu_i + u_i(t) - \sum_{j \neq i} p_{ij} u_j(t - h_{ij})$$

$$x_i^{avg} = \frac{\sum_{j=1}^{n} x_j(t - \tau_{ij})}{n}$$

$$y_i(t) = x_i(t) - x_i^{avg}$$

$$u_i(t) = -K_i \text{sat}\,(y_i(t)) \tag{5}$$

$$p_{ij} = \frac{\text{sat}\left(x_j^{avg} - x_i(t - \tau_{ji})\right)}{\sum_{i \ni i \neq j} \text{sat}\left(x_j^{avg} - x_i(t - \tau_{ji})\right)} \delta_{ij}$$

3 Simulations

The simulations here were performed using the model (1) in order to compare with the actual experimental data in the next section. Experimental procedures to determine the delay values are given in [20] and summarized in [21]. These give representative values for a Fast Ethernet network with three nodes of $\tau_{ij} = \tau = 200\,\mu$ sec for $i \neq j, \tau_{ii} = 0$, and $h_{ij} = 2\tau = 400\,\mu$ sec for $i \neq j, h_{ii} = 0$. The initial conditions for the waiting times were chosen as $x_1(0) = 0.6$, $x_2(0) = 0.4$ and $x_3(0) = 0.2$. The inputs were set as $\lambda_1 = 3\mu_1, \lambda_2 = 0, \lambda_3 = 0, \mu_1 = \mu_2 = \mu_3 = 1$. The t_{p_i}'s were taken to be equal and $p_{ij} = (1/2)\delta_{ij}$ for all i, j.

Figures 3 and 4 show the responses with the gains set as $K = 1000$ and $K = 5000$, respectively. These figures indicate that the value of the gain K has a significant effect on the response of the system. Many simulations were performed that are not presented here, and it was found that the system did not go unstable. However, for low values of the gains, the response was sluggish as in Figure 3 while for high values of the gains, the response was quite oscillatory.

To compare with the experimental results given in Figure 8 of the next section, Figure 5 shows the output responses with the gains set as $K_1 = 6667, K_2 = 4167, K_3 = 5000$, respectively.

It is important to note that these plots are of the quantities y_i in equation (1) which are the amount of waiting time *relative* to the local average and therefore can and do go negative. The actual waiting times x_i do *not* go negative.

4 Experimental Results

A parallel machine has been built to implement an experimental facility for evaluation of load balancing strategies. To date, this work has been performed for the FBI Laboratory to evaluate candidate designs of the parallel CODIS database. The design layout of the parallel database is shown in Figure 6. A root node communicates with k groups of computer networks. Each of these groups is composed of n nodes (hosts)

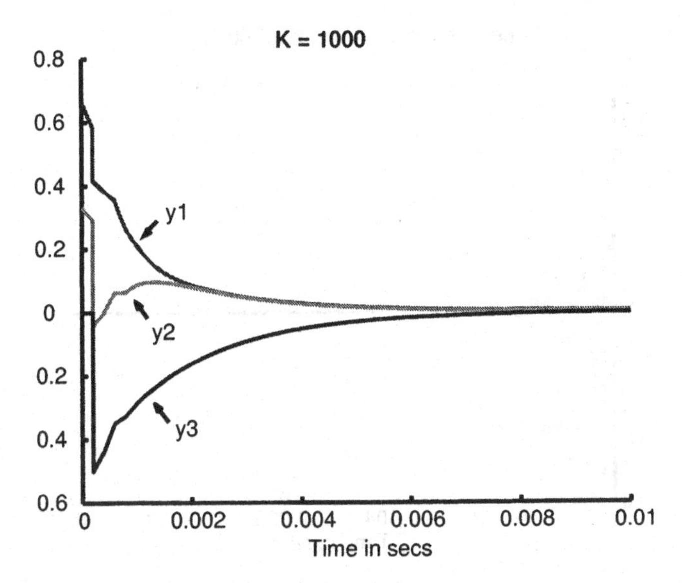

Fig. 3. Output responses with $K = 1000$.

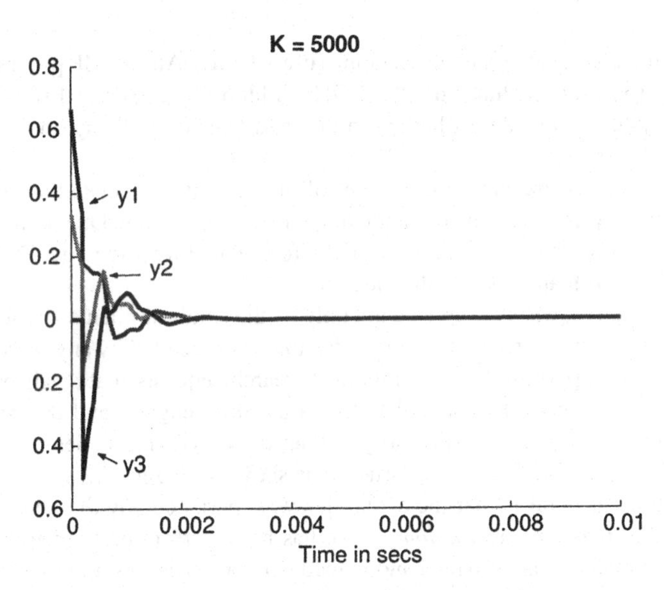

Fig. 4. Output responses with $K = 5000$.

holding identical copies of a portion of the database. (Any pair of groups correspond to different databases, which are not necessarily disjoint. A specific record, or DNA profile, is in general stored in two groups for redundancy to protect against failure of a node.) Within each node, there are either one or two processors. In the experi-

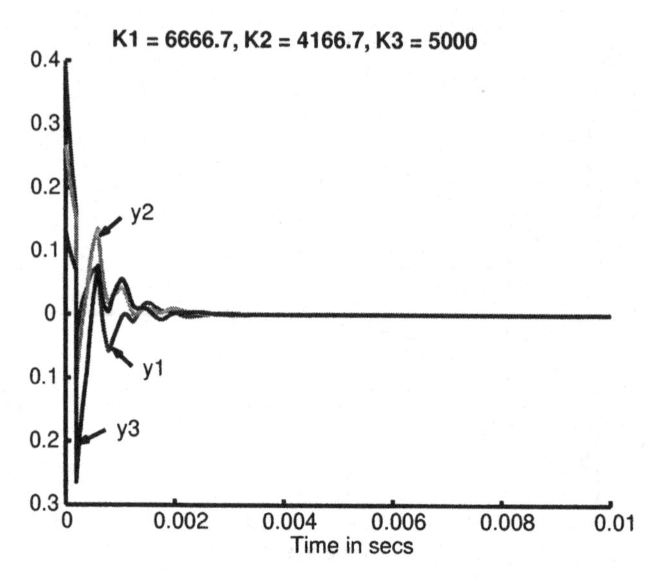

Fig. 5. Output responses with $K1 = 6666.7$; $K2 = 4166.7$; $K3 = 5000$

mental facility, the dual processor machines use 1.6 GHz Athlon MP processors, and the single processor machines use 1.33 GHz Athlon processors. All run the Linux operating system. Our interest here is in the load balancing in any one group of n nodes/hosts.

The database is implemented as a set of queues with associated search engine threads, typically assigned one per node of the parallel machine. Due to the structure of the search process, search requests can be formulated for any target DNA profile and associated with any node of the index tree.

These search requests are created not only by the database clients; the search process also creates search requests as the index tree is descended by any search thread. This creates the opportunity for parallelism; search requests that await processing may be placed in any queue associated with a search engine, and the contents of these queues may be moved arbitrarily among the processing nodes of a group to achieve a balance of the load. This structure is shown in Figure 7.

An important point is that the actual delays experienced by the network traffic in the parallel machine are *random*. Work has been performed to characterize the bandwidth and delay on unloaded and loaded network switches, in order to identify the delay parameters of the analytic models and is reported in [20] [21]. The value $\tau = 200\ \mu$sec used for simulations represents an average value for the delay and was found using the procedure described in [21]. The interest here is to compare the experimental data with that from the simulations given presented in the previous section.

To explain the connection between the control gain K and the actual implementation, recall that the waiting time is related to the number of tasks as $x_i(t) = q_i(t)t_{p_i}$

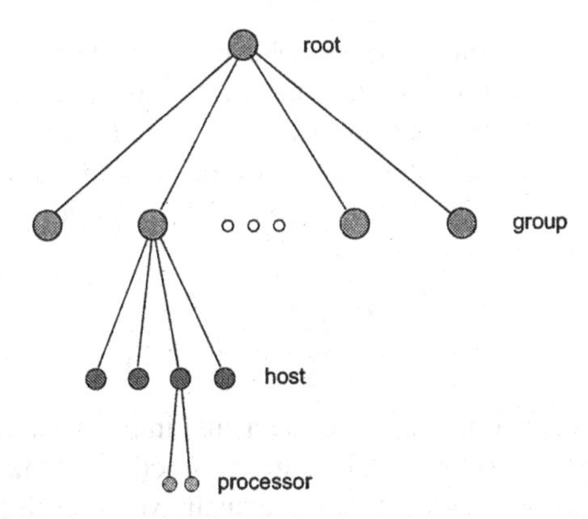

Fig. 6. Hardware structure of the parallel database. Each of the host computers in any given group have the same database. Load balancing is carried out only within a given group as they all perform the same task of searching a particular database.

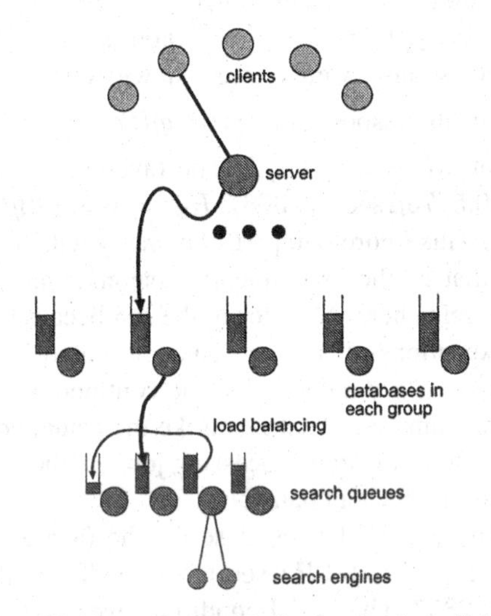

Fig. 7. A depiction of multiple search threads in the database index tree. Here the server corresponds to the "root" in Figure 6. To even out the search queues in each database group, load balancing is done between the nodes (hosts) of a group. If a node has a dual processor, then it can be considered to have two search engines for its queue.

where t_{p_i} is the average time to carry out a task. The continuous time control law is

$$u(t) = -K\,\mathrm{sat}\,(y_i(t))$$

where $u(t)$ is the rate of decrease of waiting time $x_i(t)$ per unit time. Consequently, the gain K represents the rate of reduction of waiting time per second in the continuous time model. Also, $y_i(t) = \left(q_i(t) - \left(\sum_{j=1}^{n} q_j(t-\tau_{ij})\right)/n\right) t_{p_i} = r_i(t)t_{p_i}$

where $r_i(t)$ is simply the number of tasks above the estimated (local) average number of tasks. As the interest here is the case $y_i(t) > 0$, consider $u(t) = -Ky_i(t)$. With Δt the time interval between successive executions of the load balancing algorithm, the control law says that a fraction of the queue $K_z r_i(t)$ $(0 < K_z < 1)$ is removed in the time Δt so the rate of reduction of *waiting time* is $-K_z r_i(t) t_{p_i}/\Delta t = -K_z y_i(t)/\Delta t$ so that

$$u(t) = -\frac{K_z y_i(t)}{\Delta t} \text{ or } K = \frac{K_z}{\Delta t}. \tag{6}$$

This shows that the gain K is related to the actual implementation by how fast the load balancing can be carried out and how much (fraction) of the load is transferred. In the experimental work reported here, Δt actually varies each time the load is balanced. As a consequence, the value of Δt used in (6) is an average value for that run. The average time t_{p_i} to process a task is the same on all nodes (identical processors) and is equal 10μ sec while the time it takes to ready a load for transfer is about 5μ sec . The initial conditions were taken as $q_1(0) = 60000, q_2(0) = 40000, q_3(0) = 20000$ (corresponding to $x_1(0) = q_1(0)t_{pi} = 0.06, x_2(0) = 0.04, x_3(0) = 0.02$). All of the experimental responses were carried out with constant $p_{ij} = 1/2$ for $i \neq j$.

Figure 8 is a plot of the responses $r_i(t) = q_i(t) - \left(\sum_{j=1}^{n} q_j(t - \tau_{ij})\right)/n$ for $i = 1, 2, 3$ (recall that $y_i(t) = r_i(t)t_{p_i}$). The (average) value of the gains were $(K_z = 0.5)$ $K_1 = 0.5/75\mu\,\text{sec} = 6667, K_2 = 0.5/120\mu\,\text{sec} = 4167, K_3 = 0.5/100\mu\,\text{sec} = 5000$. This figure compares favorably with Figure 5 except for the time scale being off, that is, the experimental responses are slower. The explanation for this it that the gains here vary during the run because Δt (the time interval between successive executions of the load balancing algorithm) varies during the run. Further, this time Δt is *not* modelled in the continuous time simulations, only its average effect in the gains K_i. That is, unlike the actual computer network, the continuous time model does not stop processing jobs (at the average rate t_{p_i}) while it is transferring tasks to do the load balancing.

Figure 9 shows the plots of the response for the (average) value of the gains given by $(K_z = 0.2)$ $K_1 = 0.2/125\mu\,\text{sec} = 1600$, $K_2 = 0.2/80\mu\,\text{sec} = 2500$, $K_3 = 0.2/70\mu\,\text{sec} = 2857$. The initial conditions were $q_1(0) = 60000, q_2(0) = 40000, q_3(0) = 20000$ $(x_1(0) = q_1(0)t_{pi} = 0.06, x_2(0) = 0.04, x_3(0) = 0.02)$.

Figure 10 shows the plots of the response for the (average) value of the gains given by $(K_z = 0.3)$ $K_1 = 0.3/125\mu\,\text{sec} = 2400, K_2 = 0.3/110\mu\,\text{sec} = 7273, K_3 = 0.3/120\mu\,\text{sec} = 2500$.

5 Summary and Conclusions

In this work, a load balancing algorithm was modelled by a system of nonlinear delay-differential equations. Simulations were preformed and compared with actual experimental data. The comparison indicates that the model does indeed capture dynamic behavior of the load balancing network.

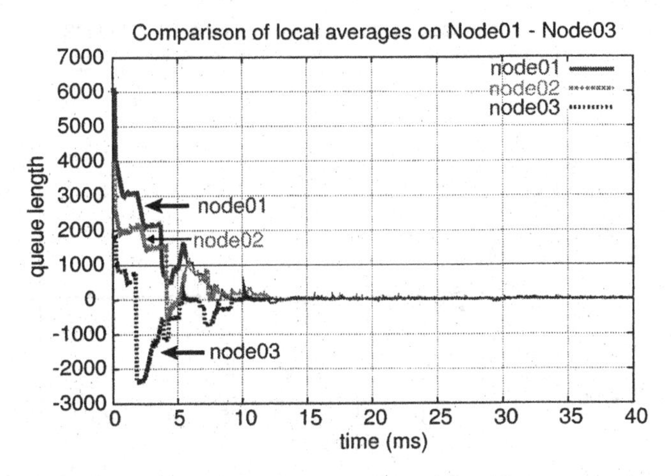

Fig. 8. Experimental response of the load balancing algorithm. The average value of the gains are $(K_z = 0.5)$ $K_1 = 6667, K_2 = 4167, K_3 = 5000$ with constant p_{ij}.

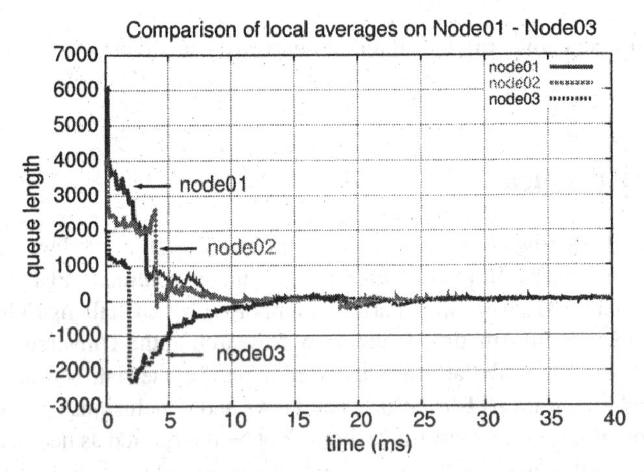

Fig. 9. Experimental response of the load balancing algorithm. The average value of the gains are $(K_z = 0.2)$ $K_1 = 1600, K_2 = 2500, K_3 = 2857$.

A consideration for future work is the fact that the load balancing operation involves processor time which is not being used to process tasks. Consequently, there is a trade-off between using processor time/network bandwidth and the advantage of distributing the load evenly between the nodes to reduce overall processing time. Another issue is that the delays in actuality are not constant and depend on such factors as network availability, the execution of the software, etc. An approach to modelling using a discrete-event / hybrid state formulation that accounts for block

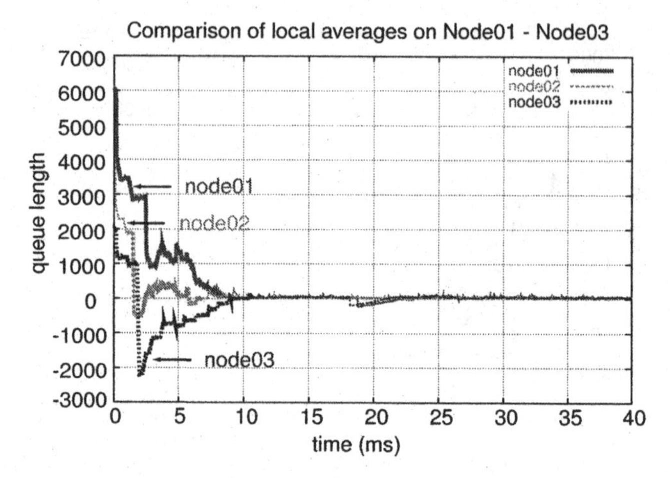

Fig. 10. Experimental response of the load balancing algorithm. The average value of the gains are $(K_z = 0.3)$ $K_1 = 2400, K_2 = 7273, K_3 = 2500$.

transfers that occur after random intervals may also be advantageous in analyzing the network.

6 Acknowledgements

The work of J.D. Birdwell, Z. Tang, and T.W. Wang was supported by U.S. Department of Justice, Federal Bureau of Investigation under contract J-FBI-98-083. Drs. Birdwell and Chiasson were also partially supported by a Challenge Grant Award from the Center for Information Technology Research at the University of Tennessee. The work of C.T. Abdallah was supported in part by the National Science Foundation through the grant INT-9818312. The views and conclusions contained in this document are those of the authors and should not be interpreted as necessarily representing the official policies, either expressed or implied, of the U.S. Government.

References

1. Abdallah, C.T., Hayat, M., Birdwell, J.D. and Chiasson, J.: "Dynamic Time Delay Models for Load Balancing," Part II: A Stochastic Analysis of the Effect of Delay Uncertainty," in *Advances in time-delay systems* (S.-I. NICULESCU, K. GU, EDS.) (this volume) (2003).
2. Abdallah, C.T., Alluri, N., Birdwell, J.D., Chiasson, J., Chupryna, V., Tang, Z. and Wang, T.: "A Linear Time Delay Model for Studying Load Balancing Instabilities in Parallel Computations," *International Journal of System Science* (2003) (to appear).
3. Abdallah, C.T., Dorato, P., Benitez-Read, J. and Byrne, R.: "Delayed positive feedback can stabilize oscillatory systems," *Proceedings of the American Control Conference*, San Francisco CA (1993) 3106-3107

4. Abdallah, C.T., Birdwell, J.D., Chiasson, J., Chupryna, V., Tang, Z. and Wang, T.: "Load Balancing Instabilities due to Time Delays in Parallel Computation," *Proceedings of the 3rd IFAC Conference on Time Delay Systems*, Santa Fe, NM (2001).
5. Altman, E. and Kameda, H.: "Equilibria for Multiclass Routing in Mult-Agent Networks," *Proceedings of the 2001 IEEE Conference on Decision and Control*, Orlando, FL, USA, December 2001.
6. Bellman, R. and Cooke, K.L.: *Differential-Difference Equations*, (Academic Press: New York, 1963).
7. Birdwell, J.D., Horn, R.D., Icove, D.J., Wang, T.W., Yadav, P. and Niezgoda, S.: "A hierarchical database design and search method for CODIS," in *Proc. Tenth International Symposium on Human Identification*, Orlando, FL, September, 1999.
8. Birdwell, J.D., Wang, T.-W. and Rader, M.: "The University of Tennessee's new search engine for CODIS," in *Proc. 6th CODIS Users Conference*, Arlington, VA, February, 2001.
9. Birdwell, J.D., Wang, T.W., Horn, R.D., Yadav, P. and Icove, D.J.: "Method of Indexed Storage and Retrieval of Multidimensional Information," in *Proc. Tenth SIAM Conference on Parallel Processing for Scientific Computation, U. S. Patent Application 09/671,304*, September, 2000.
10. Birdwell, J.D., Chiasson, J., Abdallah, C.T., Tang, Z., Alluri, N., Churpryna, V. and Wang, T.W.: "Load Balancing Instabilities due to Time Delays in Parallel Computation," *Automatica* (Submitted for Publication, 2002).
11. Birdwell, J.D., Chiasson, J., Tang, Z., Abdallah, C.T., Hayat, M., and Wang, T.W.: "Dynamic Time Delay Models for Load Balancing Part I: Deterministic Models," in *Proc. CNRS-NSF Workshop: Advances in Control of Time-Delay Systems, Paris France*, 2003.
12. Birdwell, J.D., Chiasson, J., Abdallah, C.T., Tang, Z., Alluri, N., and Wang, T.W.: "Load Balancing Instabilities due to Time Delays in Parallel Computation," submitted to the *42nd Conference on Decision and Control* (2003).
13. Cavendish, D., Mascolo, S. and Gerla, M.: "SP-EPRCA: an ATM rate based congestion control scheme based on a Smith predictor," *Preprint* (1996).
14. Chiasson, J.N., Brierley, S.D. and Lee, E.B.: "A simplified derivation of the Zeheb-Walach 2-D stability test with applications to time-delay systems," *IEEE Transactions on Automatic Control*, (1985).
15. Chiasson, J.: "A method for computing the interval of delay values for which a differential-delay system," *IEEE Transactions on Automatic Control*, **33** (1988) 1176-1178.
16. Chiasson, J. and Abdallah, C.T.: "A Test for Robust Stability of Time Delay Systems," in *Proceedings of the 3rd IFAC Conference on Time Delay Systems*, Sante Fe, NM, 2001.
17. Cooke, K.L. and Ferreira, J.M.: "Stability conditions for linear retarded functional differential equations," *Journal of Mathematical Analysis and Applications* **96** (1983).
18. Corradi, A., Leonardi, L. and Zambonelli, F.: "Diffusive load-balancing polices for dynamic applications," *IEEE Concurrency* **22** (no. **31**) (1999) 979-993.
19. Cybenko, G.: "Journal of Parallel and Distributed Computing," *IEEE Transactions on Automatic Control*, **7** (1989) 279-301.
20. Dasgupta, P., *Performance Evaluation of Fast Ethernet, ATM and Myrinet under PVM*, MS Thesis, 2001, University of Tennesse.
21. Dasgupta, P., Birdwell, J.D. and Wang, T.W.: "Timing and congestion studies under PVM," in *Proc. Tenth SIAM Conference on Parallel Processing for Scientific Computation*, Portsmouth, VA, March 2001.
22. Datko, R.: "A procedure for determination of the exponential stability of certain differential-difference equations," in *Quarterly Applied Mathematics*, **36** (1978) 279-292.

23. Diekmann, O., van Gils, S.A., Verduyn Lunel, S.M. and Walther, H.-O.: *Delay Equations* (Springer-Verlag: New York, 1995).
24. Hale, J.K. and Verduyn Lunel, S.M. : *Introduction to Functional Differential Equations*, Springer-Verlag: 1993.
25. Hertz, D., Jury, E.I. and Zeheb, E.: "Simplified analytic stability test for systems with commensurate time delays," in *IEE Proceedings, part D*, **131** (1984).
26. Hertz, D., Jury, E.I. and Zeheb, E.: "Stability independent and dependent of delay for delay differential systems," *J. Franklin Institute* (1984).
27. Kameda, H., Li, J., Kim, C. and Zhang, Y.: *Optimal Load Balancing in Distributed Computer Systems* (Springer: London, 1997).
28. Kameda, H., El-Zoghdy Said Fathy, Ryu, I. and Li, J.: "A Performance Comparison of Dynanmic versus Static Load Balancing Policies in a Mainframe," in *Proceedings of the 2000 IEEE Conference on Decision and Control*, Sydney, Australia, pp. 1415-1420, December, 2000.
29. Kamen, E.W.: "Linear systems with commensurate time delays: Stability and stabilization independent of delay," *IEEE Transactions on Automatic Control*, **27** (1982) 367-375.
30. Kleinrock, L.: *Queuing Systems Vol I : Theory* (John Wiley & Sons: New York, 1975).
31. Mascolo, S.: "Smith's Principle for Congestion Control in High-Speed Data Network," *IEEE Transactions on Automatic Control*, **45** (2000) 358-364.
32. Niculescu, S.-I.: *Delay Effects on Stability. A robust control approach* (Springer-Verlag: Heidelberg, 2001).
33. Niculescu, S.-I. and C. T. Abdallah, "Delay effects on static ouput feedback stabilization," *Proceedings of the IEEE Conf. Dec. Contr.*, Sydney, Australia (2000).
34. Ataşlar, B., Quet, P.F., Üftar, A., Özbay, H., Kang, T. and Kalyanaraman, S.: "Robust rate-based flow controllers for high-speed networks: the care of uncertain time-varying multiple time-delays," *Preprint*, 2001.
35. Petterson, B.J., Robinett, R.D. and Werner, J.C.: "Lag-stabilized force feedback damping," *Internal Report SAMD91-0194, UC-406*, 1991.
36. Power, H.M. and Simpson, R.J.: *Introduction to Dynamics and Control* (McGraw-Hill: 1978).
37. Smith, O.J.M.: "Closed Control of Loops with Dead Time," *Chemical Engineering Progress*, **57** (1957) 217-219.
38. Spies, F.: "Modeling of optimal load balancing strategy using queuing theory," *Microprocessors and Microprogramming*, **41** (1996) 555-570.
39. Wang, T.W., Birdwell, J.D., Yadav, P., Icove, D.J., Niezgoda, S. and Jones, S.: "Natural clustering of DNA/STR profiles," in *Proc. Tenth International Symposium on Human Identification*, Orlando, FL, September, 1999.
40. Willebeek-LeMair, M.H. and Reeves, A.P.: "Strategies for dynamic load balancing on highly parallel computers," *IEEE Transactions on Parallel and Distributed Systems*, (1993) 979-993.
41. Xu, C. and Lau, F.C.M.: *Load Balancing in Parallel Computers: Theory and Practice* (Kluwer: Boston, 1997).

Dynamic Time Delay Models for Load Balancing. Part II: A Stochastic Analysis of the Effect of Delay Uncertainty

Majeed M. Hayat[1], Sagar Dhakal[1], Chaouki T. Abdallah[1], J. Douglas Birdwell[2], and John Chiasson[2]

[1] ECE Dept, University of New Mexico, Albuquerque NM 87131-1356, USA
{hayat,dhakal,chaouki}@eece.unm.edu
[2] ECE Dept, University of Tennessee, Knoxville TN 37996, USA
{chiasson,birdwell}@utk.edu

Summary. In large-scale distributed computing systems, in which the computational elements are physically or virtually distant from each other, there are communication-related delays that can significantly alter the expected performance of load-balancing policies that do not account for such delays. This is a particularly significant problem in systems for which the individual units are connected by means of a shared broadband communication medium (e.g., the Internet, ATM, wireless LAN or wireless Internet). In such cases, the delays, in addition to being large, fluctuate randomly, making their one-time accurate prediction impossible. In this work, the stochastic dynamics of a load-balancing algorithm in a cluster of computer nodes are modeled and used to predict the effects of the random time delays on the algorithm's performance. A discrete-time stochastic dynamical-equation model is presented describing the evolution of the random queue size of each node. Monte Carlo simulation is also used to demonstrate the extent of the role played by the magnitude and uncertainty of the various time-delay elements in altering the performance of load balancing. This study reveals that the presence of delay (deterministic or random) can lead to a significant degradation in the performance of a load-balancing policy. One way to remedy such a problem is to weaken the load-balancing mechanism so that the load-transfer between nodes is down-scaled (or discouraged) appropriately.

1 Introduction

Effective load balancing of a cluster of computational elements (CEs) in a distributed computing system relies on accurate knowledge of the state of the individual CEs. This knowledge is used to judiciously assign incoming computational tasks to appropriate CEs, according to some load-balancing policy [1, 2]. In large-scale distributed computing systems in which the CEs are physically or virtually distant from each other, there are a number of inherent time-delay factors that can seriously alter the expected performance of the load-balancing policies that do not account for such delays. One manifestation of such time delay is attributable to the computational limitations of the individual CEs. A more significant manifestation of such delay arises from the communication limitations between the CEs. These include delays in transferring loads between CEs and delays in the communication between them. Moreover, these delay elements not only fluctuate within each CE, as the amounts of the loads to be transferred vary, but also fluctuate as a result of the uncertainty in the condition of the communication medium that connects the units. There has been an extensive research in the development of the appropriate dynamic load balancing policies. The policies have been proposed for categories such as local versus global, static versus dynamic, and centralized versus distributed scheduling [3–5]. Some of the existing approaches consider constant performance of the network while others consider deterministic communication and transfer delay. Here, we propose and investigate a dynamic load balancing scheme for distributed systems which incorporates the stochastic nature of the delay in both communication and load transfer.

To adequately model load balancing problems, several features of the parallel computation environment should be captured including: (1) the workload awaiting processing at each CE (i.e., queue size); (2) the relative performances of the CEs; (3) the computational requirements of each workload component; (4) the delays and bandwidth constraints of CEs and network components involved in the exchange of workloads, and (5) the delays imposed by CEs and the network on the exchange of measurements and information. The authors have previously developed a deterministic model, based on dynamic rate equations, describing the load-balancing dynamics and characterizing conditions for its stability [2, 6–8]. While this deterministic model is appropriate when dealing with a dedicated communication medium, it may become inadequate for cases when a shared communication medium is used whereby the delays encountered are stochastic. In this chapter, we will focus on the effect of stochastic delay on the performance of load balancing. The effect of delay is expected to be a key factor as searching large databases moves toward distributed architectures with potentially geographically distant units.

This chapter is organized as follows. In Section 2 we identify the stochastic elements of the load-balancing problem at hand and describe its time dynamics. In Section 3 we present a discrete-time queuing model describing the evolution of the random queue size of each node in the presence of delay for a typical load balancing algorithm. In Section 4 we present the results of Monte-Carlo simulations which demonstrate the extent of the role played by the uncertainty of the various time-delay elements in altering the performance of load balancing from that predicted by deterministic models, which assume fixed delays. Finally, the conclusions are given in Section 5.

2 Description of the Stochastic Dynamics

The load balancing problem in the presence of delay can be generically described as follows. Consider n nodes in a network of geographically-distributed CEs. Computational tasks arrive at each node randomly and tasks are completed according to an exponential service-time model. In a typical load-balancing algorithm, each node routinely checks its queue size against other nodes and decides whether or not to allocate a portion of its load to less busy nodes according to a predefined policy. Now due to the physical (or virtual) distance between nodes in large-scale distributed computing systems, communication and load transfer activity among them cannot be assumed instantaneous. Thus, the information that a particular node has about other nodes at any time is dated and may not accurately represent the current state of the other nodes. For the same reason, a load sent to a recipient node arrives at a delayed instant. In the mean time, however, the load state of the recipient node may have considerably changed from what was known to the transmitting node at the time of load transfer. Furthermore, what makes matters more complex is that these delays are random. For example, the communication delay is random since the state of the shared communication network is unpredictable, depending on the level of traffic, congestion, and quality of service (QoS) attributes of the network. Clearly, the characteristics of the delay depend on the network configuration and architecture, the type of communication medium and protocol, and on the overall load of the system.

Other factors that contribute to the stochastic nature of the distributed-computing problem include: 1) randomness and possible burst-like nature of the arrival of new job requests at each node from external sources (i.e., from users); 2) randomness of the load-transfer process itself, as it depends on some deterministic law that may use a sliding-window history of all other nodes (which are also random); and 3) randomness in the task completion process at each node. In the next section, we lay out a queuing model that characterizes the dynamics of the load-balancing problem described so far.

3 A Discrete-time Queuing Model with Delays

Consider n nodes (CEs), and let $Q_i(t)$ denote the number of tasks awaiting processing at the ith node at time t. Suppose that the ith node completes tasks at a rate μ_i, and new job requests are assigned to it from external sources (i.e., from external users) at a rate λ_i. Note that these incoming tasks come from sources external to the network of nodes and do not included the jobs transferred to a node from other nodes as a result of load balancing. Let the counting process $J_i(t_1, t_2)$ denote the number of such external tasks arriving at node i in the interval $(t_1, t_2]$. To capture any possible burst-like characteristics in the external-task arrivals (as each job request may involve a large number of computational tasks), we will assume that the process $J_i(\cdot,\cdot)$ is a compound Poisson process [9]. That is, $J_i(t_1, t_2) = \sum_{k:t_1<\tau_k\leq t_2} H_k$, where τ_k are the arrival times of job requests (which arrive according to a Poisson process with rate λ_i) and H_k $(k = 1, 2 \ldots)$ is an integer-valued random variable describing the number of tasks associated with the kth job request. We next address the load transfer between nodes which will allow us to describe the dynamics of the evolution of the queues.

For the ith node and at its specific load-balancing instants T_ℓ^i, $\ell = 1, 2, \ldots,$ the node looks at its own load $Q_i(T_\ell^i)$ and the loads of other nodes at randomly delayed instants (due to communication delays), and decides whether it should allocate some of its load to other nodes, according to a deterministic (or randomized, if so desired) load-balancing policy. Moreover, at times when it is not balancing its load, it may receive loads from other nodes that were transmitted at a randomly delayed instant, governed by the characteristics of the load-transfer delay. With the above description of task assignments between nodes, and with our earlier description of task completion and external-task arrivals, we can write the dynamics of the ith queue in differential form as

$$Q_i(t+\Delta t) = Q_i(t) - C_i(t, t+\Delta t) - \sum_{j\neq i} L_{ji}(t) + \sum_{j\neq i} L_{ij}(t-\tau_{ij}) + J_i(t, t+\Delta t), \quad (1)$$

where

- $C_i(t, t+\Delta t)$ is a Poisson process with rate μ_i describing the random number of tasks completed in the interval $(t, t+\Delta t]$
- τ_{ij} is the delay in transferring the load arriving to node i sent by node j, and finally
- $L_{ij}(t)$ is the load transferred from node j to node i at the time t. Note that $L_{ij}(t)$ is zero except at the load-transfer instants T^j_ℓ, $\ell = 1, 2, \ldots$, for the jth node.

Now for any $k \neq \ell$, the random load $L_{k\ell}(t)$ diverted from node ℓ to node k at a pre-specified load-transfer time t is governed by the mutual load-balancing policy a-priory agreed upon between the two nodes. This policy utilizes knowledge of the state of the ℓth (transmitting) node and the delayed knowledge of the recipient kth node as well as the dated states of all the other nodes. More precisely, we assume $L_{k\ell}(t) = g_{k\ell}(Q_\ell(t), Q_k(t-\eta_{\ell k}), \ldots, Q_j(t-\eta_{\ell j}), \ldots)$, where for any $j \neq k$, $\eta_{kj} = \eta_{jk}$ is the communication delay between the kth and jth nodes. The function $g_{k\ell}$ governs the load-balancing policy between the kth and ℓth nodes. One common example is

$$\begin{aligned} g_{k\ell}(Q_\ell(t), Q_k(t-\eta_{\ell k}), \ldots, Q_j(t-\eta_{\ell j}), \ldots) \\ = K_k p_{k\ell} \cdot \left(Q_\ell(t) - n^{-1} \sum_{j=1}^{n} Q_j(t-\eta_{\ell j}) \right) \\ \cdot u\left(Q_\ell(t) - n^{-1} \sum_{j=1}^{n} Q_j(t-\eta_{\ell j}) \right), \end{aligned} \quad (2)$$

where $u(\cdot)$ is the unit step function with the obvious convention $\eta_{ii}(t) = 0$, and K_k is a parameter that controls the "strength" or "gain" of load balancing at the kth (load distributing) node. We will refer to it henceforth as the gain coefficient. In this example, the ℓth node simply compares its load to the average over all nodes and sends out a fraction $p_{k\ell}$ of its excess load, $Q_\ell(t) - n^{-1}\sum_{j=1}^{n} Q_j(t-\eta_{\ell j})$, to the ℓth node. (Of course we require that $\sum_{k\neq\ell} p_{k\ell} = 1$.) This form of policy has been previously adopted and implemented by the authors for a cluster of CEs [1, 2]. Finally, the fractions $p_{\ell k}$ can be defined in a variety of ways. In this work they are defined as follows:

$$p_{k\ell} \triangleq \frac{1}{n-2}\left\{1 - \frac{Q_k(t-\eta_{\ell k})}{\sum_{i\neq \ell} Q_i(t-\eta_{\ell i})}\right\}. \tag{3}$$

In this definition, a node sends a larger fraction of its excess load to a node with a small load relative to all other candidate recipient nodes.

4 Simulation Results

We have developed a custom-made Monte-Carlo simulation software according to our queuing model. We utilized actual data from load-balancing experiments (conducted at the University of Tennessee) pertaining to the number of tasks awaiting processing, average communication delay, average load-transfer delay, and actual load-balancing instants [2]. In the actual experiment, the communication and load-transfer delays were minimal (due the fact that the PCs were all in a local proximity and benefited from a dedicated fast Ethernet). Thus, to better reflect cases when the nodes are geographically distant we synthesized larger delays in communication and load transfer in our simulations.

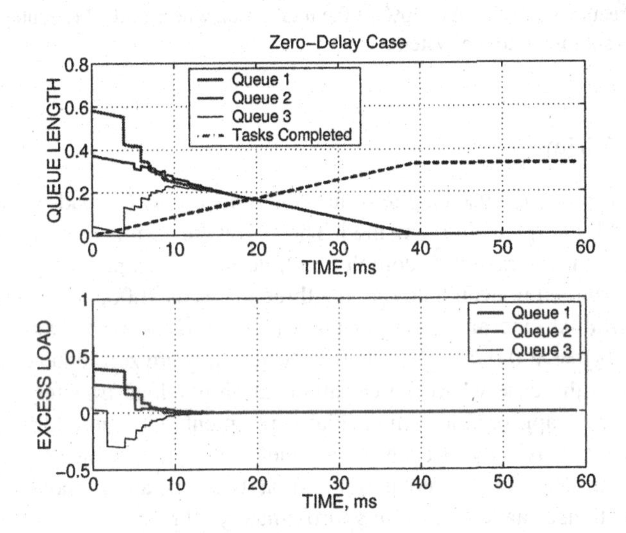

Fig. 1. Top: Queue size in the ideal case when delays are nonexistent. The queues are normalized by the total number of submitted tasks (12000 in this case). The dashed curves represent the tasks completed cumulatively in time by each node. **Bottom:** Excess queue length for each node computed as the difference between each nodes normalized queue size and the normalized queue size of the overall system. Note that the three nodes are balanced at approximately 15 ms and that all tasks are completed in approximately 39 ms.

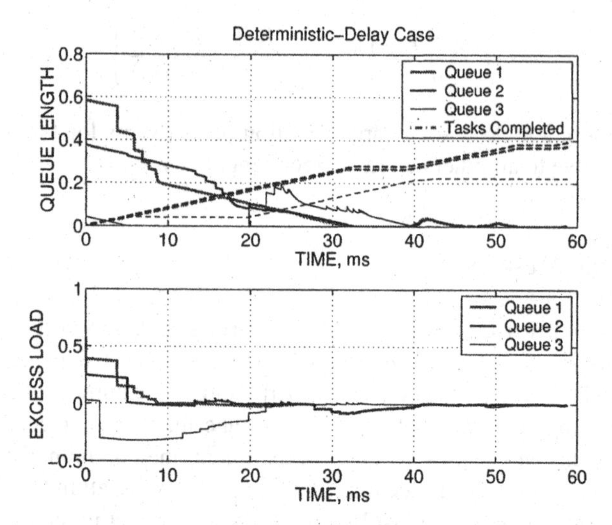

Fig. 2. Similar to Fig. 1 but with a deterministic communication and load-transfer delays of 8 ms and 16 ms, respectively. In contrast to the zero-delay case, the three nodes are balanced at approximately 60 ms and all tasks are completed shortly afterwards. Also note that nodes 2 and 3 each execute approximately 40% of the total tasks, where node 3 executes only 20% of the total tasks submitted to the system.

4.1 Effect of delay

Three CEs ($n = 3$) were used in the simulations and a standard load-balancing policy [as described by (2)] was implemented. The PCs were assumed to have equal computing power (the average task completion time was 10 μs per task), but the initial load was distributed unevenly among the three nodes as 7000, 4500, and 500 tasks, with no additional external arrival of tasks (viz., $J_1(t_1, t_2) = 7000, J_2(t_1, t_2) = 4500, J_3(t_1, t_2) = 500$ only if $t_1 = 0, 0 < t_2$ and they are zero otherwise). Figure 1 corresponds to the case where no communication nor load-transfer delays are assumed. This case approximates the actual experiment [1], where all the computers were within the proximity of each other benefiting from a dedicated fast Ethernet. Note that the system is balanced at approximately 15 ms and remains balanced thereafter until all tasks are executed in approximately 39 ms.

We next considered the presence of deterministic communication delay of 8 ms and a load transfer-delay of 16 ms. The behavior is seen in Fig. 2, where it is observed that the delay prevents load balancing to occur. For example, nodes 1 and 2 each eventually executes approximately 40% of the total tasks, whereas node 3 executes only 20% of the total tasks submitted to the system (as seen from the dashed curves in the top figure in Fig. 2). The conclusion drawn here is that the presence of delay in communication and load transfer seriously disturbs the performance of the load balancing policy, as each node utilizes *dated* information about the state of the other nodes as it decides what fraction of its load must be transferred to each of the other nodes.

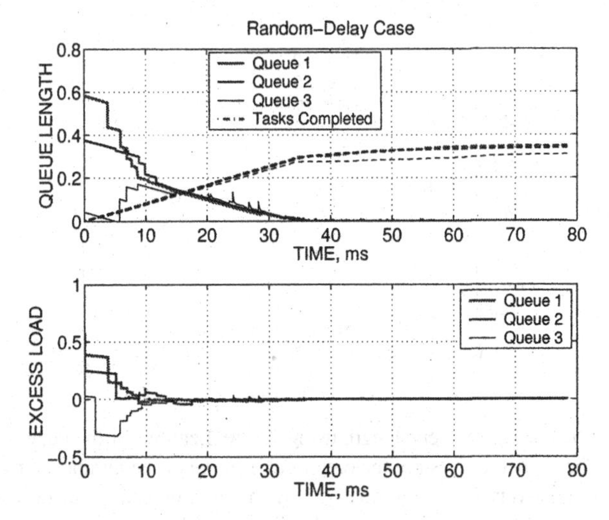

Fig. 3. In this example, the communication and load-transfer delays are assumed random with average values of 8 ms and 16 ms, respectively. Note that the performance is somewhat superior to the deterministic-delay case shown in Fig. 2.

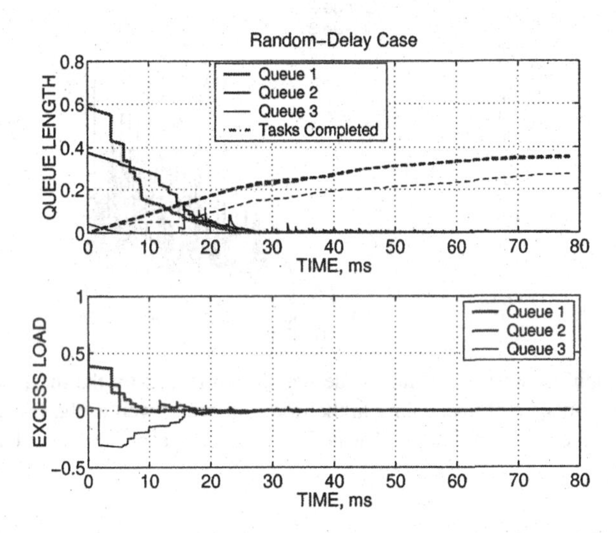

Fig. 4. Another realization of the case described in Fig. 3 showing the variability in the performance from one realization to another. Load-balancing characteristics here are inferior to those in Fig. 3.

To see the effect of the delay randomness on the load balancing performance, two representative realizations of the performance were generated and are shown in Figs. 3 and 4. The average delays were taken as in the deterministic case (i.e., 8 ms for the communication delay and 16 ms for the load-transfer delay). For simplicity

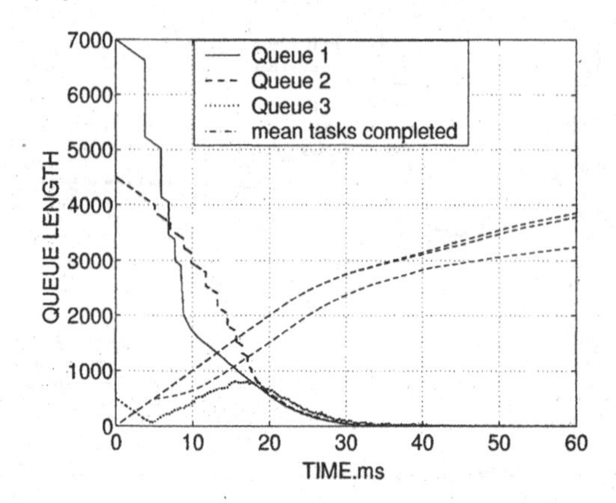

Fig. 5. The empirical average queue length using 100 realizations of the queues for each node (solid curves). The dashed curves are the empirical average of the number of tasks performed by each node cumulatively in time normalized by the total number of tasks submitted to the system. Only 87% of the total tasks are completed within 60 ms.

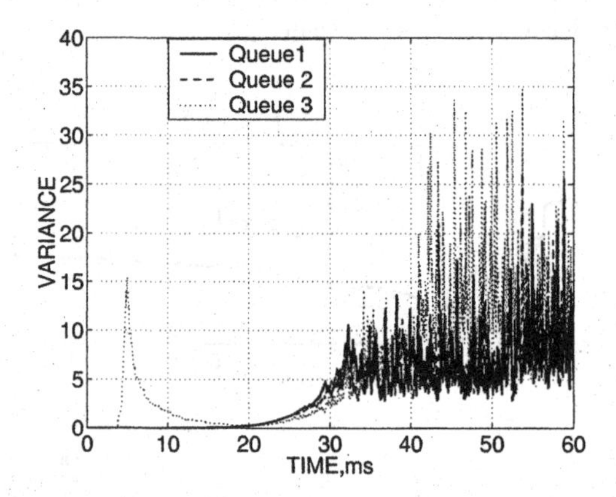

Fig. 6. The empirical variance of the queue length normalized by the mean-square values. Observe the high-degree of uncertainty in the lowest queue as well as the variability at large times, which is indicative of the fact that nodes continue to exchange tasks back and forth, perhaps unnecessarily.

and due to lack of availability of detailed information on the statistics of the delays, the delays were assumed to be uniformly-distributed in the range extending from 0 to twice their mean values. For the example considered, it turns out that the performance is sensitive to the realizations of the delays in the early phase of the load-balancing procedure. For example, it is seen from the simulation results that a deterministic (fixed) delay can lead to a more severe performance degradation than the case when the delays are assumed random (with the same mean as the deterministic case). To see the average effect of the random delay, we calculated the mean queue size and the normalized variance (normalized by the mean square) over 100 realizations of the queue sample functions, each with a different set of randomly generated delays. The results are shown in Figs. 5 and 6. It is seen from the mean behavior that the randomness in the delay actually leads, on average, to balancing characteristics (as far the excess-load is concerned) that are superior to the case when the delays are deterministic! However, there is a high level of uncertainty in the queue size, and hence in the load balancing. It is seen from Fig. 5 (dashed curves) that the average total number of tasks completed by each node continues to increase well beyond 60 ms, which is inferred from the positive slope of the dashed curves. This indicates that in comparison to the deterministic-delay case, the system requires (1) almost twice as long as the zero-delay case to complete all the tasks and (2) a longer time to complete all the tasks than the deterministic-delay case.

4.2 Interplay between delay and the gain coefficient K

We now consider the effect of varying the gain coefficient K on the performance of load balancing (assume that $K_1 = K_2 = K_3 \equiv K$). Figures 7 and 8 show the performance under two cases corresponding to a large and small gain coefficient, $K = 0.8$ and $K = 0.2$, respectively. It is seen that when $K = 0.8$, the queue lengths fluctuate more than the case when $K = 0.2$, resulting in a longer overall time to total task completion. This example shows that a "weak" load-balancing policy can outperform a "strong" policy in the presence of random delay. We will revisit this interesting observation in more detail in the next section.

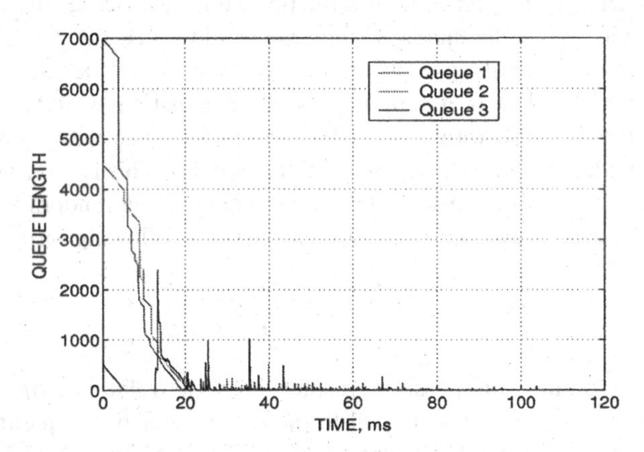

Fig. 7. Queue size versus time when the gain coefficient is $K = 0.8$, corresponding to a "strong" load-balancing policy. Notice the abundance of fluctuations in the tail of the queue in comparison to Fig. 8.

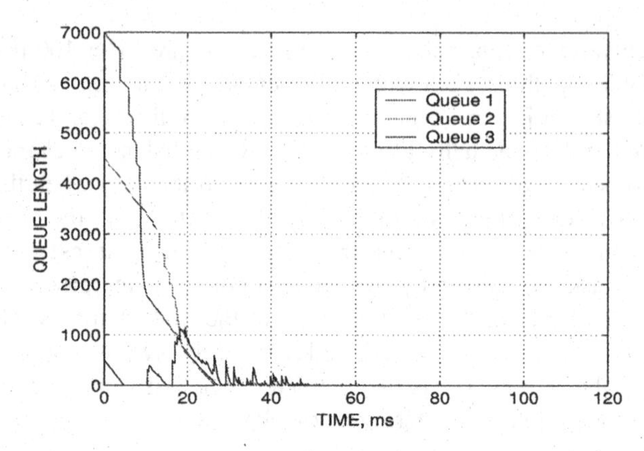

Fig. 8. Same as Fig. 7 but with $K = 0.2$, corresponding to "weak" load-balancing policy.

4.3 Load dependent delay

Clearly, the nature of the transfer delay depends on the amount of load to be transferred; a sizable load will entail, on average, a longer transfer delay than a small load. As a consequence, the load balancing policy is directly affected by the load-dependent nature of transfer delay. For example, if there is a high degree of load imbalance present at any time, it might seem tempting to redistribute big packets of data up front so as to rid the imbalance quickly. However, the tradeoff here is that the sizable load takes much longer to reach the destination node, and hence, the overall computation time will inevitably increase. Thus, we would expect the gain coefficient K to play an important role in cases when transfer delay is load dependent. Since the balancing is done frequently, it is intuitively obvious that we would be better off if we were to select K conservatively. To address this issue quantitatively, we will need to develop a model for the load-dependent transfer delay. This is done next.

We propose to capture the load-dependent nature of the random transfer delay τ_{ij} by requiring that its average value, θ_{ij}, assumes the following form

$$\theta_{ij} = d_{\min} - \frac{1 + \exp([L_{ij}(t)d\beta]^{-1})}{1 - \exp([L_{ij}(t)d\beta]^{-1})}, \tag{4}$$

where $d_{\min}$ is the minimum possible transfer delay (its value is estimated as 9 ms in this chapter), d is a constant (equal to 0.082618), and β is a parameter which characterizes the transfer delay (selected as 0.04955). Moreover, we will assume that conditional on the size of the load to be transferred, the random delay τ_{ij} is uniformly-distributed in the interval $[0, 2\theta_{ij}]$. This model assumes that up to some threshold, the delay is constant (independent of the load size) that is dependent on the capacity of the communication medium. Beyond this threshold, however, the average delay is expected to increases monotonically with the load size. The parameters d and b are selected so that the above model is consistent with the overall average delay for all the actual transfers that occurred in the simulation. We omit the details.

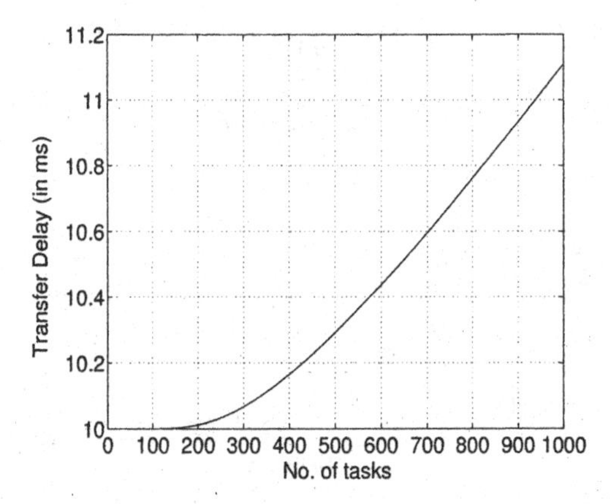

Fig. 9. The load-dependent transfer delay as a function of the load size according to the model shown in (4).

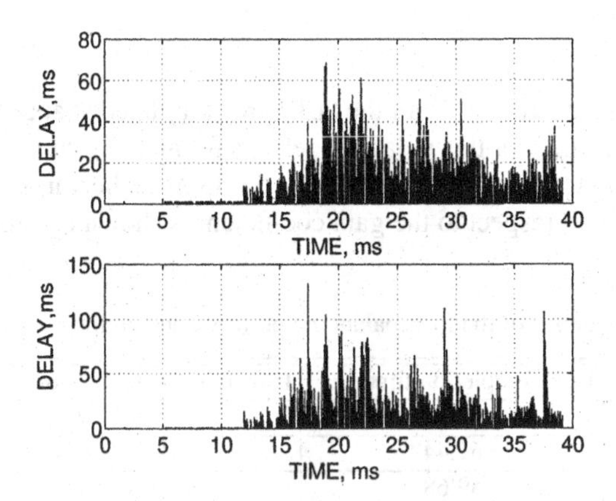

Fig. 10. Scatter-plot of the transfer-delay showing its fluctuations for a particular realization of the queues.

The load-dependent transfer delay versus the load is shown in the Fig. 9. The transfer delay for the loads sent from node 1 to node 3 (top) and from node 2 to node 3 (bottom) over the period of execution time is shown in Fig. 10. With the average communication delay being equal to 8 ms (as before) and the transfer delay made load dependent, according to the model described in (4), one realization of the load-balancing performance for $K = 0.5$ was generated and it is shown in Fig. 11. As expected, the performance deteriorates beyond the case corresponding to a fixed transfer delay. For example, we see from the figure that a load sent by node 1 at approximately 5 ms arrives at node 3 approximately 50 ms later, thereby bringing more fluctuation to the tail of the queues. The average effect (over 50 realizations) of this delay model for two different gain coefficients ($K = 0.1$ and $K = 0.9$) can be seen in Figs. 12 and 13. When $K = 0.9$, the queue is fluctuating even beyond

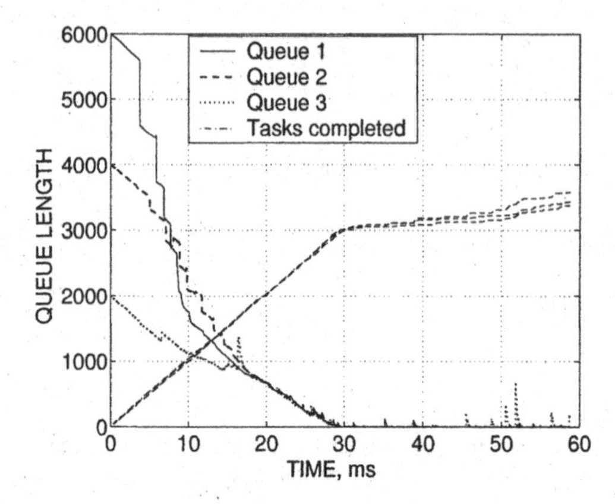

Fig. 11. Queue size and cumulative (in time) number of tasks completed (by each node) as a function of time. These graphs show that queues become more uncertain when the load-transfer delays are load dependent. ($K = 0.5$ in this example.)

$t = 80$ ms while when $K = 0.1$, all the tasks are completed at approximately 60 ms. The optimal value of K for this delay model was found to be equal to $K = 0.06$ and the overall completion time in this case was 54.85 ms. The variation of the overall completion time with respect to the gain coefficient is shown in Table 1 below.

Table 1. Dependence of the load-balancing performance on the gain coefficient K.

Gain (K)	Task completion time (ms)	Time to execute 95% of tasks (ms)
0.01	62.53	41.80
0.02	61.44	42.86
0.03	59.68	42.59
0.04	57.27	41.98
0.05	56.79	41.35
0.06	54.85	41.99
0.07	56.04	42.49
0.08	59.68	41.56
0.09	62.53	41.81
0.1	61.10	42.18
0.2	65	43.38
0.3	63.40	46.2
0.4	78.313	53.33
0.5	> 80	55.21

It is clearly seen in Fig. 13 that the required time for completing all tasks (in the system) is significantly larger than the time required to execute 95% percent of the assigned tasks. This difference increases with higher values of K. This is due to the fact that even when all the queues are *almost* depleted of tasks, they continue to execute the balancing policy. As a result, small amounts of tasks (e.g., one or two) are sent from one node to other nodes and vice versa. This unnecessary task-swapping significantly increases the transfer delay, therefore increasing the overall computational time. This phenomenon is clearly depicted in Fig. 13 where the minute fluctuations are evident near the tail of the queues.

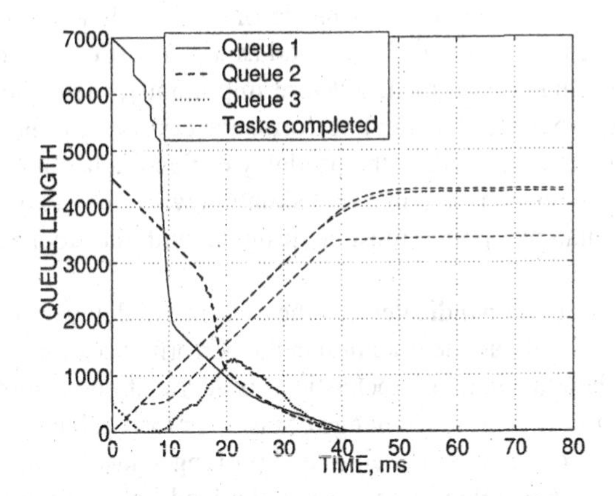

Fig. 12. Queue size versus time for the case where the transfer delay is load dependent. The gain coefficient K is 0.1. Note that the total execution time is approximately 60 ms.

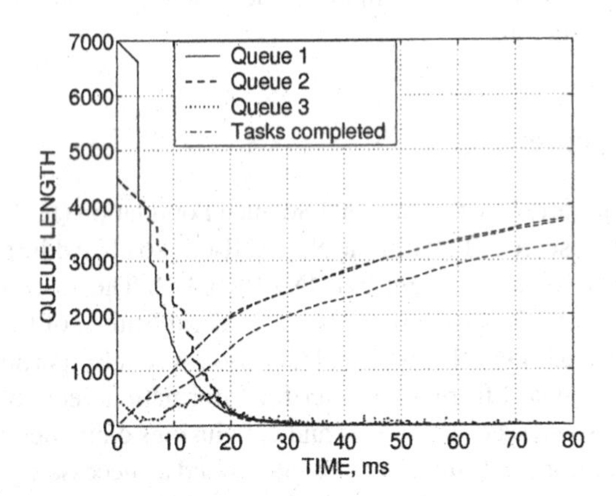

Fig. 13. Same as Fig. 12 but with $K = 0.9$. The queues fluctuate even at 80 ms.

5 Summary and Conclusions

Whenever there are tangible communication limitations between nodes in a distributed system, possibly with geographically-distant computational elements, we must take a totally new look at the problem of load balancing. In such cases, the presence of non-negligible random delays in inter-node communication and load transfer can significantly alter the expected performance of existing load-balancing strategies. The load-balancing problem must be viewed as a stochastic system, whose performance must be evaluated statistically. More importantly, the policy itself must be developed with appropriate statistical performance criteria in mind. Thus, if we design a load-balancing policy under the no-delay or fixed-delay assumptions, the policy will not perform as expected in a real situation when the delays are non-zero or random. A load-balancing policy must be designed with the stochastic nature of the delay in mind.

Monte-Carlo simulation indicates that the presence of delay (deterministic or random) can lead to a significant degradation in the performance of a load-balancing policy. Moreover, when the delay is stochastic, this degradation is worsened, leading to extended cycles of unnecessary exchange of tasks (or loads), back and forth between nodes, leading to extended overall delays and prolonged task-completion times. One way to remedy such a problem is to weaken the load-balancing mechanism so that the load-transfer between nodes is down-scaled (or discouraged) appropriately. This action makes the load-balancing policy in the presence of random delay "less reactionary" to changes in the load distribution within the system. This, in turn, reduces the sensitivity of the load-balancing process to inaccuracies in the state-of-knowledge of each node about the load distribution in the remainder of the system caused by communication limitations.

6 Acknowledgements

This work was supported by the National Science Foundation (NSF) under Information Technology Research (ITR) grant No. ANI-0312611. Additional support was also provided by NSF under grant No. INT-9818312. The work of J.D. Birdwell was supported by U.S. Department of Justice, Federal Bureau of Investigation under contract J-FBI-98-083. Drs. Birdwell and Chiasson were also partially supported by a Challenge Grant Award from the Center for Information Technology Research at the University of Tennessee. The views and conclusions contained in this document are those of the authors and should not be interpreted as necessarily representing the official policies, either expressed or implied, of the U.S. Government.

References

1. J. D. Birdwell, T. W. Wang, R. D. Horn, P. Yadav, and D. J. Icove, "Method of indexed storage and retrieval of multidimensional information," U. S. Patent Application 09/671,304 (2000)

2. C. T. Abdallah, N. Alluri, J. D. Birdwell, J. Chiasson, V. Chupryna, Z. Tang, and T. Wang "A linear time delay model for studying load balancing instabilities in parallel computations," *The International Journal of System Science*, to appear (2003)
3. T.L. Casavant and J.G. Kuhl, "A Taxonomy of Scheduling in General-Purpose Distributed Computing Systems," *IEEE Trans. Software Eng.*, vol. 14, pp. 141–154 (Feb. 1988)
4. Zhiling Lan, Valerie E. Taylor, Greg Bryan, "Dynamic Load Balancing for Adaptive Mesh Refinement Application,"*In Proc. Of ICPP'2001*, Valencia, Spain (2001)
5. G. Cybenko, "Dynamic load balancing for distributed memory multiprocessors," *IEEE Transactions on Parallel and Distributed computing*, vol. 7, pp. 279–301 (1989)
6. J. D. Birdwell, J. Chiasson, Z. Tang, T. Wang, C. T. Abdallah, and M. M. Hayat, "Dynamic time delay models for load balancing Part I: Deterministic models," *CNRS-NSF Workshop: Advances in Control of Time-Delay Systems*, Paris, France (Jan. 2003)
7. C. T. Abdallah, J.D. Birdwell, J. Chiasson, V. Churpryna, Z. Tang, and T.W. Wang "Load balancing instabilities due to Time delays in farallel computation," *Proceedings of the 3rd IFAC Conference on Time Delay Systems*, Santa Fe, NM (Dec. 2001)
8. M. M. Hayat, C. T. Abdallah, J. D. Birdwell and J. Chiasson, "Dynamic time delay models for load balancing, Part II: A stochastic analysis of the effect of delay uncertainty," *CNRS-NSF Workshop: Advances in Control of Time-Delay Systems*, Paris, France (Jan. 2003)
9. D. J. Daley and D. Vere-Jones, *An Introduction to the Theory of Point Processes* (Springer-Verlag, New York, NY 1988)

Part VII

Miscellaneous Topics

Asymptotic Properties of Stochastic Delay Systems

Erik I. Verriest

School of ECE, Georgia Institute of Technology, Atlanta, GA 30332-0250
erik.verriest@ee.gatech.edu

Summary. A gentle and elementary introduction to the theory of stochastic time delay systems is presented in this contribution. First an introduction to the stability problem in the discrete case is given. This class of systems is simpler from a dynamical point of view, as they remain finite dimensional, and provide thus a 'warm-up' for what is to come. Since continuous time stochastic systems are analyzed in the language of Itô-calculus, an elementary introduction to the latter is included to make these notes self-contained. Delay-independent and delay dependent conditions for stochastic stability are derived. Some of these are new. In the absence of equilibria, invariant distributions may still exist. Existence conditions, and the stationary Fokker-Planck equation are discussed. These results are further extended to the class of stochastic neutral systems. Finally, a new realistic design procedure is suggested for dynamic controllers in the absence of precise delay information.

1 Introduction

The past decades has seen renewed interest and a flurry of activity in stabilization and control of systems with delays [10,47]. Although a fairly recent endeavor, the literature on stochastic delay systems and stochastic functional differential equations (SFDE) in general, is already considerable, and no attempt will be made to give a complete survey of the results known to date. The main problems range from existence and uniqueness of solutions to SFDE's to stability and qualitative behavior of stochastic systems with delay, estimation in systems with delay and control of systems with delay (e.g., stabilization and optimal stochastic control). Stochastic delay systems are used to model behaviour and/or describe phenomena in epidemiology, neural networks, traffic control (freeway and communication) and in many other biological and biomedical applications.

The seminal work in stochastic stability theory originates with Bucy [6], Khasminskii [19], and Kushner [30] based on Itô stochastic differential equations and

[0] This chapter is dedicated to the memory of my mother.

(super-)martingale methods. This was followed by the work of many mathematicians: we mention among others, Arnold, Elworthy, Friedman, Gihman, Kliemann, Kolmanovskii, Ladde, Lakshmikantham, Skorohod but the enumeration of a few names can hardly begin to cover the exciting history.

Unlike deterministic system theory, there are many different notions of stochastic stability: moment stability, exponential stability, stability in probability, and distributional stability. In the context of delay systems, a good starting point is the pioneering work of Mohammed [44], Mao [34,35], Kolmanovskii, Myshkis, Nosov and Shaikhet [22–24] for the main theory. Note that besides adding substantial new material in stochastic dynamics, e.g. Lyapunov exponents, and existence of smooth densities using Malliavin calculus [1], the approach in [46] is more gentle than in [44].

Filtering issues are discussed in [22, 24, 53], but already here the literature is plenty. In pathwise stability (almost sure stability) it is understood that the solution converges pathwise to an equilibrium. This means that each sample path of the process converges to an equilibrium. If the equilibrium is unique, without loss of generality it is taken to be at the origin. Of course, this is only possible if the drift and the diffusion terms at this equilibrium are zero. Unless this is the case, as for instance in a system driven by additive noise, all one can possibly ask for is the convergence of the joint distributions to some limit distribution.

This chapter gives a brief introduction to the necessary mathematical background and surveys the problem of stability and the limiting distribution. Results for stabilization follow from the corresponding stability results, and will not be covered in this chapter. See for instance [60]. For simplicity, mainly linear or semilinear systems will be discussed. In addition, some new results are presented in sections 2, 4 and 5, dealing respectively with the stability of discrete delay systems, continuous delay systems and neutral systems.

For instance, if the delay system, $\dot{x}(t) = f(x_t, u(t))$, has a constant delay τ, with where $x_t = \{x(t+\theta); \theta \in (-\tau, 0)\}$ and $u(\cdot)$ is a random process, then one may be interested in the evolution of the joint distribution of the variables $x(t), x(t-\tau_1), x(t-\tau_N)$ with $0 < \tau_1 < \cdots \tau_N = \tau$. In a problem with discrete measurements, $y(t_k)$, this allows one to obtain formulas for the *time update*, to be combined with a *measurement update* in Bayesian fashion at the sampling times $\{t_k\}$ in order to obtain say the least squares estimates of the states.

For finite dimensional systems, the joint distribution of the state variables is given by the solution of the Fokker-Planck equation, also known as the forward Kolmogorov equation. This is a PDE for the density $\rho(x, t)$. Unfortunately, for a delay system, things are not so simple. This is because the delay system is infinite dimensional, and thus the state x_t lives in a function space. It is well known that the density of a process with respect to Lebesgue measure may not be defined. Hence more involved machinery (Radon-Nikodym derivative and Wiener measure) needs to be introduced if one wants to do things the right way.

In recent work [14, 15] this has been avoided to some extent. A finite dimensional equation for the density may still be obtained, but only in disguise. Indeed, the Fokker-Planck equation derived by Frank is actually a coupled system of (infinitely many) finite dimensional PDE's. For instance, it cannot readily be used to compute the equilibrium solution, which typically is the first quantity of interest. The moment problem [13,67] behaves similarly.

One goal in this chapter is to give finite dimensional approximations for this density. For simplicity we focus first on the linear delay system, however the ideas are readily generalized to the nonlinear case as far as setting up the Fokker-Planck equation is concerned. Solving it is another problem, but readily available numerical packages exist for solving such equations, which after all are parabolic and therefore numerically well behaved.

In the present survey we shall not focus too much on any particular aspect, but try to give a sampler. To this effect, we begin in Section 2 with the the stochastic stability for discrete delay system as a bridge between the finite and infinite dimensional theory. In Section 3, a short digression on Itô-calculus is presented in order to keep the chapter self-contained. It forms a minimal necessary background in stochastics and is explained in very simple terms. We also shed some light on the nature of the solutions of stochastic delay systems. Section 4 treats stochastic stability of continuous time stochastic delay systems, and we illustrate how delay dependent conditions may be obtained from the Lyapunov-Krasovskii theory. Existence of invariant distributions is discussed in Section 5. We also present a new approximation for the Fokker-Planck equation of a delay system. Stability results for neutral equations are presented in Section 6.

With regards to notation, we need to clarify that we opted to use identical notation, x_τ, for two different things: If it appears in the conjunction with x, then the *values* $x(t-\tau)$ and $x(t)$ will be meant respectively, thus suppressing 't'. In the other context x_t denotes the (infinite dimensional) state of the delay system at time t, as defined in [16] and already introduced above. We hope that the burden of this dual use is offset by the notational simplicity.

2 Discrete Stochastic Delay Systems

Stochastic theory is typically simpler in discrete time problems. Indeed, a stochastic sequence is fundamentally a countable collection of random variables, and a terse detailed theory of processes can be avoided. Thus let us consider, for $x_k \in X$ for all $k \geq 0$, the discrete delay system

$$x_{k+1} = f(x_{[k]}, w_k) \quad , \quad x_k = \phi_k \ , \ k \in \{-N, \ldots, 0\} = \mathcal{I}. \tag{1}$$

where w_k is a white noise sequences (independent zero mean random variables), and $x_{[k]}$ denotes $[x'_k, x'_{k-1}, \ldots, x'_{k-N}]' \in X^N$. We assume that $f(0,0) = 0$, so that $x_k \equiv 0$ is the solution to the stochastically unperturbed ($w_k \equiv 0$) equation with zero initial condition ($\phi_k = 0 \ , \ k \in \mathcal{I}$), called the null-solution. Let $\phi = \phi_{[0]}$ for simplicity of notation.

2.1 Stability definitions

The following are standard definitions [29,31]

Definition 1. *The null-solution of the stochastic system (1) is said to be*
i) Mean square stable, if $\forall\epsilon$, there exists δ such that

$$\|\phi\|_0^2 \stackrel{\text{def}}{=} \sup_{i\in\mathcal{I}} \mathbf{E}\|\phi_i^2\| < \delta \implies \mathbf{E}\|x_i\|^2 < \epsilon\,,\ \forall i \in \mathbf{Z}_+$$

ii) Asymptotically mean square stable, if
- it is mean square stable
- $\lim_{i\to\infty} \mathbf{E}x_i^2 \to 0$.

iii) Stable in probability, if $\forall\epsilon, \epsilon' > 0$, there exists $\delta > 0$ such that

$$\mathbf{P}\{\|\phi\|_0 < \delta\} = 1 \implies \mathbf{P}\{\sup \|x_i\| > \epsilon\} < \epsilon'$$

iv) Asymptotically stable in probability, if
- it is stable in probability
- $\mathbf{P}\{\lim_{i\to\infty} \|x_i\| = 0\} = 1$.

Mean square (asymptotic) stability implies (asymptotic) stability in probability [31]. The next type of stability is different in character, as the null solution may not be an equilibrium for the system.

5. Stability in distribution, if the probability distribution of the state converges (in some appropriate norm) to an invariant distribution.

The following Lyapunov stability theorems are established [25,48].

Theorem 1 (Kolmanovskii and Shaikhet, 1995). *If there exists a nonnegative functional $V : \mathcal{I} \to \boldsymbol{R}$, such that*

$$\begin{aligned} \mathbf{E}V_0 &\le c_1\|\phi\|_0^2 \\ \mathbf{E}\Delta V_i &\le -c_2\,\mathbf{E}x_i^2 \end{aligned}$$

for $c_1, c_2 > 0$, then the system is asymptotically mean square stable.

Theorem 2 (Paternoster and Shaikhet, 2000). *If there exists a functional $V : \mathcal{I} \to \boldsymbol{R}$, such that*

$$\begin{aligned} V_i &\ge c_0\,\|x_i\|^2 \\ \mathbf{E}V_0 &\le c_1\|\phi\|_0^2 \\ \mathbf{E}\Delta V_i &\le 0 \end{aligned}$$

for $c_0, c_1 > 0$, then the null solution of the system is stable in probability.

The paper [27] surveys methods for the construction of appropriate Lyapunov functionals. In it, systems with varying delays and classes of nonlinear systems are discussed. Recently, extensions for the case where the delays are random and modeled by a Markov process have been given [21]. In what follows the focus is on stochastically perturbed linear delay systems of the forms

$$x_{k+1} = Ax_k + Bx_{k-N} + Cx_k w_k, \tag{2}$$

called semilinear (linear if $C = 0$) with zero mean multiplicative noise, $\{w_k\}_{k \geq 0}$, or,

$$x_{k+1} = Ax_k + Bx_{k-N} + Cw_k + v_k, \tag{3}$$

with zero mean additive noise, $\{v_k\}_{k \geq 0}$. As in the deterministic case, the discrete time theory leads to sufficient conditions for stability and *instability*, unlike the Riccati based methods in continuous time, which only supply sufficient conditions for stability. We first define our notion of robust (in)stability in either of the above senses

Definition 2. *i) The delay-difference system (2) is* robustly (asymptotically) stable *(R(A)SS) if for any number of delay steps $N \geq 0$ the dynamical system is (asymptotically) stable.*
ii) The delay-difference system (2) is robustly unstable *if for all $N \geq 0$ the dynamical system is unstable.*

2.2 Stability with multiplicative Noise

Shaikhet studied a scalar linear stochastic delay system, $x_{k+1} = \sum_{i=0}^{n} a_i x_{k-i} + \sigma x_{k_\ell} w_k$,where the unknown but otherwise fixed delay enters only in (one) stochastic term. Such a system is readily represented in a finite dimensional matrix form $\chi_{k+1} = A\chi_k + C\chi_k w_k$, where C has rank one, for which necessary and sufficient conditions for asymptotic mean square stability are obtained [51]. In a more general case, only a partial converse is possible (i.e., necessity fails in general). Indeed, the following stochastic extensions for robust stability, shown in [17, 18, 68] for the deterministic case, are easily established:

Theorem 3. *The system with multiplicative white noise*

$$x_{k+1} = Ax_k + Bx_{k-n} + Cx_k w_k$$

is stochastically stable for all n, i.e., robustly stochastically stable (RSS) if there exist positive definite matrices P and S such that

$$M = \begin{bmatrix} A'PA - P + S + B'PB + C'PC & A'PB \\ B'PA & -S \end{bmatrix} < 0.$$

If $M > 0$ for some $P, S > 0$, then the system is not RSS.

Proof: Take the discrete Lyapunov-Krasovskii functional

$$V(\phi_{-n}, \ldots, \phi_0) = \phi_0' P \phi_0 + \sum_{i=-n}^{-1} \phi_i' Q \phi_i, \tag{4}$$

with $P > 0, Q > 0$, and express the LMI

$$\mathbf{E}\Delta V_k = \mathbf{E}\,[\, x_k' \; x_{k-n}' \,]\, M \begin{bmatrix} x_k \\ x_{k-n} \end{bmatrix},$$

upon setting $Q - B'PB = S$. Then $V_k = V(x_k)$ satisfies the conditions of Theorem 2, so that stability in probability follows. On the other hand, if $M > 0$, it follows that $\mathbf{E}V_k > \mathbf{E}V_0 > 0$ and the null solution cannot be stable. □
No conclusion is reached if M is indefinite. The previous theorem leads directly to the following Riccati conditions for robust asymptotic mean square stability (RASS).

Theorem 4. *The system*

$$x_{k+1} = Ax_k + Bx_{k-n} + Cx_k w_k \tag{5}$$

is RASS if either there exist positive definite symmetric matrices P, W, R such that

$$A'PA + B'PB + C'PC + W + A'PBW^{-1}B'PA + R = P,$$

or there exist positive definite symmetric matrices Π, Z, S such that

$$A'\Pi A + B'\Pi B + C'\Pi C + Z + B'\Pi AZ^{-1}A'\Pi B + S = \Pi.$$

Proof: The first equality is derived from the Schur complement of S in M, the second the Schur complement of the (1,1)-block in M, with suitable relabeling. Then $\mathbf{E}\, V_i^2 \leq c_2\, \mathbf{E}\, \|x_i\|^2$, where c_2 is respectively $\|R\| > 0$ or $\|S\| > 0$. Then invoke Theorem 3. □

Corollary 1. *The system (5) is RASS if either*

$$\begin{aligned} A'\Lambda^{-1}A + \Omega + R &= P - C'PC \\ B\Omega^{-1}B' + \Lambda &= P^{-1} \end{aligned}$$

for P, R, Λ and $\Omega > 0$; or,

$$\begin{aligned} B'\Sigma^{-1}B + \Gamma + S &= \Pi - C\Pi C' \\ A\Gamma^{-1}A' + \Sigma &= \Pi^{-1} \end{aligned}$$

for Π, S, Γ and $\Sigma > 0$.

The proof is similar to the deterministic case (making use of Woodbury's lemma), and omitted. The special choice $P = X, R = Q, \Lambda = \frac{1}{p}X^{-1}$, and $\Omega = qB'PB$ leads to a nice form under the condition of invertibility of B. Indeed, substitution in the corollary gives $pA'XA + qB'XB + Q = X - C'XC$ and $B\Omega^{-1}B' = X^{-1} - \frac{1}{p}X^{-1}$. Hence one obtains a generalized Lyapunov equation involving a Hölder pair:

$$\begin{aligned} \tfrac{1}{p} + \tfrac{1}{q} &= 1 \\ pA'XA + qB'XB + C'XC + Q &= X\ . \end{aligned} \tag{6}$$

However, while the positive definiteness of the solution to the Lyapunov equation $A'XA + Q = X$ is necessary and sufficient for the Schur-Cohn stability of A, the condition (6) is only sufficient for RASS.

Example: Consider the scalar difference equation $x_{i+1} = ax_i + b_{i-n} + cx_i w_i$.One finds

$$pa^2 + qb^2 + c^2 < 1$$

thus, for $|c| < 1$, a Hölder pair (p, q) can be found if (a, b) lies in the diamond bordered by

$$\frac{|a|}{\sqrt{1-c^2}} + \frac{|b|}{\sqrt{1-c^2}} = 1.$$

In addition *instability* theorems can be obtained, giving a partial converse.

Theorem 5. *The system (5) is unstable if*

$$\begin{bmatrix} A'PA - P + S + B'PB + C'PC & A'PB \\ B'PA & -S \end{bmatrix} > 0.$$

for some $P > 0$, $S > 0$.

Proof: By expressing that $\mathbf{E}\Delta V > 0$, hence V is increasing on average. □

Corollary 2. *An equivalent condition is expressed by the Riccati equation*

$$A'PA + B'PB + C'PC - S - A'PBS^{-1}B'PA - R = P$$

for some $P > 0, R > 0, S > 0$.

2.3 Stability with additive and multiplicative noise

Consider now the model

$$x_{k+1} = Ax_k + Bx_{k-n} + Cx_k w_k + v_k \tag{7}$$

where w_k and v_k are independent zero mean white noises, with respectively the variance σ^2 and covariance matrix $R \geq 0$. If the pair $(A, R^{1/2})$ is reachable, it is clear that $x_k \equiv 0$ can no longer be a solution. Hence the only meaningful questions are the one of existence of a stationary distribution (and convergence towards it), and the moment stability [31]. The first problem lies in the domain of ergodic theory.

Using the same Lyapunov functional (4), one finds with M as defined in Theorem 3,

$$\mathbf{E}\,\Delta V_k = \mathbf{E}\,[\,x_k'\;x_{k-n}'\,]\,M \begin{bmatrix} x_k \\ x_{k-n} \end{bmatrix} + \operatorname{Tr} PR. \tag{8}$$

Obviously, under the conditions of RASS in the additive noise free case, it will be true that $\mathbf{E}\,\Delta V_k \leq -c_2\mathbf{E}\|x_k|^2$, *outside* some ball centered at the origin. This guarantees the existence of a stationary distribution, π, see Tong [55].

Assuming a statistical steady state exists, we investigate its properties: Set $P_\ell \stackrel{\text{def}}{=} \mathbf{E}x_{k+\ell}x_k'$,it is then readily established that

$$P_0 = AP_0A' + BP_0B' + \sigma^2 CP_0C' + AP_NB' + BP_N'A' + R \tag{9}$$

$$P_\ell = AP_{\ell-1} + BP_{\ell-1-n} \quad , \quad \ell = 1, \ldots, n. \tag{10}$$

Since by definition $P_\ell = P'_{-\ell}$, one obtains upon setting

$$\mathcal{P} \stackrel{\text{def}}{=} \begin{bmatrix} P_0 & P_1 & \cdots & P_N \\ P_1' & P_0 & \cdots & P_{N-1} \\ \vdots & & \ddots & \vdots \\ P_N' & P_{N-1}' & \cdots & P_0 \end{bmatrix} \quad , \quad \mathcal{A} \stackrel{\text{def}}{=} \begin{bmatrix} A & \cdots & 0 & B \\ I & & & \\ & \ddots & & \\ & & I & 0 \end{bmatrix}$$

the generalized Lyapunov equation

$$\mathcal{P} = \mathcal{A}\mathcal{P}\mathcal{A}' + \sigma^2 \mathcal{C}\mathcal{P}\mathcal{C}' + \mathcal{R}, \tag{11}$$

where also we defined

$$\mathcal{C} \stackrel{\text{def}}{=} \begin{bmatrix} C & 0 & \cdots & 0 \\ 0 & 0 & & \\ \vdots & & \ddots & \vdots \\ 0 & \cdots & \cdots & 0 \end{bmatrix} \quad \text{and} \quad \mathcal{R} \stackrel{\text{def}}{=} \begin{bmatrix} R & 0 & \cdots & 0 \\ 0 & 0 & & \\ \vdots & & \ddots & \vdots \\ 0 & \cdots & \cdots & 0 \end{bmatrix}.$$

Note that if $C = 0$, and the additive noise is gaussian, then if a steady state exists, it will also be gaussian, in which case case the steady state density is completely determined by $\mathcal{P}$. If (11) is solved by an iterative scheme. convergence is guaranteed if and only if the matrix $\mathcal{A} \otimes \mathcal{A} + \sigma^2 \mathcal{C} \otimes \mathcal{C}$ is Schur-Cohn stable.

3 Continuous Time Stochastic Delay Systems

There are very many intricate details, as already the study of of finite dimensional stochastic systems is challenging and requires quite some machinery and new concepts. A full study of stochastic delay systems, involving dynamic modeling, existence and uniqueness of solutions, stability in its various guises, qualitative properties, filtering and control is out of the questions in this chapter. We refer the reader once again to monographs of Mohammed, Mao, Kolmanovskii and Myshkis, Shaikhet and many other. Additional biographical notes regarding topics not covered are found at the end of this chapter. The aim is to equip the readers quickly with some powerful tools, with the aid of which they can attack their own problems. In this sense, we will also strip also all unnecessary complications most of the time. The idea being that once the reader sees how to attack the simple problem, chances will be good that the 'usual' extensions bearing quantifiers as "uncertain", "perturbed", "time-variant", "multiple", and "distributed" will work as well. When one considers a continuous time model in the form: $\dot{x}(t) = Ax(t) + Bx(t-\tau) + [Cx(t) + Dx(t-\tau)]u(t)$,where u is "white noise", then one must realize that such a model is not well defined, since u has infinite variance. While in the linear theory, hand waving arguments can be presented to make things work; in the nonlinear realm, things get quickly out of hand. The work of Itô and many others led to a new way of thinking about stochastic models. We give a very brief glimpse of the Itô-calculus in order to make this chapter self-contained. The cogniscenti can easily skip this digression.

3.1 Digression: Itô-calculus in a nutshell

We briefly show how the limit of a random walk model defines the Wiener process (or Brownian motion). Then we shall define the quadratic variation and show its repercussions on a Taylor expansion of a function of a Wiener process, thus heuristically presenting the celebrated Itô-rule.

Random walk

We start from a random walk model: let the time step be h and the spatial step Δ. Let $\{\epsilon_i\}$ be a symmetric Bernoulli process. i.e., a sequence of random variable whose values attain only two values: +1 and -1, and each with probability $1/2$. Moreover, the random variables ϵ_i are independent. The random walk starts at the zero position, and at the k-th step (time $t = kh$), a spatial step of size Δ is taken in the forward or the backward direction, depending on whether ϵ_k was +1 or -1. Thus, if $x(t)$ denotes the position at time t, we have

$$x(kh) = \sum_{i=1}^{k} \Delta\epsilon_i.$$

The following properties are easily verified:

i) $\mathbf{E}\, x(kh) = \sum_{i=1}^{k} \Delta \mathbf{E}\, \epsilon_i = 0$
ii) $\mathrm{Var}\, x(kh) = \sum_{i=1}^{k} \mathrm{Var}\,(\Delta\epsilon_i) = \Delta^2 k.$
iii) Increments in *nonoverlapping* intervals are independent.

Wiener Process (Brownian Motion)

Consider now the limit of a random walk for $h \to 0$ and $\Delta \to 0$. Let $t = kh$ remain fixed, so that $h \to 0$ also implies that the number of steps $k \to \infty$. Thus $x(kh) = x(t)$, and by the *central limit theorem*, $x(t)$ is normally distributed, with mean zero and variance equal to $\lim_{kh=t} \Delta^2 k = t \lim \left(\frac{\Delta^2}{h}\right)$. Obviously, the process $x(t)$ will only be meaningful and nontrivial if $\frac{\Delta^2}{h} \to \sigma^2$where σ^2 is a positive constant (called the *variance parameter*). This has important repercussions: Given an arbitrary partition $\mathcal{P} = \{0 = t_0 < t_1 < \ldots < t_n = T\}$ of the interval $[0, T]$, define the *quadratic variation* over $\mathcal{P}$ of a process x, denoted $\mathrm{QV}_{\mathcal{P}}(x)$, as the sum $\sum_i [x(t_{i+1}) - x(t_i)]^2$. The quadratic variation, $\mathrm{QV}_{[0,T]}(x)$ of the process x over the interval $[0, T]$ is then defined as the limit of $\mathrm{QV}_{\mathcal{P}}(x)$ over finer and finer partitions such that $\max_k(t_{k+1} - t_k) \to 0$. It follows then from the definition of the Wiener process as the limit of a random walk that $\mathrm{QV}_{[0,T]}(x) = \lim \sum [x((k+1)h) - x(kh)]^2 = \lim \sigma \Delta^2 = \frac{t}{h}\Delta^2 = \sigma^2 t$ with probability one. Likewise, the *total variation* $\mathrm{TV}_{\mathcal{P}}(x)$, of a process x given the partition $\mathcal{P}$, is defined as the limit of the sum $\sum_i |x(t_{i+1}) - x(t_i)|$. The total variation of x in $[0, T]$ is then again the limit of $\mathrm{TV}_{\mathcal{P}}(x)$, when the partition gets finer. Now one has

$$\begin{aligned} \mathrm{QV}_{\mathcal{P}} &= \sum [x(t_{i+1}) - x(t_i)]^2 \\ &\leq \max |x(t_{i+1}) - x(t_i)| \sum |x(t_{j+1}) - x(t_j)| \\ &\leq \max |x(t_{i+1}) - x(t_i)| \, \mathrm{TV}_{\mathcal{P}} \end{aligned}$$

The left hand side converges to the quadratic variation $\sigma^2 t$, the sum on the right hand side approaches the total variation. But the maximum size of a subinterval in the partition converges to zero. Hence, the inequality can only be maintained if the total variation goes to infinity. Now when the total variation of a function x is infinite in any arbitrary subinterval, this function x is nowhere differentiable. Indeed in any finite amount of time, an infinite distance is covered by the process, giving it an infinite average speed. Hence considering white noise as a "derivative" of a Wiener process is somewhat far fetched, as the latter is nowhere differentiable. Note however that by construction, the Wiener path is continuous almost everywhere.

Itô-differentiation rule

The Wiener process $w(t)$ will be called *standard* if its variance parameter equals 1. Heuristically, the quadratic variation identity of the standard Wiener process in $[0, T]$ can be written as $\int_0^T (dw(t))^2 = t$. Hence the square $(dw)^2$ acts like dt. The effect on the Taylor expansion of $f(w(t), t)$ is

$$\mathrm{d}f(w(t), t) = \frac{\partial f}{\partial t} \mathrm{d}t + \frac{\partial f}{\partial w} \mathrm{d}w + \frac{1}{2} \left[\frac{\partial^2 f}{\partial t^2} \mathrm{d}t^2 + 2 \frac{\partial^2 f}{\partial t \partial w} \mathrm{d}t \, \mathrm{d}w + \frac{\partial^2 f}{\partial w^2} \mathrm{d}w^2 \right] + \cdots$$

By the quadratic variation property, $\mathrm{d}w^2$ acts as $\mathrm{d}t$ with probability one. The other terms inside the square bracket, and all other higher order terms have greater order than $\mathrm{d}t$, and are negligible. Hence, one obtains (heuristically) the rule

$$\mathrm{d}f(w, t) = \left(\frac{\partial f}{\partial t} + \frac{1}{2} \frac{\partial^2 f}{\partial w^2} \right) \mathrm{d}t + \frac{\partial f}{\partial w} \mathrm{d}w \tag{12}$$

In fact, a rigourous interpretation of the above is through its integral form

$$f(w(T), T) = f(0, 0) + \int_0^T \left(\frac{\partial f}{\partial t} + \frac{1}{2} \frac{\partial^2 f}{\partial w^2} \right) \mathrm{d}t + \int_0^T \frac{\partial f}{\partial w} \mathrm{d}w(t),$$

where the second integral is the non-anticipatory Itô integral, defined as a limit over a partition $\mathcal{P} = \{0 = t_0 < t_1 < \ldots < t_n = T\}$ of the interval $[0, T]$:

$$\int_0^T g(x,t)\,\mathrm{d}x(t) = \lim_{\mathcal{P}} \sum g(x(t_i),t_i)\,x[t_{i+1}) - x(t_i)].$$

This integral may be generalized for g a functional of x as long as it is non-anticipatory, i.e., g({x},t) does not depend on *future* values $x(s)$, for $s > t$.

More generally, if the state of a first order stochastic system satisfies the Itô equation: $\mathrm{d}x = f(x,t)\,\mathrm{d}t + g(x,t)\,\mathrm{d}w(t)$, then for any smooth F:

$$\mathrm{d}F(x,t) = \frac{\partial F}{\partial x}\,\mathrm{d}x + \frac{\partial F}{\partial t}\,\mathrm{d}t + \frac{1}{2}\frac{\partial^2 F}{\partial x^2}\,(\mathrm{d}x)^2.$$

With $(\mathrm{d}x)^2 \sim g(x,t)^2\,\mathrm{d}t$, we obtain

$$\mathrm{d}F(x,t) = \frac{\partial F}{\partial x} g(x,t)\,\mathrm{d}w + \left[\frac{\partial F}{\partial t} + \frac{\partial F}{\partial x} f(x,t) + \frac{1}{2}\frac{\partial^2 F}{\partial x^2}\, g(x,t)^2\right]\,\mathrm{d}t.$$

The operator $\mathcal{L}$ such that

$$\mathcal{L}F = \frac{\partial F}{\partial t} + \frac{\partial F}{\partial x} f(x,t) + \frac{1}{2}\frac{\partial^2 F}{\partial x^2}\, g(x,t)^2 \tag{13}$$

is called the *infinitesimal generator* for the process x.

3.2 Solutions of stochastic delay systems

We briefly glance over the main results on existence and uniqueness of solutions to delay stochastic differential equations and some of their properties. Only a stripped down version of the conditions will be presented here. For the intricate details, see the books of Mohammed [44] and Mao [34, 35].

Generically speaking, if a deterministic dynamical system $\dot{x} = f(t,x)$ has state space X, then a trajectory is an element from $C([0,T],X)$, the set of continuous maps from the interval $[0,T]$ to X, assigning to $t \in [0,T]$ the vector $x(t) \in X$. Stochastic theory invariably starts with the specification of the ubiquitous probability triple $(\Omega,\mathcal{B},P)$. All random variables within the theory are measurable in this (big) space. The state space of the corresponding stochastic problem defined over a probability space $(\Omega,\mathcal{B},P)$ is the set of maps $\Omega \to X$ having some prescribed "well-defined"-ness properties: $\int_\Omega \|x(\omega)\|^2 \mathrm{d}P(\omega) < \infty$. Thus $X_{\mathrm{stoch}} = \mathcal{L}^2(\Omega,X)$. In addition, there is a well defined *information structure*, technically presented by what is called a *filtration*. This is an increasing family $\{\mathcal{F}_t\}_{t\geq 0}$ of sigma-algebras $\mathcal{F}_t \subseteq \mathcal{B}$, with the property that if $t_1 < t_2$, then $\mathcal{F}_{t_1} \subseteq \mathcal{F}_{t_2}$. We say that a random variable $y : \Omega \to \mathbf{R}$ is *adapted* to $\mathcal{F}_t$, denoted $y \in \mathcal{F}_t$, if its value $y(\omega)$ can be unambiguously determined from the knowledge of the sigma algebra $\mathcal{F}_t$ at time t. For instance, if we consider the sigma-algebra generated by the successive coin tosses, then the value of the second toss is known if $t \geq 2$, but unknown at $t = 1$. We say that $f(\omega,t,x(t))$ is *non-anticipative* with respect to the filtration $\{\mathcal{F}_t\}_{t\geq 0}$ if

$f(\omega, t, x(t)) \in \mathcal{F}_t$ for all $t \geq 0$.

For a general deterministic n-dimensional functional differential system, $\dot{x} = f(t, x_t)$, with finite delay τ, using the standard notation

$$x_t \stackrel{\text{def}}{=} \{\, x(t+s) \mid s \in [-\tau, 0] \stackrel{\text{def}}{=} \mathcal{I} \,\}, \tag{14}$$

the state space is the Banach space $X_{\text{det}} = C(\mathcal{I}, \mathbf{R}^n)$ equipped with the sup-norm [16]:

$$\forall \phi \in X_{\text{det}} : \quad \|\phi\|_C = \sup_{-\tau \leq s \leq 0} |\phi(s)|,$$

where $|\cdot|$ denotes the Euclidean norm in $\mathbf{R}^n$.

In the stochastic sense, $X_{\text{stoch}} = \mathcal{L}^2(\Omega, C(\mathcal{I}, \mathbf{R}^n))$, and $\theta \in X_{\text{stoch}}$ implies $\|\theta(\omega)\|_C$ is in $\mathcal{L}^2$ with respect to Ω, i.e., $\|\theta\|_{X_{\text{stoch}}} = \left[\int_\Omega \|\theta(\omega)\|_C^2 \,\mathrm{d}P(\omega)\right]^{1/2}$, with trajectories being elements of $C\left([0,T], \mathcal{L}^2(\Omega, X_{\text{det}})\right)$.

The general stochastic functional differential equation (SFDE)

$$x(\omega)(t) = \begin{cases} \phi(\omega)(t) & t \in \mathcal{I} \\ \phi(\omega)(0) + ⨍_0^t g(s, x_s)\, \mathrm{d}z(\cdot)(s) & 0 \leq t \leq T \end{cases} \tag{15}$$

where z is a general 'noise'-process, separable into $z(\omega)(t) = \lambda(t) + Z_m(\omega)(t)$, where λ is a Lipshitz function, Z_m an $\mathcal{F}_t$-adapted martingale satisfying a growth condition, and $⨍$ indicates a stochastic integral, is studied by Mohammed in his monograph [44]. If g is Lipschitz (albeit in a 'stochasticized' form) and non-anticipatory, then the SFDE has a solution, i.e., a trajectory, $t \in [-\tau, T] \longrightarrow x_t \in \mathcal{L}^2(\Omega, C(\mathcal{I}, \mathbf{R}^n))$, such that $x_0 = \phi$. It is unique up to equivalence of stochastic processes, and its trajectory $[0,T] \to \mathcal{L}^2(\Omega, C(\mathcal{I}, \mathbf{R}^n))$ is adapted to the filtration $\{\mathcal{F}_t\}$, and almost all sample paths are continuous for almost all $\omega \in \Omega$ [44, p.36]. For $0 \leq t_1 \leq t_2 \leq T$, the solutions define a family of evolution maps

$$T_{t_2}^{T_1} : \mathcal{L}^2(\Omega, C(\mathcal{I}, \mathbf{R}^n); \mathcal{F}_{t_1}) \longrightarrow \mathcal{L}^2(\Omega, C(\mathcal{I}, \mathbf{R}^n); \mathcal{F}_{t_2})$$

and set $T_t^0 = T_t$. It is easily verified that the stochastic semigroup property $T_{t_2} = T_{t_2}^{t_1} \circ T_{t_1}$ holds for all $0 \leq t_1 \leq t_2 \leq T$. If T_t is Lipschitz and g has continuous partial derivatives with respect to its second argument, then T_t is continuously differentiable.

The Itô delay differential system with x taking values in $\mathbf{R}^n$,

$$\mathrm{d}x(t) = f(t, x_t)\,\mathrm{d}t + g(t, x_t)\,\mathrm{d}w(t), \tag{16}$$

is a special case of the previous class of systems. If f and g are globally Lipschitz and continuous then the trajectories x_t through the initial data $x_0 = \phi \in C(\mathcal{I}, \mathbf{R}^n)$ describe a Markov process on $C(\mathcal{I}, \mathbf{R}^n)$ with filtration generated by the Wiener process $w(t)$, and transition probabilities

$$p(t_1, \phi_1, t_2, B) = P\{\omega \in \Omega \mid T_{t_2}^{t_1}(\phi_1)(\omega) \in B\},$$

where B is a Borel set in the state space $C(\mathcal{I}, \mathbf{R}^n)$. Enter C_b, the Banach space of bounded continuous functions $\psi : X = C(\mathcal{I}, \mathbf{R}^n) \to \mathbf{R}$ with the sup-norm $\|\psi\|_{C_b} = \sup_{\phi \in X} |\psi(\phi)|$. For $0 \le t_1 \le t_2 \le T$, define for each $\psi \in C_b$, the map

$$P_{t_2}^{t_1}(\psi) : X = C(\mathcal{I}, \mathbf{R}^n) \longrightarrow \mathbf{R} \tag{17}$$

$$\eta \longrightarrow \mathbf{E}\left(\psi \circ T_{t_2}^{t_1}(\phi)\right) = \int_X \psi(s) p(t_1, \phi, t_2, \mathrm{d}s), \tag{18}$$

then the set $\{P_{t_2}^{t_1} ; 0 \le t_1 \le t_2 \le T\}$ is a contraction semigroup on C_b. However, unlike the non-delay case ($\tau = 0$), this contraction semigroup is not strongly continuous, thus adding further complications to the theory of SFDE's. This is a consequence of the state space X not being locally compact when $\tau > 0$. Consequently, an unbounded weak infinitesimal generator exists

$$A : \mathcal{D}(A) \subset C_b \to C_b, \tag{19}$$

$$A(\psi) = \mathrm{w-} \lim_{h \to 0+} \frac{P_t(\psi) - \psi}{h}. \tag{20}$$

Recall that a parameterized family of functions $\{\psi_t; t > 0\}$ in C_b *converges weakly* to ψ in C_b as $t \to 0$ if $\lim_{t \to 0+} \langle \psi_t, \mu \rangle = \langle \psi, \mu \rangle$ for all Borel measures μ on $C(\mathcal{I}, \mathbf{R}^n)$. There are many intricate details, for which we refer to [44]. The infinitesimal generator is needed to obtain an Itô-differentiation rule (and thus a stochastic Lyapunov theory). However a simplification is possible. Let $C_b^0 \subset C_b$ be the domain of strong continuity of $\{P_t\}_{t \ge 0}$, i.e., $P_t(\psi) \to \psi$ for all ψ in C_b^0. Introducing the class of *quasitame* functionals, Mohammed has shown that they form a weakly dense subalgebra of C_b^0, belong to $\mathcal{D}(A)$, and generate the Borel sets in C. In addition, if ψ is quasi-tame, then its value along a solution is a semi-martingale, and Itô's formula holds.

A functional $\psi : C \to \mathbf{R}$ is quasitame if there are C^∞ bounded maps $h : (\mathbf{R}^n)^k \to \mathbf{R}$, $K_j : \mathbf{R}^n \to \mathbf{R}^n$, and piecewise C^1 functions $\ell_j : \mathcal{I} \to \mathbf{R}; 1 \le j \le k$ such that for all $x \in C$:

$$\psi(x) = h\left(\int_{-\tau}^0 K_1(x(s))\ell_1(s)\,\mathrm{d}s, \ldots, \int_{-\tau}^0 K_{k-1}(x(s))\ell_{k-1}(s)\,\mathrm{d}s\,,\, x(0)\right).$$

Under appropriate conditions, regularity of the the trajectory random field is shown. While it has been established that stochastic delay systems with delay in the diffusion have very irregular behavior - continuous versions of the trajectory field never exist - regularity is preserved in a distributional sense. See [44, Chapter V].

Remark 1. Alternatively, the Delfour-Mitter Hilbert space $M_2 = \mathbf{R}^n \times L^2(\mathcal{I},$ $REri k^n)$ with the Hilbert norm $\|(x, \psi)\|_{M_2} = \left(\|x\|^2 + \int_{-\tau}^0 \|\psi(s)\|^2\,\mathrm{d}s\right)^{1/2}$ may be taken as the state space. This only requires minor modifications for the existence and uniqueness conditions, but is more appropriate in dealing the LQG problem (stochastic optimal control). An important feature of this formulation is that continuity of initial data is no longer required [22].

In the remaining part of this chapter, we will for simplicity put the emphasis on the linear (additive noise) and the semilinear (multiplicative noise) delay systems. Whereas many different criteria for stability can be obtained, see [34,35,37,45,52], we will single out methods that bear resemblance to the ubiquitous Riccati (stemming from LMI) approaches in the deterministic theory. Previous work [28,75] has recently come to our attention. A few constructive results on stabilization are mentioned. in [59,60].

Thus consider the Itô delay-differential system

$$\mathrm{d}x(t) = \left\{ Ax(t) + \sum_{i=1}^{N} B_k x(t-\tau_i) + bu(t) \right\} \mathrm{d}t + C\,\mathrm{d}w(t) \tag{21}$$

with $C \in \mathbf{R}^{n\times m}$, and $w(t)$ an m-dimensional standard Brownian motion (SBM); and

$$\begin{aligned}\mathrm{d}x(t) = &\left\{ Ax(t) + \sum_{i=1}^{k} B_i x(t-\tau_i) + bu(t) \right\} \mathrm{d}t + \\ &+ \sum_{j=1}^{N} \left\{ C^{(j)} x(t)\,\mathrm{d}w_{0j}(t) + \sum_{i=1}^{k} D_i^{(j)} x(t-\tau_i)\,\mathrm{d}w_{ij}(t) \right\}\end{aligned} \tag{22}$$

with $C^{(j)}$ and $D_i^{(j)} \in \mathbf{R}^{n\times n}$ and the $\{w_{i,j}(t)\}$ standard one dimensional Brownian motions. where A and $B_i, i = 1,\ldots N$, are all $n\times n$ matrices over $\mathbf{R}$, and $b \in \mathbf{R}^{n\times m}$.

The main point we make is that as far as stability is concerned, the complexity of the analysis is not much greater for the stochastic than for the deterministic theories, see [13,57,67]. One approach to stabilization with static and dynamic output feedback, is based on this Riccati theory [60,72]. The earlier publication assumed (quite unrealistically) exact knowledge of the delay τ. In addition some very restrictive technical conditions were required. In the later paper it was shown how to combine the Riccati method with perturbation theory, to circumvent the need for such precise information.

4 Stability of Stochastic Retarded Systems

Substantial work in the stability of stochastic time delay systems originated in the works of Kolmanovskii, Myshkis, Nosov and Shaikhet [22,23,26], Mohammed [44] and Mao [34,35], and many others. The main approach as in the deterministic case - is Lyapunov based, and centers around the martingale convergence theorem and Itô's formula [19]. Both Lyapunov-Krasovskii based methods and Razumikhin type theorems are known. The different point of view, the first using Lyapunov functionals, the second functions, is elaborated in [16] for the deterministic concepts. We shall focus on the functionals. Razumikhin-type methods are derived in [38,50].

The definitions for stability for the general Itô-model

$$dx(t) = f(t, x_t)\,dt + g(t, x_t)\,dw(t), \tag{23}$$

in continuous time are analogous to the discrete time ones. We only specify:

Definition 3. *If* $f(0) = 0$ *and* $g(0) = 0$*, then the equilibrium* $x_t \equiv 0$ *of (23) is* globally asymptotically stable in probability *if* $\forall\, s \geq 0$, *and* $\forall\, \epsilon \geq 0$,

$$\lim_{x\to 0} \mathbf{P}\{\sup_{s<t} |x_t^{s,\phi}| > \epsilon\} = 0$$
$$\mathbf{P}\{\lim_{t\to\infty} |x_t^{s,\phi}| = 0\} = 1.$$

Here $x_t^{s,x}$ *is the solution at time t for the system with initial condition* $x_t = \phi$ *when* $t = s$.

Definition 4. *The stochastic delay-differential equation (23) is* robustly (asymptotically) stochastically stable *(R(A)SS) if it is globally (asymptotically) stable for all values of the delay(s).*

The following Riccati type conditions were obtained for the continuous delay systems were obtained in [13, 67]:

Theorem 6. *The system (22) with* $N = k = 1$ *and* $w_{01} \equiv w_{11}$ *(for simplicity) is RSS if either of the following holds:*

i) There exist symmetric positive definite matrices P, R, W *such that*

$$A'P+PA+W+C'PC+D'PD+(PB+C'PD)W^{-1}(B'P+D'PC)+R=0$$

ii) There exist symmetric positive definite matrices P, R, Z *such that*

$$A'P+PA+Z+C'PC+D'PD+(B'P+D'PC)Z^{-1}(PB+C'PD)+R=0$$

Proof: Consider the Lyapunov-Krasovskii functional

$$V(x) = x'Px + \int_{t-\tau}^{t} x'(\sigma)Qx(\sigma)\,d\sigma.$$

The Itô-rule: $dV = dx'\,Px + x'P\,dx + [x'Qx - x_\tau Q x_\tau]\,dt + dx'\,P\,dx$, yields

$$\mathcal{L}V = [x'\ x_\tau'] \begin{bmatrix} A'P + PA + Q + C'PC & PB + C'PB \\ B'P + D'PC & D'PD - Q \end{bmatrix} \begin{bmatrix} x \\ x_\tau \end{bmatrix}, \tag{24}$$

from which an LMI-condition (negative definiteness of the above weight matrix) follows. The two Riccati equations follow by setting either $D'PD - Q = -W < 0$ or $Z = -(A'P + PA + Q + C'PC) > 0$. □

Remark 2. The Lyapunov-Krasovskii functional, V, used in the theorem is not quasitame, due to the presence of the (unbounded) quadratic forms, but the direct application of Itô's formula and super-martingale estimates justify its use in stability analysis.

Remark 3. Kolmanovskii and Shaikhet considered also sufficient conditions for time-varying and distributed delays [26]. It should be noted that, as in the deterministic case, the bound $\dot{\tau}(t) \leq 1$ on all delays is allowed [58]. A simple explanation is that as time proceeds, the lower boundary of the interval for the definition of the state, i.e., $t - \tau(t)$, keeps moving forward. 'The system never has to remember what it already forgot.' Hence the idea that the state is a sufficient statistic (Markovian character in the stochastic case) is conserved.

Remark 4. The results of Korenevskii [28] and Zelentsovskii [75] used a special choice of the Lyapunov functional for which a *linear* sufficient condition LMI was obtained.

Finally, we state a useful definition for delay dependent stability:

Definition 5. *The stochastic delay-differential equation is* delay-dependent robustly stochastically stable *($\overline{\tau}$-RSS) if it is globally asymptotically stable for all values of the delay(s) in* $[0, \overline{\tau})$.

4.1 A distributed system

In [12], Florchinger investigated the stability of a distributed delay stochastic system

$$\mathrm{d}x(t) = \left(Ax(t) + B \int_{t-\tau}^{t} x(s)\,\mathrm{d}s \right) \mathrm{d}t + Cx(t)\,\mathrm{d}w(t). \tag{25}$$

Using a Lyapunov-Krasovskii functional of the form

$$V(\phi) = \phi(0)'P\phi(0) + \int_0^{\tau} \int_{-s}^{0} \phi'(\theta)Q\phi(\theta)\,\mathrm{d}\theta\,\mathrm{d}s,$$

he proved that if there exist symmetric $P > 0$ and $Q > 0$ such that

$$A'P + PA + C'PC + \tau Q + \tau PBQ^{-1}B'P < 0,$$

then the system is τ-RSS in probability.

This condition is *delay dependent*, and useful for smaller delays. But, the larger τ, the more stable A should be. It is therefore impossible to conclude robust stochastic stability via this method.

An alternative is obtained by considering a larger partial state vector. Indeed, introducing for some invertible $S \in \mathbf{R}^{n \times n}$ the new state $y(t) = S^{-1} \int_{t-\tau}^{t} x(s)\,ds$, the distributed system (25) reduces to the coupled

$$\begin{bmatrix} dx(t) \\ dy(t) \end{bmatrix} = \begin{bmatrix} A & B \\ I & 0 \end{bmatrix} \begin{bmatrix} x(t) \\ y(t) \end{bmatrix} + \begin{bmatrix} 0 & 0 \\ -I & 0 \end{bmatrix} + \begin{bmatrix} x(t-\tau) \\ y(t-\tau) \end{bmatrix} + \begin{bmatrix} C & 0 \\ 0 & 0 \end{bmatrix} \begin{bmatrix} x(t) \\ y(t) \end{bmatrix} dw(t). \tag{26}$$

(For simplicity, we have let $S = I$). This system is of the 'usual' form with crisp delays. Hence the simple criterion (24), with $\mathcal{P}$ and $\mathcal{Q}$ in the form

$$\mathcal{P} = \begin{bmatrix} P_1 & P \\ P' & P_2 \end{bmatrix}, \qquad \mathcal{Q} = \begin{bmatrix} Q_1 & 0 \\ 0 & Q_2 \end{bmatrix}$$

can be applied, thus giving the LMI

$$\begin{bmatrix} W(P_1) + P' + P + Q_1 + PQ_1^{-1}P' & P_1B + A'P + P_2 + PQ_1^{-1}P_2 \\ B'P_1 + P'A + P_2 + P_2Q_1^{-1}P' & B'P + P'B + Q_2 + P_2Q_1^{-1}P_2 \end{bmatrix} < 0$$

where $W(P_1)$ is the linear term $A'P_1 + P_1A + C'P_1C$. For instance, setting $P = -I$, $P_2 = pI$ and $Q_1 = I$, while $Q_2 \to 0$ gives a criterion for the case $B_s > 0$, where B_s is the symmetric part of B. (The freedom in choice of S can be exploited in other cases). If there exists a positive definite matrix P_1 and a positive number p such that

$$pP_1 > I, \quad \text{and} \quad \begin{bmatrix} W(P_1) & P_1B - A' \\ B'P_1 - A & -2B_s + p^2I \end{bmatrix} < 0,$$

then the stochastic delay system (25) is robustly asymptotically stable. Differential delay systems driven by multi-dimensional Wiener processes, and having multiple crisp and distributed delay terms can be dealt with in the same way. A technique of reducing a more general distributed delay system with rational kernel to a system with crisp delays was discussed in [63].

4.2 Transformation method—delay-dependent condition

Inspired by Niculescu's *transformation method* [47] to obtain delay-dependent stability conditions in the noise-free case, de Souza replaced (22) (again taking $N = k = 1$ and $w_{01} \equiv w_{11}$) by

$$\begin{aligned} \mathrm{d}x(t) = \Big[(A+L)x(t) + (B-L)x(t-\tau) - L\int_{t-\tau}^{t} [Ax(\theta) + Bx(\theta-\tau)]\,\mathrm{d}\theta + \\ -L\int_{t-\tau}^{t} Cx(\theta)\,\mathrm{d}w(\theta) \Big]\,\mathrm{d}t + Cx(t)\,\mathrm{d}w(t) \end{aligned}$$

A matrix L of free parameters is introduced, which is used to optimize τ [9]. Typical for the transformation method is the change in state space. The state of the transformed system are elements in the space $C([-2\tau, 0], \mathbf{R}^n)$. He obtained:

Theorem 7 (de Souza, 2000). *The stochastic delay system (22) with $N = k = 1$ is mean square asymptotically stable for all $\tau \in [0, \overline{\tau})$ if there exist symmetric positive definite matrices P, Q and $\overline{R}$ and a matrix Y such that the following LMI holds:*

$$\begin{bmatrix} A'P+PA+Y+Y'+C'(P+\overline{R})C+Q & \overline{\tau}Y & PB-Y & A'\overline{R} \\ \overline{\tau}Y' & -\overline{R} & 0 & 0 \\ B'P-Y' & 0 & -Q & B'\overline{R} \\ \overline{R}A & 0 & \overline{R}B & -\overline{R} \end{bmatrix} < 0. \quad (27)$$

We took the liberty to combine $\overline{\tau}$ and R in de Souza's original statement to $\overline{R} = \overline{\tau}R$. Since mean square asymptotic stability implies stochastic stability, we have at once the

Corollary 3. *The stochastic delay system (22) with $N = k = 1$ is $\overline{\tau}$-RSS if there exist symmetric positive definite matrices P, Q and $\overline{R}$ and a matrix Y such that (27) holds.*

Remark 5. When $\overline{R} \overset{>}{\to} 0$ and $Y \to 0$, the RSS criterion, equivalent to $A'P + PA + C'PC + Q + PBQ^{-1}B'P < 0$ by the Schur complement, is retrieved.

We give an alternative form of de-Souza's theorem, which sheds more light on the delay dependence, and provides a *necessary condition* for the inequality (27) itself. Define first for positive definite symmetric matrices P, Q and R the forms

$$L(P) \overset{\text{def}}{=} (A' + B')P + P(A + B) + C'PC \quad (28)$$

$$X(Q, R) \overset{\text{def}}{=} C'RC + (A' + B')(BQ^{-1}B' - R^{-1})^{-1}(A + B) \quad (29)$$

$$T(P, Q, R) \overset{\text{def}}{=} (PB - Q)R^{-1}(B'P - Q). \quad (30)$$

Note that the last two matrix forms are necessarily positive semidefinite.

Theorem 8. *If there exist symmetric positive definite matrices P, Q and $\overline{R}$ such that*

$$L(P) + X(Q, R) < 0$$

then there exists $\overline{\tau} > 0$ such that the stochastic delay system is $\overline{\tau}$-RSS.

Proof: Pre- and postmultiply the LMI of de Souza's theorem respectively with

$$\begin{bmatrix} I & \overline{\tau}Y\overline{R}^{-1} & I & 0 \\ & I & 0 & 0 \\ & & I & 0 \\ & & BQ^{-1} & R^{-1} \end{bmatrix}$$

and its transpose and, since there are no restrictions on Y, let $Y = PB - Q$, to get the LMI condition

$$\begin{bmatrix} \Omega(P, Q, \overline{R}, \overline{\tau}) & 0 & 0 & A' + B' \\ 0 & -\overline{R} & 0 & 0 \\ 0 & 0 & -Q & 0 \\ A + B & 0 & 0 & \overline{R}^{-1} - BQ^{-1}B' \end{bmatrix} < 0.$$

To simplify notation, we introduced: $\Omega(P,Q,\overline{R},\overline{\tau}) = (A'+B')P + P(A+B) + C'(P+\overline{R})C + \overline{\tau}^2(PB-Q)\overline{R}^{-1}(B'P-Q)$. This LMI is obviously satisfied if $\overline{R}^{-1} < BQ^{-1}B'$ and $L(P) + X(Q,\overline{R}) + \overline{\tau}^2 T(P,Q,\overline{R}) < 0$. Clearly, if $L(P) + X(Q,\overline{R}) < 0$ there exists a $\overline{\tau} > 0$ such that the above inequality holds for all $0 < \tau < \overline{\tau}$. In turn this implies the stability for all $0 \leq \tau < \overline{\tau}$. □

The *linear* form $(A'+B')P + P(A+B) + C'PC$ is instrumental in obtaining stability dependent on the delay. Indeed, the following is obtained.

Corollary 4. *If there exists a symmetric positive definite matrix P such that*

$$L(P) = (A'+B')P + P(A+B) + C'PC < 0$$

then there exist $\overline{\tau} > 0$ such that the stochastic delay system (22) ($N = k = 1$)is stochastically stable for all $0 \leq \tau < \overline{\tau}$.

Proof: Without loss of generality, we may assume (rescaling if necessary) that $L(P) < -I$, For arbitrary $S > 0$, the matrix $R = (S^{-1} + BQ^{-1}B')^{-1} > 0$ is well defined. Note that $R^{-1} > BQ^{-1}B'$, and by Woodbury's lemma

$$R = S - SB(Q + B'SB)^{-1}B'S.$$

Hence

$$X(Q,R) = \tilde{X}(Q,S) = C'SC - C'SB(Q+B'SB)^{-1}B'SC + (A'+B')S(A+B).$$

Choose $S = sI$ and $Q = qI$ to get

$$\tilde{X}(sI,qI) = [C'C + (A'+B')(A+B)]s - C'B(qI + B'Bs)^{-1}B'Cs^2 > 0.$$

If the ratio $q/s = \rho$ is kept fixed,

$$\tilde{X}(sI,\rho sI) = [C'C + (A'+B')(A+B)]s - C'B(\rho I + B'B)^{-1}B'Cs,$$

can be made arbitrarily small by making $s > 0$ sufficiently small. In particular, for any $\epsilon > 0$ we have $s_\epsilon > 0$ such that $\tilde{X}(s_\epsilon I, \rho s_\epsilon I) = s_\epsilon X_o(\rho) < \epsilon I$. It follows that $L(P) + X(Q,R) < -(1-\epsilon)I$. Now, consider

$$T(P,Q,R) = (PB - Q)R^{-1}(B'P - Q)$$

which, for the chosen matrices, evaluates to:

$$T_o(P,\rho,s) = (PB - \rho sI)(BB' + \rho I)(B'P - \rho sI)\frac{1}{s}$$

It follows that a bound $\overline{\tau}$ is given by letting $L(P) + X(Q,R) + \overline{\tau}^2 T(P,Q,R) = 0$, or,

$$\overline{\tau}(\rho,s) \geq \sqrt{\frac{(1-\epsilon)s}{\|PB - \rho sI)(BB' + \rho I)(B'P - \rho sI)\|}} > 0. \qquad \square$$

The condition $L(P) < 0$ implies also that the stochastic system

$$dx = (A + B)x\,dt + C dw(t),$$

i.e. the delay system with $\tau = 0$ is stochastically stable. Since it is also a necessary condition [5], we get:

Corollary 5 (Continuity in the delay). *If the stochastic system*

$$dx = (A + B)x\,dt + C\,dw(t)$$

is stochastically stable, then the stochastic delay system

$$dx(t) = [Ax(t) + Bx(t - \tau)]\,dt + Cx(t)\,dw(t)$$

remains stochastically stable for sufficiently small values of the delay.

Remark 6. A necessary condition for the LMI to hold is that $A + B$ is a Hurwitz matrix. However, this is not necessary for the stability (with probability one) of the null solution of the stochastic delay system itself, but it is necessary for the stability of the moments. Indeed sample equations may possess stable equilibrium solutions with probability one, even though all moments become unbounded [29].

Iff $A + B$ is a Hurwitz matrix, and $w(t)$ has variance parameter σ^2 instead of 1, then the linear form $L(P) = (A' + B')P + P(A + B) + \sigma^2 C'PC$ has a positive definite solution for sufficiently small σ. Hence for a (nonstandard) Wiener process with *sufficiently small* variance parameter, the stochastic delay free system remains stable. and by corollary 5 also the stochastic delay system for sufficiently small noise intensity and delay.

Remark 7. Now we show that the condition in the corollary 4 is not vacuus. Indeed, for

$$A = \begin{bmatrix} 1 & 1 \\ 0 & 1 \end{bmatrix}, B = \begin{bmatrix} -2 & -1 \\ 0 & -2 \end{bmatrix}, C = \gamma \begin{bmatrix} 1 & 0 \\ 0 & 1 \end{bmatrix},$$

it is obvious that for all positive definite P and Q the quadratic form $A'P + PA + Q + PBQ^{-1}B'P$ cannot be negative definite. Hence robust stochastic stability cannot be concluded. However, the choice

$$P = \begin{bmatrix} 1 & 0 \\ 0 & 1 \end{bmatrix}, R = \begin{bmatrix} 1 & 0 \\ 0 & 1 \end{bmatrix}, Q = \frac{1}{2}B'B = \begin{bmatrix} 2 & 1 \\ 1 & 2.5 \end{bmatrix},$$

leads to $L(P) = (\gamma^2 - 2)I$, $X(Q, R) = (\gamma^2 + 1)I$, and $T(P, Q, R) = (PB - Q)S^{-1}(B'P - Q) = \begin{bmatrix} 20 & 13 \\ 13 & 21.25 \end{bmatrix}$. The LMI matrix is negative definite for $\gamma < 1/\sqrt{2}$ and then a bound on the delay is given by

$$\overline{\tau}(\gamma) = \frac{1}{64}\sqrt{2(165 - \sqrt{10841})(1 - 2\gamma^2)} = .17241\,\sqrt{1 - 2\gamma^2}.$$

Note that in the scalar case with positive b, the existence of a bound $\overline{\tau}$ implies that $2(a + b) + c^2 < 0$. But this is equal to $2a + c^2 + \inf_{w>0}(w + b^2/w)$. Hence there exist $p = 1, w = b > 0$ and $r > 0$ such that $2ap + c^2p + w + p^2b^2/w + r = 0$. By theorem 6 the stochastic delay system (with $D = 0$) is then RSS. Thus if this scalar delay system is $\overline{\tau}$-RSS for some $0 < \overline{\tau} < \infty$, it is stochastically stable for *all* $\tau > 0$.

5 Invariant Distributions

Scheutzow [54] considered stochastic delay equations of the form:

$$\mathrm{d}x(t) = f(x_t)\,\mathrm{d}t + \mathrm{d}w(t)$$

For arbitrary initial data $\psi \in C$, he showed that all bounded Borel sets B in the state space C are revisited infinitely often either with probability one or with probability zero. In the first case call $(x_t)_{t\geq 0}$ recurrent, and in the second case transient. If recurrent, there exists an invariant measure (but not necessarily a probability measure). He derived sufficient criteria on f for the existence of an invariant probability measure. Some systems with non-unit diffusion may be transformed by change of variables to the above form. A noteworthy example is the generalized delayed logistic equation.

$$\mathrm{d}x(t) = [k_1 - k_2 x^{k_3}(t-1)]x(t)\,\mathrm{d}t + k_4 x(t)\,\mathrm{d}w(t).$$

It is shown that the conditions $k_1 > (k_4^2/2)$, and $k_2, k_3, k_4 > 0$ guarantee a unique invariant probability measure.

Mohammed proved that whenever the deterministic equation, $\dot{x}(t) = f(x_t)$, with $f : C(\mathcal{I}, \mathbf{R}^n) \to \mathbf{R}^n$ a continuous *linear* map, is globally asymptotically stable, then a unique invariant Gaussian distribution exists for $\mathrm{d}x(t) = f(x_t)\,\mathrm{d}t + G\,\mathrm{d}w(t)$, and is globally asymptotically stable [44, p.217]. The result is based on a stochastic version of the variation of parameters formula and a splitting of the state space $C = \mathcal{U} \oplus \mathcal{S}$ [16]. The (unstable) subspace, $\mathcal{U}$, is finite dimensional, and the stable one, $\mathcal{S}$, is closed. The splitting is invariant under the semigroup $(T_t)_{t\geq 0}$.

Da Prato and Zabczyk [8] studied the linear equation

$$\mathrm{d}x(t) = \left[Ax(t) + \sum_{i=1}^{p} B_i(t-\tau_i)\right]\mathrm{d}t + C\,\mathrm{d}w(t) \tag{31}$$

using the dissipativity method developed in their monograph. Let

$$\rho = \sup\left\{\mathrm{Re}\,\lambda \mid \det\left[\lambda I - \sum_{i=0}^{p} e^{-\lambda\tau_i} B_i\right] = 0\right\}. \tag{32}$$

If $\rho < 0$, then there exists a unique invariant measure for the system (31). Conversely, if an invariant measure exists for (31), and the matrix $\left[\lambda I - \sum_{i=0}^{p} e^{-\lambda\tau_i} B_i \mid C\right]$ has full rank for all $\lambda \in \mathbf{C}$, then $\rho < 0$. Note that this converse involves a reachability condition.

5.1 Existence for all delays

For simplicity, let us restrict the rest of the section to the additive noise, single ($p = 1$) delay model (31)(generalization is straightforward). An invariant distribution exists if there exists a positive definite $V(x)$ vanishing at 0, and such that $\mathcal{L}V < 0$ outside some ball centered at O.

Noting that: $\mathcal{L}V = x'[A'P + PA + Q]x - x_\tau' Q x_\tau + 2x'PBx_\tau + \operatorname{Tr} C'PC$, one obtains (see [67]),

Theorem 9. *The system (31) (p = 1) has an invariant distribution for all τ if there exists $P, Q, R > 0$ such that*

$$A'P + PA + Q + PBQ^{-1}B'P + R = 0.$$

The condition of the theorem implies also the robust stability of the theorem of the (deterministic) system $\dot{x}(t) = Ax(t) + Bx(t - \tau)$.

5.2 Correlation

Various properties can be derived for the stationary distribution with specific τ. For instance the *correlation matrix* is obtained as follows: Fix $\tau = 1$ (one may always rescale the time), and let $t > s$. Then

$$x(t) = e^{A(t-s)}x(s) + \int_s^t e^{A(t-\zeta)}[Bx(\zeta - 1) + Cu(\zeta)]\, d\zeta.$$

Postmultiply by $x'(s)$ and take expectations. If the initial condition satisfies $\mathbf{E}\,\phi(\theta) = 0$ for $\theta \in [-\tau, 0]$, then the process $x(t)$ has zero mean and $R(t,s) = \mathbf{E}\,x(t)x'(s)$ satisfies the integral equation

$$R(t,s) = e^{A(t-s)}R(s,s) + \int_s^t e^{A(t-\zeta)}BR(\zeta - 1, s)\, d\zeta.$$

Set $s = t - \theta$, and take the limit for $t \to \infty$, letting $\overline{R}(\theta) \stackrel{\text{def}}{=} \lim_{t\to\infty}(t, t-\theta)$,

$$\overline{R}(\theta) = e^{A\theta)}\overline{R}(0) + \int_0^\theta e^{A(\theta-\zeta)}B\overline{R}(\zeta - 1)\, d\zeta$$

Differentiate

$$\frac{d\overline{R}(\theta)}{d\theta} = A\overline{R}(\theta) + B\overline{R}(\theta - 1) \quad , \quad \forall\theta > 0$$

Also, symmetry gives:

$$\overline{R}(\theta) = \overline{R}(-\theta)'$$

The case $\theta = 0$ is special, and can be obtained by "squaring up" $x(t)$ and taking expectations, followed by differentiation. The results are summarized in the following

Theorem 10. *If the deterministic system*

$$\dot{x}(t) = Ax(t) + Bx(t - 1)$$

is asymptotically stable, then the stochastic additive noise system

$$\mathrm{d}x = [Ax(t) + Bx(t-1)]\,\mathrm{d}t + C\,\mathrm{d}w$$

has an invariant distribution with mean 0 and covariance $\overline{R}(0)$. The sample paths exhibit temporal correlation $\overline{R}(\theta)$ satisfying

$$\begin{aligned} A\overline{R}(0) + \overline{R}(0)A' + \overline{R}(1)B' + B\overline{R}(1) + CC' &= 0 \\ \tfrac{d\overline{R}(\theta)}{d\theta} = A\overline{R}(\theta) + B\overline{R}(\theta - 1) \quad &, \quad \forall\theta > 0 \\ \overline{R}(-\theta) = \overline{R}(\theta)' \quad &, \quad \forall\theta > 0 \end{aligned}$$

5.3 Fokker-Planck approximation

For finite dimensional systems, the stationary density can be obtained as the equilibrium solution to the Fokker-Planck (or forward Kolmogorov) equation. This is a scalar PDE governing the evolution of the probability density of the (finite dimensional) state. Let us fix the ideas again on the simple scalar toy model, thus avoiding only notational complexity. Given the stochastic delay system $\mathrm{d}x = [Ax(t)+Bx(t-\tau)]\,\mathrm{d}t + b\,\mathrm{d}w(t)$, consider the $(N+1)$-dimensional approximation by simple discretization, where $\delta = \tau/N$.

$$\mathrm{d}\chi(t) = \frac{1}{\delta}\begin{bmatrix} A\delta & 0 & \cdots & 0 & B\delta \\ 1 & -1 & & & \\ & \ddots & \ddots & & \\ & & \ddots & \ddots & \\ & & & 1 & -1 \end{bmatrix} \chi(t)\,\mathrm{d}t + \begin{bmatrix} b \\ 0 \\ \vdots \\ \vdots \\ 0 \end{bmatrix} \mathrm{d}w(t) \tag{33}$$

or

$$\mathrm{d}\chi(t) = \frac{1}{\delta}[\mathcal{A} - I]\chi(t)\,\mathrm{d}t + \mathcal{B}\,\mathrm{d}w(t) \tag{34}$$

with

$$\mathcal{A} = \begin{bmatrix} 1 + A\delta & 0 & \cdots & 0 & B\delta \\ 1 & 0 & & & \\ & \ddots & \ddots & & \\ & & \ddots & \ddots & \\ & & & 1 & 0 \end{bmatrix},$$

a matrix in the companion form. Note that $x(t) = e_1'\chi(t) = \chi_1(t)$, the first component of $\chi(t)$. In this model, one has also

$$\dot{x}(t - (k-1)\delta) \approx \frac{1}{\delta}[x(t - k\delta) - x(t - (k-1)\delta)]$$

from which the approximation of the N remaining components of $\chi(t)$ follow: The $(k+1)$st component of the state $\chi(t)$ is an approximation for $x(t - k\delta)$. The deterministic part of this approximation is asymptotically stable if $\mathcal{A}$ has its spectrum to the left of $1 + \delta$. The eigenvalues of $\mathcal{A}$ are the roots of the characteristic polynomial

$$a_N(z) = z^{N+1} - (1 + A\delta)z^N - B\delta.$$

For fixed N, there is a root of multiplicity N at 0, and a single root at -1. Temporarily neglecting the relation between N and δ, the influence of δ on the roots of this polynomial may be studied with the classical root locus method. When generalizing to vector delay systems, the multivariable extension of the root locus, involving branch points and Riemann surfaces, needs to be used [49]. Thus,

$$\frac{1}{\delta} = \frac{Az^N + B}{z^N(z-1)}$$

The (0 degree) root locus branches leave the poles (N at the origin, one at 1), and move towards the zeros, which are located in a Butterworth pattern. For small δ, the departure from the single pole at 1 is critical for the stability. Obviously, $|B/A| < 1$ is a sufficient condition for stability for this model, regardless of N. Hence in the limit, the stability of the delay system is guaranteed for all values of τ.

The Fokker-Planck equation associated with this approximate (N+1)-dimensional state model is

$$\frac{\partial}{\partial t}\rho^{(N)}(\chi, t) = -\nabla\rho^{(N)}(\chi, t)\, \mathcal{A}\chi + \frac{b^2}{2}\frac{\partial^2}{\partial\chi_1^2}\rho^{(N)}(\chi, t). \tag{35}$$

This expands to

$$\begin{aligned}\frac{\partial}{\partial t}\rho^{(N)}(\chi, t) = &-\frac{\partial}{\partial\chi_1}\rho^{(N)}(\chi, t)\,(A\chi_1 + B\chi_{N+1}) - \frac{\partial}{\partial\chi_2}\rho^{(N)}(\chi, t)\,\frac{\chi_1 - \chi_2}{\delta} - \\ &\cdots - \frac{\partial}{\partial x_N}\rho^{(N)}(\chi, t)\,\frac{\chi_N - \chi_{N+1}}{\delta} + \frac{b^2}{2}\frac{\partial^2}{\partial\chi_1^2}\rho^{(N)}(\chi, t).\end{aligned}$$

If the deterministic part of the system is asymptotically stable, then the solution, $\rho^{(N)}(\chi, t)$ converges for $t \to \infty$ to a steady state solution, $\overline{\rho}^{(N)}(\chi)$. As we are here dealing with a linear system in gaussian noise, the limiting distribution must be gaussian. Hence it suffices to compute the mean and the covariance matrix.

Introducing for the original delay system in the steady state,

$$\begin{aligned}&\overline{\rho}_{(N)}(\chi_1, \ldots, \chi_{N+1})\, d\chi_1 \cdots d\chi_{N+1} \\ &\quad \stackrel{\text{def}}{=} \Pr\{x(t) \in (\chi_1, \chi_1 + d\chi_1), \cdots, x(t - N\delta) \in (\chi_{N+1}, \chi_{N+1} + d\chi_{N+1}),\end{aligned}$$

it follows that the covariance matrix must have a symmetric Toeplitz structure. This is not consistent with the above approximation. However, remark that this can be remedied by adding noise terms in the last N rows of the approximating dynamics. This is heuristically justifiable by noting that the mean value theorem implies the existence of $\xi_k \in [0, \delta]$ such that

$$\dot{x}(t - k\delta + \xi_k) = \frac{1}{\delta}[x(t - (k-1)\delta) - x(t - k\delta)].$$

The difference $\dot{x}(t-k\delta+\xi_k)-\dot{x}(t-k\delta)$ is the requisite perturbation for the above approximating model. Once the requisite perturbation terms are in place, we postulate their conservation in the model to compute the transient solution via the finite dimensional approximation.

While these ideas may be extended to the nonlinear case, the Fokker-Planck equation itself remains a linear PDE. These approximations may be valuable in the applications to neural networks with delay, applications in epidemiology, applications to synchronization and applications in traffic models.

6 Stochastic Stability of Neutral Systems

The analysis of stochastic stability for neutral systems and conditions for existence of invariant distributions also received considerable attention recently. Existence and uniqueness of solutions to general stochastic neutral equations are found in Kolmanovskii and Nosov [23], where also stability and asymptotic stability are discussed. The problem is further discussed in Kolmanovskii and Myshkis [22]. Mao obtained general criteria for exponential stability using various techniques in a series of papers [36,39,41].

In this section we focus again on the linear stochastic neutral systems

$$\mathrm{d}x(t)=[Ax(t)+Bx(t-\tau)]\,\mathrm{d}t+C\,\mathrm{d}x(t-\tau)+\Gamma x\,\mathrm{d}w(t) \tag{36}$$

Guided by the Lyapunov-Krasovskii methods for the deterministic case [62,69–71] we construct new stochastic versions, based on different choices of Lyapunov functionals.

6.1 First criterion

Consider first the Lyapunov-Krasovskii functional, $V: C(\mathcal{I},\mathbf{R}^n)\to\mathbf{R}$, defined by

$$V(\psi)=\psi'(0)P\psi(0)+\int_{-\tau}^{0}\psi(t)'Q\psi(t)\,\mathrm{d}t+\int_{-\tau}^{0}\mathrm{d}\psi'(t)\,R\,\mathrm{d}\psi(t), \tag{37}$$

recalling that $\mathbf{E}\,\mathrm{d}x'R\mathrm{d}x$ is of order $\mathrm{d}t$, the second integral contributes really a non-zero term. Applying the Itô-differentiation rule: (here, and in what follows, x_τ denotes the value of x at the delayed time argument)

$$\begin{aligned}\mathrm{d}V=&\;2[Ax+Bx_\tau]'Px\,\mathrm{d}t+2x'PC\,\mathrm{d}x_\tau+\mathrm{d}x_\tau'\,C'PC\,\mathrm{d}x_\tau+x'\Gamma'P\Gamma x\,\mathrm{d}t+\\&+[x'Qx-x_\tau'Qx_\tau]\,\mathrm{d}t+\mathrm{d}x'\,R\,\mathrm{d}x-\mathrm{d}x_\tau'\,R\,\mathrm{d}x_\tau\end{aligned}$$

Substituting $\mathrm{d}x'\,R\,\mathrm{d}x$ by $\mathrm{d}x_\tau'\,C'RC\,\mathrm{d}x_\tau+x'\Gamma'R\Gamma x\,\mathrm{d}t$ gives for $\mathbf{E}\,\mathrm{d}V$ the quadratic form,

$$\mathbf{E}\,\mathrm{d}V = [x'\sqrt{\mathrm{d}t}\;,\; x'_\tau\sqrt{\mathrm{d}t}\;,\; \mathrm{d}x'_\tau] \cdot \begin{bmatrix} A'P+PA+Q+\Gamma'(P+R)\Gamma & PB & PC \\ B'P & -Q & 0 \\ C'P & 0 & C'(P+R)C-R \end{bmatrix} \begin{bmatrix} x\sqrt{\mathrm{d}t} \\ x_\tau\sqrt{\mathrm{d}t} \\ \mathrm{d}x_\tau \end{bmatrix}.$$

Note that Schur-Cohn stability of C is a necessary condition for negative definiteness of this quadratic form. If $\Gamma = 0$, we obtain the deterministic criterion

$$\begin{bmatrix} A'P+PA+Q & PB \\ B'P & -Q \end{bmatrix} < 0,$$

since the term $dx'R\,dx$ is now of second order in dt and is set to zero. This differs from the deterministic Lyapunov-Krasovskii functional in [62] where we defined $V_{\mathrm{det}}(\psi) = \psi'(0)P\psi(0) + \int_{-\tau}^{0} \psi'(t)Q\psi(t)\,\mathrm{d}t + \int_{-\tau}^{0} \dot{\psi}'(t)\,N\,\dot{\psi}(t)\,\mathrm{d}t$.

Theorem 11. *The linear stochastic neutral equation (36) is RASS if there exists symmetric positive definite matrices P, Q and R such that either of the following are negative definite:*

$$i)\; \begin{cases} C'(P+R)C-R \quad \text{and} \\ \begin{bmatrix} A'P+PA+Q+\Gamma'(P+R)\Gamma+PC(R-C'(P+R)C)^{-1}C'P & PB \\ B'P & -Q \end{bmatrix} \end{cases} \tag{38}$$

$$ii)\; \begin{bmatrix} A'P+PA+Q+PBQ^{-1}B'P+\Gamma'(P+R)\Gamma & PC \\ C'P & C'(P+R)C-R \end{bmatrix} \tag{39}$$

Proof: The LMI's are the Schur complements of respectively the (3,3) and the (2,2) blocks in the above 3×3 blockmatrix in the expression of $\mathcal{L}V$. □

Corollary 6. *The linear stochastic neutral equation is stochastically stable if there exists symmetric positive definite matrices P, Q and R such that $R - C'(P+R)C > 0$ and*

$$A'P+PA+Q+PBQ^{-1}B'P+\Gamma'(P+R)\Gamma+PC(R-C'(P+R)C)^{-1}C'P < 0. \tag{40}$$

Proof: By taking Schur complements in either of the LMI's in Theorem 11. □

6.2 Second stability criterion

Here we start from a stochastic varaiant of the Lyapunov-Krasovskii functional in the form presented in [70, 71]. Let

$$\overline{V}(\psi) = [\psi(0) - C\psi(-\tau)]'P[\psi(0) - C\psi(-\tau)] + \\ + \int_{-\tau}^{0} \psi'(t)Q\psi(t)\,\mathrm{d}t + \int_{-\tau}^{0} \mathrm{d}\psi'(t)\,R\,\mathrm{d}\psi(t), \tag{41}$$

The Itô-differentiation rule leads now to:

$$\begin{aligned}
\mathrm{d}\overline{V} &= 2(\mathrm{d}x - C\,\mathrm{d}x_\tau)'P(x - Cx_\tau) + (\mathrm{d}x - C\,\mathrm{d}x_\tau)'P(\mathrm{d}x - C\,\mathrm{d}x_\tau) + \\
&\quad +(x'Qx - x_\tau' Q x_\tau)\,\mathrm{d}t + \mathrm{d}x' R\,\mathrm{d}x - \mathrm{d}x_\tau' R\,\mathrm{d}x_\tau \\
\mathcal{L}\,\mathrm{d}\overline{V} &= 2[Ax + Bx_\tau]'P(x - Cx_\tau)\,\mathrm{d}t + x'\Gamma' P \Gamma x\,\mathrm{d}t + \\
&\quad +[x'Qx - x_\tau' Q x_\tau]\,\mathrm{d}t + \mathrm{d}x'\,R\,\mathrm{d}x - \mathrm{d}x_\tau'\,R\,\mathrm{d}x_\tau \\
&= 2[Ax + Bx_\tau]'P(x - Cx_\tau)\,\mathrm{d}t + x'\Gamma' P \Gamma x\,\mathrm{d}t + \\
&\quad +[x'Qx - x_\tau' Q x_\tau]\,\mathrm{d}t + \mathrm{d}x_\tau' C'\,RC\mathrm{d}x_\tau + x'\Gamma' R \Gamma x\,\mathrm{d}t - \mathrm{d}x_\tau'\,R\,\mathrm{d}x_\tau \\
&= 2[Ax + Bx_\tau]'P(x - Cx_\tau)\,\mathrm{d}t + x'\Gamma'(P+R)\Gamma x\,\mathrm{d}t + \\
&\quad +[x'Qx - x_\tau' Q x_\tau]\,\mathrm{d}t + \mathrm{d}x_\tau'[C'\,RC - R]\mathrm{d}x_\tau
\end{aligned}$$

This yields for $\mathcal{L}\,\mathrm{d}\overline{V}$ the sum of the quadratic forms $\mathrm{d}x_\tau'[C'RC - R]\,\mathrm{d}x_\tau$, and

$$[\,x'\,,\,x_\tau'\,]\begin{bmatrix} A'P + PA + Q + \Gamma'(P+R)\Gamma & PB - A'PC \\ B'P - C'PA & -B'PC - C'PB - Q \end{bmatrix}\begin{bmatrix} x \\ x_\tau \end{bmatrix}\mathrm{d}t.$$

This proves

Theorem 12. *The stochastic system (36) is RASS if C is Schur Cohn stable and there exist positive definite matrices P, Q, R such that the following holds:*

$$M = \begin{bmatrix} A'P + PA + Q + \Gamma'(P+R)\Gamma & PB - A'PC \\ B'P - C'PA & -B'PC - C'PB - Q \end{bmatrix} < 0$$

Note that if we let $\Gamma = 0$, the LMI

$$\begin{bmatrix} A'P + PA + Q & PB - A'PC \\ B'P - C'PA & -B'PC - C'PB - Q \end{bmatrix} < 0, \tag{42}$$

results. But $\Gamma = 0$ means that the system is deterministic, and thus the above is a new criterion for asymptotic stability of a deterministic linear neutral system. Mao obtained in [39] the LMI (42) for $P = I$ as one condition for almost sure exponential stability for the neutral system (36), the other conditions are $||C|| < 1$, and a nonlinear and time varying stochastic perturbation term, satisfying a Lipschitz and linear growth condition.

Remark 8. Note that the LMI given in [70] for the deterministic case is

$$\begin{bmatrix} A'P + PA + S & P(AC + B) + SC \\ (B' + C'A')P + C'S & C'SC - S \end{bmatrix} < 0$$

6.3 Application: multiple commensurate delays

The neutral stability criterion may aid in obtaining less conservative bounds for stochastic stability of delay equations with commensurate delays. Indeed, when the sufficient condition for RSS is used for the multi-delay system, (the result given in section 4 is readily generalized) the information on the precise relation between the two delays is lost. However, treating the system as a single delay neutral system, and using the neutral RASS criterion, a less conservative result may be expected. The deterministic case was explored in [65]. For instance, the commensurate delay equation

$$\mathrm{d}x = (Ax + Bx_\tau + Cx_{2\tau})\,\mathrm{d}t + \Gamma x\,\mathrm{d}w$$

can be augmented to a neutral equation

$$\begin{bmatrix} \mathrm{d}x \\ \mathrm{d}x_\tau \end{bmatrix} = \left(\begin{bmatrix} A & M_1 \\ 0 & M_2 \end{bmatrix} \begin{bmatrix} x \\ x_\tau \end{bmatrix} + \begin{bmatrix} B - M_1 & C \\ -M_2 & 0 \end{bmatrix} \begin{bmatrix} x_\tau \\ x_{2\tau} \end{bmatrix} \right) \mathrm{d}t + \\ + \begin{bmatrix} 0 & 0 \\ I & 0 \end{bmatrix} \begin{bmatrix} \mathrm{d}x_\tau \\ \mathrm{d}x_{2\tau} \end{bmatrix} + \begin{bmatrix} \Gamma & 0 \\ 0 & 0 \end{bmatrix} x\,\mathrm{d}w$$

This is of the single-delay form

$$\mathrm{d}\chi = [\mathsf{A}\chi + \mathsf{B}\chi_\tau]\,\mathrm{d}t + \mathsf{C}d\chi_\tau + \mathsf{G}\chi\,\mathrm{d}w,$$

where two free parameter matrices M_1 and M_2 are introduced. This additional freedom may aid in the search for a feasible LMI.

7 Conclusions

We explored some aspects of the stochastic delay differential and delay difference equations. We briefly looked at some of the intricate details a deeper study of the subject is involved with. We surveyed some of the results on stability of delay systems. Some new material was included in this chapter, relating to the stability of discrete and continuous linear stochastic systems. This was given in terms of the existence of solutions to certain Riccati-type equations (or LMI's). A sufficient condition for the existence of a unique invariant distribution is related to the same Riccati equation. We also obtained new results for the stability of neutral systems and discussed briefly the Fokker-Planck equation.

8 Bibliographical Notes

We have left the general discussion of filtering and control of stochastic time delay systems and some aspects of their dynamical properties largely untouched. Noteworthy are the extensions to La Salle's theorem by Mao [40, 42], the computation of Lyapunov exponents, the deeper theory of invariant densities [2, 11], the approach by dissipation methods [8], and the numerical approximation and simulation of stochastic delay systems [7,20,73]. A very important class of systems are the ones with random delay and Markovian jump parameters. We refer to the recent monograph [4] and recent work of Mao and collaborators [43, 74], and Mahmoud and Shi [32, 33]. Page restrictions also prohibited us from presenting applications of stochastic delay systems in traffic control and the biosciences. (e.g., prey-predator models, epidemiology [3] and dynamics of cancer). We hope further research in these areas will be stimulated.

Acknowledgement: The author is indebted to Professor S.-E.A. Mohammed for useful comments, and to Professor A. Tannenbaum for 'freeing up some time'.
Support of the collaborative NSF-CNRS grants INT-9818312 and INT-0129062, is hereby also gratefully acknowledged.

References

1. D. Bell and S.-E. A. Mohammed, "The Malliavin Calculus and Stochastic Delay Equations," *J. Functional Analysis*, Vol. 99, 1991, pp. 75-99.
2. D. Bell and S.-E. A. Mohammed, "Smooth Densities for Degenerate Stochastic Delay Equations with Hereditary Drift," *Annals of Probability*, Vol. 23, No. 4, 1995, pp. 1875-1894.
3. E. Beretta, V.B. Kolmanovskii and L.E. Shaikhet, "Stability of Epidemic Model with Time Delays Influenced by Stochastic Perturbation," *Mathematics and Computers in Simulation*, Vol. 45, 1998, pp. 269-277.
4. E.-K. Boukas and Z.-K. Liu, *Deterministic and Stochastic Time Delay Systems*, Birkhäuser, 2002.
5. S. Boyd, L. El Ghaoui, E. Feron and V. Balakrishnan, *Linear Matrix Inequalities in System and Control Theory* SIAM Studies in Applied Mathematics, 1994.
6. H.J. Bucy, "Stability and Positive Supermartingales," *J. Differential Equations*, Vol. 1, 1965, pp. 151-155.
7. M.-H. Chang, "Discrete Approximation of Nonlinear Filtering for Stochastic Delay Equations," *Stochastic Analysis and Appl.*, Vol. 5, No. 3, 1987, pp. 267-298.
8. G. Da Prato and J. Zabczyk, *Ergodicity for Infinte Dimensional Systems*, Cambridge University Press, 1996.
9. C. de Souza, "Stability and Stabilization of Linear State-Delayed Systems with Multiplicative Noise," *Proceedings 2nd IFAC Workshop on Linear Time Delay Systems*, Ancona, Italy, pp. 21-26, September 2000.
10. L. Dugard and E.I. Verriest (Eds.), *Stability and Control of Time Delay Systems*, Springer-Verlag Lecture Notes in Control and Information Sciences, Vol. 228, 1998.
11. M. Ferrante, C. Rovira and M. Sanz-Solé, "Stochastic Delay Equations with Hereditary Drift: Estimates of Density,", *J. Functional Analysis*, 177, pp. 138-177, 2000.
12. P. Florchinger, "Stability of some linear stochastic systems with delays, *Proc. 40th IEEE Conf. Dec. & Control*, Orlando, FL, Dec. 2001, pp. 4744-4745.
13. P. Florchinger and E.I. Verriest , Stabilization of nonlinear stochastic systems with delay feedback, *Proc. 32nd IEEE Conf. Dec. & Control*, San Antonio, TX, Dec. 1993, pp. 859-860.
14. T.D. Frank, "Multivariate Markov Processes for Stochastic Systems with Delays: Applications to the Stochastic Gompertz Model with Delay," *Phys. Rev. E*, Vol. 66, Aug. 2002.
15. T.D. Frank and P.J. Beek, "Stationary solutions of linear stochastic delay differential equations: Applications to biological systems," *Phys. Rev. E*, Vol. 64, 021917-1-12, 2001.
16. J. Hale and S.M. Verduyn-Lunel, *Introduction to Functional Differential Equations*, Springer-Verlag, 1993.

17. A. F. Ivanov and E. I. Verriest, "Robust Stability of Delay-Difference Equations," in *Systems and Networks: Mathematical Theory and Applications*, U. Helmke, R. Mennicken, and J. Saurer, (eds.), University of Regensburg, pp. 725-726, 1994.
18. A. F. Ivanov and E. I. Verriest, "Stability and Existence of Periodic Solutions in Discrete Delay Systems," Research Report 32/98 School of Information Technology and Mathematical Sciences, University of Ballarat, Victoria, Australia, December 1998.
19. R.Z. Khasminskii, *Stochastic Stability of Differential Equations*, Sijthoff and Noordhoff, Alphen aan den Rijn, 1980.
20. J.R. Klauder and W.P. Petersen, "Numerical Integration of Multiplicative-Noise Stochastic Differential Equations," *SIAM J. Numer. Anal.*, Vol. 22, No. 6, 1985, pp. 1153-1166.
21. V.B. Kolmanovskii, T.L. Maizenberg and J.-P. Richard, "Mean Square Stability of Difference Equations with a Stochastic Delay," *Nonlinear Analysis,* 52 (2003) pp. 795-804.
22. V.B. Kolmanovskii and A. Myshkis, *Applied Theory of Functional Differential Equations*, Kluwer Academic Publishers, Dordrecht, 1992.
23. V.B. Kolmanovskii and V.R. Nosov, *Stability of Functional Differential Equations*, Academic Press, 1986.
24. V.B. Kolmanovskii and L.E. Shaikhet, *Control of Systems with Aftereffect,* AMS 1991.
25. V.B. Kolmanovskii and L.E. Shaikhet, "General Methods of Lyapunov Functionals Construction for Stability Investigations of Stochastic Difference Equations," in *Dynamical Systems and Applications*, Vol. 4, Agarwal (Edt.), pp. 397-439, World Scientific, 1995.
26. V.B. Kolmanovskii and L.E. Shaikhet, "Matrix Riccati Equations and Stability of Stochastic Linear Systems with Nonincreasing Delays," *Functional Differential Equations*, Vol. 4, No. 3-4, 1997, pp. 279-293.
27. V.B. Kolmanovskii and L.E. Shaikhet, "Construction of Lyapunov Functionals for Stochastic Herditary Systems: A Survey of Some Recent Results," *Mathematical and Computer Modelling*, 36 (2002), pp. 691-716.
28. D.G. Korenevskii, "Stability with Probability 1 of Solution of Systems of Linear Itô Stochastic Differential-Difference Equations," Ukrainian mathematical Journal, Vol. 39, No, 1, pp. 26-30.
29. F. Kozin, "A survey of stability odf stochastic systems," *Automatica*, Vol. 5, 1969, pp. 95-112.
30. H.J. Kushner, "On the Stability of Processes Defined by Stochastic Difference-Differential Equations," *J. Differential Equations*, Vol. 4, No. 3, pp. 424-443, 1968.
31. F. Ma, "Stability Theory of Stochastic Difference Systems,"in *Probabilistic Analysis and Related Topics*, Vol. 3, A.T. Bharucha-Reid (Edt.), Academic Press, 1983, pp. 127-160.
32. M.S. Mahmoud and P. Shi, "Robust Kalman filtering for continuous time-lag systems with Markovian jump parameters," *IEEE Trans. Auto. Control,* Vol. 50, No. 1, (2003) pp. 98-105.
33. M.S. Mahmoud and P. Shi, "Output feedback stabilization and disturbance attenuation of time-delay jumping systems," *IMA Journal of Mathematical Control and Information*, Vol. 20, (2003) pp. 179-199.
34. X. Mao, (1991) *Stability of stochastic differential equations with respect to semimartingales*, Pitman Research Notes in Mathematics Series, Vol. 251, Longman, Essex.
35. X. Mao, (1994) *Exponential Stability of Stochastic Differential Equations*. M. Dekker, New York.
36. X. Mao, "Exponential Stability in Mean Square of Neutral Stochastic Differential Functional Equations," *Systems & Control Letters*, 26, 1995, pp. 245-251.
37. Mao, X. (1996). "Robustness of exponential stability of stochastic delay equations," *IEEE Trans. Automatic Control*, **41**, No. 3, 442-447.

38. Mao, X. (1997). "Almost Sure Exponential Stability of Neutral Differential Difference Equations with Damped Stochastic Perturbations," *SIAM J. Mathematical Analysis*, Vol. 28, No. 2, pp. 389-401.
39. Mao, X. (1997). "Almost Sure Exponential Stability of Neutral Differential Difference Equations with Damped Stochastic Perturbations," *J. Math. Anal. Appl.* Vol. 212, 1997, pp. 554-570.
40. X. Mao, "LaSalle-Type Theorems for Stochastic Differential Delay Equations,", J. Math. Anal. Appl., Vol. 236, pp. 350-369, 1999.
41. X. Mao, "Asymptotic Properties of Neutral Stochastic Differential Delay Equations," *Stochastics and Stochastic Reports*, Vol. 68, No. 3-4, pp. 273-295, 2000.
42. X. Mao, "A Note on the LaSalle-Type Theorems for Stochastic Differential Delay Equations,", J. Math. Anal. Appl., Vol. 268, pp. 125-142, 2002.
43. X. Mao, "Exponential stability of stochastic delay interval systems with markovian systems," *IEEE Transactions on Automatic Control*, Vol. 47, No. 10, 2002, pp. 1604-1612.
44. Mohammed, S.-E. A. (1984). *Stochastic functional differential equations*, Pitman.
45. S.-E.A. Mohammed, "Stability of Linear Delay Equations under Small Noise,", *Proc. Edinburgh Math. Soc.*, Vol. 29, pp. 233-254, 1986.
46. S.-E. A. Mohammed, "Stochastic Differential Systems with Memory: Theory, Examples and Applications," in *Stochastic Analysis and Related Topics VI*, Birkhäuser, 1998, pp. 1-77.
47. S.-I Niculescu, *Delay Effects on Stability* Springer-Verlag Lecture Notes in Control and Information Sciences, Vol. 269, 2002.
48. B. Paternoster and L. Shaikhet, "About Stability of Nonlinear Stohastic Difference Equations," *Appl. Math. Lett.*, Vol. 13, pp. 27-32, 2000.
49. I. Postlethwaite and A.G.J. MacFarlane, *A Complex Variable Approach to the Analysis of Linear Multivariable Feedback Systems,* Springer-Verlag, Lecture Notes in Control and Information Sciences, No. 12, 1979.
50. M. Reiss, `http://www.mathematik.hu-berlin.de/ reiss`, *Stochastic Delay Differential Systems*, February 2003.
51. L.E. Shaikhet, "Necessary and Sufficient Conditions of Asymptotic Mean Square Stability for Stochastic Linear Difference Equations," *Appl. Math. Lett.*, Vol. 10, No. 3, pp. 111-115, 1997.
52. L.E. Shaikhet, "Stability in Probability of Nonlinear Stochastic Systems with Delay," *Mathematical Notes,* Vol. 57, Nos. 1-2, 1995, pp. 103-106.
53. L.E. Shaikhet and M.L. Shafir, "Linear Filtering of Solutions of Stochastic Integral Equations in Non-Gaussian Case," *Problems of Control and Information Theory,* Vol. 18, No. 6, 1989, pp. 421-434.
54. M. Scheutzow, "Qualitative Behaviour of Stochastic Delay Equations with a Bounded Memory," *Stochastics*, Vol. 12, pp. 41-80, 1984.
55. H. Tong, *Non-linear Time Series* Oxford University Press, 1990.
56. C. Tudor and M. Tudor, "On Approximation of Solutions for Stochastic Delay Equations," *Stud. Cerc. Mat.* Vol. 39, No. 3, pp. 265-274.
57. E.I. Verriest,"Stabilization of Deterministic and Stochastic Systems with Uncertain Time Delays," *Proceedings of the 33d IEEE Conference on Decision and Control*, Orlando, FL, Dec. 1994, pp. 3829-3834.
58. E.I. Verriest, "Robust Stability of Time-Varying Systems with Unknown Bounded Delays," *Proceedings of the 33rd IEEE Conference on Decision and Control*, Orlando, FL, pp. 417-422, December 1994.

59. E.I. Verriest, "Stability and Stabilization of Stochastic Systems with Distributed Delays," *Proceedings of the 34th IEEE Conference on Decision and Control*, New Orleans, LA, pp. 2205-2210, December 1995.
60. E.I. Verriest, "Robust Stability and Stabilization of Deterministic and Stochastic Time-Delay Systems *European Journal of Automation*, Vol. 31, No. 6, pp. 1015-1024, October 1997.
61. E.I. Verriest, "Stochastic Stability of Feedback Neural Networks," *Proceedings of the IFAC Conference on Large Scale Systems (LSS'98)*, pp. 444-457, Patras, Greece, July 1998.
62. E.I. Verriest, "Robust Stability of Differential-Delay Systems," *Zeitschrift für Angewandte Mathematik und Mechanik*, pp. S1107-S1108, 1998.
63. E.I. Verriest, "Linear Systems with Rational Distributed Delay: Reduction and Stability," *Proceedings of the 1999 European Control Conference*, DA-12, Karlsruhe, Germany, September 1999.
64. E.I. Verriest, "Robust Stability and Stabilization: from Linear to Nonlinear," *Proceedings 2nd IFAC Workshop on Linear Time Delay Systems*, Ancona, Italy, pp. 184-195, September 2000.
65. E.I. Verriest, "Perturbation and Interpolation Approach to Stability of Functional Differential Systems," *Proceedings of the 36th annual Conference on Information Systems and Sciences* (CISS), Princeton, NJ, March 20-22, 2002.
66. E.I. Verriest, "Stability of Systems with State-Dependent and Random Delays," *IMA Journal of Mathematical Control and Information*, Vol. 19, pp. 103-114, 2002.
67. E.I. Verriest and P. Florchinger, "Stability of stochastic systems with uncertain time delays," *Systems & Control Letters* Vol. 24, No. 1, 1995, pp. 41-47.
68. E.I. Verriest and A.F. Ivanov, "Robust Stability of Delay-Difference Equations," *Proceedings of the 34th IEEE Conference on Decision and Control*, New Orleans, LA, pp. 386-391, December 1995.
69. E.I. Verriest and S.-I. Niculescu, "Delay-Independent Stability of Linear Neutral Systems: A Riccati Equation Approach," *Proceedings of the European Control Conference*, TU-A D4 1-5, Brussels, Belgium, July 1-4, 1997.
70. E.I. Verriest and S.-I. Niculescu, "Delay-Independent Stability of Linear Neutral Systems: A Riccati Equation Approach," in *Stability and Control of Time-Delay Systems* L. Dugard and E.I. Verriest (Eds.), Springer-Verlag, Lecture Notes on Control and Information Sciences, Vol. 228, 1998, pp. 92-100.
71. E. I. Verriest and S.-I. Niculescu, "On Stability Properties of Some Class of Linear Neutral Systems," *MTNS 1998*, Padua, Italy, in *Mathematical Theory of Networks and Systems*, Beghi, Finesso and Picci, (eds.) Il Poligrafo, Padova, pp. 559-562, 1998.
72. E.I. Verriest, O. Sename and P. Pepe, "Robust Observer-Controller for Delay-Differential Systems," accepted *41th IEEE Conference on Decision and Control*, Las Vegas, NV, December, 2002, TuP07-3.
73. H. Yoo, "Semi-Discretization of Stochastic Partial Differential Equations on R^1 by a Finite-Difference Method," *Mathematics of Computation*, 1999, pp. 1-14.
74. C. Yuan and X. Mao, "Asymptotic stability in distribution of stochastic differential equations with Markovian switching," *Stochastic Processes and their Applications*, Vol. 103, (2003), pp. 277-291.
75. A.L. Zelentsovskii, "Stability with Probability 1 of Solutions of Systems of Linear Stochastic Differential-Difference Ito Equations," *Ukrainian Math. J.* 43, 123-126, 1991.

Stability and Dissipativity Theory for Nonnegative and Compartmental Dynamical Systems with Time Delay

Wassim M. Haddad[1] and VijaySekhar Chellaboina[2]

[1] School of Aerospace Engineering, Georgia Institute of Technology, Atlanta, GA 30332-0150 wm.haddad@aerospace.gatech.edu
[2] Mechanical and Aerospace Engineering, University of Missouri, Columbia, MO 65211 ChellaboinaV@missouri.edu

Summary. Nonnegative and compartmental dynamical system models are derived from mass and energy balance considerations that involve dynamic states whose values are nonnegative. These models are widespread in engineering and life sciences and typically involve the exchange of nonnegative quantities between subsystems or compartments wherein each compartment is assumed to be kinetically homogeneous. However, in many engineering and life science systems, transfers between compartments are not instantaneous and realistic models for capturing the dynamics of such systems should account for material in transit between compartments. Including some information of past system states in the system model leads to infinite-dimensional delay nonnegative dynamical systems. In this chapter we present necessary and sufficient conditions for stability of nonnegative and compartmental dynamical systems with time delay. Specifically, asymptotic stability conditions for linear and nonlinear as well as continuous-time and discrete-time nonnegative dynamical systems with time delay are established using linear Lyapunov-Krasovskii functionals. Furthermore, we develop new notions of dissipativity theory for nonnegative dynamical systems with time delay using linear storage functionals with linear supply rates. These results are then used to develop general stability criteria for feedback interconnections of nonnegative dynamical systems with time delay.

1 Introduction

Modern complex engineering systems are highly interconnected and mutually interdependent, both physically and through a multitude of information and communication networks. By properly formulating these systems in terms of subsystem interaction and energy/mass transfer, the dynamical models of many of these systems can be derived from mass, energy, and information balance considerations that involve dynamic states whose values are nonnegative. Hence, it follows from physical considerations that the state trajectory of such systems remains in the nonnegative orthant of the state space for nonnegative initial conditions. Such systems are commonly referred to as *nonnegative dynamical systems* in the literature [1, 2]. A subclass of nonnegative dynamical systems are *compartmental systems* [2–5]. Compartmental systems involve dynamical models that are characterized by conservation laws (e.g., mass and energy) capturing the exchange of material between coupled macroscopic subsystems known as compartments. Each compartment is assumed to be kinetically homogeneous; that is, any material entering the compartment is instantaneously mixed with the material of the compartment. The range of applications of nonnegative systems and compartmental systems is not limited to complex engineering systems. Their usage includes biological and physiological systems, chemical reaction systems, queuing systems, large-scale systems, stochastic systems (whose state variables represent probabilities), ecological systems, economic systems, demographic systems, telecommunication systems, transportation systems, power systems, heat transfer systems, and structural vibration systems, to cite but a few examples. A key physical limitation of such systems is that transfers between compartments are not instantaneous and realistic models for capturing the dynamics of such systems should account for material, energy, or information in transit between compartments [5]. Hence, to accurately describe the evolution of the aforementioned systems, it is necessary to include in any mathematical model of the system dynamics some information of the past system states. This of course leads to (infinite-dimensional) delay dynamical systems [6, 7].

In this chapter we develop necessary and sufficient conditions for stability of time-delay nonnegative and compartmental dynamical systems. Specifically, using *linear* Lyapunov-Krasovskii functionals we develop necessary and sufficient conditions for asymptotic stability of linear nonnegative dynamical systems with time delay. The consideration of a linear Lyapunov-Krasovskii functional leads to a *new* Lyapunov-like equation for examining stability of time delay nonnegative dynamical systems. The motivation for using a linear Lyapunov-Krasovskii functional follows from the fact that the (infinite-dimensional) state of a retarded nonnegative dynamical system is nonnegative and hence a linear Lyapunov-Krasovskii functional is a valid candidate Lyapunov-Krasovskii functional. For a time delay compartmental system, a linear Lyapunov-Krasovskii functional is shown to correspond to the total mass of the system at a given time plus the integral of the mass flow in transit between compartments over the time intervals it takes for the mass to flow through the intercompartmental connections.

Next, exploiting the input-output properties related to conservation, dissipation, and transport of mass and energy in nonnegative and compartmental dynamical systems, we develop a *new* notion of classical dissipativity theory [8] for nonnegative dynamical systems with time delay. Specifically, using linear storage functionals with linear supply rates we develop sufficient conditions for dissipativity of nonnegative dynamical systems with time delay. The motivation for using linear storage functionals and linear supply rates follows from the fact that the (infinite-dimensional) state as well as the inputs and outputs of retarded nonnegative dynamical systems are nonnegative. The consideration of linear storage functionals and linear supply rates leads to new Kalman-Yakubovich-Popov equations for characterizing dissipativity of nonnegative systems with time delay. For a time delay compartmental system, a linear storage functional is shown to correspond to the total mass of the system at a given time plus the integral of the mass flow in transit between compartments over the time intervals it takes for the mass to flow through the intercompartmental connections. In this case dissipativity implies that the total system mass transport is equal to the supplied system flux minus the expelled system flux. Finally, using the concepts of dissipativity for retarded nonnegative dynamical systems, we develop feedback interconnection stability results for nonnegative systems with time delay. In particular, general stability criteria are given for Lyapunov and asymptotic stability of feedback nonnegative dynamical systems with time delays.

2 Notation and Mathematical Preliminaries

In this section we introduce notation, several definitions, and some key results concerning linear nonnegative dynamical systems [2, 3] that are necessary for developing the main results of this chapter. Specifically, $\mathcal{N}$ denotes the set of nonnegative integers, $\mathbb{R}$ denotes the reals, and $\mathbb{R}^n$ is an n-dimensional linear vector space over the reals with the maximum modulus norm $\|\cdot\|$ given by $\|x\| = \max_{i=1,\ldots,n} |x_i|$, $x \in \mathbb{R}^n$. For $x \in \mathbb{R}^n$ we write $x \geq\geq 0$ (resp., $x >> 0$) to indicate that every component of x is nonnegative (resp., positive). In this case we say that x is *nonnegative* or *positive*, respectively. Likewise, $A \in \mathbb{R}^{n\times m}$ is *nonnegative*[1] or *positive* if every entry of A is nonnegative or positive, respectively, which is written as $A \geq\geq 0$ or $A >> 0$, respectively. Let $\overline{\mathbb{R}}_+^n$ and $\mathbb{R}_+^n$ denote the nonnegative and positive orthants of $\mathbb{R}^n$; that is, if $x \in \mathbb{R}^n$, then $x \in \overline{\mathbb{R}}_+^n$ and $x \in \mathbb{R}_+^n$ are equivalent, respectively, to $x \geq\geq 0$ and $x >> 0$. Finally, $\mathcal{C}([a,b], \mathbb{R}^n)$ denotes a Banach space of continuous functions mapping the interval $[a,b]$ into $\mathbb{R}^n$ with the topology of uniform convergence. For a given real number $\tau \geq 0$ if $[a,b] = [-\tau, 0]$ we let $\mathcal{C} = \mathcal{C}([-\tau,0], \mathbb{R}^n)$ and designate the norm of an element ϕ in $\mathcal{C}$ by $|||\phi||| = \sup_{\theta\in[-\tau,0]} \|\phi(\theta)\|$. If $\alpha, \beta \in \mathbb{R}$ and $x \in \mathcal{C}([\alpha - \tau, \alpha + \beta], \mathbb{R}^n)$, then for every $t \in [\alpha, \alpha+\beta]$, we let $x_t \in \mathcal{C}$ be defined by $x_t(\theta) = x(t+\theta)$, $\theta \in [-\tau, 0]$. The following definition introduces the notion of a nonnegative function.

Definition 1. *Let $T > 0$. A real function $u : [0,T] \to \mathbb{R}^m$ is a* nonnegative *(resp.,* positive*)* function *if $u(t) \geq\geq 0$ (resp., $u(t) >> 0$) on the interval $[0,T]$.*

The next definition introduces the notion of essentially nonnegative matrices.

Definition 2 ([2, 3]). *Let $A \in \mathbb{R}^{n\times n}$. A is* essentially nonnegative *if $A_{(i,j)} \geq 0$, $i,j = 1,\ldots,n$, $i \neq j$.*

Next, we present a key result for linear nonnegative dynamical systems

$$\dot{x}(t) = Ax(t), \quad x(0) = x_0, \quad t \geq 0, \tag{1}$$

where $x(t) \in \mathbb{R}^n$, $t \geq 0$, and $A \in \mathbb{R}^{n\times n}$ is essentially nonnegative. The solution to (1) is standard and is given by $x(t) = e^{At}x(0)$, $t \geq 0$. The following lemma proven in [3] (see also [2]) shows that A is essentially nonnegative if and only if the state transition matrix e^{At} is nonnegative on $[0, \infty)$.

[1] In this chapter it is important to distinguish between a square nonnegative (resp., positive) matrix and a nonnegative-definite (resp., positive-definite) matrix.

Lemma 1. *Let $A \in \mathbb{R}^{n\times n}$. Then A is essentially nonnegative if and only if e^{At} is nonnegative for all $t \geq 0$. Furthermore, if A is essentially nonnegative and $x_0 \geq\geq 0$, then $x(t) \geq\geq 0$, $t \geq 0$, where $x(t)$, $t \geq 0$, denotes the solution to (1).*

Next, we consider a subclass of nonnegative systems; namely, compartmental systems.

Definition 3. *Let $A \in \mathbb{R}^{n\times n}$. A is a* compartmental matrix *if A is essentially nonnegative and $\sum_{i=1}^{n} A_{(i,j)} \leq 0$, $j = 1, 2, \ldots, n$.*

If A is a compartmental matrix, then the nonnegative system (1) is called an *inflow-closed compartmental system* [2, 5]. Recall that an inflow-closed compartmental system possesses a dissipation property and hence is Lyapunov stable since the total mass in the system given by the sum of all components of the state $x(t)$, $t \geq 0$, is nonincreasing along the forward trajectories of (1). In particular, with $V(x) = e^{\mathrm{T}}x$, where $e \triangleq [1, 1, \cdots, 1]^{\mathrm{T}}$, it follows that $\dot{V}(x) = e^{\mathrm{T}}Ax = \sum_{j=1}^{n}\left[\sum_{i=1}^{n} A_{(i,j)}\right] x_j \leq 0$, $x \in \overline{\mathbb{R}}_{+}^{n}$. Furthermore, since $\mathrm{ind}(A) \leq 1$, where $\mathrm{ind}(A)$ denotes the index of A, it follows that A is semistable; that is, $\lim_{t\to\infty} e^{At}$ exists. Hence, all solutions of inflow-closed linear compartmental systems are convergent. For details of the above facts see [2, 3].

3 Stability Theory for Nonnegative Dynamical Systems with Time Delay

In this chapter we consider linear time-delay dynamical systems $\mathcal{G}$ of the form

$$\dot{x}(t) = Ax(t) + A_{\mathrm{d}}x(t-\tau) + Bu(t), \quad x(\theta) = \phi(\theta), \quad -\tau \leq \theta \leq 0, \quad t \geq 0, \tag{2}$$

$$y(t) = Cx(t) + Du(t), \tag{3}$$

where $x(t) \in \mathbb{R}^n$, $u(t) \in \mathbb{R}^m$, $y(t) \in \mathbb{R}^l$, $t \geq 0$, $A \in \mathbb{R}^{n\times n}$, $A_{\mathrm{d}} \in \mathbb{R}^{n\times n}$, $B \in \mathbb{R}^{l\times n}$, $C \in \mathbb{R}^{l\times n}$, $D \in \mathbb{R}^{l\times m}$, $\tau > 0$, and $\phi(\cdot) \in \mathcal{C} = \mathcal{C}([-\tau, 0], \mathbb{R}^n)$ is a continuous vector-valued function specifying the initial state of the system. Note that the state of (2) at time t is the *piece of trajectories* x between $t-\tau$ and t, or, equivalently, the *element* x_t in the space of continuous functions defined on the interval $[-\tau, 0]$ and taking values in $\mathbb{R}^n$; that is, $x_t \in \mathcal{C}([-\tau, 0], \mathbb{R}^n)$. Hence, $x_t(\theta) = x(t+\theta), \theta \in [-\tau, 0]$. Furthermore, since for a given time t the piece of the trajectories x_t is defined on $[-\tau, 0]$, the uniform norm $|||x_t||| = \sup_{\theta\in[-\tau,0]}\|x(t+\theta)\|$ is used for the definitions of Lyapunov and asymptotic stability of (2). For further details see [6, 9]. Finally, note that since $\phi(\cdot)$ and $u(\cdot)$ are continuous it follows from Theorem 2.1 of [6, p. 14] that there exists a unique solution $x(\phi)$ defined on $[-\tau, \infty)$ that coincides with ϕ on $[-\tau, 0]$ and satisfies (2) for $t \geq 0$. The following definition is needed for the main results of this section.

Definition 4. *The linear time delay dynamical system $\mathcal{G}$ given by (2) is* nonnegative *if for every $\phi(\cdot) \in \mathcal{C}_+$, where $\mathcal{C}_+ \triangleq \{\psi(\cdot) \in \mathcal{C} : \psi(\theta) \geq\geq 0, \theta \in [-\tau, 0]\}$, and $u(t) \geq\geq 0$, $t \geq 0$, the solution $x(t)$, $t \geq 0$, to (2) and the output $y(t)$, $t \geq 0$, are nonnegative.*

Proposition 1. *The linear time delay dynamical system $\mathcal{G}$ given by (2) is nonnegative if and only if $A \in \mathbb{R}^{n\times n}$ is essentially nonnegative and $A_{\mathrm{d}} \geq\geq 0$, $B \geq\geq 0$, $C \geq\geq 0$, and $D \geq\geq 0$.*

Proof. It follows from Lagrange's formula that the solution to (2) is given by

$$\begin{aligned} x(t) &= e^{At}x(0) + \int_0^t e^{A(t-\theta)}[A_{\mathrm{d}}x(\theta-\tau) + Bu(\theta)]\mathrm{d}\theta \\ &= e^{At}\phi(0) + \int_{-\tau}^{t-\tau} e^{A(t-\tau-\theta)}A_{\mathrm{d}}x(\theta)\mathrm{d}\theta + \int_0^t e^{A(t-\theta)}Bu(\theta)\mathrm{d}\theta. \end{aligned} \quad (4)$$

Now, if A is essentially nonnegative it follows from Lemma 1 that $e^{At} \geq\geq 0$, $t \geq 0$, and if $\phi(\cdot) \in \mathcal{C}_+$, $A_{\mathrm{d}} \geq\geq 0$, $B \geq\geq 0$, $C \geq\geq 0$, and $D \geq\geq 0$, it follows that

$$x(t) = e^{At}\phi(0) + \int_{-\tau}^{t-\tau} e^{A(t-\tau-\theta)}A_{\mathrm{d}}x(\theta)\mathrm{d}\theta + \int_0^t e^{A(t-\theta)}Bu(\theta)\mathrm{d}\theta \geq\geq 0, \quad t \in [0,\tau) \quad (5)$$

and $y(t) \geq\geq 0$ for all $t \in [0,\tau)$. Alternatively, for all $\tau < t$,

$$x(t) = e^{A\tau}x(t-\tau) + \int_0^{\tau} e^{A(\tau-\theta)}A_{\mathrm{d}}x(t+\theta-2\tau)\mathrm{d}\theta + \int_0^t e^{A(t-\theta)}Bu(\theta)\mathrm{d}\theta \quad (6)$$

and hence, since $x(t) \geq\geq 0$, $t \in [-\tau,\tau)$, it follows that $x(t) \geq\geq 0$, $t \in [\tau, 2\tau)$. Repeating this procedure iteratively it follows that $x(t) \geq\geq 0$, $t \geq 0$, and hence $y(t) \geq\geq 0$, $t \geq 0$, which implies that $\mathcal{G}$ is nonnegative.

Conversely, suppose $\mathcal{G}$ is nonnegative. Now, note that with $u(0) = 0$, $y(0) = Cx(0)$ and, since $y(0) \geq\geq 0$ for all $x(0) \in \overline{\mathbb{R}}_+^n$, it follows that $C \geq\geq 0$. Next, with $x(0) = 0$, $y(0) = Du(0)$ and, since $y(0) \geq\geq 0$ for all $u(0) \in \overline{\mathbb{R}}_+^l$, it follows that $D \geq\geq 0$. Now, let $\phi(\theta) = 0, -\tau \leq \theta \leq 0$, and let $u(t) = \delta(t-\hat{t})\hat{u}$, $t, \hat{t} \in [0,\tau)$, where $\hat{u} \geq\geq 0$. In this case, since $x(\hat{t}) = B\hat{u} \geq\geq 0$ for all $\hat{u} \in \overline{\mathbb{R}}_+^m$ it follows that $B \geq\geq 0$. Furthermore, with $u(t) \equiv 0$, $\phi(\theta) = 0$, $-\tau \leq \theta \leq 0$, $x(t) = e^{At}\phi(0)$, $t \in [0,\tau)$, and hence it follows from Lemma 1 that if $x(t) \geq\geq 0$, $t \geq 0$, for all $\phi(0) \in \overline{\mathbb{R}}_+^n$, then A is essentially nonnegative. Finally, suppose, *ad absurdum*, A_{d} is not nonnegative, that is, there exist $I, J \in \{1, 2, ..., n\}$ such that $A_{\mathrm{d}(I,J)} < 0$. Let $u(t) = 0$, $t \geq 0$, and let $\{v_n\}_{n=1}^{\infty} \subset \mathcal{C}_+$ denote a sequence of functions such that $\lim_{n\to\infty} v_n(\theta) = e_J\delta(\theta - \eta + \tau)$, where $0 < \eta < \tau$ and $\delta(\cdot)$ denotes the Dirac delta function. In this case, it follows from (4) that

$$x_n(t) = e^{A\eta}v_n(0) + \int_0^{\eta} e^{A(\eta-\theta)}A_{\mathrm{d}}x(\theta-\tau)\mathrm{d}\theta, \quad (7)$$

which implies that $x(\eta) = \lim_{n\to\infty} x_n(\eta) = e^{A\eta}A_{\mathrm{d}}e_J$. Now, by choosing η sufficiently small it follows that $x_I(\eta) < 0$ which is a contradiction. □

For the remainder of this chapter, we assume that A is essentially nonnegative and A_{d}, B, C, and D, are nonnegative so that for every $\phi(\cdot) \in \mathcal{C}_+$, the linear time delay dynamical system $\mathcal{G}$ given by (2), (3) is nonnegative. Next, we present necessary and sufficient conditions for asymptotic stability for the linear undisturbed (i.e., $u(t) \equiv 0$) time delay nonnegative dynamical system (2). Note that for addressing the stability of the zero solution of a time delay nonnegative system, the usual stability definitions given in [6] need to be slightly modified. In particular, stability notions for nonnegative dynamical systems need to be defined with respect to relatively open subsets of $\overline{\mathbb{R}}_+^n$ containing the equilibrium solution $x_t \equiv 0$. For a similar definition see [2]. In this case, standard Lyapunov-Krasovskii stability theorems for nonlinear time delay systems [6] can be used directly with the required sufficient conditions verified on $\mathcal{C}_+$.

Theorem 1. *Consider the linear undisturbed (i.e., $u(t) \equiv 0$) nonnegative time delay dynamical system $\mathcal{G}$ given by (2) where $A \in \mathbb{R}^{n\times n}$ is essentially nonnegative and $A_\mathrm{d} \in \mathbb{R}^{n\times n}$ is nonnegative. Then $\mathcal{G}$ is asymptotically stable for all $\tau \in [0,\infty)$ if and only if there exist $p, r \in \mathbb{R}^n$ such that $p >> 0$ and $r >> 0$ satisfy*

$$0 = (A + A_\mathrm{d})^\mathrm{T} p + r. \tag{8}$$

Proof. To prove necessity, assume that the linear undisturbed (i.e., $u(t) \equiv 0$) time delay dynamical system $\mathcal{G}$ given by (2) is asymptotically stable for all $\tau \in [0,\infty)$. In this case, it follows that the linear nonnegative dynamical system

$$\dot{x}(t) = (A + A_\mathrm{d})x(t), \quad x(0) = x_0 \in \overline{\mathbb{R}}_+^n, \quad t \geq 0, \tag{9}$$

or, equivalently, (2) with $\tau = 0$ and $u(t) \equiv 0$, is asymptotically stable. Now, it follows from Theorem 3.2 of [2] that there exists $p >> 0$ and $r >> 0$ such that (8) is satisfied. Conversely, to prove sufficiency, assume that (8) holds and consider the candidate Lyapunov-Krasovskii functional $V : \mathcal{C}_+ \to \mathbb{R}$ given by

$$V(\psi) = p^\mathrm{T}\psi(0) + \int_{-\tau}^{0} p^\mathrm{T} A_\mathrm{d}\psi(\theta)\mathrm{d}\theta, \quad \psi(\cdot) \in \mathcal{C}_+.$$

Now, note that $V(\psi) \geq p^\mathrm{T}\psi(0) \geq \alpha\|\psi(0)\|$, where $\alpha \triangleq \min_{i\in\{1,2,\ldots,n\}} p_i > 0$. Next, using (8), it follows that the Lyapunov-Krasovskii directional derivative along the trajectories of (2) with $u(t) \equiv 0$ is given by

$$\dot{V}(x_t) = p^\mathrm{T}\dot{x}(t) + p^\mathrm{T}A_\mathrm{d}[x(t) - x(t-\tau)] = p^\mathrm{T}(A + A_\mathrm{d})x(t) = -r^\mathrm{T}x(t) \leq -\beta\|x(t)\|,$$

where $\beta \triangleq \min_{i\in\{1,2,\ldots,n\}} r_i > 0$ and $x_t(\theta) = x(t+\theta)$, $\theta \in [-\tau, 0]$, denotes the (infinite-dimensional) state of the time delay dynamical system $\mathcal{G}$. Now, it follows from Corollary 3.1 of [6, p. 143] that the linear nonnegative time delay dynamical system $\mathcal{G}$ is asymptotically stable for all $\tau \in [0,\infty)$. □

Remark 1. The results presented in Proposition 1 and Theorem 1 can be easily extended to systems with multiple delays of the form

$$\dot{x}(t) = Ax(t) + \sum_{i=1}^{n_\mathrm{d}} A_{\mathrm{d}i}x(t - \tau_i), \quad x(\theta) = \phi(\theta), \quad -\overline{\tau} \leq \theta \leq 0, \quad t \geq 0, \tag{10}$$

where $x(t) \in \mathbb{R}^n$, $t \geq 0$, $A \in \mathbb{R}^{n\times n}$ is essentially nonnegative, $A_{\mathrm{d}i} \in \mathbb{R}^{n\times n}$, $i = 1,\ldots,n_{\mathrm{d}}$, is nonnegative, $\overline{\tau} = \max_{i\in\{1,\cdots,n_{\mathrm{d}}\}} \tau_i$, and $\phi(\cdot) \in \{\psi(\cdot) \in \mathcal{C}([-\overline{\tau},0],\mathbb{R}^n) : \psi(\theta) \geq\geq 0,\ \theta \in [-\overline{\tau},0]\}$. In this case, (8) becomes

$$0 = (A + \sum_{i=1}^{n_{\mathrm{d}}} A_{\mathrm{d}i})^{\mathrm{T}} p + r, \tag{11}$$

which is associated with the Lyapunov-Krasovskii functional

$$V(\psi) = p^{\mathrm{T}}\psi(0) + \sum_{i=1}^{n_{\mathrm{d}}} \int_{-\tau_i}^{0} p^{\mathrm{T}} A_{\mathrm{d}i}\psi(\theta)\mathrm{d}\theta. \tag{12}$$

Similar remarks hold for the nonlinear extension of Theorem 1 presented below and the discrete-time results addressed in Remark 2.

Remark 2. An analogous theorem to Theorem 1 can be derived for discrete-time, nonnegative time delay dynamical systems $\mathcal{G}$ of the form

$$x(k+1) = Ax(k) + A_{\mathrm{d}}x(k-\kappa), \quad x(\theta) = \phi(\theta), \quad -\kappa \leq \theta \leq 0, \quad k \in \mathcal{N}, \tag{13}$$

where $x(k) \in \mathbb{R}^n$, $k \in \mathcal{N}$, $A \in \mathbb{R}^{n\times n}$ and $A_{\mathrm{d}} \in \mathbb{R}^{n\times n}$ are nonnegative, $\kappa \in \mathcal{N}$, and $\phi(\cdot) \in \mathcal{C}_+$, where $\mathcal{C}_+ \triangleq \{\psi(\cdot) \in \mathcal{C}(\{-\kappa,\cdots,0\},\mathbb{R}^n) : \psi(\theta) \geq\geq 0,\ \theta \in \{-\kappa,\cdots,0\}\}$ is a vector sequence specifying the initial state of the system. In this case, $\mathcal{G}$ is asymptotically stable for all $\kappa \in \{0,\cdots,\overline{\kappa}\}$, where $\overline{\kappa} > 0$, if and only if there exist $p, r \in \mathbb{R}^n$ such that $p >> 0$ and $r >> 0$ satisfy

$$p = (A + A_{\mathrm{d}})^{\mathrm{T}} p + r. \tag{14}$$

The Lyapunov-Krasovskii functional $V : \mathcal{C}_+ \to \mathbb{R}$ used to prove this result is given by

$$V(\psi) = p^{\mathrm{T}}\psi(0) + \sum_{\theta=-\kappa}^{-1} p^{\mathrm{T}} A_{\mathrm{d}}\psi(\theta), \quad \psi(\cdot) \in \mathcal{C}_+. \tag{15}$$

A similar remark holds for Theorem 2 below.

Next, we present a nonlinear extension of Proposition 1 and Theorem 1. Specifically, we consider nonlinear time delay dynamical systems $\mathcal{G}$ of the form

$$\dot{x}(t) = Ax(t) + f_{\mathrm{d}}(x(t-\tau)), \quad x(\theta) = \phi(\theta), \quad -\tau \leq \theta \leq 0, \quad t \geq 0, \tag{16}$$

where $x(t) \in \mathbb{R}^n$, $t \geq 0$, $A \in \mathbb{R}^{n\times n}$, $f_{\mathrm{d}} : \mathbb{R}^n \to \mathbb{R}^n$ is locally Lipschitz and $f_{\mathrm{d}}(0) = 0$, $\tau \geq 0$, and $\phi(\cdot) \in \mathcal{C}$. Once again, since $\phi(\cdot)$ is continuous, existence and uniqueness of solutions to (16) follow from Theorem 2.3 of [6, p. 44]. Nonlinear time delay systems of the form given by (16) arise in the study of physiological and biomedical systems [10], ecological systems [11], population dynamics [12], as well as neural Hopfield networks [13]. For the nonlinear time delay dynamical system (16), the definition of nonnegativity holds with (2) replaced by (16). The following definition is needed for our next result.

Definition 5. *Let* $f_{\mathrm{d}} = [f_{\mathrm{d}1}, \cdots, f_{\mathrm{d}n}]^{\mathrm{T}} : \mathcal{D} \to \mathbb{R}^n$, *where* $\mathcal{D}$ *is an open subset of* $\mathbb{R}^n$ *that contains* $\overline{\mathbb{R}}_+^n$. *Then* f_{d} *is* nonnegative *if* $f_{\mathrm{d}i}(x) \geq 0$, *for all* $i = 1, \ldots, n$, *and* $x \in \overline{\mathbb{R}}_+^n$.

Proposition 2. *Consider the nonlinear time delay dynamical system* $\mathcal{G}$ *given by (16). If* $A \in \mathbb{R}^{n\times n}$ *is essentially nonnegative and* $f_{\mathrm{d}} : \mathbb{R}^n \to \mathbb{R}^n$ *is nonnegative, then* $\mathcal{G}$ *is nonnegative.*

Proof. The proof is identical to the proof of Proposition 1. □

Next, we present sufficient conditions for asymptotic stability for nonlinear nonnegative dynamical systems given by (16).

Theorem 2. *Consider the nonlinear nonnegative time delay dynamical system* $\mathcal{G}$ *given (16) where* $A \in \mathbb{R}^{n\times n}$ *is essentially nonnegative,* $f_{\mathrm{d}} : \mathbb{R}^n \to \mathbb{R}^n$ *is nonnegative, and* $f_{\mathrm{d}}(x) \leq\leq \gamma x$, $x \in \overline{\mathbb{R}}_+^n$, *where* $\gamma > 0$. *If there exist* $p, r \in \mathbb{R}^n$ *such that* $p >> 0$ *and* $r >> 0$ *satisfy*

$$0 = (A + \gamma I_n)^{\mathrm{T}} p + r, \tag{17}$$

then $\mathcal{G}$ *is asymptotically stable for all* $\tau \in [0, \infty)$.

Proof. Assume that (17) holds and consider the candidate Lyapunov-Krasovskii functional $V : \mathcal{C}_+ \to \mathbb{R}$ given by

$$V(\psi) = p^{\mathrm{T}}\psi(0) + \int_{-\tau}^{0} p^{\mathrm{T}} f_{\mathrm{d}}(\psi(\theta))\mathrm{d}\theta, \quad \psi(\cdot) \in \mathcal{C}_+.$$

Now, note that $V(\psi) \geq p^{\mathrm{T}}\psi(0) \geq \alpha\|\psi(0)\|$, where $\alpha \triangleq \min_{i\in\{1,2,\ldots,n\}} p_i > 0$. Next, using (17), it follows that the Lyapunov-Krasovskii directional derivative along the trajectories of (16) is given by

$$\begin{aligned}
\dot{V}(x_t) &= p^{\mathrm{T}}\dot{x}(t) + p^{\mathrm{T}}[f_{\mathrm{d}}(x(t)) - f_{\mathrm{d}}(x(t-\tau))] \\
&= p^{\mathrm{T}}(Ax(t) + f_{\mathrm{d}}(x(t))) \\
&\leq p^{\mathrm{T}}Ax(t) + \gamma p^{\mathrm{T}}x(t) \\
&= -r^{\mathrm{T}}x(t) \\
&\leq -\beta\|x(t)\|,
\end{aligned}$$

where $\beta \triangleq \min_{i\in\{1,2,\ldots,n\}} r_i > 0$. Now, it follows from Corollary 3.1 of [6, p. 143] that the nonlinear nonnegative time delay dynamical system $\mathcal{G}$ is asymptotically stable for all $\tau \in [0, \infty)$. □

Remark 3. The structural constraint $f_{\mathrm{d}}(x) \leq\leq \gamma x$, $x \in \overline{\mathbb{R}}_+^n$, where $\gamma > 0$, in the statement of Theorem 2 is naturally satisfied for many compartmental dynamical systems. For example, in nonlinear pharmacokinetic models [14] the transport across biological membranes may be facilitated by carrier molecules with the flux described by a saturable from $f_{\mathrm{d}i}(x_i, x_j) = \phi_{\max}[(x_i^\alpha/(x_i^\alpha + \beta) - (x_j^\alpha/(x_j^\alpha + \beta)]$, where x_i, x_j are the concentrations of the ith and jth compartments and $\phi_{\max}$, α, and β are model parameters. This nonlinear intercompartmental flow model satisfies the structural constraint of Theorem 2.

4 Dissipativity Theory for Nonnegative Dynamical Systems with Time Delay

In this section, we present sufficient conditions for dissipativity for the linear time delay nonnegative dynamical system (2), (3). Recall that a function $s : \mathbb{R}^m \times \mathbb{R}^l \to \mathbb{R}$ is called a *supply rate* [8] if it is locally integrable; that is, for all input-output pairs $u \in \mathbb{R}^m$ and $y \in \mathbb{R}^l$, $s(\cdot,\cdot)$ satisfies $\int_{t_1}^{t_2} |s(u(\sigma), y(\sigma))| \mathrm{d}\sigma < \infty$, $t_1, t_2 \geq 0$. The following definition is needed for the next result.

Definition 6. *The nonnegative dynamical system (2), (3) is* dissipative with respect to the supply rate $s : \overline{\mathbb{R}}_+^m \times \overline{\mathbb{R}}_+^l \to \mathbb{R}$ *if there exists a C^0 nonnegative-definite functional $V_\mathrm{s} : \mathcal{C}_+ \to \mathbb{R}$, called a* storage functional, *such that the* dissipation inequality

$$V_\mathrm{s}(x_t) \leq V_\mathrm{s}(x_{t_1}) + \int_{t_1}^{t} s(u(\sigma), y(\sigma)) \mathrm{d}\sigma, \tag{18}$$

is satisfied for all $t_1, t \geq 0$, where $x(t)$, $t \geq 0$, is the solution to (2) with $\phi(\cdot) \in \mathcal{C}_+$ and $u(t) \in \overline{\mathbb{R}}_+^m$. The nonnegative dynamical system (2), (3) is strictly dissipative with respect to the supply rate $s : \overline{\mathbb{R}}_+^m \times \overline{\mathbb{R}}_+^l \to \mathbb{R}$ *if the dissipation inequality (18) is strictly satisfied.*

Theorem 3. *Let $q \in \mathbb{R}^l$ and $r \in \mathbb{R}^m$. Consider the nonnegative dynamical system $\mathcal{G}$ given by (2), (3) where $A \in \mathbb{R}^{n\times n}$ is essentially nonnegative, $A_\mathrm{d} \geq\geq 0$, $B \geq\geq 0$, $C \geq\geq 0$, and $D \geq\geq 0$. If there exist $p \in \overline{\mathbb{R}}_+^n$, $l \in \overline{\mathbb{R}}_+^n$ (resp., $l \in \mathbb{R}_+^n$), and $w \in \overline{\mathbb{R}}_+^m$ such that*

$$0 = (A + A_\mathrm{d})^\mathrm{T} p - C^\mathrm{T} q + l, \tag{19}$$
$$0 = B^\mathrm{T} p - D^\mathrm{T} q - r + w, \tag{20}$$

then $\mathcal{G}$ is dissipative (resp., strictly dissipative) with respect to the supply rate $s(u, y) = q^\mathrm{T} y + r^\mathrm{T} u$.

Proof. Suppose that there exist $p \in \overline{\mathbb{R}}_+^n$, $l \in \overline{\mathbb{R}}_+^n$, and $w \in \overline{\mathbb{R}}_+^m$ such that (19) and (20) hold. Then, with storage functional

$$V_\mathrm{s}(\psi) = p^\mathrm{T}\psi(0) + \int_{-\tau}^{0} p^\mathrm{T} A_\mathrm{d}\psi(\theta)\mathrm{d}\theta, \quad \psi(\cdot) \in \mathcal{C}_+, \tag{21}$$

it follows that, for all $x_t \in \mathcal{C}_+$ and $u \in \overline{\mathbb{R}}_+^m$,

$$\begin{aligned} \dot{V}_\mathrm{s}(x_t) &= p^\mathrm{T}\dot{x}(t) + p^\mathrm{T} A_\mathrm{d}[x(t) - x(t-\tau)] \\ &= p^\mathrm{T}[Ax(t) + A_\mathrm{d}x(t-\tau) + Bu(t)] + p^\mathrm{T} A_\mathrm{d} x(t) - p^\mathrm{T} A_\mathrm{d} x(t-\tau) \\ &= p^\mathrm{T}(A + A_\mathrm{d})x(t) + p^\mathrm{T} B u(t) \\ &= q^\mathrm{T} C x(t) - l^\mathrm{T} x(t) + q^\mathrm{T} D u(t) + r^\mathrm{T} u(t) - w^\mathrm{T} u(t) \\ &\leq q^\mathrm{T} y(t) + r^\mathrm{T} u(t), \end{aligned} \tag{22}$$

which implies that $\mathcal{G}$ is dissipative with respect to the supply rate $s(u,y) = q^{\mathrm{T}}y + r^{\mathrm{T}}u$. Finally, in the case where $l \in \mathbb{R}^n_+$, the inequality in (22) is strict and hence $\mathcal{G}$ is strictly dissipative with respect to the supply rate $s(u,y) = q^{\mathrm{T}}y + r^{\mathrm{T}}u$. □

The result presented in Theorem 3 can be easily extended to systems with multiple and distributed state delays. Specifically, consider the multiple point-wise delay system $\mathcal{G}$ given by

$$\dot{x}(t) = Ax(t) + \sum_{i=1}^{n_{\mathrm{d}}} A_{\mathrm{d}i}x(t-\tau_i) + Bu(t), \;\; x(\theta) = \phi(\theta), \;\; -\overline{\tau} \le \theta \le 0, \;\; t \ge 0, \tag{23}$$

$$y(t) = Cx(t) + Du(t), \tag{24}$$

where $x(t) \in \mathbb{R}^n$, $t \ge 0$, $A \in \mathbb{R}^{n\times n}$ is essentially nonnegative, $A_{\mathrm{d}i} \in \mathbb{R}^{n\times n}$, $i = 1,...,n_{\mathrm{d}}$, is nonnegative, $\overline{\tau} = \max_{i\in\{1,...,n_{\mathrm{d}}\}}\tau_i$, and $\phi(\cdot) \in \overline{\mathcal{C}}_+ \triangleq \{\psi(\cdot) \in \mathcal{C}([-\overline{\tau},0],\mathbb{R}^n) : \psi(\theta) \ge\ge 0, \theta \in [-\overline{\tau},0]\}$. In this case, with (19) replaced by

$$0 = (A + \sum_{i=1}^{n_{\mathrm{d}}} A_{\mathrm{d}i})^{\mathrm{T}}p - C^{\mathrm{T}}q + l \tag{25}$$

and storage functional

$$V_{\mathrm{s}}(\psi) = p^{\mathrm{T}}\psi(0) + \sum_{i=1}^{n_{\mathrm{d}}} \int_{-\tau_i}^{0} p^{\mathrm{T}}A_{\mathrm{d}i}\psi(\theta)\mathrm{d}\theta, \quad \psi(\cdot) \in \overline{\mathcal{C}}_+, \tag{26}$$

it can be shown using a similar construction as in the proof of Theorem 3 that $\mathcal{G}$ given by (23), (24) is dissipative with respect to the supply rate $s(u,y) = q^{\mathrm{T}}y + r^{\mathrm{T}}u$.

Alternatively, for pure distributed delay systems of the form

$$\dot{x}(t) = Ax(t) + \int_{-\tau}^{0} A_{\mathrm{d}}x(t+\sigma)\mathrm{d}\sigma + Bu(t), \;\; x(\theta) = \phi(\theta), \;\; -\tau \le \theta \le 0, \;\; t \ge 0, \tag{27}$$

$$y(t) = Cx(t) + Du(t), \tag{28}$$

it can also be shown, with (19) replaced by

$$0 = (A + \tau A_{\mathrm{d}})^{\mathrm{T}}p - C^{\mathrm{T}}q + l \tag{29}$$

and storage functional

$$V_{\mathrm{s}}(\psi) = p^{\mathrm{T}}\psi(0) + \int_{-\tau}^{0}\int_{\sigma}^{0} p^{\mathrm{T}}A_{\mathrm{d}}\psi(\theta)\mathrm{d}\theta\mathrm{d}\sigma, \quad \psi(\cdot) \in \mathcal{C}_+, \tag{30}$$

that (27), (28) is dissipative with respect to the supply rate $s(u,y) = q^{\mathrm{T}}y + r^{\mathrm{T}}u$.

Finally, we show that linear compartmental dynamical systems with time delays [5] are a special case of the linear nonnegative time delay systems (2), (3). To see this, for $i = 1,\ldots,n$, let $x_i(t)$, $t \ge 0$, denote the mass (and hence a nonnegative quantity) of the ith subsystem of the compartmental system shown in Figure 1, let

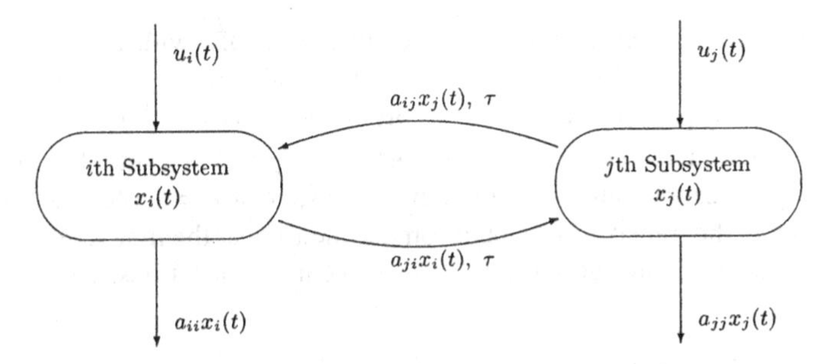

Fig. 1. Linear compartmental interconnected subsystem model with time delay.

$a_{ii} \geq 0$ denote the loss coefficient of the ith subsystem, let $\phi_{ij}(t-\tau)$, $i \neq j$, denote the net mass flow (or flux) from the jth subsystem to the ith subsystem given by $\phi_{ij}(t-\tau) = a_{ij}x_j(t-\tau) - a_{ji}x_i(t)$, where the transfer coefficient $a_{ij} \geq 0$, $i \neq j$, and τ is the fixed time it takes for the mass to flow from the jth subsystem to the ith subsystem, and let $u_i(t) \geq 0$, $t \geq 0$, denote the input mass flux to ith compartment. For simplicity of exposition we have assumed that all transfer times between compartments are given by τ. The more general multiple delay case can be addressed as shown for the system $\mathcal{G}$ given by (23), (24). Now, a mass balance for the whole compartmental system yields

$$\dot{x}_i(t) = -(a_{ii} + \sum_{j=1, i\neq j}^{n} a_{ji})x_i(t) + \sum_{j=1, i\neq j}^{n} a_{ij}x_j(t-\tau) + u_i(t), \quad t \geq 0, \quad i = 1, \ldots, n, \tag{31}$$

or, equivalently,

$$\dot{x}(t) = Ax(t) + A_{\mathrm{d}}x(t-\tau) + u(t), \quad x(\theta) = \phi(\theta), \quad -\tau \leq \theta \leq 0, \quad t \geq 0, \tag{32}$$

where $x(t) = [x_1(t), \cdots, x_n(t)]^{\mathrm{T}}$, $u(t) = [u_1(t), \cdots, u_n(t)]^{\mathrm{T}}$, $\phi(\cdot) \in \mathcal{C}_+$, and for $i, j = 1, \ldots, n$,

$$A_{(i,j)} = \begin{cases} -\sum_{k=1}^{n} a_{ki}, & i = j \\ 0, & i \neq j \end{cases}, \quad A_{\mathrm{d}(i,j)} = \begin{cases} 0, & i = j \\ a_{ij}, & i \neq j \end{cases}. \tag{33}$$

Note that A is essentially nonnegative and A_{d} is nonnegative. Furthermore, note that $A + A_{\mathrm{d}}$ is a compartmental matrix and

$$(A + A_{\mathrm{d}})x = (J_n(x) - D(x))e, \tag{34}$$

where $J_n(x)$ is a skew-symmetric matrix function with $J_{n(i,i)}(x) = 0$ and

$$J_{n(i,j)}(x) = a_{ij}x_j - a_{ji}x_i, \qquad i \neq j,$$

and

$$D(x) = \mathrm{diag}[a_{11}x_1, a_{22}x_2, \ldots, a_{nn}x_n] \geq\geq 0, \qquad x \in \overline{\mathbb{R}}_+^n.$$

To show that all compartmental systems of the form (32) with measured outputs corresponding to material outflows $y = D(x)e = [a_{11}x_1, a_{22}x_2, ..., a_{nn}x_n]^{\mathrm{T}}$ are dissipative with respect to the special supply rate $s(u,y) = \mathrm{e}^{\mathrm{T}}u - \mathrm{e}^{\mathrm{T}}y$, consider the storage functional $V_{\mathrm{s}}(\psi) = \mathrm{e}^{\mathrm{T}}\psi(0) + \mathrm{e}^{\mathrm{T}}\int_{-\tau}^{0} A_{\mathrm{d}}\psi(\theta)\mathrm{d}\theta$. Note that the storage functional $V_{\mathrm{s}}(\psi)$ captures the total mass of the system at $t = 0$ plus the integral of the mass flow in the transit between the compartments over the time intervals it takes for the mass to flow through the intercompartmental connections. Now, it follows that

$$\begin{aligned} \dot{V}_{\mathrm{s}}(x_t) &= \mathrm{e}^{\mathrm{T}}\dot{x}(t) + \mathrm{e}^{\mathrm{T}}A_{\mathrm{d}}[x(t) - x(t-\tau)] + \mathrm{e}^{\mathrm{T}}u(t) \\ &= \mathrm{e}^{\mathrm{T}}[(A + A_{\mathrm{d}})x(t) + u(t)] \\ &= \mathrm{e}^{\mathrm{T}}u(t) - \mathrm{e}^{\mathrm{T}}y(t) + \mathrm{e}^{\mathrm{T}}J_n(x(t))e \\ &= \mathrm{e}^{\mathrm{T}}u(t) - \mathrm{e}^{\mathrm{T}}y(t), \quad x_t \in \mathcal{C}_+, \end{aligned} \tag{35}$$

which shows that all compartmental systems with outputs $y = D(x)e$ are dissipative with respect to the supply rate $s(u,y) = \mathrm{e}^{\mathrm{T}}u - \mathrm{e}^{\mathrm{T}}y$. Note that in the case where the system is *closed*; that is, $u(t) \equiv 0$ and $y(t) \equiv 0$, $\dot{V}_{\mathrm{s}}(x_t) = 0$, $x_t \in \mathcal{C}_+$, which corresponds to conservation of mass in the system.

5 Feedback Interconnections of Nonnegative Dynamical Systems with Time Delay

In this section we consider feedback interconnections of dynamical systems with time delay. We begin by considering the nonnegative dynamical system $\mathcal{G}$ given by (2), (3) with the nonnegative dynamical feedback system $\mathcal{G}_{\mathrm{c}}$ given by

$$\dot{x}_{\mathrm{c}}(t) = A_{\mathrm{c}}x_{\mathrm{c}}(t) + A_{\mathrm{dc}}x_{\mathrm{c}}(t-\tau) + B_{\mathrm{c}}u_{\mathrm{c}}(t), \quad x_{\mathrm{c}}(\theta) = \phi_{\mathrm{c}}(\theta), \quad -\tau \le \theta \le 0, \quad t \ge 0, \tag{36}$$

$$y_{\mathrm{c}}(t) = C_{\mathrm{c}}x_{\mathrm{c}}(t), \tag{37}$$

where $A_{\mathrm{c}}, A_{\mathrm{dc}} \in \mathbb{R}^{n_{\mathrm{c}} \times n_{\mathrm{c}}}$, $B_{\mathrm{c}} \in \mathbb{R}^{n_{\mathrm{c}} \times m_{\mathrm{c}}}$, $C_{\mathrm{c}} \in \mathbb{R}^{l_{\mathrm{c}} \times n_{\mathrm{c}}}$, A_{c} is essentially nonnegative, $A_{\mathrm{dc}} \ge\ge 0$, $B_{\mathrm{c}} \ge\ge 0$, $C_{\mathrm{c}} \ge\ge 0$, and $\phi_{\mathrm{c}}(\cdot) \in \mathcal{C}_{\mathrm{c}+} \triangleq \{\psi_{\mathrm{c}}(\cdot) \in \mathcal{C}([-\tau, 0], \mathbb{R}^{n_{\mathrm{c}}}) : \psi_{\mathrm{c}}(\theta) \ge\ge 0,\ \theta \in [-\tau, 0]\}$. Note that the delay amount in the feedback system $\mathcal{G}_{\mathrm{c}}$ need not be the same as the delay amount in the dynamical system $\mathcal{G}$. The assumption of equal delay amounts in $\mathcal{G}$ and $\mathcal{G}_{\mathrm{c}}$ is made for convenience only. If this were not the case, the closed-loop system has the form given by (10) and thus can easily be addressed. For the following result we assume that $\mathrm{row}_i(C) \neq 0$, $i = 1, \ldots, n$, and $\mathrm{row}_i(C_{\mathrm{c}}) \neq 0$, $i = 1, \ldots, n_{\mathrm{c}}$, where $\mathrm{row}_i(\cdot)$ denotes the ith row operator.

Theorem 4. *Let $q \in \mathbb{R}^l$, $r \in \mathbb{R}^m$, $q_{\mathrm{c}} \in \mathbb{R}^{l_{\mathrm{c}}}$, and $r_{\mathrm{c}} \in \mathbb{R}^{m_{\mathrm{c}}}$. Consider the linear nonnegative dynamical systems $\mathcal{G}$ and $\mathcal{G}_{\mathrm{c}}$ given by (2), (3) and (36), (37), respectively. Assume that $\mathcal{G}$ is dissipative with respect to the linear supply rate $s(u,y) = q^{\mathrm{T}}y + r^{\mathrm{T}}u$ and with a linear storage functional $V_{\mathrm{s}}(\psi) = p^{\mathrm{T}}\psi(0) + \int_{-\tau}^{0} p^{\mathrm{T}}A_{\mathrm{d}}\psi(\theta)\mathrm{d}\theta$,* *where* $p >> 0$, *and assume that* $\mathcal{G}_c$ *is dissipative with respect to the linear supply rate* $s(u_c, y_c) = q_c^T y_c + r_c^T u_c$ *and with a linear storage functional* $V_{sc}(\psi_c) = p_c^T \psi_c(0) + \int_{-\tau}^{0} p_c^T A_{dc} \psi_c(\theta) d\theta$, *where* $p_c >> 0$. *Then the following statements hold:*

i) If there exist a scalar $\sigma > 0$ *such that* $q + \sigma r_c \leq\leq 0$ *and* $r + \sigma q_c \leq\leq 0$, *then the positive feedback interconnection of* $\mathcal{G}$ *and* $\mathcal{G}_c$ *is Lyapunov stable.*
ii) If there exist a scalar $\sigma > 0$ *such that* $q + \sigma r_c << 0$ *and* $r + \sigma q_c << 0$, *then the positive feedback interconnection of* $\mathcal{G}$ *and* $\mathcal{G}_c$ *is asymptotically stable.*

Proof. Note that the positive feedback interconnection of $\mathcal{G}$ and $\mathcal{G}_c$ is given by $u = y_c$ and $u_c = y$ so that the closed-loop dynamics of $\mathcal{G}$ and $\mathcal{G}_c$ is given by

$$\begin{bmatrix} \dot{x}(t) \\ \dot{x}_c(t) \end{bmatrix} = \begin{bmatrix} A & BC_c \\ B_c C & A_c + B_c D C_c \end{bmatrix} \begin{bmatrix} x(t) \\ x_c(t) \end{bmatrix} + \begin{bmatrix} A_d & 0 \\ 0 & A_{dc} \end{bmatrix} \begin{bmatrix} x(t-\tau) \\ x_c(t-\tau) \end{bmatrix},$$
$$\begin{bmatrix} x(\theta) \\ x_c(\theta) \end{bmatrix} = \begin{bmatrix} \phi(\theta) \\ \phi_c(\theta) \end{bmatrix}, \quad -\tau \leq \theta \leq 0, \quad t \geq 0. \tag{38}$$

Now, note that $\tilde{A} \triangleq \begin{bmatrix} A & BC_c \\ B_c C & A_c + B_c D C_c \end{bmatrix}$ is essentially nonnegative and $\tilde{A}_d \triangleq \begin{bmatrix} A_d & 0 \\ 0 & A_{dc} \end{bmatrix}$ is nonnegative. Hence, the closed-loop system is nonnegative and thus $\tilde{x}(t) \triangleq [x^T(t)\ x_c^T(t)]^T \geq\geq 0$, $t \geq 0$. Next, consider the Lyapunov-Krasovskii functional $V : \mathcal{C}_+ \to \mathbb{R}$ given by

$$\begin{aligned} V(\tilde{\psi}) &= V_s(\psi) + \sigma V_{sc}(\psi_c) \\ &= p^T \psi(0) + \int_{-\tau}^{0} p^T A_d \psi(\theta) d\theta + \sigma p_c^T \psi_c(0) + \sigma \int_{-\tau}^{0} p_c^T A_{dc} \psi_c(\theta) d\theta, \\ &\qquad \tilde{\psi} = [\psi^T\ \psi_c^T]^T \in \tilde{\mathcal{C}}_+, \end{aligned}$$

where $\tilde{\phi} \triangleq \left[\phi^T\ \phi_c^T\right]^T \in \tilde{\mathcal{C}}_+ \triangleq \{[\psi^T\ \psi_c^T]^T : \psi(\cdot) \in \mathcal{C}_+,\ \psi_c(\cdot) \in \mathcal{C}_{c+}\}$. Now, note that

$$V(\tilde{\psi}) \geq p^T \psi(0) + p_c^T \psi_c(0) = \tilde{p}^T \tilde{\psi}(0) \geq \alpha \|\tilde{\psi}(0)\|,$$

where $\tilde{p} \triangleq \left[p^T\ p_c^T\right]^T$ and $\alpha \triangleq \min_{i \in \{1,2,\ldots,n+n_c\}} \tilde{p}_i > 0$. Next, it follows that the Lyapunov-Krasovskii directional derivative along the trajectories of (38) is given by

$$\begin{aligned} \dot{V}(\tilde{x}_t) &\leq q^T y(t) + r^T u(t) + \sigma(q_c^T y_c(t) + r_c^T u_c(t)) \\ &= (q + \sigma r_c)^T y(t) + (r + \sigma q_c)^T u(t) \\ &= (q + \sigma r_c)^T C x(t) + [(q + \sigma r_c)^T D C_c + (r + \sigma q_c)^T C_c] x_c(t) \\ &= -\tilde{r}^T \tilde{x}(t) \\ &\leq -\beta \|\tilde{x}(t)\|, \end{aligned}$$

where $\tilde{r} \triangleq -\left[(q + \sigma r_c)^T C,\ [(q + \sigma r_c)^T D + (r + \sigma q_c)^T] C_c\right]^T$, $\beta \triangleq \min_{i \in \{1,2,\ldots,n+n_c\}} \tilde{r}_i \geq 0$, and $\tilde{x}_t(\theta) = \tilde{x}(t+\theta)$, $\theta \in [-\tau, 0)$, denotes the (infinite-dimensional) state of time delay dynamical system. Now, it follows from Corollary 3.1 of [6, p. 143] that the closed-loop linear nonnegative time delay dynamical system (38) is Lyapunov stable which proves *i*).

Finally, to show *ii*) note that if $q + \sigma r_c << 0$ and $r + \sigma q_c << 0$, then $\beta > 0$. Hence, it follows from from Corollary 3.1 of [6, p. 143] that the closed-loop linear nonnegative time delay dynamical system (38) is asymptotically stable. □

6 Conclusion

Nonnegative and compartmental models are widely used to capture system dynamics involving the interchange of mass and energy between homogeneous subsystems or compartments. In this chapter, necessary and sufficient conditions for asymptotic stability of nonnegative dynamical systems with time delay were given. Furthermore, we developed sufficient conditions for dissipativity with linear storage functionals and linear supply rates for nonnegative dynamical systems with time delay. Finally, general stability criteria were given for Lyapunov and asymptotic stability of feedback interconnections of retarded nonnegative dynamical systems. Analogous dissipativity results for discrete-time nonnegative dynamical systems with time delay are given in [15].

Acknowledgement

This research was supported in part by the Air Force Office of Scientific Research under Grant F49620-03-1-0178 and the National Science Foundation under Grant ECS-0133038.

References

1. L. Farina and S. Rinaldi, *Positive Linear Systems: Theory and Applications.* New York, NY: John Wiley & Sons, 2000.
2. W. M. Haddad, V. Chellaboina, and E. August, "Stability and dissipativity theory for nonnegative dynamical systems: A thermodynamic framework for biological and physiological systems," in *Proc. IEEE Conf. Dec. Contr.*, (Orlando, FL), pp. 442–458, 2001.
3. D. S. Bernstein and D. C. Hyland, "Compartmental modeling and second-moment analysis of state space systems," *SIAM J. Matrix Anal. Appl.*, vol. 14, pp. 880–901, 1993.
4. W. Sandberg, "On the mathematical foundations of compartmental analysis in biology, medicine and ecology," *IEEE Trans. Circuits and Systems*, vol. 25, pp. 273–279, 1978.
5. J. A. Jacquez, *Compartmental Analysis in Biology and Medicine, 2nd ed.* Ann Arbor: University of Michigan Press, 1985.
6. J. K. Hale and S. M. VerduynLunel, *Introduction to Functional Differential Equations.* New York: Springer-Verlag, 1993.
7. S. I. Niculescu, *Delay Effects on Stability: A Robust Control Approach.* New York: Springer, 2001.

8. J. C. Willems, "Dissipative dynamical systems part I: General theory," *Arch. Rational Mech. Anal.*, vol. 45, pp. 321–351, 1972.
9. N. N. Krasovskii, *Stability of Motion*. Stanford: Stanford University Press, 1963.
10. S. A. Campbell and J. Belar, "Multiple-delayed differential equations as models for biological control systems," in *Proc. World Math. Conf.*, pp. 3110–3117, 1993.
11. K. Gopalsamy, *Stability and Oscillations in Delay Differential Equations of Population Dynamics*. Kluwer Academic Publishers, 1992.
12. Y. Kuang, *Delay Differential Equations with Applications in Population Dynamics*. Boston: Academic Press, 1993.
13. C. W. Marcus and R. M. Westervelt, "Stability of analog neural networks with delay," *Phys. Rev.*, vol. 34, pp. 347–359, 1989.
14. W. M. Haddad, T. Hayakawa, and J. M. Bailey, "Nonlinear adaptive control for intensive care unit sedation and operating room hypnosis," in *Proc. Amer. Contr. Conf.*, (Denver, CO), pp. 1808–1813, June 2003.
15. W. M. Haddad, V. Chellaboina, and T. Rajpurohit, "Dissipativity theory for nonnegative and compartmental dynamical systems with time delay," in *Proc. Amer. Contr. Conf.*, (Denver, CO), pp. 857–862, June 2003.

A

List of Contributors

Chaouki T. Abdallah
Electrical and Computer Engineering
University of New Mexico
Albuquerque, NM 87131, USA

Anuradha Annaswamy Mechanical Engineering, Room 3-461
Massachusetts Inst. of Tech.
Cambridge, MA 02139-4307, USA

Marco Ariola
Dipartimento di Informatica e Sistemistica
Università degli Studi di Napoli Federico II
Napoli, ITALY

Lotfi Belkoura
LAIL, Universite des Sciences et Technologies de Lille
France

Alfredo Bellen
Dipartimento di Scienze Matematiche
Università degli Studi di Trieste
34100 Trieste, Italy

J. Douglas Birdwell
ECE Dept, University of Tennessee
Knoxville TN 37996, USA

Pierre-Alexandre Bliman
INRIA, Rocquencourt BP 105
78153 Le Chesnay cedex, France

Catherine Bonnet
INRIA, Rocquencourt, Domaine de Voluceau
B.P. 105, 78153 Le Chesnay cedex, France

Pedro Castillo
Heudiasyc-UTC, UMR CNRS 6599, B.P. 20529
Compiègne, France

VijaySekhar Chellaboina
Mechanical and Aerospace Engineering
University of Missouri, Columbia, MO, USA

Jie Chen
Department of Electrical Engineering
Marlan and Rosemary Bourns College of Engineering
University of California
Riverside, CA 92521, USA

John Chiasson
ECE Dept, University of Tennessee
Knoxville TN 37996, USA

Michel Dambrine
LAIL, Ecole Centrale de Lille, France

Richard Datko
Georgetown University
Department of Mathematics
Washington, DC 200257-1233, USA

Sagar Dhakal
Electrical and Computer Engineering
University of New Mexico
Albuquerque, NM 87131, USA

Jean-Michel Dion
Laboratoire d'Automatique de Grenoble (INPG CNRS UJF)
ENSIEG, BP 46, 38402, St. Martin d'Hères, FRANCE

Luc Dugard
Laboratoire d'Automatique de Grenoble (INPG CNRS UJF)
ENSIEG, BP 46, 38402, St. Martin d'Hères, FRANCE

Alejandro Dzul
División de Estudios de Posgrado e Investigación
Instituto Tecnológico da la Laguna
27000 Torreón, Coahuila, México

Mehmet Önder Efe
Collaborative Center of Control Science
Department of Electrical Engineering
The Ohio State University
Columbus, OH 43210, U.S.A.

Koen Engelborghs
K.U. Leuven, Department of Computer Science
Celestijnenlaan 200A, 3001 Heverlee, Belgium

Anas Fattouh
Automatic Laboratory of Aleppo
Faculty of Electrical and Electronic Engineering
University of Aleppo, Aleppo, Syria

Emilia Fridman
Department of Electrical Engineering
Tel-Aviv University
Ramat-Aviv, Tel-Aviv 69978, Israel

Pedro Garcia Gil
Dept. of Systems Engineering and Control
Universidad Politecnica de Valencia
P. O. Bax 22012, E-46071 Valencia, Spain

Keqin Gu
Department of Mechanical and Industrial Engineering
Southern Illinois University at Edwarsville
Edwardsville, IL 62026, USA

Wassim M. Haddad
School of Aerospace Engineering
Georgia Institute of Technology
Atlanta, GA 30332-0150, USA

Jack K. Hale
School of Mathematics
Georgia Institute of Technology
Atlanta, GA 30332, USA

Majeed M. Hayat
Electrical and Computer Engineering
University of New Mexico
Albuquerque, NM 87131, USA

Vladimir L. Kharitonov
Department of Automatic Control
CINVESTAV-IPN, A.P. 14-740
Mexico, D.F., Mexico

Jean Jacques Loiseau
Institut de Recherche en Communications et Cybernétique de Nantes
UMR CNRS 6597, Ecole Centrale de Nantes, BP 92101
44 321 Nantes Cedex 03, France

James Louisell
Department of Mathematics
Colorado State University—Pueblo
Pueblo, CO 81001, USA

Rogelio Lozano
Heudiasyc-UTC, UMR CNRS 6599, B.P. 20529
Compiègne, France

Sjoerd M. Verduyn Lunel
Mathematisch Instituut
Universiteit Leiden, P.O. Box 9512
2300 RA Leiden, The Netherlands

Tatyana Luzyanina
K.U. Leuven, Department of Computer Science
Celestijnenlaan 200A, 3001 Heverlee, Belgium

Frédéric Mazenc
INRIA Lorraine, Projet CONGE, ISGMP Bât A
Ile du Saulcy, 57045 Metz Cedex 01, France

Wim Michiels
K.U. Leuven, Department of Computer Science
Celestijnenlaan 200A, 3001 Heverlee, Belgium

Sabine Mondié
Departamento de Control Automático
CINVESTAV-IPN, Av. IPN 2508, A.P. 14-740
07300 México, D.F., México

Silviu-Iulian Niculescu
HEUDIASYC (UMR CNRS 6599)
Université de Technologie de Compiègne
Centre de Recherche de Royallieu, BP 20529
60205, Compiègne, cedex, France.

Nejat Olgac
191 Auditorium Rd., Engineering II Building
University of Connecticut
Storrs, CT 06269-3139, USA

Yuri Orlov
CICESE Research Center
Electronics and Telecom Dpt., San Diego, USA

Hitay Özbay
Department of Electrical and Electronics Engineering
Bilkent University, Bilkent, Ankara, TR-06800, Turkey
on leave from The Ohio State University, USA

Jonathan R. Partington
University of Leeds
School of Mathematics
Leeds LS2 9JT, U.K.

Dan Popescu
Department of Automatic Control
University of Craiova
A.I.Cuza Str. No. 13, RO 1100 Craiova, ROMANIA

Rabah Rabah
IRCCyN UMR 6597, 1 rue de la Noë, BP 92101
F-44321, Nantes Cedex 3, France.

Alexandr V. Rezounenko
Department of Mechanics and Mathematics
Kharkov University, 4 Svobody sqr.
Kharkov, 61077, Ukraine.

Jean-Pierre Richard
LAIL, Ecole Centrale de Lille, France

Salvador A. Rodriguez
Laboratoire d'Automatique de Grenoble (INPG CNRS UJF)
ENSIEG, BP 46, 38402, St. Martin d'Hères, FRANCE

Dirk Roose
K.U. Leuven, Department of Computer Science
Celestijnenlaan 200A, 3001 Heverlee, Belgium

Vladimir Răsvan
Department of Automatic Control
University of Craiova
A.I.Cuza Str. No. 13, RO 1100 Craiova, ROMANIA

Olivier Sename
LAG, ENSIEG-BP 46
38402 Saint Martin d'Hères Cedex, France

Rifat Sipahi
191 Auditorium Rd., Engineering II Building
University of Connecticut
Storrs, CT 06269-3139, USA

Grigory M. Sklyar
Mathematics, University of Szczecin, 70–451 Szczecin
Wielkopolska 15, Poland.

Zhong Tang
ECE Dept, University of Tennessee
Knoxville TN 37996, USA

Damia Taoutaou
HEUDIASYC (UMR CNRS 6599)
Université de Technologie de Compiègne
Centre de Recherche de Royallieu, BP 20529
60205, Compiègne, cedex, France.

Sophie Tarbouriech
LAAS-CNRS, 7 Avenue du Colonel Roche
31077 Toulouse cedex 4, FRANCE

Erik I. Verriest
School of ECE, Georgia Institute of Technology
Atlanta, GA 30332-0250, USA

Tsewei Wang
ChE Dept, University of Tennessee
Knoxville TN 37996, USA

Xin Yuan
Collaborative Center of Control Science
Department of Electrical Engineering
The Ohio State University
Columbus, OH 43210, U.S.A.

Marino Zennaro
Dipartimento di Scienze Matematiche
Università degli Studi di Trieste
34100 Trieste, Italy

Index

Editorial Policy

§1. Volumes in the following three categories will be published in LNCSE:

i) Research monographs
ii) Lecture and seminar notes
iii) Conference proceedings

Those considering a book which might be suitable for the series are strongly advised to contact the publisher or the series editors at an early stage.

§2. Categories i) and ii). These categories will be emphasized by Lecture Notes in Computational Science and Engineering. **Submissions by interdisciplinary teams of authors are encouraged.** The goal is to report new developments – quickly, informally, and in a way that will make them accessible to non-specialists. In the evaluation of submissions timeliness of the work is an important criterion. Texts should be well-rounded, well-written and reasonably self-contained. In most cases the work will contain results of others as well as those of the author(s). In each case the author(s) should provide sufficient motivation, examples, and applications. In this respect, Ph.D. theses will usually be deemed unsuitable for the Lecture Notes series. Proposals for volumes in these categories should be submitted either to one of the series editors or to Springer-Verlag, Heidelberg, and will be refereed. A provisional judgment on the acceptability of a project can be based on partial information about the work: a detailed outline describing the contents of each chapter, the estimated length, a bibliography, and one or two sample chapters – or a first draft. A final decision whether to accept will rest on an evaluation of the completed work which should include

- at least 100 pages of text;
- a table of contents;
- an informative introduction perhaps with some historical remarks which should be accessible to readers unfamiliar with the topic treated;
- a subject index.

§3. Category iii). Conference proceedings will be considered for publication provided that they are both of exceptional interest and devoted to a single topic. One (or more) expert participants will act as the scientific editor(s) of the volume. They select the papers which are suitable for inclusion and have them individually refereed as for a journal. Papers not closely related to the central topic are to be excluded. Organizers should contact Lecture Notes in Computational Science and Engineering at the planning stage.

In exceptional cases some other multi-author-volumes may be considered in this category.

§4. Format. Only works in English are considered. They should be submitted in camera-ready form according to Springer-Verlag's specifications.
Electronic material can be included if appropriate. Please contact the publisher.
Technical instructions and/or TEX macros are available via
http://www.springeronline.com/sgw/cda/frontpage/0,10735,5-111-2-71391-0,00.html
The macros can also be sent on request.

General Remarks

Lecture Notes are printed by photo-offset from the master-copy delivered in camera-ready form by the authors. For this purpose Springer-Verlag provides technical instructions for the preparation of manuscripts. See also *Editorial Policy.*

Careful preparation of manuscripts will help keep production time short and ensure a satisfactory appearance of the finished book.

The following terms and conditions hold:

Categories i), ii), and iii):
Authors receive 50 free copies of their book. No royalty is paid. Commitment to publish is made by letter of intent rather than by signing a formal contract. Springer-Verlag secures the copyright for each volume.

For conference proceedings, editors receive a total of 50 free copies of their volume for distribution to the contributing authors.

All categories:
Authors are entitled to purchase further copies of their book and other Springer mathematics books for their personal use, at a discount of 33,3 % directly from Springer-Verlag.

Addresses:

Timothy J. Barth
NASA Ames Research Center
NAS Division
Moffett Field, CA 94035, USA
e-mail: barth@nas.nasa.gov

Michael Griebel
Institut für Angewandte Mathematik
der Universität Bonn
Wegelerstr. 6
53115 Bonn, Germany
e-mail: griebel@iam.uni-bonn.de

David E. Keyes
Department of Applied Physics
and Applied Mathematics
Columbia University
200 S. W. Mudd Building
500 W. 120th Street
New York, NY 10027, USA
e-mail: david.keyes@columbia.edu

Risto M. Nieminen
Laboratory of Physics
Helsinki University of Technology
02150 Espoo, Finland
e-mail: rni@fyslab.hut.fi

Dirk Roose
Department of Computer Science
Katholieke Universiteit Leuven
Celestijnenlaan 200A
3001 Leuven-Heverlee, Belgium
e-mail: dirk.roose@cs.kuleuven.ac.be

Tamar Schlick
Department of Chemistry
Courant Institute of Mathematical
Sciences
New York University
and Howard Hughes Medical Institute
251 Mercer Street
New York, NY 10012, USA
e-mail: schlick@nyu.edu

Springer-Verlag, Mathematics Editorial IV
Tiergartenstrasse 17
69121 Heidelberg, Germany
Tel.: *49 (6221) 487-8185
e-mail: peters@springer.de

Lecture Notes in Computational Science and Engineering

Vol. 1 D. Funaro, *Spectral Elements for Transport-Dominated Equations.* 1997. X, 211 pp. Softcover. ISBN 3-540-62649-2

Vol. 2 H. P. Langtangen, *Computational Partial Differential Equations.* Numerical Methods and Diffpack Programming. 1999. XXIII, 682 pp. Hardcover. ISBN 3-540-65274-4

Vol. 3 W. Hackbusch, G. Wittum (eds.), *Multigrid Methods V.* Proceedings of the Fifth European Multigrid Conference held in Stuttgart, Germany, October 1-4, 1996. 1998. VIII, 334 pp. Softcover. ISBN 3-540-63133-X

Vol. 4 P. Deuflhard, J. Hermans, B. Leimkuhler, A. E. Mark, S. Reich, R. D. Skeel (eds.), *Computational Molecular Dynamics: Challenges, Methods, Ideas.* Proceedings of the 2nd International Symposium on Algorithms for Macromolecular Modelling, Berlin, May 21-24, 1997. 1998. XI, 489 pp. Softcover. ISBN 3-540-63242-5

Vol. 5 D. Kröner, M. Ohlberger, C. Rohde (eds.), *An Introduction to Recent Developments in Theory and Numerics for Conservation Laws.* Proceedings of the International School on Theory and Numerics for Conservation Laws, Freiburg / Littenweiler, October 20-24, 1997. 1998. VII, 285 pp. Softcover. ISBN 3-540-65081-4

Vol. 6 S. Turek, *Efficient Solvers for Incompressible Flow Problems.* An Algorithmic and Computational Approach. 1999. XVII, 352 pp, with CD-ROM. Hardcover. ISBN 3-540-65433-X

Vol. 7 R. von Schwerin, *Multi Body System SIMulation.* Numerical Methods, Algorithms, and Software. 1999. XX, 338 pp. Softcover. ISBN 3-540-65662-6

Vol. 8 H.-J. Bungartz, F. Durst, C. Zenger (eds.), *High Performance Scientific and Engineering Computing.* Proceedings of the International FORTWIHR Conference on HPSEC, Munich, March 16-18, 1998. 1999. X, 471 pp. Softcover. 3-540-65730-4

Vol. 9 T. J. Barth, H. Deconinck (eds.), *High-Order Methods for Computational Physics.* 1999. VII, 582 pp. Hardcover. 3-540-65893-9

Vol. 10 H. P. Langtangen, A. M. Bruaset, E. Quak (eds.), *Advances in Software Tools for Scientific Computing.* 2000. X, 357 pp. Softcover. 3-540-66557-9

Vol. 11 B. Cockburn, G. E. Karniadakis, C.-W. Shu (eds.), *Discontinuous Galerkin Methods.* Theory, Computation and Applications. 2000. XI, 470 pp. Hardcover. 3-540-66787-3

Vol. 12 U. van Rienen, *Numerical Methods in Computational Electrodynamics.* Linear Systems in Practical Applications. 2000. XIII, 375 pp. Softcover. 3-540-67629-5

Vol. 13 B. Engquist, L. Johnsson, M. Hammill, F. Short (eds.), *Simulation and Visualization on the Grid.* Parallelldatorcentrum Seventh Annual Conference, Stockholm, December 1999, Proceedings. 2000. XIII, 301 pp. Softcover. 3-540-67264-8

Vol. 14 E. Dick, K. Riemslagh, J. Vierendeels (eds.), *Multigrid Methods VI.* Proceedings of the Sixth European Multigrid Conference Held in Gent, Belgium, September 27-30, 1999. 2000. IX, 293 pp. Softcover. 3-540-67157-9

Vol. 15 A. Frommer, T. Lippert, B. Medeke, K. Schilling (eds.), *Numerical Challenges in Lattice Quantum Chromodynamics.* Joint Interdisciplinary Workshop of John von Neumann Institute for Computing, Jülich and Institute of Applied Computer Science, Wuppertal University, August 1999. 2000. VIII, 184 pp. Softcover. 3-540-67732-1

Vol. 16 J. Lang, *Adaptive Multilevel Solution of Nonlinear Parabolic PDE Systems.* Theory, Algorithm, and Applications. 2001. XII, 157 pp. Softcover. 3-540-67900-6

Vol. 17 B. I. Wohlmuth, *Discretization Methods and Iterative Solvers Based on Domain Decomposition.* 2001. X, 197 pp. Softcover. 3-540-41083-X

Vol. 18 U. van Rienen, M. Günther, D. Hecht (eds.), *Scientific Computing in Electrical Engineering.* Proceedings of the 3rd International Workshop, August 20-23, 2000, Warnemünde, Germany. 2001. XII, 428 pp. Softcover. 3-540-42173-4

Vol. 19 I. Babuška, P. G. Ciarlet, T. Miyoshi (eds.), *Mathematical Modeling and Numerical Simulation in Continuum Mechanics.* Proceedings of the International Symposium on Mathematical Modeling and Numerical Simulation in Continuum Mechanics, September 29 - October 3, 2000, Yamaguchi, Japan. 2002. VIII, 301 pp. Softcover. 3-540-42399-0

Vol. 20 T. J. Barth, T. Chan, R. Haimes (eds.), *Multiscale and Multiresolution Methods.* Theory and Applications. 2002. X, 389 pp. Softcover. 3-540-42420-2

Vol. 21 M. Breuer, F. Durst, C. Zenger (eds.), *High Performance Scientific and Engineering Computing.* Proceedings of the 3rd International FORTWIHR Conference on HPSEC, Erlangen, March 12-14, 2001. 2002. XIII, 408 pp. Softcover. 3-540-42946-8

Vol. 22 K. Urban, *Wavelets in Numerical Simulation.* Problem Adapted Construction and Applications. 2002. XV, 181 pp. Softcover. 3-540-43055-5

Vol. 23 L. F. Pavarino, A. Toselli (eds.), *Recent Developments in Domain Decomposition Methods.* 2002. XII, 243 pp. Softcover. 3-540-43413-5

Vol. 24 T. Schlick, H. H. Gan (eds.), *Computational Methods for Macromolecules: Challenges and Applications.* Proceedings of the 3rd International Workshop on Algorithms for Macromolecular Modeling, New York, October 12-14, 2000. 2002. IX, 504 pp. Softcover. 3-540-43756-8

Vol. 25 T. J. Barth, H. Deconinck (eds.), *Error Estimation and Adaptive Discretization Methods in Computational Fluid Dynamics.* 2003. VII, 344 pp. Hardcover. 3-540-43758-4

Vol. 26 M. Griebel, M. A. Schweitzer (eds.), *Meshfree Methods for Partial Differential Equations.* 2003. IX, 466 pp. Softcover. 3-540-43891-2

Vol. 27 S. Müller, *Adaptive Multiscale Schemes for Conservation Laws.* 2003. XIV, 181 pp. Softcover. 3-540-44325-8

Vol. 28 C. Carstensen, S. Funken, W. Hackbusch, R. H. W. Hoppe, P. Monk (eds.), *Computational Electromagnetics.* Proceedings of the GAMM Workshop on "Computational Electromagnetics", Kiel, Germany, January 26-28, 2001. 2003. X, 209 pp. Softcover. 3-540-44392-4

Vol. 29 M. A. Schweitzer, *A Parallel Multilevel Partition of Unity Method for Elliptic Partial Differential Equations.* 2003. V, 194 pp. Softcover. 3-540-00351-7

Vol. 30 T. Biegler, O. Ghattas, M. Heinkenschloss, B. van Bloemen Waanders (eds.), *Large-Scale PDE-Constrained Optimization.* 2003. VI, 349 pp. Softcover. 3-540-05045-0

Vol. 31 M. Ainsworth, P. Davies, D. Duncan, P. Martin, B. Rynne (eds.) *Topics in Computational Wave Propagation.* Direct and Inverse Problems. 2003. VIII, 399 pp. Softcover. 3-540-00744-X

Vol. 32 H. Emmerich, B. Nestler, M. Schreckenberg (eds.) *Interface and Transport Dynamics.* Computational Modelling. 2003. XV, 432 pp. Hardcover. 3-540-40367-1

Vol. 33 H. P. Langtangen, A. Tveito (eds.) *Advanced Topics in Computational Partial Differential Equations.* Numerical Methods and Diffpack Programming. 2003. XIX, 658 pp. Softcover. 3-540-01438-1

Vol. 34 V. John, *Large Eddy Simulation of Turbulent Incompressible Flows.* Analytical and Numerical Results for a Class of LES Models. 2004. XII, 261 pp. Softcover. 3-540-40643-3

Vol. 35 E. Bänsch, *Challenges in Scientific Computing - CISC 2002.* Proceedings of the Conference *Challenges in Scientific Computing*, Berlin, October 2-5, 2002. 2003. VIII, 287 pp. Hardcover. 3-540-40887-8

Vol. 36 B. N. Khoromskij, G. Wittum, *Numerical Solution of Elliptic Differential Equations by Reduction to the Interface.* 2004. XI, 293 pp. Softcover. 3-540-20406-7

Vol. 37 A. Iske, *Multiresolution Methods in Scattered Data Modelling.* 2004. XII, 182 pp. Softcover. 3-540-20479-2

Vol. 38 S.-I. Niculescu, K. Gu, *Advances in Time-Delay Systems.* 2004. XIV, 446 pp. Softcover. 3-540-20890-9

For further information on these books please have a look at our mathematics catalogue at the following URL: `www.springeronline.com/series/3527`

Texts in Computational Science and Engineering

Vol. 1 H. P. Langtangen, *Computational Partial Differential Equations.* Numerical Methods and Diffpack Programming. 2nd Edition 2003. XXVI, 855 pp. Hardcover. ISBN 3-540-43416-X

Vol. 2 A. Quarteroni, F. Saleri, *Scientific Computing with MATLAB.* 2003. IX, 257 pp. Hardcover. ISBN 3-540-44363-0

For further information on these books please have a look at our mathematics catalogue at the following URL: `www.springeronline.com/series/5151`

Druck und Bindung: Strauss GmbH, Mörlenbach